P9-AGU-985

Advanced RFID Systems, Security, and Applications

Nemai Chandra Karmakar
Monash University, Australia

Managing Director: Lindsay Johnston
Editorial Director: Joel Gamon
Book Production Manager: Jennifer Romanchak
Publishing Systems Analyst: Adrienne Freeland
Development Editor: Hannah Abelbeck
Assistant Acquisitions Editor: Kayla Wolfe
Typesetter: Alyson Zerbe
Cover Design: Nick Newcomer

Published in the United States of America by
Information Science Reference (an imprint of IGI Global)
701 E. Chocolate Avenue
Hershey PA 17033
Tel: 717-533-8845
Fax: 717-533-8661
E-mail: cust@igi-global.com
Web site: http://www.igi-global.com

Library of Congress Cataloging-in-Publication Data

Advanced RFID systems, security, and applications / Nemai Chandra Karmakar, editor.
p. cm.
Includes bibliographical references and index.
Summary: "This book features a comprehensive collection of research provided by leading experts in both academia and industries, providing state-of-the-art development on RFID"-- Provided by publisher.
ISBN 978-1-4666-2080-3 (hardcover) -- ISBN 978-1-4666-2081-0 (ebook) -- ISBN 978-1-4666-2082-7 (print & perpetual access) 1. Radio frequency identification systems. I. Karmakar, Nemai Chandra, 1963-
TK6570.I34A38 2013
006.2'45--dc23
2012013148

British Cataloguing in Publication Data
A Cataloguing in Publication record for this book is available from the British Library.

All work contributed to this book is new, previously-unpublished material. The views expressed in this book are those of the authors, but not necessarily of the publisher.

To my parents, Haridhan Karmakar and Raju Bala Karmakar

Editorial Advisory Board

List of Reviewers

Table of Contents

Section 2
Middleware

Section 3
Anti-Collision Protocol

Section 4
Applications

Detailed Table of Contents

Remote technologies are changing our way of life. The radio frequency identification (RFID) system is a new technology which uses the open air to transmit information. This information transmission needs to be protected to provide user safety and privacy. Business will look for a system that has fraud resilience to prevent the misuse of information to take dishonest advantage. The business and the user need to be assured that the transmitted information has no content which is capable of undertaking malicious activities. Public awareness of RFID security will help users and organizations to understand the need for security protection. Publishing a security guideline from the regulating body and monitoring implementation of that guideline in RFID systems will ensure that businesses and users are protected. This chapter explains the importance of security in a RFID system and will outline the protective measures. It also points out the research direction of RFID systems.

Security and privacy protection are very critical requirements for the widespread deployment of RFID technologies for commercial applications. In this chapter, the authors first present the security and privacy requirement of any commercial system, and then highlight the security and privacy threats that are unique to an RFID system. The security and privacy preserving protocols for RFID system proposed in literature are elaborately discussed, analyzing their strengths, vulnerabilities, and implementation issues. The open research challenges that need further investigation, especially with the rapid introduction of diverse RFID applications, are also presented.

Radio Frequency Identification (RFID) refers to wireless technology that is used to seamlessly and automatically track various amounts of items around an environment. This technology has the potential to improve the efficiency and effectiveness of tasks such as shopping and inventory saving commercial organisations both time and money. Unfortunately, the wide scale adoption of RFID systems have been hindered due to issues such as false-negative and false-positive anomalies that lower the integrity of captured data. In this chapter, we propose the utilisation three highly intelligent classifiers, specifically a Bayesian Network, Neural Network and Non-Monotonic Reasoning, to handle missing, wrong and duplicate observations. After discovering the potential from using Bayesian Networks, Neural Networks and Non-Monotonic Reasoning to correct captured data, we decided to improve upon the original approach by combining the three methodologies into an integrated classifier. From our experimental evaluation, we have shown the high results obtained from cleaning both false-negative and false-positive anomalies using each of our concepts, and the potential it holds to enhance physical RFID systems.

Counterfeiting affects many different sectors of the world trade, including the pharmaceutical and the aerospace industries, and, therefore, its impact is not only of financial nature but can also have fatal consequences. This chapter introduces a new robust RFID system with enhanced hardware-enabled authentication and anti-counterfeiting capabilities. The system consists of two major components, namely the near-field certificates of authenticity (NF-CoAs), which complement typical RFID tags and serve as authenticity vouchers of the products they are attached to, and a microcontroller-enabled, low-power and low-cost reader. The high entropy and security of this framework stem from the unique, conductive, and dielectric, physical structure of the certificate instances and the highly complex electromagnetic effects that take place when such a certificate is brought in the reactive near-field area of the reader's antenna array. In particular, the reader's main task is to accurately extract the 5 to 6 GHz near-field response (NF fingerprint) of the NF-CoAs. The characterization of the reader's components, with an emphasis on the accuracy achieved, is provided. Rigorous performance analysis and security test results, including uniqueness among different instances, repeatability robustness for same instance and 2D to 3D projection attack resistance, are presented and verify the unique features of this technology. Rendering typical RFID tags physically unique and hard to near-exactly replicate by complementing them with NF-CoAs can prove a valuable tool against counterfeiting.

high implementation cost, especially the cost of tags for applications involving item-level tagging. This has resulted in the development of chipless RFID tags which cost much less than conventional chip-based tags. Despite the much lower tag cost, the uptake of chipless RFID system in the market is still not as widespread as predicted by RFID experts. This chapter explores the value-added applications of chipless RFID system to promote wider adoption. The chipless technology's technical and operational characteristics, benefits, limitations and current uses will also be examined. The merit of this chapter is to contribute fresh propositions to the promising applications of chipless RFID to increase its adoption in the industries that are currently not (or less popular in) utilising it, such as retail, logistics, manufacturing, healthcare, and service sectors.

Radio Frequency Identification (RFID) is going to play a crucial role as auto-identification technology in a wide range of applications such as healthcare, logistics, supply chain management, ticketing, et cetera. The use of electromagnetic waves to identify, trace, and track people or goods allows solving many problems related to auto-identification devices based on optical reading (i.e. bar code). Currently, high interest is concentrated on the use of Radio Frequency (RF) solutions in healthcare and pharmaceutical supply chain, in order to improve drugs flow transparency and patients' safety. Unfortunately, there is a possibility that drug interaction with electromagnetic fields (EMFs) generated by RF devices, such as RFID readers, deteriorate the potency of bioactive compounds. This chapter proposes an experimental multidisciplinary approach to investigate potential alterations induced by EMFs on drug molecular structure and performance. To show the versatility of this approach, some experimental results obtained on two biological pharmaceuticals (peptide hormone-based) are discussed.

Sleep apnea is a severe, potentially life-threatening condition that requires immediate medical attention. In this chapter, a novel wireless sleep apnea monitoring system is proposed to avoid uncomfortable sleep in an unfamiliar sleep laboratory in traditional PSG-based wired monitoring systems. In wireless sleep apnea monitoring system, signal propagation paths may be affected by fading because of reflection, diffraction, energy absorption, shadowing by the body, body movement, and the surrounding environment. To combat the fading effect in WBSN, the MIMO technology is introduced in this chapter. In addition, the presented active RFID based system is composed of two main parts. The first is an on-body sensor system; the second is a reader and base station. In order to minimize the physical size of the on-body sensors and to avoid interference with 2.4 GHz wireless applications, the system is designed to operate in the 5.8 GHz ISM band. Each on-body sensor system consists of a physiological signal detection circuit, an analogue-to-digital convertor (ADC), a microcontroller (MCU), a transceiver, a channel selection bandpass filter (BPF), and a narrow band antenna.

Preface

Radio Frequency Identification (RFID) is a wireless data capturing technique for automatic identification, tracking, security surveillance, and logistic and supply chain management. In the modern time RFID is one of the top ten technologies that have tremendously impacted society. RFID offers flexibility in operation and higher data capacity compared to that for optical barcodes. Therefore, RFID has gained momentum to be used in all possible applications. The most visible application of RFID is Electronic Article Surveillance (EAS) in superstores. Expensive items are tagged so that the unpaid items give warning signals at the entry and exit points of stores. EAS is a 1-bit tag, can only respond to 'yes' or 'no' situation. More expensive and high capacity tags carry much more useful information that an optical barcode can offer. Therefore, RFIDs offer not only flexibility and capacity but also item level tagging, tracking, and surveillance. However, the bottleneck of mass deployment of RFIDs for low cost item tagging is the cost of the tag. The cost of the conventional RFIDs has been decreasing day by day. However, there is a limit due to the silicon chip attached to the tag. These chips are application specific integrated circuits (ASICs) and the price of the chip can be tens of cents. To alleviate this cost problem, researchers are envisaging alternatives such as chipless tags. Thin film transistors (TFTs) on organic substrates and fully printable tags are the two commercially viable solutions that can compete with optical barcodes in mass implementations. If the cost of the tags can be reduced to less than a cent, the tags will find many potential applications. This book has addressed the most recent development of chipless and conventional tags—their systems and applications.

According to the respected research institute IDTechEx (2009), chipless RFID tags will occupy more that 60% market share of RFID markets within a few years time if the tags can be made less than a cent. The market of RFID technology surpassed $5B in 2009 and is projected to be more than $25bn in 2018. The accelerated pace of RFID tags, middleware, and reader development will address many technological challenges as well as provide many new solutions in printing techniques, algorithm and reader architectures. Anti-collision will also play an important role in mass deployments of chipless RFID technology in multi-faceted applications. Chipless RFID will change the culture of the way we do transactions in our businesses and livings. Like chipped RFIDs, chipless RFIDs have the capability to provide flexibility in operations with its salient features of non-line-of-sight (NLOS) and all-weather reading capability without much human intervention. It has the potential to replace trillions of optical barcodes printed each year. Therefore, many research activities on chipless RFID tags have been conducted not only in academia but also in industries. In this regard, printed electronics technologies shall play the vital role. Again according to the market analyses by IDTechEx, a few hundred industries are engaged in printed electronics for identifications, tagging and telecommunications markets. As quoted by Harrop, Reuter, & Das (2009), "This organic and printed electronics is growing to become a $300 billion market and, in 2007 alone, many factories came on stream to make "post silicon" transistors, displays and solar cells using thin films and, increasingly, printing. Most of the action is taking place in East Asia, Europe and North America......*There is also the prospect of replacing 5-10 trillion barcodes yearly with printed RFID that is more versatile, reliable and has a lower cost of ownership.*"

RFID is an emerging technology which has been going through various developmental phases in terms of technological developments and businesses (applications), the potential as well as the challenges are huge. As for the example of the implementation of the RFID in Monash University's Library above, the bottleneck is the cost of the tag and its mass deployment. The answer to the problem lies in the development of new materials and printing technologies which can appropriately address the problem and bring forth a sustainable solution in terms of economy and technological advancements.

When the tags become dumb, the reader should be smart. The smartness will come from the smart signal capturing capabilities from the dumb tags and the post-processing of the returned echoes which are the signals from the uniquely identifiable tags. Significant advancement has been made in the new design of RFID systems and detection techniques of RFID tags, discrimination of tagged items and protocols developed for wireless sensors network applied to RFID systems. The book includes a full section on these topics.

As an enabling technology, RFID encompasses multiple disciplines. Similar to radar technology, RFID is a multi-disciplinary technology which encompasses a variety of disciplines: (i) RF and microwave engineering, (ii) RF and digital integrated circuits, (iii) antenna design, and (iv) signal processing software and computer engineering. The latter encodes and decodes analog signals into meaningful codes for identification. According to Lai et al (2005), "The fact that RFID reading operation requires the combined interdisciplinary knowledge of RF circuits, antennas, propagation, scattering, system, middleware, server software, and business process engineering is so overwhelming that it is hard to find one single system integrator knowledgeable about them all. .... In view of the aforesaid situation, this present invention (RFID system) seeks to create and introduce novel technologies, namely redundant networked multimedia technology, auto-ranging technology, auto-planning technology, smart active antenna technology, plus novel RFID tag technology, to consolidate the knowledge of all these different disciplines into a comprehensive product family." The book has incorporated these multi-disciplinary contents in seven different sections: (i) Security, (ii) Middleware, (iii) Anti-collision Protocol, and (iv) Applications.

Due to the flexibility and numerous advantages of RFID systems compared to barcodes and other identification systems available so far, RFIDs are now becoming a major player in retail sectors and government organisations. Patronization of the RFID technology by organisations such as Wal-Mart, K-Mart, the USA Department of Defense, Coles Myer in Australia and similar consortia in Europe and Asia has accelerated the progress of RFID technology significantly in the new millennium. As a result, significant momentum in the research and development of RFID technology has developed within a short period of time. The RFID market has surpassed the billion dollar mark recently (Das & Harrop, 2006), and this growth is exponential, with diverse emerging applications in sectors including medicine and health care, agriculture, livestock, logistics, postal deliveries, security and surveillance and retail chains. The book includes application in tracing systems on the integrity of pharmaceutical products, near field authentication, monitoring system for sleep apnoea diagnosis in wireless body sensor network (WBSN) using active RFID and MIMO technology, chipless RFID based temperature and partial discharge (PD) detection sensors and finally, wireless sensors network and their applications in RFIDs.

Today, RFID is being researched and investigated by both industry and academic scientists and engineers around the world. Recently, a consortium of the Canadian RFID industry has put a proposal to the Universities Commission on the education of fresh graduates with knowledge about RFID (GTA, 2007). The Massachusetts Institute of Technology (MIT) has founded the AUTO-ID centre to standardize RFID, thus enabling faster introduction of RFID into the mainstream of retail chain identification and asset management (McFarlane & Sheffi, 2003; Karkkainen & Ala-Risku, 2003). The synergies of implementing and promoting RFID technology in all sectors of business and day to day life have overcome the boundaries of country, organisation and discipline.

As a wireless system, RFID has undergone close scrutiny for reliability and security (EPCglobal, Inc., 2006). With the advent of new anti-collision and security protocols, efficient antennas and RF and microwave systems, these problems are being delineated and solved. Smart antennas have been playing a significant role in capacity and signal quality enhancement for wireless mobile communications, mobile ad-hoc networks and mobile satellite communications systems. Smart antennas are used in RFID readers where multiple antennas and associated signal processing units are easy to implement (Lai et al, 2005). Even multiple antennas are proposed in RFID tags to improve reading rate and accuracy (Ingram, 2003).

Besides the contributions from outside, this book editor's research group at Monash University have contributed significantly in the physical layer development of RFID reader architectures for chipped and chipless RFID tag systems, RFID smart antennas, wireless sensor network protocols for RFID, and anti-collision algorithm. The research group has been supported by the Australian Research Council's Discovery Project Grants *DP665523: Chipless RFID for Barcode Replacement* and *DP110105606: Electronically Controlled Phased Array Antenna for Universal UHF RFID Applications*; the Australian Research Council's Linkage Project Grants *LP0989652: Printable, Multi-Bit RFID for Banknotes*; *LP0776796: Radio Frequency Wireless Monitoring in Sleep Apnoea (particularly for paediatric patients)*; LP0669812*: Investigation into improved wireless communication for rural and regional Australia*; *LP0991435: Back-scatter based RFID system capable of reading multiple chipless tags for regional and suburban libraries*; and *LP0989355: Smart Information Management of Partial Discharge in Switchyards using Smart Antennas*; and finally, Victorian Department of Innovation, Industry & Regional Development (DIIRD) Grant: *Remote Sensing Alpine Vehicles Using Radio Frequency Identification (RFID) Technology* within the Department of Electrical and Computer Systems Engineering, Monash University from 2006 to date. The dedication of former postgraduate students Drs. Sushim Mukul Roy, Stevan Preradovic and Isaac Balbin and current research staff and PhD students under my supervision has brought the chipless RFID tag and reader system as the viable commercial products for Australian polymer banknotes, library database management systems, access cards, remote sensing of faulty power apparatuses in switchyards and the wireless monitoring of sleep apnoea patients. The RFID and smart sensor related research projects supported by Australian Research Council's Discovery and Linkage Projects and Victoria Government are worth approximately three million dollars. More than twenty researchers have been working in various aspects of these projects. The book contains three chapters on our research findings in the above research topics.

The dramatic growth of the RFID industry has created a huge market opportunity. Patronization from Wal-Mart alone has triggered their more than two thousand suppliers to implement RFID system for their products and services. The motto is to track the goods, items and services from their manufacturing point until the boxes are crashed once the goods are sold. Thus industries can track every event in their logistics and supply chain management and make sound plan for efficient operations and business transactions. The RFID system providers are searching all possible technologies that can be implemented in the existing RFID system (Gen2 becomes a worldwide standard) that can be made cheap, can be implemented to provide high accuracy in multiple tags reading with minimum errors and extremely low false alarm rate, location finding of tags for inventory control and asset tracking. Employing smart information management system in the reader presents an elegant way to improve the performance of the RFID system. The book has covered many technical aspects of these requirements.

The book aims to provide the reader with comprehensive information with the recent development of chipless and conventional RFID systems both in the physical layer development and the software algorithm and protocols. To serve the goal of the book the book features thirteen chapters authored by the leading experts in both academia and industries. They offer in depth descriptions of terminologies and concepts relevant to the RFID systems—the security issues of chipped and chipless tags, development of chipless RFID tags and reader system to address authentications, middleware and applications.

I continuously collect and read books on RFIDs. These books are readily available from on-line book shops such as Amazon.com. Every scientific book publisher has a series of book on RFIDs and their applications in governance, pharmaceuticals, logistics, supply chain managements, retails and original part manufacturing. These books mainly report specific applications, introduce fundamental issues, and gather information on RFIDs, specific technical details that are commonly available from other resources. This book aims to come out of the convention approach of reporting the technology. The book presents the most recent technological development from renowned researchers and scientists from academia and industries. Therefore, a comprehensive coverage of definitions of important terms of RFID systems and how the RFID technology is evolving into a new phase of development can be found in the book. The book covers the state-of-the-art development on RFID in recent years. Seven scientists from five large to medium size industries including Microsoft research Center, USA, Securency Intl. Pty. Ltd, Australia, Unique Microwave Design, Australia, Tata Consultancy Services, India, and fifty academic researchers from Australia, Italy, Mexico, Taiwan, Ukraine, UK and USA, have contributed chapters in the book. Therefore, the book not only delivers the emerging development in a total package of chipless and conventional RFID systems but also provides diversities in topics. The rich contents of the book will benefit the RFID technologist, planners, policy makers, educators, researchers and students. Many universities and tertiary educational institutions teach RFID in certificate, diploma, undergraduate and graduate levels. This book can be served as a textbook or a companion book and a very useful reference for students and researchers in all levels.

The book can be best used as a complete reference guide if an expert wants to design a complete RFID system using either a chipless or a conventional radio frequency identification system. The beneficiaries of the book are the specialists of specific disciplines such system aspects on detection, discrimination, sensor network protocols, security issues and design of security protocols and systems. The readers of the book can maximize their knowledge on a systematic middleware and enterprise software planning, anti-collision protocol designs for multiple tag and reader scenarios such as warehouses, manufacturing plants, supply chain managements, and pharmaceuticals. If some experts and executives want to implement RFID in a particular system in their organizations, they are encouraged to read the last few chapters on design and implementation of RFIDs and RFID based sensors in various emerging applications. Each section is rich with new information and research results to cater for the needs of specialists in system as well as specific components of the RFID.

In the book utmost care has been paid to keep the sequential flow of information related to the various aspects as mentioned above on the RFID system and its emerging development. Hope that the book will serve as a good reference of RFID and will pave the ways for further motivation and research in the field.

Nemai Chandra Karmakar
Monash University, Australia
March 11, 2012

REFERENCES

Das, R., & Harrop, P. (2006). RFID forecasts, players & opportunities 2006–2016. London, UK: IDTechEx. Retrieved September 11, 2007, from http://www.idtechex.com/products/en/view.asp?productcategoryid=93

EPCglobal. (2005, January). *EPCTM radio-frequency identity protocols class-1 generation-2 UHF RFID: Protocol for communications at 860 MHz - 960 MHz,* Version 1.0.9.

GTA. (2007, January). *RFID industry group RFID applications training and RFID deployment lab request*. Background paper.

Harrop, P., Reuter, S., & Das, R. (2009). *Organic and printed electronics in Europe*. London, UK: IDTechEx.

IDTechEx. (2009). *RFID forecasts, players and opportunities 2009-2019: Executive summary and col-lusions*. London, UK: Author.

Ingram, M. A. (2003, January 21). *Smart reflection antenna system and method.* (US patent no. US 6,509,836, B1).

Karkkainen, M., & Ala-Risku, T. (2003). *Automatic identification – Applications and technologies.* Logistics Research Network 8th Annual Conference, London UK, September 2003.

Lai, K. Y. A., Wang, O. Y. T., Wan, T. K. P., Wong, H. F. E., Tsang, N. M., Ma, P. M. J., et al. (2005, 22 July). *Radio frequency identification (RFID) system.* (European Patent Application EP 1 724 707 A2).

McFarlane, D., & Sheffi, Y. (2003). The impact of automatic identification on supply chain operations. *International Journal of Logistics Management*, *14*(1), 407–424. doi:10.1108/09574090310806503

Section 1
Security

Chapter 1
Security Risks/Vulnerability in a RFID System and Possible Defenses

Morshed U. Chowdhury
Deakin University, Australia

Biplob R. Ray
Melbourne Institute of Technology, Australia

ABSTRACT

Remote technologies are changing our way of life. The radio frequency identification (RFID) system is a new technology which uses the open air to transmit information. This information transmission needs to be protected to provide user safety and privacy. Business will look for a system that has fraud resilience to prevent the misuse of information to take dishonest advantage. The business and the user need to be assured that the transmitted information has no content which is capable of undertaking malicious activities. Public awareness of RFID security will help users and organizations to understand the need for security protection. Publishing a security guideline from the regulating body and monitoring implementation of that guideline in RFID systems will ensure that businesses and users are protected. This chapter explains the importance of security in a RFID system and will outline the protective measures. It also points out the research direction of RFID systems.

INTRODUCTION

RFID is a wireless technology used to identify an object. Generally, it has three main components: a tag, a reader and a back-end. This chapter assumes that the back-end has enough computational power to operate under existing security protection such as cryptography, hash algorithm, etc. It will discuss secure tag data transmission between the back-end and the tag, malware protection from malicious tag content and the infecting of a clean tag by another infected tag. A tag uses the open air to transmit data via radio frequency (RF). It is also weak in computational capability. RFID automates information collection regarding an individual's location and actions which could be abused by hackers, retailers and even the government.

DOI: 10.4018/978-1-4666-2080-3.ch001

The tag can be "promiscuous," that is, it will communicate with any reader. The reader can query any tag and gather information which makes users vulnerable to information exposure and location privacy threat. The tag might contain different information sets based on its implication. Commonly, a tag might contain a product code or object code, patient identification code, credit card information, passport number, etc. Exposing part or all of this information could put someone into a life threatening situation.

Retailers are initiating processes for collecting customer information to make their business processes more efficient. This opens up the possibility to identify a customer by attaching a tag. The system can use the information to locate the tag bearer. In the supply chain a competitor might use information gathered from various tag fields. The tag contains a very small amount of information but this information might be sufficient to take unfair advantage of competitors. The tag format and the process of reading data from a tag by reader are very important in understanding the security concerns of a RFID system. There are two popular tag formats available for users; an ISO tag format and an EPC Global tag format. These tags contain different fields which relate to the business entity. The EPC tag format is shown in Table 1.

The tag data format shown in Table 1 is a General Identifier (GID-96) 96-bit EPC tag format, and helps an application to identify an object. The EPCglobal Gen 2 tag encodes a header field, EPC manager, object class and serial number. The header field defines the overall length and format of the values of tag fields. The EPC manager identifies the company associated with the EPC.

Table 1. EPC-96 bits tag data format

HEADER	EPC MANAGER	OBJECT CLASS	SERIAL NUMBER
8 bits	28 bits	24 bits	36 bits

The object class number refers to the exact type of product being identified. The serial number field identifies the serial number of the product itself. There are two different sizes of Tag: the 64-bit scheme and the 96-bit scheme. The data is stored in a tag using a particular binary encoding. The EPC scheme is then translated using a tag data translation technique (TDT), to a uniform resource identifier (URI). The URI is assembled to form a uniform resource name (URN) notation to represent the identity of a tag (EPCglobal, 2011).

The ISO/IEC 15693-3 tag format is shown in Table 2. The 64 bit Unique Identification Number (UID) has four fields: 40 bit unique serial number for products, 8 bits data to specify tag type, 8 bits to store the manufacturer code and the last 8 bits for ISO specific code for a particular tag model. The ISO specific code should be E0 to ISO/IEC 15693-3.

The intruder can use various fields such as the Object Class and the EPC manager for standard database matching. Later, information can be used to take an illegal business advantage from competitors. To protect the data transmission and the tag, we have to protect security properties (Ray, Chowdhury et al, 2010) such as:

- Data Confidentiality
- Anti-Cloning
- Availability
- Indistinguishability
- Forward Security
- Backward Security
- Tag Tempering
- Reader Authenticity

Tag data can be sensitive in terms of the security of a human being, an animal and objects. We need to make sure that the system has a proven security protocol to protect all possible security properties before we deploy RFID in a sensitive data exchange chain. Scientists are working hard to establish a security layer standard for all types

Table 2. ISO-64 bits tag data format

ISO SPECIFIC CODE FOR TAG MODEL	MANUFACTURER CODE	TAG TYPE	SERIAL NUMBER
8 bits	8 bits	8 bits	40 bits

of RFID uses. To date, we have several standards applicable to some specific applications. We need to be aware of these standards so that we can judge before using this technology. There are a number of well-established RFID security and privacy threats (Rieback, Crispo et al, 2006) such as:

- **Sniffing:** RFID tags are designed to be read by any generic reader device which might occur without the knowledge of the tag bearer and/or over a large distance. This could trigger the skimming of passport data and credit card information theft.
- **Tracking:** This might happen in strategic locations to track individuals. A reader could collect the information and look for a unique identity. The problem arises when individuals are tracked involuntarily. These might leave schools children, senior citizens and company employees vulnerable to illegal tracking.
- **Spoofing:** Intruders can create a cloned tag by properly formatting an existing tag which can then be used for identity theft. Researchers from Johns Hopkins University and RSA cloned an RFID transponder, using a sniffed and decrypted identifier. They were able to use that transponder to buy gasoline.
- **Reply Attack:** A RFID relay device could be used to intercept and transmit RFID queries. These can exploit payment systems using tag, building access control, etc. However, existing authentication protocols are effective to stop a reply attack at present.
- **Malware:** The RFID tag is capable of storing small amounts of data but Melanie R. Rieback et el (Rieback et al,2006) proved that RFID malware, RFID worms and RFID viruses are a reality. The RFID malware can exploit RFID system components such as back-end databases and generic protocols.
- **Rogue Reader:** Attacks based on malicious RFID readers are also a serious concern which can manipulate, corrupt and hijack the connection between an authentic reader and a tag (Konidala, 2006). The rogue reader can obtain personal information about a user's preferences, medical information and other sensitive data.

Moreover, a RFID system is vulnerable to DoS attack, privacy, traceability and non-repudiation.

RFID middleware transmits the massive amount of data in the RFID system to the upper systems/back-end server as services. The middleware must process the incoming RFID data intelligently to integrate into the business application. Major challenges of this layer are collision in singulation, data redundancy and data noise. This part of the RFID system can be used for a host of malicious activities which might disrupt the entire system. The tag might have a pointer to a specific code in middleware to activate malicious action.

Backend server has a business application which receives process data from middleware as input and displays output as requested to the business application by the user. The tag search technique is one of the major challenges in the backend. The various malicious attacks such as:

SQL injection, code insertion, buffer overflows and database infection can use middleware and backend to exploit the RFID system.

This chapter covers the background of existing security protocols for a RFID system, the use of protocols in different business cases, recommendations and the future of RFID security. The purpose of this chapter is to describe the security risks /vulnerabilities businesses and users need to carefully evaluate before using or implementing. It will explain the existing security protocols and the way these protocols are mitigating existing threats. Readers will know the implementation techniques and challenges of existing RFID security protocols in the business model. Moreover, it has a guide to future research directions to improve RFID system protection. User awareness can reduce the risk of using an emerging technology. This chapter is describing the past, present and future issues of RFID security evaluation. In addition, it will make readers knowledgeable regarding the best way to protect themselves and their business while using a RFID system.

BACKGROUND

Protecting user data and location privacy are two of the most important security requirements for the RFID tag. There are schemes such as the kill command feature (EPCglobal, 2011), blocker tag (EPCglobal, 2011) and Faraday cage (Gildas and Oechslin, 2005), which are hardware-based schemes. In "kill command," the reader sends a code to the tag that turns the tag off permanently. In RFID deployment in a store environment, RFID tags will be in their products, in their packaging, and each checkout counter will usually have a RFID reader that can kill the tags. As the same time, companies will normally place kiosks near the exits of stores, so consumers who wish to can kill the tags. Products will have symbols indicating that the package contains a tag and retailers may remove the tags at the checkout if consumers want them removed. However, in this scheme, we cannot reuse the tag, even though this scheme is providing good security. Researchers at MIT and RSA Laboratories have developed an interesting solution, the RFID blocker tag. Basically, any time it comes within the range of a reader, the blocker tag will emit a garbage signal that overrides any normal signals it encounters. The blocker tag can also be set so that it only blocks a certain subset of tags, therefore it will not interfere with non-consumer tags such as shipping and inventory tags. The researchers suggest that the tags be used to set up "privacy zones" where people can feel safe from outside readers. The blocker tag's privacy zone can be a stationary area like the home or office, or it can be attached to the person's body, thereby creating a personal RFID dead-zone. A blocker tag is a good way to provide security but it disables the legitimate user who wants to collect information. The Faraday Cage tag can be shielded from scrutiny. It uses a metal or foil-lined container that is impenetrable to radio-frequency waves. An RFID tag in a Faraday cage is effectively unreadable. The Faraday cage encloses the tag and, as a result, the tag cannot hear the request. It is also possible to jam the reading of RFID tags with devices that broadcast powerful, disruptive radio signals. Such jamming devices, however, will in most cases violate government regulations for radio emissions. For most applications which need the public appearance of a tag, we cannot use the Faraday cage. In some specific situations we can enclose the tag using a Faraday cage.

There are concepts relying on universal re-encryption such as Golle et al. (Golle et al, 2004), which can be undertaken without the knowledge of public keys. Saito et al. (Saito et al, 2004) discovered an attack on the Golle et al. protocol and suggested two enhanced versions based on it. In the first one, the author modified the operation carried out by the tag and in the second one re-encryption is carried out by the tag itself instead of the reader. All protocols based on universe re-encryption will produce previously re-encrypted data as an output of the tag of the next session. The attacker can link each session and trace the tag which makes it weak against eavesdropping.

The researcher also worked to discover a cryptographic solution for RFID security but those cryptographic primitives were too computationally intensive for passive RFID tags. RSA Laboratories has elaborated a framework for "minimalist cryptography" in RFID tags, that is, achievement of the goals of cryptography under the special constraints posed by RFID. One of the key ideas in this work is the application of pseudonyms to help enforce privacy in RFID tags. In a nutshell, a tag may carry multiple, random-looking names. Each time it is queried, the tag releases a different name. In principal, only a valid verifier can tell when two different names belong to the same tag. Of course, an adversary could query a tag multiple times to harvest all names so as to defeat the scheme. There is also some special enhancements of these frameworks. First, tags release their names only at a certain (suitably slow) prescribed rate. Second, pseudonyms can be refreshed by authorized readers. However, a method for implementing a strong cryptography primitive in low cost RFID tags is still to be discovered.

Many researchers used hash function and bitwise operation to implement mutual authentication between the back-end and the tag. These protocols have contributed the most to making a general purpose tag secure. We can categorize these protocols on the basis of their use of static and dynamic tag ID, number of times of random number generation and use of hash function. Some of these existing protocols just use a bitwise operation which is less computationally intensive and perfect for EPC passive tags but those protocols are breakable using brute force attack or cryptanalysis.

There are protocols using hash function for secure data transmission such as Weis et al. and Ohkubo et al., etc. Weis et al suggested that tag send h(ID||r), including a random number r to respond to the reader's request of tag's ID. Scientists proved that this protocol is vulnerable to attacks based on the tempering of the tag (Avoine and Oechslin, 2005). In the Ohkubo et al. protocol, the i-th tag sends $G(H^{k-1}(s^1_i))$ for the k-th response where G and H are different hash functions but this protocol is inefficient for a big system as to find the ID of the tag, the back-end database must search all of the hash chain of each tag.

Some researchers such as Henrici et al. (Henrici et al,2004), Molnar et al(Molnar et al,2004), Lee et al.(Lee et al,2005) established mutual authentication schemes between the backend and the tag. Mutual authentication is a technique where the tag and reader both verify each other to make sure they are authorized for each other. In Henrici et al.'s protocol the share secret ID between the back-end and the tag is updated upon successful identification but this protocol has a database desynchronization problem. In a desynchronization situation the tag can be easily traced. Molnar et al.'s secret shared information based authentication does not have this desynchronization problem by fixing the ID but it can be cryptographically weak against tempering with the tag since the ID is fixed. Lee el al. suggested a protocol which solves the desynchronised problem by maintaining a previous ID in the back-end. In this scheme hash ID is continually identical and the ID is updated by XORing the previous ID with a random number from the reader. As a result an attacker cannot trace the tag and/or any previous event of tag respectively. However, this protocol did not satisfy forward and backward security. Moreover, this protocol is also not protected from DoS attack because in this protocol the reader is not authenticated. As a result an intruder can use a generic reader to interrogate the tag. This might make the tag useless. Ray et al suggested an enhanced version of Lee's protocol where they mitigated the forward security and the backward security and prevented a DoS attack. Moreover, this protocol also introduced the mechanism for reader authenticity.

RFID malware is a new threat which has made a huge impact on our traditional way of thinking about RFID security. RFID exploits, RFID virus and RFID worms are a reality as proven by Melanie R. Rieback et al (Rieback et al, 2006). The tag contains less data however the backend contains a large amount of source code, generic protocols and facilities which are used for RFID installation and implementation. Researchers have proved that various attacks such as SQL injection, code insertion, buffer overflows, and viral self replication, self-referential commands, Quines, database infection, payload activation and infection of a new tag, are all possible using the tag with the present RFID system. There are some countermeasures proposed to protect the RFID system from malware which are: bounds checking, sanitization of the input, disabling of backend scripting languages, limiting the database permissions, segregating users, using parameter binding, isolating the RFID middleware server and reviewing the source code. However, these are not implemented on the RFID system and do not protect all those types of malware attacks mentioned above. Those measures are not sufficient to protect middleware and the backend server from payload activation, self replicating viruses, self-referential commands, etc. In the literature there are techniques for malware detection systems such as static function call analysis, data mining method and model checking. However, they are too complex to implement for a low power and low storage technology like the RFID system. Ray et al (Ray, Chowdhury et al, 2010) implemented a sanitization technique in the RFID system but it can only check commands in the command dictionary. Moreover, it is making the tag security clearance process slower as it will check all of the tag against the given database.

Researchers implement modified WLAN security protocols in the RFID system such as the V. Alarcon-Aquino et al (Alarcon-Aquino et al,2008) modified WEP (wired equivalent privacy) protocol to provide authentication and encryption for the RFID system. The authentication mechanism is mutual authentication based on a three way handshaking model, which authenticates both the reader and the tag in the communication protocol. The cipher algorithm, based on a symmetric-key cryptosystem, is Rivest Cipher 4(RC4) implemented. It modified the existing WEP protocol to make it more secure in terms of message privacy. This system still has the shortcomings of RC4 cipher and as the RFID system is resource limited, it is not possible to implement encryption algorithms such as DES or AES.

CASE STUDY TO UNDERSTAND SECURITY RISKS/VULNERABILITIES FOR A RFID SYSTEM AND POSSIBLE DEFENCES

Let us take two implementation scenarios to elaborate on the security risks and vulnerability discussed earlier. This section will plot the supply chain scenario and Centralized Intelligent Healthcare Systems to understand the importance of RFID security.

Supply Chain

A supply chain involves few entities in the chain; these are the supplier, manufacturer, distributor, retailer and end customer. The supplier supplies the raw material to manufacture. The manufacturer produces goods in a factory using production facilities. The distributer buys goods from the manufacturer in large scale. The retailer may buy goods from the distributor and use tools to provide a service to a wholesaler and/or customers. At the end point the customer buys the product from a retailer by going to a store or using a channel such as e-commerce etc. Figure 1 shows supply chain entities and their flow.

Conventionally, a manual process or barcode is used to record the movement of goods from one phase to another. Each entity of those phases

Figure 1. Generic retail supply chain

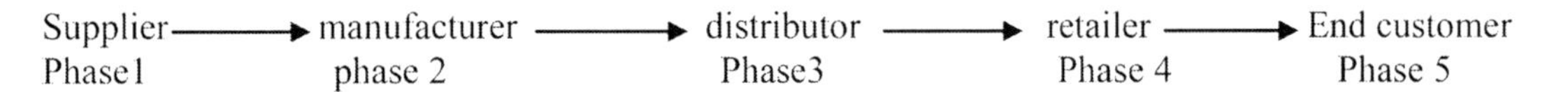

is in a different geographic location. There are many disadvantages in a conventional record process:

- Human error in calculation
- Wastage of human resources
- Misunderstanding between characters in different phases and dishonest human beings
- Time consuming

For the above reasons a manual process or barcode might not be an efficient way to create a supply chain. Let us introduce the RFID system to this supply chain where each object starting in Phase 1 will be attached to a RFID tag and the details of the data will be stored in the back-end database server. WML (Wireless Mark-up Language) could be used to make sure the updating of product movement can be accessed from any location in which they are situated. Object movement will be recorded by the reader in each location and back-end will be attached to a WML compatible mainframe which will keep up to date data of goods movement in different phases. Figure 2 shows a supply chain process using RFID in stage one. Other phases will be connected using the internet via a secure channel.

What are the uses of a secure RFID system in the example in Figure 2? We are assuming the supply chain is using a passive tag GEN-2 as this is one of the cheapest tags. To ensure the expected outcome, the RFID system needs to provide the following:

1. The information transmission between the reader and the tag of the object need to be protected from interceptors: licking of information during transmission can help unauthorized people to find the authentic tags location, partial information, provide a great deal of raw material to do crypto analysis and brute force attack. The intruder can also use the information to generate a reply attack.
2. The security protocols need to ensure that that the readers are all authentic readers: a malicious reader might create a DoS attack to use for "door knob rattling". It can create a fake database in the back-end and authentication protocol to read legitimate tags. Tags should not allow unauthentic readers to expose the location of its user by identifying a legitimate tag (this is called forward and backward security).
3. The system needs to make sure that the reader reads only legitimate tags: the system needs to verify authentic tags to stop malicious content infiltrating the system and cloning the tag.
4. There is a need to protect tags from cloning: an intruder might tamper with the tag and use this information to make a clone of that tag. The tag might have the same data structure to a legitimate tag, as a result the reader may become confused and let the tag into the system.
5. Infection of a tag by an infected tag: an infected tag can infect a clean tag causing it to pass malicious content into the back-end which might spread the infection. This will

make the total RFID system useless, as all tags will be infected over time. This makes all tags useless and the back-end will be infected.

6. Last but not least, an authentic reader should not allow malicious content or malicious pointing code from the tag into the back-end: the reader should have a mechanism to verify the content of the tag. The literature shows that an authentic tag could be infected by malware which might trigger phenomena such as RFID phishing and various malicious activities on back end servers.

To meet these security objectives we need to have security protocols which will protect information, the user and the back-end. One of the most recent mutual authentication protocols to provide security was Ray et. al's (Ray, Chowdhury et al, 2010) protocol. The use of protocol (Ray, Chowdhury et al, 2010) in a supply chain scenario would provide a secure business environment.

Ray et al developed a framework to implement a security layer between the EPC low level reader protocol (LLRP) and the discovery, configuration and initialization (DCI) protocol. This framework used two different security sub layers to make sure data are verified before passing to the authentication stage. This will increase the performance and accuracy of the authentication protocol. The framework is shown in Figure 3.

This framework can provide the secure business operation discussed in the supply chain scenario. The data verification layer used a linear search with a data dictionary to find the malicious content, which is time consuming. However, it satisfies all of the security properties of a RFID tag. In the data verification it is using a data

Figure 2. Supply chain automated object tracing flow

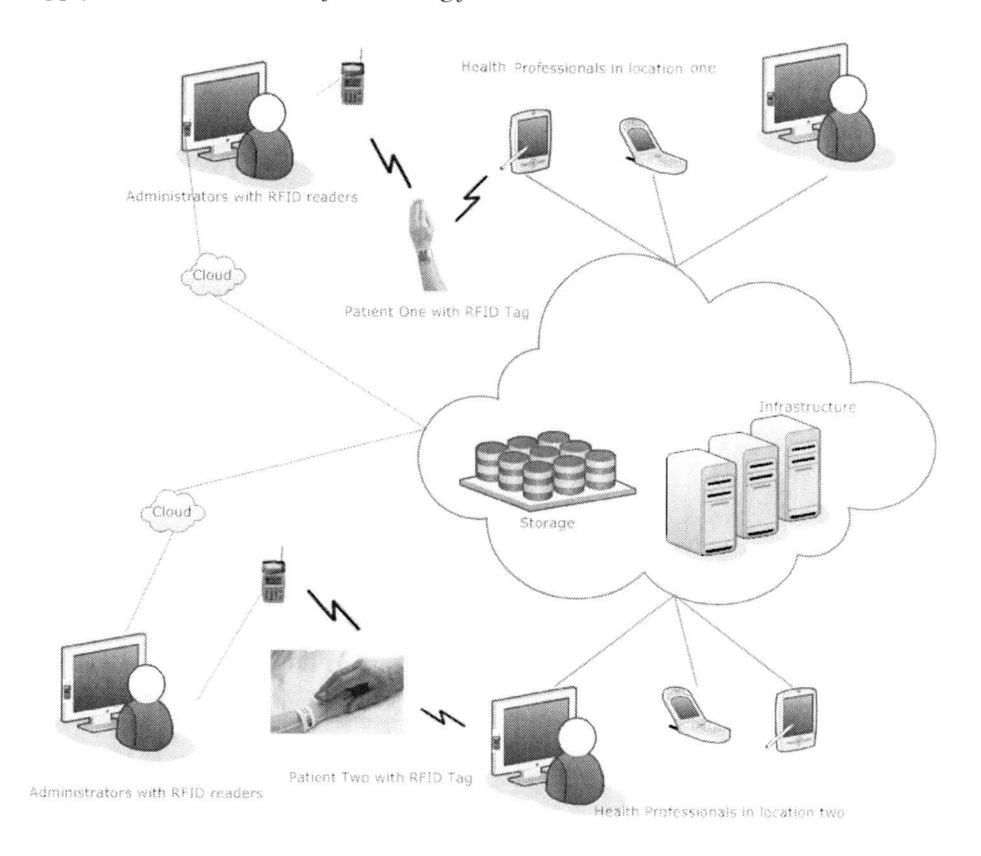

Figure 3. Framework developed by Ray, et al

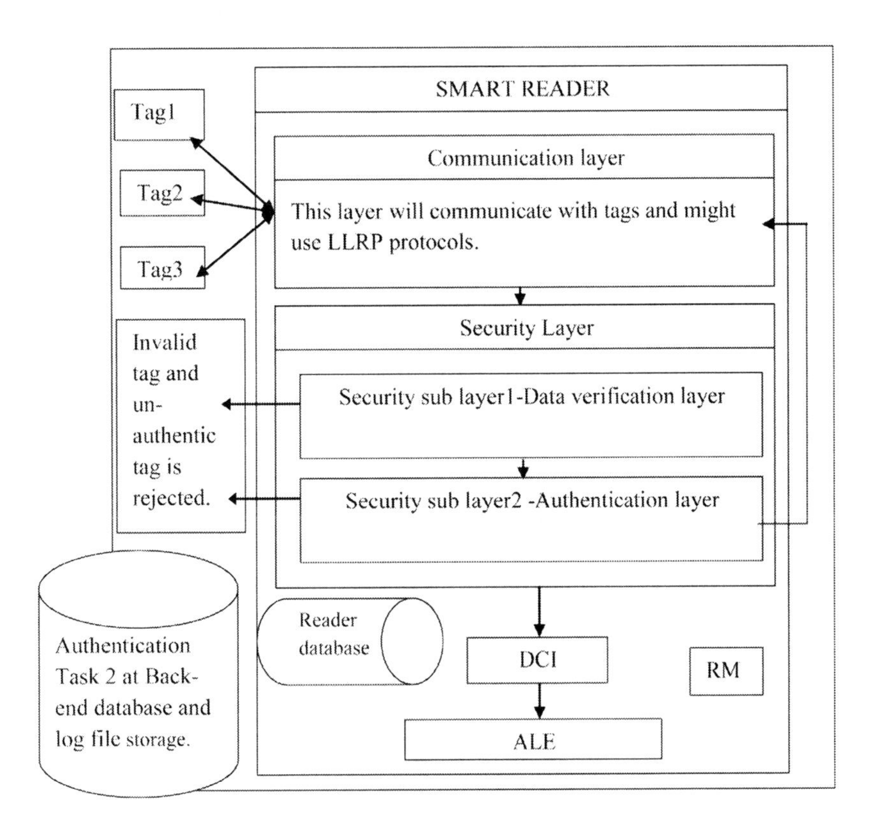

sanitization technique to match the correct data in each field. It is using a reader to store a small database which has EPC binary encoding, ASCII table and a command dictionary. A command dictionary will store all the available malicious commands that might harm the back-end. The tags with clean data proceed to the authentication layer (sub layer 2) where it will use a mutual authentication protocol to ensure the authenticity of reader and tag. Let us discuss how this authentication protocol is providing authenticity in each aspect of supply chain.

This authentication protocol has three stages.

- **Stage One:** Initiates where a reader generates and saves two new pseudorandom numbers r_1 and r_{bit} by utilizing LAMED. (The LAMED is a lightweight PRNG which is compliant with the standards and successfully passes several batteries of very demanding randomness tests (ENT, DIEHARD, NIST, and SEXTON). It can be implemented with slightly less than 1.6 K gates, and pseudo-random numbers can be generated each 1.8 ms.) It also uses its own unique hardware identification key i to generate i' which is a hash function of r_1 and i combined. This protocol uses an integer hash algorithm. The R sends i' and r_{bit} to tag. The tag also generates a new pseudorandom number r_2 by utilizing LAMED and k' using a circular shift rotation (a circular shift (or bitwise rotation) is a shift operator that shifts all bits of its operand), then the tag generates r_3.

$$\mathbf{r}_3 \leftarrow \mathrm{h}\left(\mathbf{r}_2 \oplus \mathrm{k}' \oplus \mathbf{i}'\right) \quad (1)$$

The tag then sends r_2 and r_3 to the reader. The r_1, r_2, r_3, r_{bit}, i' are used for verification by the backend database, supplied by the reader. The process of verification occurs in stage two.

- **Stage Two:** In this stage, the backend uses the data sent by the tag to the reader and matches the reader's ID and tag's ID in the database. If the backend can find both in the database then it moves to stage three, otherwise it discards all data. This back-end data matching happens when using a linear search. This is an area where this protocol could be enhanced.
- **Stage Three:** In this stage, the protocol updates the tag ID to stop cloning of the tag and stores the new ID in the database to overcome desynchronization problem. It also stores a log of any unauthentic reader to prevent a DoS attack.

The above framework meets all the requirements we need to protect the supply chain from all security concerns as follows:

- The reader's ID is verified to make sure an unauthentic reader cannot read the data from the tag which is protecting the tag's data from interceptors. This is providing the reader with authenticity property assurance. An interceptor might send a signal with random numbers and reader ID, however only an authentic reader can obtain the information in the tag.
- The protocol is not transmitting the tag's ID through air interface; it is using a circular key shift rotation, hash and random number to wrap it up, which is protecting forward and backward security, cloning and location privacy. The intruder might use a hardware device to intercept the air interface transmission but the capture will not reveal any meaningful information to intruders.
- The attacker cannot tamper with the tag as it will change the tag ID in each successive transmission; this will also have a special field in the database to protect the system from data desynchronization.
- It will protect the supply chain from DoS attack to a certain extent as it is keeping a log of unauthorized readers.
- The intruder cannot gain an illegal business benefit because the meaningful information regarding the object is always wrapped up using hash and circular key shift rotation.
- Mutual authentication is making sure the tag and reader is legitimate. The data verification sub layer is making sure that information transmitted from the tag is not malicious. This ensures non-repudiation property. The tag will not accept the transmission unless it finds some authentic reader which will prevent the infection of a legitimate tag by an infected tag.
- This protocol makes sure data is verified which protects valuable system components from malicious tag content such as RFID worms, RFID viruses, etc.

This protocol can be enhanced by introducing a faster security clearance mechanism for a trusted tag which will be a great benefit to a busy business environment such as a supply chain. This can be achieved using a security clearance handoff while a trusted tag does not need to follow the literal process of security clearance.

The framework set out in Figure 3, with its security layer and its protocols, will give a security assurance to a business seeking to take the big step of using RFID to automate its business processes. Let us discuss another case in the light of security requirements and the framework in Figure 3.

Centralized Intelligent Healthcare Systems (CIHS)

CIHS is a system which maintains the centralized data of patients for a country. Patients will have a tag attached ID which will be synchronized with a database. It mainly consists of a patient tag (i.e. wristband), a reader and health center IT systems (i.e. ICDTS). Each unique patient tag can be passive, semi-passive or active. Passive patient tags can be used for both reading/writing capabilities by the reader and do not need internal power (i.e. battery). They are energized by the reader through radio waves. A patient's basic information is stored in the back-end server for processing data. The patient database can also be linked through the Internet into other health center's databases for retrieving patient's past history. Figure 4 shows the business flow of the CIHS system. In this system, medical data will be transmitted via radio frequency and cloud. This system will give fully diagnosed, up to date data for each citizen upon request by an authorized user and the patient together. Medical data is sensitive information for each individual patient and must be kept confidential for patient safety.

Involvement of the RFID system makes global patient information access automation possible in all types of mobile devices. However, security concerns need to be addressed to assure users that their privacy is protected. The security concerns involved with CIHS are:

- **Privacy of Information:** The RFID system should assure users that information transmission through air will not expose their privacy. The information transmission between reader and the tag of the object needs to be protected from interceptors. Protocols need to make sure that crypto analysis and brute force attacks are not possible on information captured illegally.

Figure 4. Business flow of centralized intelligent healthcare systems

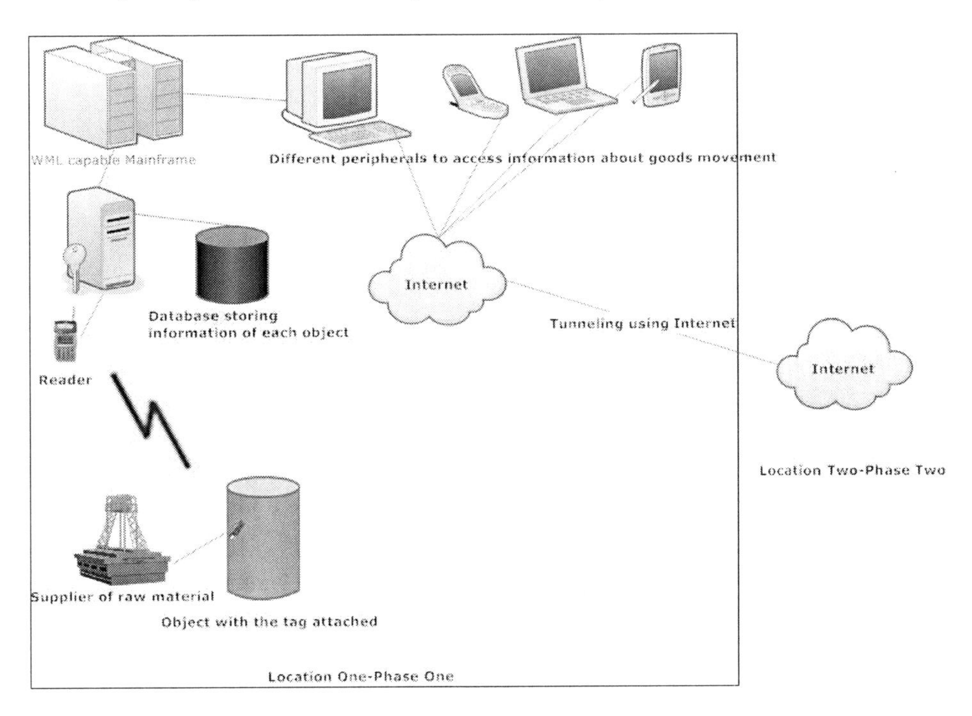

- **Traceability:** The system needs to guarantee that the patient's information cannot be traced. It also needs to ensure forward and backward security otherwise an unauthorized body might try to assume the patient's medical condition.
- **Reader's Authenticity:** Ensuring the reader's authenticity will reduce the possibility of information exposure, once a medical condition has been identified. Exposed information can be used for illegal purpose, for example, a patient might have a phobia or fear of a specific condition. An advantage taker can utilize that information against the patient.
- **Detection of Malicious Content:** Malicious code or the misfeed of information can put a patient's life in danger. A RFID system needs to ensure, that it is capable of checking the integrity of information and detecting malicious content in the tag.
- **Cloning and Tampering Protection:** The tag has to be tamper resistant which will protect patients from identity theft.
- **Infection of Other Writeable Remote Technologies by an Infected Tag:** Medical facilities are full of remote technologies. Infection of those technologies by an infected tag might shut down health facilities. An infected tag also can infect a clean tag.

In the above cases, the privacy of information, identity theft possibility and filtering of malicious content is a high priority to resolve. However, other security needs are also a concern. CIHS is a global centralized system so any protocol should be capable of handling a distributed system in heterogeneous locations. The protocol we described for a supply chain might be used for the health system too. However, this protocol needs to consider the distributed system facility, faster dictionary search and faster database search. Ray et al's protocol checks a reader's ID in the database as well as the Tag's ID which increases the time of the business process. Passive tags need a faster approach which will use a hand off technique to accelerate the security check of a secure tag.

There are many uses of RFID systems in industry and there are many more yet to be discovered. Most systems have some common security requirements. Tags must not compromise the privacy of their holders. Information should not be leaked to unauthorized readers, nor should it be possible to build long-term tracking associations between tags and holders. To prevent tracking, holders should be able to detect and disable any tags they carry. Publicly available tag output should be randomized or easily modifiable to avoid long-term associations between tags and holders. Private tag contents must be protected by access control and, if interrogation channels are assumed insecure, encryption. Both tags and readers should trust each other. Spoofing either party should be difficult. Besides providing an access control mechanism, mutual authentication between tags and readers also provides a measure of trust. Fault induction or power interruption should not compromise protocols or open windows to hijack attempts. Both tags and readers should be resistant to r man-in-the-middle attacks such as session hijacking, replay attacks, etc.

Some systems are sought for their special security features, such as the implementation of RFID in passports which requires heavy encryption of data. These special types of requirements need the special build of tags which makes tags expensive. Keeping that in mind, some researchers developed a new data structure of a tag which will ensure security for different purposes. Yan Lian and Rong Chunming (Yan et al, 2008) developed a new data structure using identity-based cryptography; this can be used to provide a digital signature along with authentication. Table 3 shows the new data structure which includes three parts: the serial number, encrypted hash code and product information of the item. The serial number is a random number

Table 3. New data structure of RFID tag

Data Structure		
Serial Number	Encrypted Hash Code	Product Information

that is allocated by the system as the identity of the tag. The serial number of each tag should be unique to this system and this serial number is used as the public key of this tag. The encrypted hash code is generated using the private key of this tag and using the hash function. The unique public key for each tag will generate a unique private key; as a result it will get a unique hash code. The third part of the data structure is product information, which includes different kinds of information according to different requirements.

In some RFID applications a digital signature is required. Yan Ling et al's system is capable of providing this. The main problem for this type of RFID system is key management. The system will have a large key base to look for as each tag will have two keys.

RECOMMENDATIONS AND FUTURE RESEARCH DIRECTIONS

The RFID system brings about a revolution in object identification but rushing immature technology in industry might have a negative effect. Researchers are working to establish a well established security layer for security assurance. Low cost tags have some mature authentication protocols but data encryption needs more research for low cost tags. Malware is a new threat. Security protocols for RFID systems need some data validation techniques which will provide faster security clearance for low cost RFID tags.

A tag might be read by the system more than once if the readers are scattered across various business locations. This might lead to the wastage of resources and also create redundancy problems in the supply chain. Security protocols should have mechanisms to identify those tags which had successful clearance.

There are many existing security protocols available, but choosing the right one for the right application is very important. RFID applications hold the potential to process data relating to an identified or identifiable natural person, a natural person being identified directly or indirectly. They can process personal data stored on the tag such as a person's name, birth date or address or biometric data or data connecting a specific RFID item number to personal data stored elsewhere in the system. Furthermore, the potential exists for this technology to be used to monitor individuals through their possession of one or more items that contain an RFID item number. Because of its potential to be both ubiquitous and practically invisible, particular attention to privacy and data protection issues is required in the deployment of RFID. Consequently, privacy and information security features should be built into RFID applications before their widespread use.

RFID will be able to deliver its numerous economic and social benefits if effective measures are in place to safeguard personal data protection, privacy and the associated ethical principles that are central to the debate on public acceptance of RFID. For the benefit of our society, system stakeholders should, especially in this initial phase of RFID implementation, make further efforts to ensure that RFID applications are monitored and the rights and freedoms of individuals are respected.

A regulatory body or government should take steps towards a policy framework for clarification and guidance which can guide industry to provide data protection and the privacy aspects of RFID applications through one or more recommended security protocol. A framework developed at community level for conducting privacy and data protection impact assessments will ensure that the provisions of this recommendation are followed coherently across the nation. The development of such a framework should build on existing practices and experiences gained. Social awareness of the technology helps the general public to find the right way to manage the use of an emerging technology such as RFID. RFID applications have implications for the general public, such as electronic ticketing in public transport and require appropriate protective measures. RFID applications that affect individuals by processing, for example, biometric identification data or health-related data, are especially critical with regard to information security and privacy and therefore require specific attention. Raising awareness among the public and small and medium-sized enterprises (SMEs) about the features and capabilities of RFID will help allow this technology to fulfill its economic promise while at the same time mitigating the risks of it being used to the detriment of the public interest, thus enhancing its acceptability.

An assessment of the privacy and data protection impacts carried by the operator prior to the implementation of an RFID application will provide the information required for appropriate protective measures. Such measures will need to be monitored and reviewed throughout the lifetime of the RFID application.

While the existing security protocol designs partially satisfy some of the desired security properties, further developments are required for secure implementation of a RFID system. One prospective area of research is the further development and implementation of low cost cryptographic primitives. These include hash functions, random number generators and both symmetric and public key cryptographic functions. There is hardware based authentication for securing a RFID system. Low cost hardware implementation must minimize circuit area and power consumption without adversely affecting computation time. This book dedicated other chapters to describing hardware based authentication in detail. RFID security may benefit from both improvements to existing systems and from new designs. More expensive RFID devices already offer symmetric encryption and public key algorithms. Adaptation of these algorithms for the low-cost passive RFID devices should be a reality in a matter of years. Protocols utilizing these cryptographic primitives must be resilient to power interruption and fault induction. Compared to smart cards, RFID tags possess more vulnerability to these types of attacks. Protocols must account for disruption of wireless channels or communication hijack attempts. Tags themselves must recover from power loss or communication interruption without compromising security.

There should be more research dedicated to data mining as the more data that is read by a reader makes more it difficult to provide security. An RFID system might have ten scattered readers and a tag might accept the communication from them periodically. The system might have an uncountable number of tags passing in the range. Redundant data makes it hard for security protocols to provide security assurance in real time.

CONCLUSION

RFID is an emerging object identification technology which raises various security concerns for users and industries. People now routinely carry radio frequency identification (RFID) tags - in passports, driver's licenses, credit cards, and other identifying cards - from which nearby RFID readers can access privacy-sensitive information. The problem is that people are often unaware of security and privacy risks associated with RFID, probably because the technology remains largely invisible and uncontrollable for the individual.

There are many authentication protocols for various levels of tag. Expensive tags with medium computational power are capable of using some of the mature RFID security protocols. However, these protocols need more enhancements in terms on objects global identification using a distributed system which is heterogeneously scattered. Low cost RFID tag's security protocols are emerging and will soon be mature.

However, RFID virus, RFID worm and RFID exploits are yet need to be embedded with these protocols. Standard regulation for RFID implementation in the application is required to be published. There should be a standard guide of RFID security compliance to make sure users and society are protected.

There are different layers stated by EPCGlobal for various stages of a RFID system. These layers have different standard protocols to operate. However, there are no security layers and standards stated by standardized organizations. Popularity of RFID technology will bring attention to lure attackers. RFID systems need to deploy standard security layers and protocols to provide a secure RFID environment for businesses and users.

REFERENCES

Alarcon-Aquino, V., Dominguez-Jimenez, M., & Ohms, C. (2008). Design and implimentation of a security layer for RFID. *Journal of Applied Research and Technology*, *6*(2), 69–83.

Avoine, G. (2005). *Radio frequency identification: adversary model and attacks on existing protocols.* Technical Report LASEC-REPORT-2005-001. Lausanne, Switzerland: EPFL. Retrieved from http://www.epcglobalinc.org/standards/tds

Avoine, G., & Oechslin, P. (2005). RFID traceability: A multilayer problem. *Financial Cryptography, 2005*, 125–140.

Golle, P., Jakobsson, M., Juels, A., & Syverson, P. (2004). Universal re-encryption for mixnets. *The Cryptographers' Track at the RSA Conference-CT-RSA* (pp. 163-178)

Henrici, D., & MÄuller, P. (2004). Hash-based enhancement of location privacy for radio-frequency identification devices using varying identifiers. *IEEE International Workshop on Pervasive Computing and Communication Security- PerSec* (pp.149-153)

Konidala, M., & Divyan, K. W.-S. (2006). *Security assessment of EPCglobal architecture framework* (pp. 13–16). AUTO-ID Labs.

Lee, S.-M., Hwang, Y. J., Lee, D. H., & Lim, J. I. (2005). Efficient authentication for low-cost RFID systems. *International Conference on Computational Science and its Applications* (pp. 619-627)

Molnar, D., & Wagner, D. (2004). Privacy and security in library RFID: Issues, practices, and architectures. *ACM Conference on Computer and Communications Security - ACM CCS* (pp. 210-219)

Ray, B., Chowdhury, M. U., & Pham, T. (2010). Mutual authentication with malware protection for RFID system. *Annual International Conference on Information Technology Security* (pp. I-24 - I-29)

Rieback, M. R., Crispo, B., & Tanenbaum, A. S. (2006). Is your cat infected with a computer virus? In *Proceedings of PerCom* (pp.169-179)

Rieback, M. R., Simpson, P. N. D., Crispo, B., & Tanenbaum, A. S. (2006). RFID malware: Design principles and examples. *Pervasive and Mobile Computing*, 405–426. doi:10.1016/j.pmcj.2006.07.008

Saito, J., Ryou, J.-C., & Sakurai, K. (2004). *Enhancing privacy of universal re-encryption scheme for RFID tags* (pp. 879–890). Embedded and Ubiquitous Computing. doi:10.1007/978-3-540-30121-9_84

Yan, L., & Rong, C. (2008). Strengthen RFID tags security using new data structure. *International Journal of Control and Automation*, *1*, 51–58.

Chapter 2
Security and Privacy in RFID Systems

Joarder Kamruzzaman
Monash University, Australia

A. K. M. Azad
Monash University, Australia

Nemai Chandra Karmakar
Monash University, Australia

Gour C. Karmakar
Monash University, Australia

Bala Srinivasan
Monash University, Australia

ABSTRACT

Security and privacy protection are very critical requirements for the widespread deployment of RFID technologies for commercial applications. In this chapter, the authors first present the security and privacy requirement of any commercial system, and then highlight the security and privacy threats that are unique to an RFID system. The security and privacy preserving protocols for RFID system proposed in literature are elaborately discussed, analyzing their strengths, vulnerabilities, and implementation issues. The open research challenges that need further investigation, especially with the rapid introduction of diverse RFID applications, are also presented.

1. INTRODUCTION

Radio frequency identification (RFID) technology has seen rapid evolution in recent years with innovative applications in various domains. RFID tags are now as small as the size of a rice grain and have built-in microchip, antenna and memory. Tags can be either active or passive. Passive and semi-active tags harness power from the reader to communicate, while active tags use their own battery power and can communication over greater range. In recent times, development of chipless RFID tags, which have no chip, memory or power circuit, has also shown remarkable prog-

DOI: 10.4018/978-1-4666-2080-3.ch002

ress. RFID tags are expected to replace barcodes and will enable automatic reading of many tags simultaneously attached to any object. This offers enormous opportunity for business organizations to use RFID to increase productivity and lower operating costs. RFID has already found application in supply chain, pharmaceutical industries, military, healthcare, emergency management, etc. The U.S. Department of Defense has set a requirement that all its shipments be equipped with RFID tags (Li and Ding, 2007). Business share of RFID application was estimated to reach US$3 billion by 2010 (Bashir et. al., 2011).

Although RFID as a technology is very promising and can offer enormous benefit, the biggest concern on RFID use is security and privacy. There are many ways how security and privacy breach can occur in RFID system. An RFID system would consist of readers, tags and back end systems and interfaces. An attack by a malicious party can occur at any of these components, but by attack on RFID system we would mainly mean attack on readers and tags while attack on back end system could also be other forms of attack on information systems. However, some types of attack on readers and tags may cause de-synchronization of tags/readers' state with those recorded in back end system. Some examples of privacy and security breaches include: an attack on RFID tag attached to a patients' body may prevent reading of the tag by legitimate reader and delay the patient's care which may turn into catastrophic situation depending on patient's condition; an attack on readers and tags in a retailer may prevent normal operation of the store and disrupt inventory and stock replenishment task causing financial loss; and unauthorized tracking of objects in bags or wallet being carried by a person may compromise his/her personal security and privacy.

Since computer security is high in the agenda for both government and private organizations, and people are increasing becoming aware of privacy violation and their rights for protection, security and privacy issues are perhaps the biggest barrier for wide-spread adoption of this technology in every sphere of our life, starting from retailer to healthcare to military (Zuo, 2010). It is estimated that Wal-Mart's in-store implementation will generate about 7 terabytes of RFID data per day (Mao, William & Sanchez, 2008). This means any security breach in RFID system will cost Wal-Mart millions of dollars. Military applications will suffer even greater consequence, including the possibility of loss of human life. All these underline the enormous importance of ensuring security in RFID system, without which prolific deployment of RFID system will not accepted and hence, not materialized. In this chapter, we review the current status of research towards providing security and privacy in RFID system and highlight the open research challenges.

2. RFID SYSTEM

Figure 1 shows a three-layer architecture for an RFID system where the bottom RFID network layer manages one or more physical RFID networks consists of tags, readers and base stations, the top most application layer is for user applications and/or RFID based services, and the integration and interfaces between these two layers are done in the RFID middleware. In the following we briefly describe the components in the RFID network layer and middleware.

In an RFID network, an RFID reader scans a tag (or multiple tags concurrently) and transmits the tag identifier to the tag information server (TIS)/base station (BS). The TIS/BS maintains attributes (e.g., product code, manufacturer code, product specific information, date time, location, etc.) for each tag identifier and transmits the attributes to the RFID task manager (TM). In some applications, information flow might be bidirectional also, i.e., TIS/BS might also send some control information to the readers which in turn based on the instruction received generate control information and disseminate them to the tags. For

Figure 1. Functional block diagram of RFID system

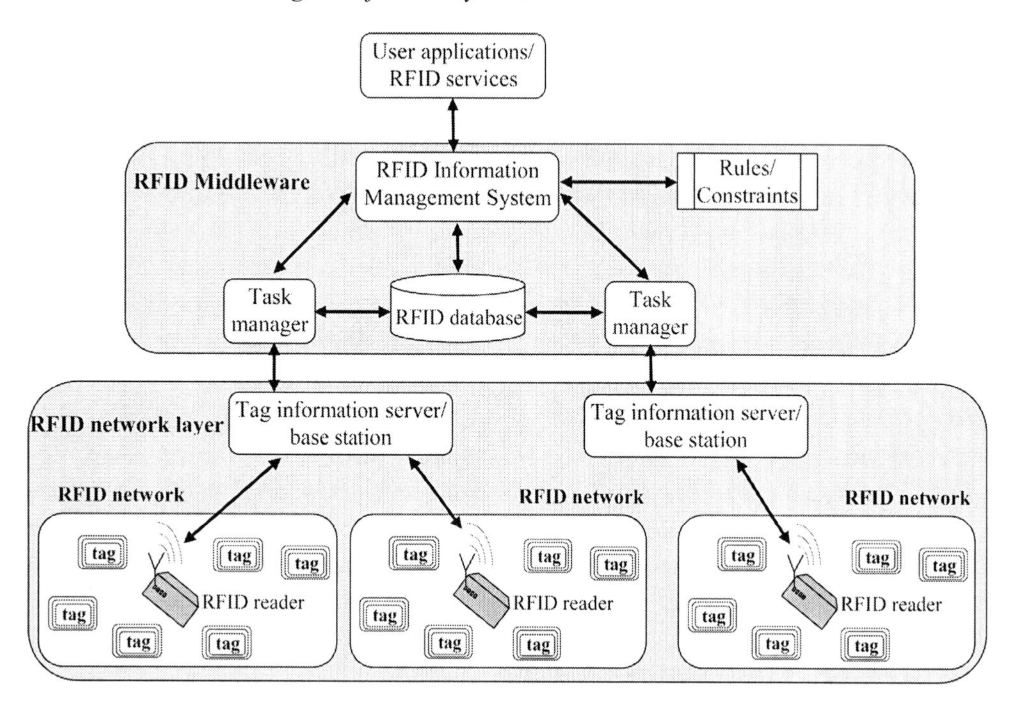

example, based on the information received from the tags RFID task manager might need to increase or decrease the tag reading rate and such instruction is propagated to the RFID reader through the TIS/BS, also to reduce collisions during multiple tags reading reader might need to command the tags to reduce their response rate, etc.

The RFID task manager is required to interwork with the RFID network layer. This component receives the tag information from the RFID network and stores it to the RFID Database. If necessary, the RFID task manager can be assigned a task by the RFID information management system (IMS). The RFID task manager can transform a task into one or more RFID subtasks, each of which is transferred to an RFID reader (or a group of readers) to locate objects specified by the TIS/BS.

The rules/constraints repository maintains a list of rules mapping tables for tags and readers. The IMS accesses this repository to retrieve the information required to construct a task for RFID networks or to return the answers to queries from applications or users. For instance, when a new object enters the RFID network and its information is delivered to the TM, which in turn notify the IMS.

The IMS then look the rules repository for any specific constraints listed for this specific tag and triggers a new task for readers according to the rules. Another example may be that once user/application asks for specific operation the IMS looks into the rules repository for any restrictions/constraints on such operation by the user and respond accordingly.

The IMS handles incoming RFID data by accessing the RFID database. For each query request from users, the IMS provides advanced processing techniques such as data aggregation, filtering, and mapping. When it receives a query about historic data from an application, the IMS accesses the RFID DB and the rule repository and processes the relevant data to return results. When the query requires the present or future data, it look ups the rule repository and dispatch tasks to the RFID network through the TM. As requested by the tasks, the RFID network reports the requested data to the RFID DB and the IMS process them to generate response to the user query.

The security and privacy issues are present at all stages of communications, namely, between tags and readers, readers and TIS/BS, TIS/BS and TM, TM and IMS, and IMS and user applications. This chapter focuses only on the issues during the tags and readers communications.

3. SECURITY AND PRIVACY REQUIREMENTS

3.1 Security Requirements

In ensuring information security, in general, any given security mechanism will address one or more of the following requirements. These requirements have been defined by the International Organization for Standardization (ISO) in X.800 standard. The main requirements for a secured system are as below:

- **Confidentiality:** Confidentiality of data transmission is one of the fundamental security requirements for organizations as well as individuals. The ISO has defined confidentiality as "ensuring that information is accessible only to those authorized to have access" (ISO, 2005). This means that information should only be disclosed to "the intended entity" and data must be protected from unauthorized disclosure. Since RFID transmission is easily interceptable, meeting this security requirement is more difficult in wireless network. Confidentiality can be ensured by using proper access control and by using cryptosystems for securing data transmissions.
- **Integrity:** The ISO has defined integrity as "safeguarding the accuracy and completeness of information and processing methods" (ISO, 2005). In secured information systems, this means that data has to be protected from unauthorized modification. Data should be received as sent (e.g., reading a tag) with no duplication, insertion, modification or replays. If any violation of integrity is detected, whether by the introduction of error or hostile activity, that should be detected and notified. In many cases, we may need a system that will recover from the violation.
- **Availability:** The ISO has defined "availability" as "ensuring that authorized users have access to information and associated assets when required" (ISO, 2005). This means that the system should be able to serve its intended purpose to users whenever required. In RFID system, malfunctioning of the readers by an attack means tags cannot be read when needed, and therefore, service interruption will occur.
- **Authentication:** It is the assurance that the communicating entity is the one that it claims to be. Authentication of both parties has to be established at the time of communication initiation and also to be maintained during the communication, i.e., during communication another malicious entity should not be able to impersonate as one of the legitimate entities. X.800 defines two types of authentication services. One is peer entity authentication where two entities implement the same protocol in different system and authentication takes place during the connection phase. The other is data origin authentication (e.g., email) where no prior interaction takes place between the entities.
- **Access Control:** This is the ability to restrict access to the host or information system, e.g., the underlying system that governs the RFID system. An entity attempting to gain access must be authenticated before access permission is granted and needs to be tailored according to the role the entity play in the overall system.

- **Non-Repudiation:** An entity taking participation in a communication should not be able to latter deny having participation in the communication. Although this requirement is an integral part of many secure systems, e.g., ecommerce, it has less relevance to RFID system.

3.2 Privacy Requirement

"Privacy" is an abstract notion which varies with time and society. People often think of privacy as some kind of right. As Roger Clark puts it, "Unfortunately, the concept of a 'right' is a problematical way to start, because a right seems to be some kind of absolute standard. What's worse, it's very easy to get confused between legal rights, on the one hand, and natural or moral rights, on the other" (Roger Clark, 2011). He has defined privacy as "Privacy is the interest that individuals have in sustaining a 'personal space,' free from interference by other people and organizations." Again, privacy may have many dimensions, e.g.:

- **Privacy of Person**: Usually refer to physical and biological privacy.
- **Privacy of Personal Behavior:** Usually refer to one's preferences, practices, choices and habits, e.g., religious and political belief, sexual orientation, etc.
- **Privacy of Personal Communications**: Usually refer to one's or organization's right to communicate with peers without being monitored or eavesdropped by others. For example, this may invade privacy of someone's location, giving the notion of location privacy in case of a person carrying RFID tag.
- **Privacy of Personal Data**: Data on someone should not be automatically available to others, or used and analyzed by others without the formal consent of that person. This gives the notion of data privacy, e.g., personal data embedded into a tag associated with a person can be revealed.

The last two points of the above introduce the notion of information privacy, a term coined by Roger Clark. Combining the privacy of personal communications and of personal data, "Information Privacy is defined as the interest an individual has in controlling, or at least significantly influencing, the handling of data about themselves" (Roger Clark, 2011).

Sometimes a person's one interest may conflict with another interest of own self or interest of another person within a group and there may be a need to reach a balance among completing interest. For example, we have to disclose some personal medical information to subscript health insurance. Disclosure of disease information of individual may help research bringing societal benefit. Overall privacy protection is a process of finding appropriate balances between privacy and multiple competing interests.

The Centre for Democracy and Technology (http://www.cdt.org/) believes that people should have "the ability to take control of their personal information on line and make informed, meaningful choices about the collection, use and disclosure of personal information." About twenty years ago the OECD formulated some fundamental principles for the protection of privacy (OECD, 2011). Of course, once we accept such principles of information privacy, the question that then arises is: how do we ensure them or enforce them?

4. SECURITY AND PRIVACY THREATS IN RFID SYSTEM

RFID makes identification easy and mass deployment is possible at a cheap cost, however, its effective use requires well planned and controlled deployment. While deploying for a specific application (ie., supply chain management, library, healthcare, etc.) we have to consider the potential risk of security and privacy violation, at least the system has be to designed keeping security and privacy issues in mind with an aim to minimize the risks.

4.1 Security Threats

Following Igure and William (Igure & William, 2008), Deng et al. (Deng, Li & Feng, 2008) has classified RFID security threats into two levels. In the first level, threats are classified according to the communication layers at which the attack may be launched. At the second level, threats are classified according to system-specific attacks. In the following, the various possible attacks in RFID system at these levels are presented.

4.1.1 Threats of Physical Layer

The threats of the physical layer include RF Eavesdropping, Jamming and Cloning. They violate mainly the electromagnetic properties (the RF signal) in the physical layer, such as the frequency and the carrier clock cycle.

4.1.1.1 Eavesdropping

Wireless medium is inherently less secure, mainly due to the communication medium which is open to the intruders. Since wireless transmission is of broadcast nature and easily interceptable, meeting the security requirement is more difficult in wireless network than wireline network. The physical security countermeasure is far less ineffective in this case. Interception of data means compromising proprietary information, network IDs and passwords, configuration data and encryption keys. The signal between the RFID tags and reader can be eavesdropped posing a security risk. An attacker equipped with RF antennas can scan through the whole range of frequency spectrum and use wireless packet analyzer to record the signal and analyze the data communication between the tag and the reader. The passive RFID system is especially vulnerable to eavesdropping. They rely on simple modulation scheme using narrowband frequencies which make them easy to be eavesdropped.

4.1.1. 2 Jamming

Another way an attacker may use to disturb the communication between readers and tags is the transmission of jamming signals. Jamming is different from interference in the sense that jamming is a deliberate attempt to reduce signal to noise ratio with an intention to disrupt an ongoing or scheduled communication between parties. It was extensively used during World War II and cold war to mislead enemy pilots and to block broadcast of foreign radio stations. In jamming, an attack device (i.e. a radio-based voltage-controller oscillator (Jong, Helmond, & Koot, 2005) broadcasting RF signals blocks radio waves to reach the RFID-tagged objects, and thereby disrupts the normal operation of any nearby RFID system (Xiao, Boulet, & Gibbons, 2007). Though any un-modulated carrier would cause disruption of tag-reader communications, other forms of modulation are commonly used in conjunction with jamming signal to maximize disruption effect (Deng, Li & Feng, 2008). In cases where RFIDs are used in medical and emergency applications, any hindrance to read data from RFID tags or corruption of data may lead to disastrous consequences.

4.1.1.3 Cloning or Mimicking of Tags

An RFID tag emits a unique number known as the EPC number when interrogated by the reader. Tags many contain other data, but low cost tags are very simple and may contain only EPC. Such tags can be widely used in commercial products on store shelves. A malicious reader or other devices may scan those tags and later create exact copies of those. The "cloning" (ie., copying) of a tag is the creation of an exact logical copy that is indistinguishable from the original tag. Once cloned the RFID readers will treat the original tag and its cloned copy in the same way such that both appear to be authentic.

In many applications, a RFID tag's primary use is the unique identification of an object (e.g., a tag attached to an expensive item), and in those cases cloning of tags should be prevented to avoid forgery of items. In case of healthcare, fake or low quality medicine can be sold at a high price compromising or complicating someone's treatment process. Other examples where cloning must be prevented are access control, forgery of banknote, etc.

For many applications an exact copy of a tag or its functionality may not be required (Henrici 2008). The reason is, in most of the times, a simple reading of EPC number or basic data will be needed for normal operation of the RFID system, while other data or functionality will be rarely used. This means that an attacker with simple cloning may inflict considerable harm to the system that depends on RFID.

4.1.1.3 Side Channel Analysis (SCA)

Some researchers think that side channel analysis would potentially be the most serious threat to RFID tags (Duc et al., 2009). Side channel analysis exploits the knowledge gained from the physical implementation of a cryptographic algorithm. For example, timing information, power consumption, electromagnetic radiation can provide an extra source of information that may help to expose secret data stored in RFID tags for the analysis. A non-invasive attack can be launched by gathering information from the fluctuations of the electromagnetic field emanated by a device while performing a cryptographic operation (Oswald, Kasper & Paar, 2011).

4.1.2 Threats of Communication Layer

One problem that occurs at the communication layer is when a reader scans multiple tags at the same time or multiple readers in close proximity read multiple tags. When multiple tags are energized by the reader simultaneously, their respective reflected signals come back to the reader at almost the same time. Then reader is then unable to differentiate between those signals to get individual tag's data. This issue becomes a real concern when a large number of tags are read together.

The main threat at the communication layer is the 'Collision' attack. In this case the attacker interferes with the mechanism that a reader applies to identify a single tag for communication. Any interference above a certain level from nearby readers or other RF radio transmitters may prevent a reader from discovering and polling tags. An attacker can install an unauthorized RF transmitter or tags with malicious intention. This RF transmitter will interfere with the legitimate reader or unauthorized tags will responds to the reader as the same as other legitimate tags are responding creating deliberate collision. Collision can also lead to the Denial of Sevice (DoS) attack.

4.1.3 Threats of Application Layer

The following major threats exist at the application layer.

4.1.3.1 Spoofing

In spoofing, a RFID reader is fooled into believing that the data it has received is from a legitimate tag. A portable reader might covertly read and record a data transmission from a tag that could contain the tag's ID and when this data transmission is retransmitted, it appears to be a valid tag. The tag's ID can also be written in appropriately formatted data on blank RFID tags. In that case the forged tag masquerades as a valid tag and thereby gains products and accesses services available to someone else's ID (Xiao, Boulet, & Gibbons, 2007; Peris-Lopez et al, 2006). This kind of attack can be used to retag extensive item in supermarket, masquerade digital passports, and fool contact-less payment systems and building access control stations. Though many people consider 'spoofing' as 'cloning,' but there is shuttle difference between them. In spoofing what we have is 'a copy of the tag's transmission,' not a clone of the tag. A portable reader in a bag can copy the transmission and then retransmit.

Researchers have tried to assess difference scenarios where 'spoofing' attack may be launched. An experiment by Johns Hopkins University researchers recently cloned a cryptographically-protected Texas Instruments digital signature transponder, which they used to buy gasoline and unlock a DST-based car immobilization system (Rieback, Crispo, & Tanenbaum, 2006).

4.1.3.2 Replay

In a replay attack, an attacker records a legitimate tag's response to a reader's interrogation and later broadcasts the recorded response to impersonate that tag to a reader. A typical example of this kind of attack is to gain unauthorized access to restricted areas by simply replaying the exact radio signal that was used by a legitimate tag to access permission. In 2008, a reply attack on plastic tickets with an embedded RFID chip used in Netherlands public transport was reported. One particular attack was on a single-use RFID ticket and it allowed the attacker to use the same ticket again and again (http://www.cs.vu.nl/~ast/ov-chip-card/). Another attack was reported with UK's RFID-enabled license plate, *e-plates*. An attacker can record the radio signals that contain encrypted ID and payment information when a car's license plate is scanned. The attacker can later replay the same information to avoid toll-ways charges and congestion fees.

4.1.3.3 Denial of Service (DoS)

DoS is one category of attacks on RFID systems. An attacker attempts to prevent the target tag from receiving legitimate services. One kind of DoS attack is desynchronization. If an attacker can break the synchronization between the information stored in a tag and reader, the tag can't be conventionally authenticated, rendering the tag useless (Lee, 2005). In desynchronization attack the shared secret values among the tag and the back-end server are made inconsistent by an attacker. Then, the tag and back-end server cannot recognize each other.

4.2 Privacy Threats

RFID use raises concern about two types of privacy concern as described below:

1. **Location Privacy:** A person's location privacy is compromised if a tag ID associated with that person is spotted at a particular reader's location. Such tags may not have unique ID. A certain combinations of non-unique tags may still form unique constellations of items that can be used to identify an individual. If a person's car is detected at the parking space of a hospital or of a market place, then one can infer (not prove) the location and the activity of that person.
2. **Data Privacy:** When a tag carries not only the identifier, but also a person's personal details (like name, contact, financial and medical information, etc.) data privacy may be compromised. In absence of strong security protection, eavesdropping or a malicious reader can fool a tag to reveal such data. A special case of data privacy is the disclosure of the products that a person is carrying, e.g., the types and brands of clothing one is wearing, the items in one's shopping bag, or even the furniture in a house (Langheinrich, 2009). Disclosing this information is tan-

tamount to invading someone's personal space. Even when the actual identity of the victim remains unknown, knowing a person's possession may put him/her into jeopardy, on the top of invading privacy.

Langheinrich (Langheinrich, 2009) noted that loss of privacy may create fear among customer of using RFID. Apart from revealing personal data directly or indirectly through clandestine scanning as described above, several clandestine scans can be pooled to track a person's movements compromising location privacy. Fabian et al. also mentioned the vulnerability of the commercial product information network to data disclosure attacks (Fabian, Gunther & Spiekermann, 2005). Andrew Kantor, a columnist for USA Today, has this observation: "A department store's RFID system recognizes that you're carrying an item you bought there last week. Now it knows who you are. And if there are readers scattered about, it knows where you're going. Come home to a phone call, 'Mr. Kantor – we noticed you were shopping for a television...'" (Kantor, 2003). Similarly Forbes Magazine commented: "As the shopper enters the store, scanners identify her clothing by the tags embedded in her pants, shirt and shoes. The store knows where she bought everything she is wearing (Schoenberger, 2002). These scenarios certainly depict a frightening aspect of privacy violation.

Langheinrich (2009) also noted the potential increase of criminal activities, like "Sophisticated thieves walk by homes with RFID readers to get an idea of what's inside. Slightly less sophisticated thieves do the same thing in a parking lot, scanning car trunks" (Kantor, 2003) and "Using mobile readers, future pickpockets could find out how much cash someone would carry" (Zeidler, 2003). Potential criminal activities are not only confined to burglary: "In the future, there will be this very tiny microchip embedded in the envelope or stamp. You won't be able to shred it because it's so small...Someone will come along and read my garbage and know every piece of mail I received" (Roberti, 2003).

5. RFID SECURITY PROTOCOLS

5.1 Authentication Protocols

Important concerns associated with the RFID technology are security and privacy of the tag content. Indeed, it is pretty much easy for anybody with technical skills to set up a device for reading the tag content. Nevertheless, to reserve user privacy, only authorized RFID readers should be enabled to access the tag content. At the same time, legal RFID readers would like to be sure that the tags they are reading are authentic and have not been counterfeit. An authentication protocol, which grants access to the tag content only to a legitimate reader and, at the same time, guarantees the reader of the identity of the tag, is therefore required. Based on the computational cost and the operations supported on tags, authentication protocols can be divided in classes: full-fledged class, simple class, lightweight class and ultralightweight class.

The protocols (Juels, Molner & Wagner, 2005; Kumar & Paar, 2006) belonging to the full-fledged class support cryptographic functions like hashing, encryption, and even public key algorithms on tags. One of the main applications of these full-fledged protocols is E-passport (Juels, Molner & Wagner, 2005).

The tags in the protocols of the simple class should support random number function and hash functions but not encryption functions/public key algorithms. Examples are like (Chien, 2006; Molnar & Wagner, 2004; Weis et al, 2004; Yang et al, 2005), where Chien (Chien, 2006) and Avoine et al. (Avoine, Dysli, & Oechslin, 2005) had reported the weaknesses of the schemes such as the secret key disclosure problem and the violation of anonymity.

The lightweight class of RFID authentication protocols does not require hashing function on tags; for example, the EPCglobal Class-1 Gen-2 RFID tag (EPCglobal 2011) supports Pseudo-Random Number Generator (PRNG) and Cyclic Redundancy Code (CRC) checksum but not hashing function. The protocols (Duc et al, 2006; Juels 2005; Karthikeyan & Nesterenko, 2005) belong to this class, where the scheme (Juels 2005) did not take the eavesdropping and privacy issues into consideration, and Chien and Chen (Chien, & Chen, 2007) had reported the DOS attack, replay attack, tracking attack and spoofing tag problem on the schemes (Duc et al, 2006), (Karthikeyan & Nesterenko, 2005), respectively. The HB-series (Hopper & Blum, 2001; Bringer, Chabanne & Dottax, 2006; Piramuthu, 2006) can also be classified into this class, since they demand the support of random number function but not hash function on tags. Hopper and Blum (Hopper & Blum, 2001), based on the LPN problem, first proposed the HB protocol to defect the passive attacker. Later, the HB protocol was successively attacked and improved by its sister works (Bringer, Chabanne & Dottax, 2006; Piramuthu, 2006). Actually, the HB-series cannot be regarded as complete, since these protocols only consider the authentication of tags. They neglected the issue of the authentication of the readers, the tracking problem, and the anonymity issue, and even the privacy of the tag identification.

Recently, Peris-Lopez et al. proposed a series of ultralightweight class of RFID authentication protocols (Peris-Lopez et al 2006; Hernandez-Castro, Estevez-Tapiador & Ribagorda, 2006a, b), where the tags involve only simple bit-wise operations like XOR, AND, OR, and addition mod 2m. These schemes are very efficient, and they only require about 300 gates. Unfortunately, Li and Wang (Li & Wang, 2007) and Li and Deng (Li & Deng, 2007), respectively, reported the de-synchronization attack and the full-disclosure attack on these protocols. Moreover, (Chien, 2007) finds that the previous schemes (Peris-Lopez et al 2006; Hernandez-Castro, Estevez-Tapiador & Ribagorda, 2006a, b) only provided weak authentication and weak integrity protection, which make them vulnerable to various attacks, instead proposes a new ultralightweight RFID authentication protocol, namely, Strong Authentication and Strong Integrity (SASI) for low-cost RFIDs, and the scheme should withstand all the possible attacks.

Since ultralightweight protocols require significantly low resource in RFID tag compared to full-fledged, simple and lightweight class of authentication protocols, they are best suited for RFID authentication. In the rest of this section we review such an ultralightweight authentication protocol, namely, SASI protocol, as seen in Figure 2. (Chien, 2007).

- **SASI Protocol:** The protocol involves three entities: tag, reader, and backend server. The channel between the reader and the backend server is assumed to be secure, but that between the reader and the tag is susceptible to all the possible attacks. Each tag has a static identification (ID), and pre-shares a pseudonym (IDS) and two keys K1|K2 with the backend server. The length of each of ID|IDS|K1|K2 is 96 bits. To resist the possible de-synchronization attack, each tag actually keeps two entries of (IDS, K1, K2): one is for the *old* values and the other is for the *potential next* values. The protocol consists of three stages: tag identification phase, mutual authentication phase, and pseudonym updating and key updating phase.
    - **Tag Identification:** In each protocol instance, the reader may probe the tag twice or once in the tag identification phase, depending on the tag's IDS is found or not. The reader first sends "*hello*" message to the tag, and the tag will respond with its *potential next* IDS. The reader uses the tag's response IDS to find a matched entry in the database, and goes to the mu-

Figure 2. The SASI protocol

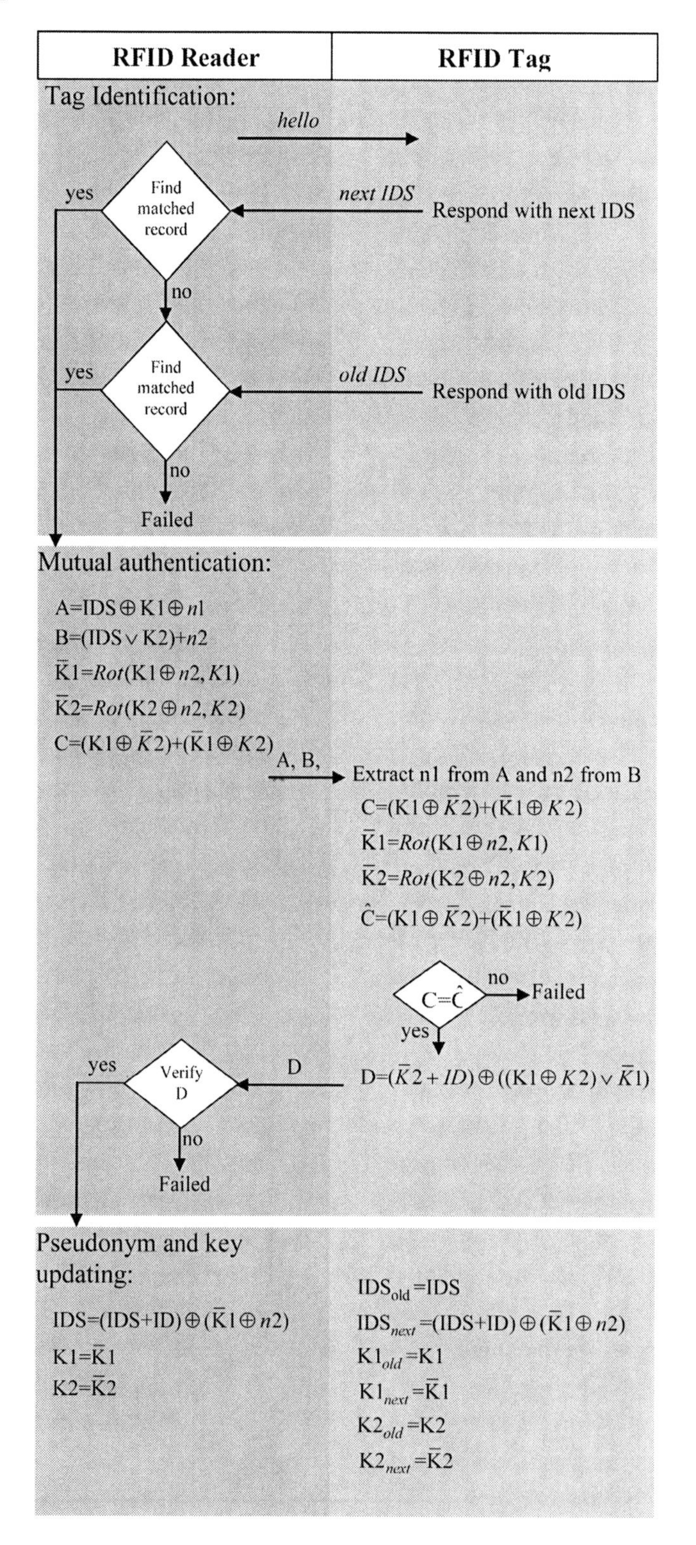

tual authentication phase if a matched entry is found; otherwise, it probes again and the tag responds with its *old* IDS.

- **Mutual Authentication:** In the mutual authentication phase, the reader and the tag authenticate each other, and they, respectively, update their local pseudonym and the keys after successful authentication. The reader uses IDS to find a matched record in the database. It could be the *potential next* IDS or the *old* IDS of the tag. It then uses the matched values and two generated random integers $n1$ and $n2$ to compute the values A, B, and C (the calculation equations are specified in Figure 2). From A||B||C, the tag first extracts $n1$ from A, extracts $n2$ from B, computes $\bar{K}1$ and $\bar{K}2$ and then verifies the value of C. If the verification succeeds, then it computes the response value D. Upon receiving D, the reader uses its local values to verify D.
- **Pseudonym and Key Updating:** After successful authentication, the tag stores the matched values to the entry (IDS_{old}, $K1_{old}$, $K2_{old}$) and stores the updated values to the entry (IDS_{next}, $K1_{next}$, $K2_{next}$). In addition, the reader updates the local pseudonym and keys as specified in Figure 2.

From above protocol description it is clear that the random number generator is required on the reader only, and the tags only involve simple bitwise operations like bitwise XOR $(\oplus)$, , bitwise OR $(\vee)$, bitwise AND $(\wedge)$, addition mod 2m (+), and left rotation *Rot*(*x*,*y*). Here, *Rot*(*x*,*y*) is defined to left rotate the value of x with y bits.

The protocol is examined against a number of security features and possible attacks and found to be very robust compared to its counterfeit other ultralightweight RFID authentication protocols such as LMAP, M²AP, EMAP (Peris-Lopez et al 2006; Hernandez-Castro, Estevez-Tapiador & Ribagorda, 2006a, b; Li & Wang, 2007; Li & Deng, 2007), and the result is shown in Table 1.

5.2 Anti-Collision Protocols

RFID systems consist of a reading device called a reader, and one or more tags. The reader is typically a powerful device with sufficient memory and computational resources. On the other hand, tags vary significantly in their computational capabilities. They range from dumb passive tags, which respond only at reader commands, to smart active tags, which have an on-board microcontroller, transceiver, memory, and power supply. Among tag types, passive ones are emerging to be a popular choice for large scale deployments due to their low cost.

Collision due to simultaneous tag responses is one of the key issues in RFID systems (Finkenzeller, 2003). Figure 3 shows the collision problem when multiple tags are present within the reading zone of a reader. Without an anti-collision protocol, the replies from these tags would collide and thereby prolong their identification. It results in wastage of bandwidth, energy, and increases identification delays. To minimize collisions, RFID readers must use an anti-collision protocol. To this end, this section reviews state-of-the-art anti-collision protocols, and provides a comparison of the different approaches used to minimize collisions, and hence help reduce identification delays. Such review will be of great importance to researchers and designers that are building RFID systems.

Anti-collision protocols can be categorized into, space division multiple access (SDMA), frequency division multiple access (FDMA), code

Table 1. A simple comparison of ultralightweight authentication protocols (Chien, 2007)

	LMAP	M²AP	EMAP	SASI
Resistance to de-synchronization attacks	No	No	No	Yes
Resistance to disclosure attacks	No	No	No	Yes
Privacy and anonymity	No	No	No	Yes
Mutual authentication and forward secrecy	No	No	No	Yes
Total messages for mutual authentications	$4L$	$5L$	$5L$	$4L$
Memory size on tag	$6L$	$6L$	$6L$	$7L$
Memory size for each tag on server	$6L$	$6L$	$6L$	$4L$
Operation types on tag	$\oplus, \wedge, \vee, +$	$\oplus, \wedge, \vee, +$	$\oplus, \wedge, \vee$	$\oplus, \wedge, \vee, +, Rot$

L denotes the bit length of one pseudonym or one key

division multiple access (CDMA), and time division multiple access (TDMA) (Finkenzeller, 2003). Briefly, SDMA protocols spatially separate the channel using directional antennas or multiple readers to identify tags. They, however, are expensive and require intricate antenna designs. On the other hand, FDMA protocols involve tags transmitting in one of several predefined frequency channels; thus, requiring a complex receiver at the reader. Lastly, systems based on CDMA require tags to multiply their ID with a pseudo-random sequence (PN) before transmission. Unfortunately, CDMA based systems are expensive and power hungry. TDMA protocols constitute the largest group of anti-collision protocols, and hence the focus on this section. These protocols can be classified as reader-driven, also called Reader-talk-first (RTF), and tag driven also called Tag-talk-first (TTF) respectively. Most applications use RTF protocols, which can be further classified into Aloha and tree based protocols. The basic idea behind RTF is that tags remain quiet until specifically addressed or commanded by a reader. On the other hand, TTF procedures function asynchronously. This means a TTF tag announces itself to the reader by transmitting its ID in the presence of a reader. Tags driven procedures are slow as compared to RTF procedures (Tang, & He, 2007).

5.2.1 ALOHA Based Anti-Collision Protocols

Pure Aloha (PA). In PA based RFID systems, a tag responds with its ID randomly after being energized by a reader. It then waits for the reader to reply with, i) a positive acknowledgment (*ACK*), indicating its ID has been received correctly, or ii) a negative acknowledgment (*NACK*), meaning a collision has occurred. If two or more tags transmit, a complete or partial collision occurs [46], which tags then resolve by backing off randomly before retransmitting their ID. Pure Aloha based systems have several variants (Burdet, 2004; Klair, Chin & Raad, 2007):

Figure 3. The tag collision problem

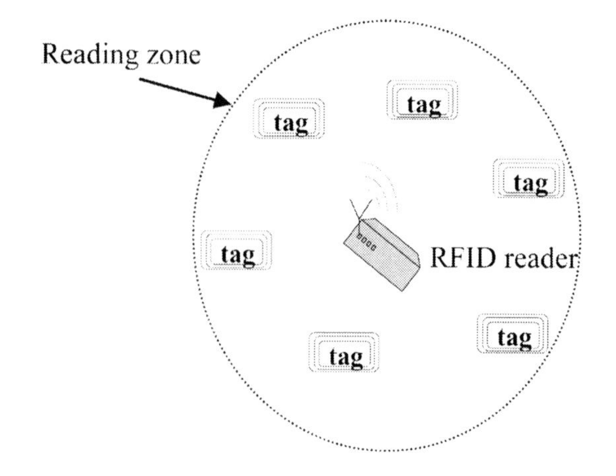

- **PA with Muting:** After identification, the reader silences read tags using the "mute" command. When muting is used, the number of tags in a reader's interrogation zone is reduced, hence, muting has the effect of reducing the offered load to the reader after each successful identification.
- **PA with Slow Down:** Instead of being muted, a tag can be instructed using a "slow down" command to reduce its rate of transmissions, hence decreasing the probability of collision.
- **PA with Fast Mode:** A "silence" command is sent by the reader once it has detected the start of a tag transmission. This command has the effect of stopping other tags from transmitting. Tags are allowed to transmit again after the reader has sent an ACK command or until their waiting timer expires.

Slotted Aloha (SA). In Slotted Aloha (SA) based RFID systems, tags transmit their ID in synchronous time slots. If there is a collision, tags retransmit after a random delay. The collision occurs at slots boundary only, hence there are no partial collisions (Schwartz, 1988). Slotted Aloha also has numerous variants (Burdet, 2004; Klair, Chin & Raad, 2007; Schwartz, 1988):

- **SA with Muting/Slow Down:** The principle operation is similar to PA with muting/slow down, but operates in a slotted manner.
- **SA with Early End:** If no transmission is detected at the beginning of a slot, the reader closes the slot early. Two commands are used: start-of-frame (SOF) and end-of-frame (EOF). The former is used to start a reading cycle, and the latter is used by the reader to close an idle slot early.

Framed Slotted Aloha (FSA). In PA and SA based systems, a tag with a high response rate will frequently collide with potentially valid responses from other tags. Therefore, basic FSA (BFSA) protocols mandates that each tag responds only once per frame. Note, the term "basic" refers to the frame size being fixed throughout the reading process. BFSA has four variants: 1) BFSA-non muting, 2) BFSA-muting, 3) BFSA-non-muting-early-end, and 4) BFSA-muting-early end. In BFSA-non muting, a tag is required to transmit its ID in each read round. BFSA non-muting suffers from an exponential increase in identification delay when the number of tags is higher than the frame size (Cha & Kim, 2005). For BFSA-muting, the number of tags reduces after each read round, since tags are silenced after identification. When a read round is collision free, the reader concludes that all tags have been identified successfully. BFSA-non-muting-early-end and BFSA-muting-early-end variants incorporate the early-end feature. Specifically, the reader closes a slot early if no response is detected at beginning of a slot.

Dynamic Frame Slotted Aloha (DFSA). FSA protocols with variable frame sizes are called dynamic framed slotted Aloha (DFSA) [44]. Similar to BFSA, DFSA operates in multiple rounds, and it can also incorporate the early-end feature. The key difference, however, is that in each read round, the reader uses a tag estimation function to vary its frame size (Cha & Kim, 2005). A tag estimation function calculates the number of tags based on feedback from a reader's frame, which include the number of slots filled with zero ($c0$), one ($c1$) and multiple tag responses (ck). This information is then used by the function to obtain a tag estimate, and hence the optimal frame size N for a given round. Here, the optimal frame size is one which promises the maximum system efficiency and minimum identification delay (Vogt, 2002; Zhen, Kobayashi & Shimizu, 2005).

5.2.2 Tree Based Anti-Collision Protocols

Tree based protocols were originally developed for multiple access arbitration in wireless systems (Hush & Wood, 1998). These protocols are able to single out and read every tag, provided each tag has a unique ID. All tree based protocols require tags to have muting capability, as tags are silenced after identification. Tree based protocols can be classified into: 1) Tree splitting (TS), 2) Query tree (QT), and 3) Binary search (BS) algorithms. In addition, there are a new branch of hybrid protocols for tag reading that combine the advantages of tree and Aloha protocols.

1. **Tree Splitting (TS):** TS protocols operate by splitting responding tags into multiple subsets using a random number generator (Hush & Wood, 1998; Myung & Lee, 2005). We present two algorithms in this category. These subsets become increasingly smaller until they contain one tag. Identification is achieved in a sequence of timeslots. Each tag has a random binary number generator. In addition, each tag maintains a counter to record its position in the resulting tree. Tags with a counter value of zero are considered to be in the transmit state, otherwise tags are in the wait or sleep state. After each timeslot, the reader informs tags whether the last timeslot resulted in a collision, single or no response. If there was a collision, each tag in the transmit state generates a random binary number b and adds the number to its current counter value. On the other hand, tags in the wait state increment their counter by one. In the case of idle or single response, tags in the wait state decrement their counter by one. After identification, tags enter the sleep state. Figure 4 shows an example of tree splitting algorithm with 4 tags.
2. **Query Tree Algorithms:** In TS protocols, tags require a random number generator and a counter to track their tree position, thus making them costly and computationally complex. Query tree algorithms overcome these problems by storing tree construction information at the reader, and tags only need to have a prefix matching circuit (Law, Lee & Siu, 2000; Myung, Lee & Shih, 2006). The reader transmits a query q, and tags with a matching prefix reply to the reader. Collision occurs when multiple tags have the same prefix. In this case, the reader forms a new query by appending q with a binary 0 or 1. The reader then repeats the reading process using the augmented query. Figure 5 illustrates the QT algorithm with an example.
3. **Binary Search (BS):** BS algorithm (Finkenzeller, 2003) involves the reader transmitting a serial number to tags, which they then compare against their ID. Those tags with ID equal to or lower than the serial number respond. The reader then monitors tags reply bit by bit using Manchester coding, and once a collision occurs, the reader splits tags into subsets based on collided bits. Figure 6 depicts a reader using BS to read five tags.

Initially, the reader starts reading with the maximum possible tag ID value, i.e., 111. Tags with an ID value less than 111 respond, resulting in the reply XXX. This indicates all three bits have experienced a collision. The reader then transmits another query by replacing the most significant collided bit with 0, and sets the other bits to 1, i.e., the new query becomes 011. This subsequent query solicits the response 0XX. The reader then sends the query 001. Only tag 001 have ID lower than 001 and therefore it is identified successfully. After that, the reader restarts the reading with query 111.

Figure 4. The basic tree splitting algorithm

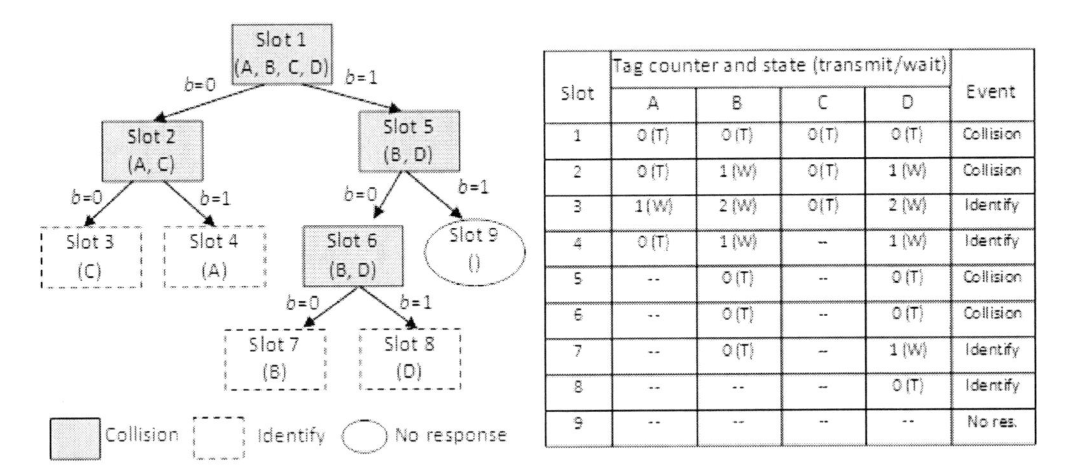

Slot	Tag counter and state (transmit/wait)				Event
	A	B	C	D	
1	0 (T)	0 (T)	0 (T)	0 (T)	Collision
2	0 (T)	1 (W)	0 (T)	1 (W)	Collision
3	1 (W)	2 (W)	0 (T)	2 (W)	Identify
4	0 (T)	1 (W)	--	1 (W)	Identify
5	--	0 (T)	--	0 (T)	Collision
6	--	0 (T)	--	0 (T)	Collision
7	--	0 (T)	--	1 (W)	Identify
8	--	--	--	0 (T)	Identify
9	--	--	--	--	No res.

Figure 5. The query tree algorithm

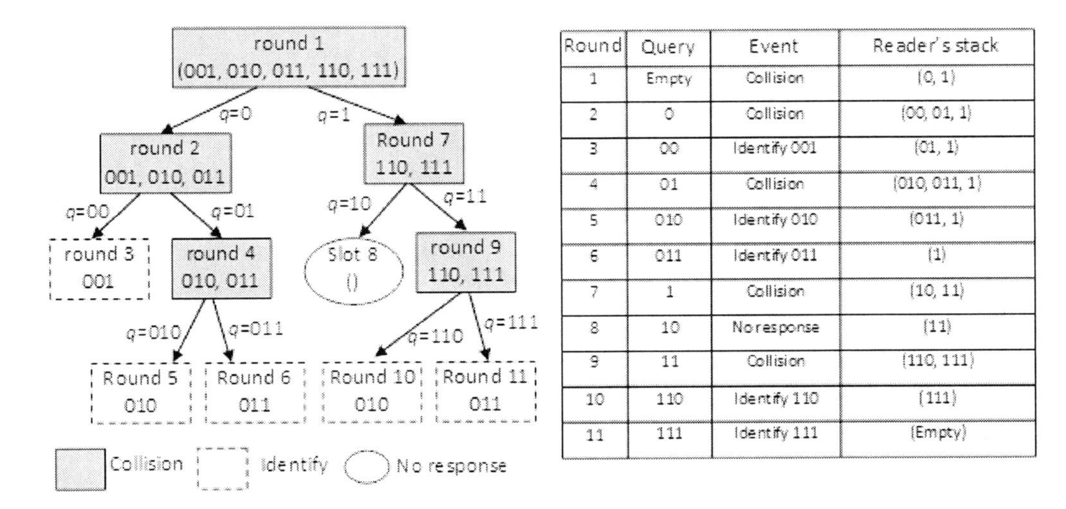

Round	Query	Event	Reader's stack
1	Empty	Collision	{0, 1}
2	0	Collision	{00, 01, 1}
3	00	Identify 001	{01, 1}
4	01	Collision	{010, 011, 1}
5	010	Identify 010	{011, 1}
6	011	Identify 011	{1}
7	1	Collision	{10, 11}
8	10	No response	{11}
9	11	Collision	{110, 111}
10	110	Identify 110	{111}
11	111	Identify 111	{Empty}

Figure 6. The binary search anti-collision algorithm

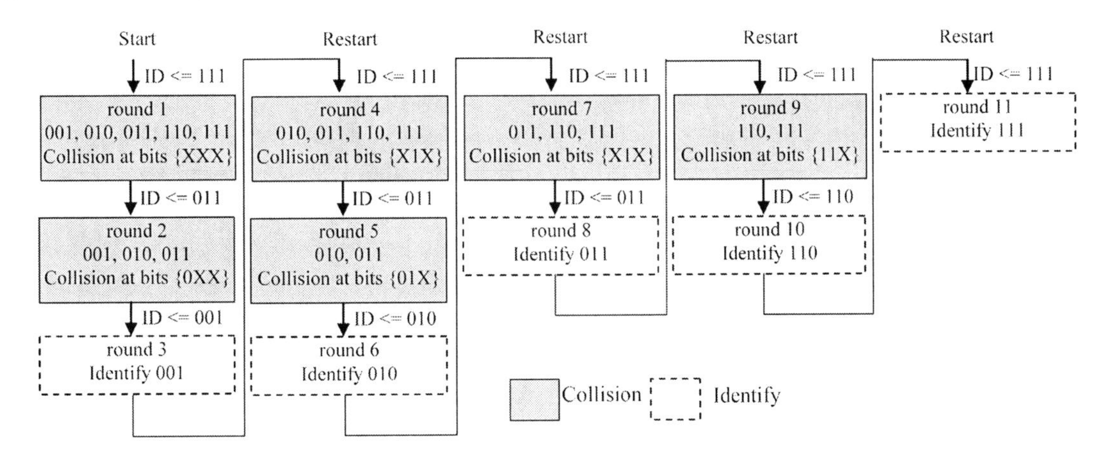

5.3 Privacy Preserving Protocol

In RFID tag identification, privacy is measured by an attacker's ability to trace tags by means of their responses. One of the earliest works on privacy preserving was done by Conti et al. (Conti et al, 2007). They proposed a privacy preserving protocol with forward security, named RIPP-FS (RFID Identification, Privacy Preserving protocol with Forward Secrecy). The proposed protocol is based on hash chains and the authors claimed it to provide key secrecy, privacy, forward secrecy and DoS resiliency. The key steps in the underlying protocol are:

- **Tag Side:** Change the reply string every time the tag is queried. Every tag shares a unique private symmetric key with the server. Before deploying the tags, the server computes a long enough one-way hash function chain. Those hash functions are used to update the shared key and send the HMAC value calculated using the updated key.
- **Reader Side:** Use a Lamport hash chain (Haller, 1994) to prove the identity of the reader. To make sure that the reader is a genuine one, the tag must verify both the time and the key received. Each tag computes the number of iterations occurred from its last reply, as the difference from its own stored time and the received reader time. A negative or zero value for the difference means that an illegal reader is trying to query, reusing a previous message, since the server time should always increase.
- After a successful authentication, use a key updating mechanism in order to allow the reader to identify the tags replying a legal query. In order to identify the tags, the server maintains some key lookup tables that contains the triplet [Tag, updated key (at time t) and HMAC vale (at t)]. When the reader at time t collects a set of HAMC values, it simply looks-up its tables for time t and identifies all the tags that answered.

One drawback of RIPP-FS is that it is only possible to make finite number of hash chain and at certain point all values of the chain are revealed and then tag reading can't be continued. A simple solution could be to broadcast to every tag a new seed for a new chain in an encrypted form.

The authors later proposed a faster and efficient version of RIPP-FS and named it FastRIPP (Conti et al, 2007b) that significantly reduces some extra computations required on the tag side. This improvement is achieved by an efficient hash traversal amortization technique that uses fractal hash sequence representation (Jakobsson, 2002). FastRIPP supports all the security features offered by RIPP-FS and reduces the computational complexity from $O(m)$ to $O(\log^2 m)$ for hash chain computation, m being the length of the chain.

Schemes like the above rely on the existence of a back-end database that serves as a central server. Each tag shares a unique secret key with the server. Once the reader queries a tag, the tag uses its secret key to generate a response to the reader, which then forwards the response to the server for verification. However, a reliable server cannot be always guaranteed to be operational all the time. Liu and Hu (Liu & Hu, 2008) proposed a protocol named ROAD which supports offline authentication and privacy protection with DoS resilience without the use of a central server. The scheme makes use of a trusted third party Citification Authority (CA) to assign a unique identifier to each tag and the reader. ROAD performs a 'challenge and response' dialogue before sending the tag secret to the reader. The reader sends a request and all tags, which upon receiving the request respond to the reader. The reader then searches its list to identify a tag from all those tags. A slightly different version of this protocol is ROAD$^+$. ROAD$^+$ differs from ROAD in that the former provides targeted tag authentication whereby the reader issues a query such that only the targeted tag can interpret and the tag replies in such a manner that only an authenticated reader can understand. Since only the authenticated reader can understand, privacy is preserved.

In Juels & Pappu (2003), the authors proposed a RFID banknote protection scheme which enables a note to be tracked by the legal law agency and verified by the merchant. However, a later study by Avoine (Avoine, 2004) reports attacks that severely compromise the privacy of the banknotes' bearers without physical contact with notes, including pick pocketing attack and attack that makes a banknote sleep and dead. Yang et al. (2008) proposed a modified version of the work by Juels and Pappu (2003) that enhances the privacy of banknote bearers. The protocol proposed in (Yang et al., 2008) used two access-keys instead of one access key used in (Juels & Pappu, 2003). An access key (D_M) is the same as in (Juels & Pappu, 2003) used for a merchant and an additional key (D_L) is a global key for law enforcement agency. A central bank can create different D_L for the different banknotes and deliver this global key to law enforcing authority with the authority to track the notes through EPC network (EPC global, 2011; Leong & Ng, 2004). Like ROAD, Yang et al. (2008) adopted a challenge-response strategy to avoid stealing an access-key to prevent attacks based on the disclosure of an access-key. However, implementations of this scheme require more memory in the RFID tag.

In the above protocols, upon receiving a tag's response, the reader is required to perform a search of all tags in the system listed in the backend database. Although this class of protocols provide privacy protection in tag identification, their practical implementation is not scalable. Alomair et al. (2010) proposed a scalable privacy preserving protocol that allows constant time identification. In this protocol, each tag has an internal counter c, and is preloaded with a unique secret pseudonym ψ, and a secret key k; ψ and k are updated each time a successful mutual authentication takes place. The reader first generates a random nonce r and sends it to the tag. On receiving r, the tag then computes $h(\psi,c)$ and $r\text{c}=\text{h}(0,\psi,c,k,r)$, and increment the counter. Using $h(\psi,c)$, the reader accesses the database to identify the tag and fetch its relevant information (eg., ψ and k), and a new pseudonym ψ' is generated to update the tag. With $r\text{c}$, the reader authenticates the tag by confirming its knowledge of the secret key, k, obtained from the database. The reader then sends some response strings that enable the tag to authenticate the reader. In this way, the protocol utilizes resources (i.e., databases) that are already available in RFID systems to improve identification efficiency.

Batina et al. (2010) proposed a privacy preserving scheme based on the 'grouping' concept. The concept of RFID grouping was first introduced by Juel (2004). The motivation behind grouping was originated from the fact that some applications may require the proof that two or more related entities are presented together. For example, hardware components that need to be shipped together, an airline ticket linked to passport and boarding pass, books on same topic are shelved together in library, etc. Batina et al. (2010) has extended the grouping concept to include privacy and forward security. It has been shown that public-key cryptography is needed to achieve a certain level of privacy protection (Vaudenay, 2007). In addition to the reader(s), the protocol employs a trusted verifier. The tags and the readers are engaged in the protocol to construct the grouping proof. The reader coordinates the execution of the protocol, collect the grouping proof and forward it to the verifier. The verifier is a trusted one and relies on the public-private key system and employs Elliptic Curve Cryptography (ECC). The proposed grouping-proof protocol inherits the security properties of the ID-transfer protocol (Lee et al., 2010) which is designed to provide secure entity authentication. Following the privacy protection attributes of the ID-transfer protocol against a narrow-strong adversary, untraceability can even be guaranteed if the challenges of the ID-transfer protocol are controlled by the adversary (Batina et al., 2010). As a direct consequence, the EC-based grouping-proof protocol is also narrow-strong privacy-preserving. However, the protocol did not consider the attack where an adversarial reader scans two non-compromised tags, and forwards the grouping proof at a later time to the verifier.

Lin et al. (2010) proposed an RFID-based children tracking scheme for large amusement parks, called REACT that preserves a child's identity privacy, unlinkable location privacy and forward security. In this scheme each child wears a passive tag and there are various checkpoints equipped with RFID readers in the park which are connected to storage nodes and control center. When a child passes by an RFID reader, the reader processes the tag information and forwards it to a storage node. When a child's location is requested, it can be retrieved by querying the storage node. Privacy of child's identity, location and tracking query to control center is achieved by using cryptographic keys and exchanging strings that are comprised of timestamps and hash functions. The study presents simulation results that show that the proposed scheme is able to quickly locate a lost child's location.

5.4 Countermeasure for SCA

Designing efficient countermeasures to thwart SCAs is a real challenge and no effective solution has been developed yet. Most notably, types of SCA attacks are Simple Power Analysis (SPA), Differential Power Analysis (DPA) and differential electromagnetic analysis (DEMA). Oren and Shamir (2007) showed a successful analysis result which can be used to extract kill password remotely from a UHF EPC tag (EPCglobal, 2011). The authors suggested common countermeasures to prevent the analysis with some examples. In (Luo & Fei, 2011), the authors proposed a systematic method for evaluating resilience of cryptographic algorithms and reveals inherent algorithmic properties related to side-channel attacks. Though the paper does not discuss SCA attack specifically in relation to RFID, the proposed method is useful for devising countermeasure for SCA attack on RFID. Plos (2008) investigated EM analysis on UHF RFID tag.

6. RESEARCH CHALLENGES IN RFID SECURITY

Even though research on RFID security and privacy protection has advanced in recent years, but it is not still adequate. For practical deployment of RFID in wide scale many issues need to be resolved. Also there are almost no works done for security in chipless RFID. The followings are some of the research challenges in RFID security as noted by Duc et al. (2009):

- Cryptography plays the core role in security protocol as described above. There are very strong and well-proven cryptographic techniques for wired networks, and in recent times cryptographic algorithms for wireless networks have also been significantly improved. However, RFID tags have very limited processing power and therefore, for practical deployment, we need cryptographic algorithms that are lightweight and fast, but at the same time possess strong cryptographic properties. The lack of the study on lightweight cryptographic primitives includes design of new primitives, analysis of the security of new primitives and their efficient implementation (Duc et al., 2009). Though few lightweight cryptographic algorithms are proposed in literature, much research needs to be done to evaluate them through cryptanalysis and investigate new algorithms for stronger protection. As new types of attacks are reported, research need to focus to counter those attacks. Another important aspect that plays vital roles in cryptography is the regeneration of pseudorandom number so that it generation cannot be predicted. Further research need to done to generate random number that satisfies the requirement of RFID. As Duc et al. pointed out the EPCglobal Gen2 standard specifies a requirement for its built-in pseudorandom number generator (PRNG) that the probability of generating a 16-bit number is bounded by $[0.8 \times 2^{-16}, 1.25 \times 2^{-16}]$. However, there is no known proposed scheme or implementation for such a PRNG except the work done by Peris-Lopez et al. (2009).

- Duc et al. (2009) stressed that the cryptographic task that is proposed in various research works must be based on practical assumption so that its implementation becomes feasible. To support forward security, most of the authentication and privacy preserving protocols employ key-evolving protocol to update the secret key at the tag and the server/reader. They conclude that "it is impossible to realize a robust interactive key-evolving protocol that is secure against de-synchronization of secret (which may imply the loss of forward security)".
- An appropriate security model for RFID is essential for developing solution of security and privacy. There are shortcomings of the known security model. For example, no security model exists for multiple tag scanning. Also security models, e.g., Universal Composable Framework (Le et al., 2007) and Vaudenay's model (Vaudenay, 2007) do not consider the RFID reader as a party of a RFID system which is not a reflection of the actual system, because mutual authentication and maintain synchronization state is very much need for make the system secured.
- There have been only very few works done on location privacy (e.g., the work by Lin et al., 2010) which is a very crucial privacy issue as discussed in Section 4.2 and further investigations are needed to ensure privacy before people feel confident about using RFID tag. In side channel analysis attack on RFID, though few works have been done with SPA and DPA type attack, to the best of the authors' knowledge no work has yet been done on DEMA type attack.

7. CONCLUSION

Although RFID technology is not new, in recent times RFID has found diverge and innovative applications in business, supply chain management, industry, sports, animal farms, library, and asset and object tracking, etc. RFID technology is evolving and more and more innovation applications may encompass may aspects of our everyday life. Therefore, the security and privacy concern of RFID deployment has drawn much research attention in recent times. This chapter presents an overview of the issues on security and privacy concern and discuss the most recent security and privacy protocols, specific to RFID system, and analysis their strength and vulnerabilities. New challenges and further research directions are identified to meet the evolving need of RFID system.

REFERENCES

Alomair, B., Clark, A., Cuellart, J., & Poovendran, R. (2010). Scalable RFID systems: A privacy-preserving protocol with constant-time identification. In *the Proceedings of the IEEEI IFIP International Conference on Dependable Systems & Networks* (DSN), (pp. 1-10).

Avoine, G. (2004). Privacy issues in RFID banknote protection schemes. In *Proceedings of the 6th International Conference on Smart Card Research and Advanced Applications*, (pp. 33-48).

Avoine, G., Dysli, E., & Oechslin, P. (2005). Reducing time complexity in RFID systems. *Proceedings of the 12th Annual Workshop Selected Areas in Cryptography* (*SAC*).

Bashir, A. K., Chauhdary, S. H., Shah, S. C., & Park, M.-S. (2011). Mobile RFID and its design issues. *IEEE Potential*, *30*(4), 34–38. doi:10.1109/MPOT.2011.940230

Batina, L., Lee, Y. K., Stefaan, S., Singelee, D., & Verbauwhede, I. (2010). Privacy-preserving ECC-based grouping proofs for RFID. In *Proceedings of the 13th International Conference on Information Security* (ISC'2010), (pp. 159-165).

Bringer, J., Chabanne, H., & Dottax, E. (2006). HB++: A lightweight authentication protocol secure against some attacks. *Proceedings of IEEE International Conference on Pervasive Service, Workshop Security, Privacy and Trust in Pervasive and Ubiquitous Computing*, 2006.

Burdet, L. A. (2004). *RFID multiple access methods*. Technical Report. Retrieved from http://www.vs.inf.ethz.ch/edu/SS2004 /DS/reports/06 rfid-mac report.pdf

Cha, J.-R., & Kim, J.-H. (2005). Novel anti-collision algorithms for fast object identification in RFID system. *The 11th International Conference on Parallel and Distributed Systems*, Korea, (pp. 63–67).

Chien, H.-Y. (2006). Secure access control schemes for RFID systems with anonymity. *Proceedings of the 2006 International Workshop Future Mobile and Ubiquitous Information Technologies* (*FMUIT* '06), 2006.

Chien, H.-Y. (2007). SASI: A new ultralightweight RFID authentication protocol providing strong authentication and strong integrity. *IEEE Transactions on Dependable and Secure Computing*, *4*(4).

Chien, H.-Y., & Chen, C.-H. (2007). Mutual authentication protocol for RFID conforming to EPC class 1 generation 2 standards. *Computer Standards & Interfaces*, *29*(2), 254–259. doi:10.1016/j.csi.2006.04.004

Clark, R. (2011). *Dataveillance and information privacy*. Retrieved from http://www.rogerclarke.com/DV/

Conti, M., Pieto, R. D., Mancini, L. V., & Spognardi, A. (2007a). RIPP-FS: An RFID identification, privacy preserving protocol with forward secrecy. In *the Proceedings of the Fifth IEEE International Conference on Pervasive Computing and Communications Workshops* (PerComW'07), (pp. 229-234).

Conti, M., Pieto, R. D., Mancini, L. V., & Spognardi, A. (2007b). FastRIPP: RFID privacy preserving protocol with forward secrecy and fast resynchronisation. *Proceedings of the 33rd Annual Conference of the IEEE Industrial Electronics Society*, (pp. 52-57).

Deng, Z., Li, J., & Feng, B. (2008). A taxonomy model of RFID security threats. In *the Proceedings of the IEEE International Conference on Communication Technology*, ICCT 2008, (pp. 765-768).

Duc, D. N., Lee, H., Konidala, D. M., & Kim, K. (2009). Open issues in RFID security. In *the Proceedings of the International Conference on Internet Technology and Secured Transactions*, ICITST 2009, (pp. 1-5).

Duc, D. N., Park, J., Lee, H., & Kim, K. (2006). Enhancing security of EPCglobal Gen-2 RFID tag against traceability and cloning. *Proceedings of the 2006 Symposium on Cryptography and Information Security*, 2006.

EPCglobal. (2011). *EPCglobal: The EPCglobal network: Overview of design, benefits and security.* Retrieved from http://www.epcglobalinc.org

Fabian, B., Gunther, O., & Spiekermann, S. (2005). *Security analysis of the object name service for RFID.* First International Workshop on Security, Privacy and Trust in Pervasive and Ubiquitous Computing, July 2005.

Finkenzeller, K. (2003). *RFID handbook: Fundamentals and applications in contactless smart cards and identification.* John Wiley and Sons Ltd, 2003.

Haller, N. M. (1994). The S/KEY one-time password system. *Proceedings of the Symposium on Network and Distributed System Security*, (pp. 151–157).

Henrici, D. (2008). *Security and privacy in large-scale RFID systems - Challenges and solutions.* Msc Thesis, University of Kaiserslautern, Germany.

Hernandez-Castro, J. C., Estevez-Tapiador, J. M., & Ribagorda, A. (2006a). EMAP: An efficient mutual authentication protocol for low-cost RFID tags. *Proceedings OTM Federated Conference and Workshop: IS Workshop.*

Hernandez-Castro, J. C., Estevez-Tapiador, J. M., & Ribagorda, A. (2006b). M2AP: A minimalist mutual-authentication protocol for low-cost RFID tags. *Proceedings of the International Conference on Ubiquitous Intelligence and Computing* (*UIC*'06), (pp. 912-923).

Hopper, N. J., & Blum, M. (2001). Secure human identification protocols. *Proceedings of the Seventh International Conference on Theory and Application of Cryptology and Information Security*, (pp. 52-66).

Hush, D. R., & Wood, C. (1998). Analysis of tree algorithms for RFID arbitration. *The IEEE International Symposium on Information Theory*, (pp. 107–107).

Igure, V. M., & Williams, R. D. (2008). Taxonomies of attacks and vulnerabilities in computer systems. *IEEE Communications Surveys & Tutorials*, *10*(1), 6–19. doi:10.1109/COMST.2008.4483667

ISO. (2005). *ISO/IEC 17799: Information technology – Security techniques – Code of practice for information security management.* International Organization for Standardization.

Jakobsson, M. (2002). Fractal hash sequence representation and traversal. In *Proceedings of the 2002 IEEE International Symposium on Information Theory* (ISIT '02), (pp. 437-444).

Jong, R. D., Helmond, D. J. V., & Koot, M. (2005). *An exploration of RFID technology.* Technical report, 2005.

Juels, A. (2004). "Yoking-Proofs" for RFID Tags. In *Proceedings of the Second IEEE Annual Conference on Pervasive Computing and Communications Workshops* (Per-comw '04), (pp. 138–143).

Juels, A. (2005). Strengthening EPC tag against cloning. *Proceedings of the ACM Workshop Wireless Security* (*WiSe* '05), (pp. 67-76).

Juels, A., Molner, D., & Wagner, D. (2005). Security and privacy issues in e-passports. *Proceedings of the First International Conference Security and Privacy for Emerging Areas in Communication Networks* (*SecureComm* '05), 2005.

Juels, A., & Pappu, R. (2003). Squealing Euros: Privacy protection in RFID–enabled banknotes. *Financial Cryptography 2003. LNCS, 2742*, 103–121.

Kantor, A. (2003, December 19). Tiny transmitters give retailers, privacy advocates goose bumps. *USAToday.com.*

Karthikeyan, S., & Nesterenko, M. (2005). RFID security without extensive cryptography. *Proceedings of the Third ACM Workshop Security of Ad Hoc and Sensor Networks*, (pp. 63-67).

Klair, D. K., Chin, K.-W., & Raad, R. (2007). *An investigation into the energy efficiency of pure and slotted Aloha based RFID anti-collision protocols*. IEEE International Symposium on a World of Wireless, Mobile and Multimedia Networks (IEEE WoWMoM'07), 2007.

Kumar, S. S., & Paar, C. (2006). Are standards compliant elliptic curve cryptosystems feasible on RFID? *Proceedings Workshop RFID Security*, July 2006.

Langheinrich, M. (2009). A survey of RFID privacy approaches. *Personal and Ubiquitous Computing*, *13*(6), 413–421. doi:10.1007/s00779-008-0213-4

Law, C., Lee, K., & Siu, K.-Y. (2000). Efficient memoryless protocol for tag identification (extended abstract). *Proceedings of the 4th International Workshop on Discrete Algorithms and Methods for Mobile Computing and Communications*, (pp. 75–84).

Le, T. V., Burnmester, M., & Medeiros, B. (2007). Universally composable and forward secure RFID authentication and authenticated key exchange. In *the Proceedings of the 2nd ACM Symposium on Information, Computer and Communications Security*, (pp. 242-252).

Lee, S. (2005). *Mutual authentication of RFID system using synchronized secret information*. MSc Thesis, School of Engineering, Information and Communications University, 2005.

Lee, Y. K., Batina, L., Stefaan, S., Singelee, D., & Verbauwhede, I. (2010). Low-cost untraceable authentication protocols for RFID (extended version). In S. Wetzel, C. N. Ro-taru, & F. Stajano (Eds.), *Proceedings of the 3rd ACM Conference on Wireless Network Security* (WiSec '10), (pp. 55–64).

Leong, K. S., & Ng, M. L. (2004). *A simple EPC enterprise model*. Auto-ID Labs Workshop, Zurich 2004. Retrieved from http://www.m-lab.ch

Li, T., & Deng, R. H. (2007). Vulnerability analysis of EMAP-An efficient RFID mutual authentication protocol. *Proceedings of the Second International Conference on Availability, Reliability, and Security (AReS* '07), 2007.

Li, T., & Wang, G. (2007). Security analysis of two ultra-lightweight RFID authentication protocols. *Proceedings of the 22nd IFIP TC-11 International Information Security Conference*, May 2007.

Li, Y., & Ding, X. (2007). *Protecting RFID communications in supply chains*. Presented at the 2nd ACM Symposium on Information, Computer & Communication Security, Singapore, 2007.

Lin, X., Lu, R., Kwan, D., & Shen, X. (2010). REACT: An RFID-based privacy-preserving children tracking scheme for large amusement parks. *Computer Networks*, *54*, 2744–2755. doi:10.1016/j.comnet.2010.05.005

Liu, F., & Hu, L. (2008). ROAD: An RFID offline authentication, privacy preserving protocol with DoS resilience. In *Proceedings of the 2008 IFIP International Conference on Network and Parallel Computing*, (pp. 139-146).

Luo, Q., & Fei, Y. (2011). Algorithmic collision analysis for evaluating cryptographic systems and side-channel attacks. In *the Proceedings of the IEEE International Symposium on Hardware-Oriented Security and Trust* (HOST), (pp. 75-80).

Molnar, D., & Wagner, D. (2004). Privacy and security in library RFID: Issues, practices, and architectures. *Proceedings of the Conference on Computer and Communication Security (CCS'04)*, (pp. 210-219).

Myung, J., & Lee, W. (2005). Adaptive binary splitting: A RFID tag collision arbitration protocol for tag identification. *IEEE BROADNETs*, (pp. 347–355).

Myung, J., Lee, W., & Shih, T. (2006). An adaptive memoryless protocol for RFID tag collision arbitration. *IEEE Transactions on Multimedia*, *8*(5), 1096–1101. doi:10.1109/TMM.2006.879817

OCED. (2011). *The OECD principles*. Retrieved from http://www.anu.edu.au/people/Roger.Clarke/DV/OECDPs.html

Oren, Y., & Shamir, A. (2007). Remote password extraction from RFID tags. *IEEE Transactions on Computers*, *56*(9), 1292–1296. doi:10.1109/TC.2007.1050

Oswald, D., Kasper, T., & Paar, C. (2011). *Side-channel analysis of cryptographic RFIDs with analog demodulation*. RFIDSec'11, June 2011.

Peris-Lopez, P., Hernández-Castro, J. C., Estévez-Tapiador, J. M., & Ribagorda, A. (2006). RFID systems: A survey on security threats and proposed solutions. *PWC*, *2006*, 159–170.

Peris-Lopez, P., Hernandez-Castro, J. C., Estevez-Tapiador, J. M., & Ribagorda, A. (2009). LAMED A PRNG for EPC class-1 generation-2 RFID specification. *Computer Standards & Interfaces*, *31*(1), 88–97. doi:10.1016/j.csi.2007.11.013

Piramuthu, S. (2006). HB and related lightweight authentication protocols for secure RFID tag/reader authentication. *Proceedings of CollECTeR Europe Conference*, June 2006.

Plos, T. (2008). Susceptibility of UHF RFID tags to electromagnetic analysis. *CT-RSA 2008*. *LNCS*, *4964*, 288–300.

Rieback, M. R., Crispo, B., & Tanenbaum, A. S. (2006). *The evolution of RFID security* (pp. 62–69). Pervasive Computing.

Roberti, M. (2003). Big brother's enemy. *RFID Journal*, July 2003.

Schoenberger, C. R. (2002). The internet of things. *Forbes Magazine, 6*.

Schwartz, M. (1988). *Telecommunication networks protocols, modeling and analysis*. USA: Addison-Wesley.

Tang, Z., & He, Y. (2007). Research of multi-access and anti-collision protocols in RFID systems. *IEEE International Workshop on Anticounterfeiting, Security, Identification*, China, (pp. 377–380).

Tingting Mao, T., Williams, J., & Sanchez, A. (2008). Interoperable internet scale security framework for RFID networks. In *the Proceedings of the IEEE 24th International Conference on Data Engineering Workshop*, ICDEW'08.

Vaudenay, S. (2007). On privacy models for RFID. In *Advances in Cryptology* (ASI-ACRYPT'07), *Lecture Notes in Computer Science, LNCS 4833*, (pp. 68–87). Springer-Verlag.

Vogt, H. (2002). Multiple object identification with passive RFID tags. *The IEEE International Conference on Man and Cybernetics*, (pp. 6–13).

Weis, S. A., Sarma, S. E., Rivest, R. L., & Engels, D. W. (2004). *Security and privacy aspects of low-cost radio frequency identification systems. Security in Pervasive Computing* (pp. 201–212). Springer.

Xiao, Q., Boulet, C., & Gibbons, T. (2007). RFID security issues in military supply chains. *ARES*, *2007*, 599–605.

Yang, C.-N., Chen, J.-R., Chiu, C.-Y., Wu, G.-C., & Wu, C.-C. (2009). Enhancing privacy and security in RFID-enabled banknotes. In *the Proceedings of the IEEE International Symposium on Parallel and Distributed Processing with Applications*, (pp. 439-444).

Yang, J., Park, J., Lee, H., Ren, K., & Kim, K. (2005). Mutual authentication protocol for low-cost RFID. *Proceedings of the Ecrypt Workshop RFID and Lightweight Crypto*, 2005.

Zeidler, M. (2003, January 8). RFID: Der Schnuffelchip im Joghurtbecher. *Monitor-Magazine*.

Zhen, B., Kobayashi, M., & Shimizu, M. (2005). Framed Aloha for multiple RFID objects identification. *IEICE-Transactions on Communications. E (Norwalk, Conn.)*, *88-B*, 991–999.

Zuo, Y. (2010). Survivable RFID systems: Issues, challenges, and techniques. *IEEE Transactions on Systems, Man and Cybernetics. Part C, Applications and Reviews*, *40*(4), 406–418. doi:10.1109/TSMCC.2010.2043949

KEY TERMS AND DEFINITIONS

Access Control: Provides protection measure against unauthorized access to a resource.

Anti-Collision Protocol: Used in RFID tag reading so that multiple tags can be read simultaneously without error.

Authentication: Security services that confirm the identities of one or more of the entities connected to one or more of the other entities.

Confidentiality: Security services that provide protection of data from unauthorized disclosure and traffic analysis.

Cryptography: Deals with the techniques that ensure confidentiality and authenticity of the data.

Denial of Service (DoS): A form of attack where all resources of an entity (e.g., servers) become unavailable to the legitimate users of that resource.

Eavesdropping: The interception of data which may result in compromising proprietary information, network IDs and passwords, configuration data and encryption keys.

Encryption: The techniques that convert data into a form that is not interpretable without the necessary secret information used for encryption.

Integrity: Security services that ensure that messages are received as sent, with no duplication, insertion, modification, reordering or replays.

Jamming: Occurs when a malicious user deliberately emanates a signal from a wireless device in order to block legitimate wireless signals.

Privacy: The right of an individual to have control on who would have access to someone's data and how the data are used.

Privacy Preserving Protocol: Designed to preserve privacy by thwarting attempts by attackers to gain information about someone's location or data.

Chapter 3
The Evolution of Intelligent Classifiers into an Integrated Approach to Correct RFID Anomalies

Peter Darcy
Institute of Integrated and Intelligent Systems, Griffith University, Australia

Bela Stantic
Institute of Integrated and Intelligent Systems, Griffith University, Australia

Abdul Sattar
Institute of Integrated and Intelligent Systems, Griffith University, Australia

ABSTRACT

Radio Frequency Identification (RFID) refers to wireless technology that is used to seamlessly and automatically track various amounts of items around an environment. This technology has the potential to improve the efficiency and effectiveness of tasks such as shopping and inventory saving commercial organisations both time and money. Unfortunately, the wide scale adoption of RFID systems have been hindered due to issues such as false-negative and false-positive anomalies that lower the integrity of captured data. In this chapter, we propose the utilisation three highly intelligent classifiers, specifically a Bayesian Network, Neural Network and Non-Monotonic Reasoning, to handle missing, wrong and duplicate observations. After discovering the potential from using Bayesian Networks, Neural Networks and Non-Monotonic Reasoning to correct captured data, we decided to improve upon the original approach by combining the three methodologies into an integrated classifier. From our experimental evaluation, we have shown the high results obtained from cleaning both false-negative and false-positive anomalies using each of our concepts, and the potential it holds to enhance physical RFID systems.

DOI: 10.4018/978-1-4666-2080-3.ch003

INTRODUCTION

Radio Frequency Identification technology uses readers to scan various environments to discover tags attached to items that are needed to be tracked. The readers will send out an electro-magnetic pulse that certain tags, especially the passive variety, will use the power to transmit its unique identifier. The mass amounts of tag IDs are then passed through middleware to filter out potential invalid observations before storing this information in a centralised data warehouse. Currently, RFID has been integrated into inventory management, baggage tracking at airports, automatic road tolls, pet identification and many other applications. The commercial sector has benefitted from these applications as it provides additional security and integrity, and saves both time and money in the management items. While a wide array of RFID integrations exist, certain issues prevent it from obtaining the maximum potential of applications it could be utilised in.

There are four major issues associated with RFID systems, the low level nature of the observations, the anomalies that lower the integrity of the captured data, the huge volume intake of readings and the ever-increasing complexities of spatial and temporal properties. The low level nature of the observations hinders the applications as the recorded information is useless without referential data to identify the meaning of the identifiers. The anomalous readings are categorized as either false-positives, observations captured that do not reflect the actual events that transpired, and false-negatives in which tagged information is not present in the data warehouse when it was present physically. The huge volume of information captured is problematic for capturing devices as it will flood the storage and result in the need to either periodically replace memory or lose observational data. With the exponential growth of technology over time, and the integration of portable readers, the spatial and temporal information associated with observations continue to also rise in complexity. In this research, we focus on proposing techniques to correct RFID anomalies by either eliminating wrong and duplicate false-positive records, or restoring missing false-negative observations.

Specifically, we have done this by proposing a novel concept that utilises intelligent classifiers to correct both false-negative and false-positive RFID anomalies. These three classifiers include a Bayesian Network, Neural Network and Non-Monotonic Reasoning which have all proven to achieve high cleaning accuracies when applied in this context. From observing the potential to accurately and intelligently clean highly ambiguous anomalies, we decided to evolve our concept to globally fuse these classifiers into one integrated approach. We accomplish this by introducing three fusion methods, namely a probabilistic Bayesian Network, a deterministic Non-Monotonic Reasoning engine and an unbiased Majority Rules algorithm. From our extensive experimental evaluation, we have found that each of our approaches obtains highly accurate results and we conclude that this cleaning approach could significantly improve the integrity of captured information.

BACKGROUND

RFID systems refer to a system that automatically identifies multiple amounts of tagged objects within a certain proximity to a reader. Although the potential benefits of cheap and durable passive tagging architectures are great, ambiguous anomalies such as missed readings hinder the world-wide adoption of this technology. High level intelligence engines, such as Bayesian Networks, Neural Networks, or Non-Monotonic reasoning, may be harnessed to provide the resolution to highly ambiguous scenarios such as missing data to increase the integrity of the data set.

Radio Frequency Identification

Radio Frequency Identification (RFID) utilises radio transmissions between a reader and identifying tags to wirelessly locate objects automatically. Items are fitted with tags which are interrogated by readers resulting in the return of the unique Electronic Product Code (EPC) (Chawathe, 2004). Unfortunately, there are several issues associated with the RFID architecture, specifically the passive tag system. There are two persistent anomalies found within recorded data sets that are continually introduced. This may be attributed to various factors such as collision, detuning or water/metallic interference among tags (Floerkemeier, 2004). These anomalies are false positive readings where the data captured did not exist in reality, and false negative readings where data is not present in the data set but is required to be. Of these two anomalies, false negative errors may be considered the hardest to correct as the data is not recorded into the database minimising the contextual information needed to correctly impute what readings may have originally been present. It has been estimated that only 60%-70% of recordings have been estimated to be captured within any RFID architecture (Jeffery, 2006). Due to the unique nature of the observations and the ambiguous issues that arise, there is an opportunity to correct these anomalies as they appear in the database using data mining techniques that examine the Tag ID, Reader ID and Timestamp.

Bayesian Network

Bayesian Networks refer to a network designed to find the highest probable solution to any given problem. This is usually performed by finding the product of evidence found in a situation and comparing it with other possible causes until the greatest probable outcome is found. When expressing the mechanics of any Bayesian Network, there are three common mediums: a joint distribution equation, an influence diagram and a Bayesian Network table. To demonstrate how a Bayesian Network operates, we have developed an example scenario in which a network is developed to determine the cause of a tree falling down (human or nature) when given such attributes as council markings and weather. The specific rules for this example are that there is a very high chance for the council cut down the tree for this coupled with fine weather. However, if the weather is stormy, there is less chance of the tree being cut down by a human.

Equation 1 will calculate the probability of $X_1 - X_n$ using a Bayesian network:

$$P\left(X_1,\ldots,X_n\right)=\prod_{i=1}^{n}P\left(X_i|\left(X_1,\ldots,X_{i-1}\right)\right) \quad (1)$$

The mathematical equation is a formula designed to express the process utilised to determine the percentage of likelihood of a cause being true. As seen in Equation 1, the probability from X_1 to X_n is equal to all the products on which X_i is dependent. Equation 2 illustrates the mathematical equation applied to the tree falling down scenario. In this Equation, T_1 represents the cutting down of the tree event which may be the result of human interaction or natural causes, C represents a tree being marked by the council to be lopped and W represents the weather.

Equation 2 will calculate the probability of the tree being cut down by either the council or the weather using a Bayesian network:

$$P\left(T_1|C_iW\right)=P\left(C|W\right)*P(W) \quad (2)$$

All information relating to the Bayesian Network is then placed within a table which can then be queried by a program. The table consists of the evidence vs. the causes in which a percentage is given to each case for the true and false outcomes in each scenario. From this table, all the percentages are multiplied together and a percentage score is given to each of the causes. A Bayesian Network will then conclude that the most probable cause is the cause with the highest achieving percentage.

Neural Network

An Artificial Neural Network is a classifier designed to emulate the learning behaviour of the brain. It does this by creating a fixed amount of Neurons which are trained to deliver a certain output when fed various input. The entire process has actually been based on the biological neuron. Dendrites will receive information which is passed to the cell body whose objective is to pass the information into the axon when certain requirements are met and, thus, to dendrites of other neurons via the synapse connection. The crucial difference between a digital Neuron and its biological counterpart is that there is a computational limit to the amount of hidden units that may be present within a network. Unfortunately, technology has not advanced enough to effectively and efficiently emulate the amount of Neurons the human brain possesses, which is estimated to be between 10 billion and 1 trillion (Williams, 1988).

Algorithm 1 is for the hard limiter activation function utilised within a neural network:

```
Input: Neuron output
Output: Activation function output
If The value is greater than 0 then
 |    Set output to +1.
else
 |    Set output to -1.
end
```

The Artificial Neural Network consists of three main layers: the Input Layer, Hidden Layer(s) and the Output Layer (McCulloch 1943). The processes include receiving inputs which are modified at a central sum area. The Neuron will then apply an activation function such as the hard limiter or sigmoidal functions, which are displayed in Algorithm 1 and Equation 3 respectively, to derive a value for the output. With regards to training the network, there are several techniques available such as the Back-propagation (Rumelhart, 1986), (Blumenstein, 2007) and Genetic Algorithms (Holland, 1975), (Rooij, 1996) which are both considered as leaders with regards configuring Artificial Neural Networks.

Equation 3 will calculate the probability of the tree being cut down by either the council or the weather using a Bayesian network:

$$f(x) = 1 / \left(1 + \exp\left(-x\right)\right) \tag{3}$$

When attempting to configure the Neural Network weights, one method that exists is to utilise training algorithms. Two dominant training algorithms that have been proven to excel in network training are the Back-Propagation Algorithm and the Evolutionary Neural Network. Back-Propagation relies on the concept of training the network by propagating error back through the network via modifying the weights after the output has been calculated (Rumelhart, 1986). The algorithm uses either a predetermined limited amount of iterations or the Root-Mean-Square error threshold of the calculated output as stopping criteria (Blumenstein, 2007).

The Evolutionary Neural Network training algorithm in contrast to Back-Propagation utilises the theory of genetic evolution to train the network weights. Similar to the genetic algorithm process of training a Bayesian Network, all the weights are added into a chromosome as genes to be manipulated according to the fittest output obtained (Holland 1975). The weights are initialised as small random numbers which are checked for either obtaining a high enough score or if the amount of generation limit has been reached. In the case that neither of the stopping criteria has been met, the algorithm will examine each chromosome in relation to achieving the correct output. A certain amount of the unfit chromosomes are then destroyed within the population to be replaced with child chromosomes of two of the fitter chromosomes. The child chromosomes can be created with different means such as one-point, two-point or uniform crossover (Rooij, 1996). Mutation will then be applied to a certain small percent of the population to assure that the network avoids problems such as network paralysis or local minima (Cha, 2008).

Non-Monotonic Reasoning

Non-Monotonic Reasoning (NMR) refers to a deterministic logic used to decipher the solution when given a number of relevant pieces of evidence. NMR is set apart from classic Monotonic Reasoning in that in contrast of arriving at one conclusion for any given problem, NMR will consider a number of solutions and will eliminate existing solutions or add additional solutions as extra information which is readily available. In particular, we have investigated the Clausal Defeasible Logic (CDL) as the proof algorithm to arrive at a conclusion as it has been designed specifically to be implemented in a computer (Antoniou, 2006). For example, it is widely known that mammals do not usually lay eggs. There is an exception to this rule, however, as certain species such as the platypus do lay eggs and is still a mammal. It is important to note the relation represented as a semi circle which provides priority to the most clockwise entity, in this case, the Platypus rule in this example. This relation is known as a "Priority Relation"; in a typical scenario, priority is given to the rule which is more specific and will always reach the conclusion before other rules.

A language called Decisive Programming Language (DPL) proposed by Billington (2005) has been employed to illustrate scenarios which use CDL. Within DPL, several symbols are used to represent different relationships of the entities preceding the relation, the antecedent, and the entities positioned subsequently after the relation, the conclusion. The first symbol is the Strict Rule Relation which is represented as "→". It dictates that this rule is certain with no possible ambiguity involved. The second symbol is the Defeasible Rule Symbol "⇒" denoting a relationship in which it is defeasible to say the former entity will result in the latter entity. The third symbol is the Warning rule symbol "⇝" which describes when the former entity cannot disprove the latter (usually the negative of the latter). Other symbols which are used include the Priority relation ">" which dictates that the former rule is greater than the latter rule, and the negative symbol "~"' which turns the following variable into its negative counterpart. The CDL Engine would illustrate the above scenario as the following observations and rules:

```
Platypus(x) → Mammals(x).
R1: Mammals(x) ⇒ Do Not Lay Eggs(x).
R2: Platypus(x) ⇒ ~Do Not Lay
Eggs(x).
R2 > R1.
```

From the rules illustrated above, the CDL Reasoning Engine can ascertain that a platypus is a mammal and, therefore, must not lay eggs. However, since the second rule which states that for a platypus, the negative of "Do Not Lay Eggs," is defeasible, it may be stated that a platypus will lay eggs. Although it is true that one conclusion must be drawn for any given situation, Clausal Defeasible Logic, has several levels of confidence, represented in formulae, which may be used to obtain a different correct answer dependent upon the amount of ambiguity allowed. CDL is very similar to Plausible Logic (PL) in the sense that it uses the same inputs to arrive at a conclusion. However, unlike PL, CDL has refined itself to include five core formulae rather than three. These formulae include:

- **μ:** This formula uses only certain information to obtain its conclusion.
- **π:** This formula allows conclusions in which ambiguity is propagated.
- **β:** This formula does not allow any ambiguity to be used in obtaining its conclusion.
- **α:** A formula in which any conjunction of the π and β formulae are used to reach its conclusion.
- **δ:** The disjunction of π and β are used to draw conclusions.

As discussed by (Billington, 2007), other strengths that set CDL apart from other reasoning algorithms are its ability to uses team defeat, failure-by-looping and discovering the loops in a given reasoning system within a set number of steps. This Logical engine has already been tested and implemented in two different scenarios to allow a robot dog to play soccer and inside a robot designed as a means of alarming individuals of an emergency in an elderly care situation (Billington, 2008).

Related Work

In past literature, popular previously proposed concepts include sliding window filtering, rule-based cleaning and event extraction to correct captured data before or after is has been stored in the database. Sliding Windows, as proposed by Jeffery (2006), utilise adaptive windows to smooth the readings. The rule-based approach allows users to define rules that will be used to correct all anomalies found (Rao, 2006). Probabilistic Event Extraction refers to the process of taking the previously observed historical high-level events and probabilistically finding others from the low-level captured data (Khoussainova, 2008). The major issues with these techniques are that the filter does not have enough information at the edge to correct highly ambiguous anomalies; the user-defined rule-based approach may be too generalised and lack the intelligence to adequately deal with the anomalies; and the probabilistic events may accidently introduce artificial anomalies due to its probabilistic nature. To improve upon these issues, we wish to present an approach that is deferred to take advantage of all readings, and has a high integration of deterministic techniques to provide high intelligence that may outperform probabilistic methodologies in certain situations.

FALSE-NEGATIVE MANAGEMENT

In this chapter, we propose an advanced data analysis methodology coupled with high level intelligence to correctly decipher the likeliest candidates of observations to be returned into the data set. This will include an outline the motivation and scenario considered in this work followed by a description of the system architecture of our approach. These discussions will be followed by the database structure that houses the RFID observations and all assumptions made towards our methodology. We will then present the results obtained from our experimental evaluation before summarising our findings.

Motivation

False Negative anomalies are hazardous to all applications that utilise RFID as it prevents the recording of data lowering the overall integrity of the whole data set. Within previous work, we have established that there is large potential for the fusion of thorough analysis coupled with high level intelligence to adequately replace missed readings. Additionally, we have also put forth a preliminary analysis of a simplistic CDL logic engine using high level analysis which resulted in high cleaning rates. We have decided to enhance the cleaning rules using three different classifiers, a Bayesian Network, Neural Network and Non-Monotonic Reasoning, and compare each approach to determine which intelligent classifier obtains the highest accuracy.

The scenario which we have intended our approach would be used for would include an enclosed static environment where tracking items is essential. Within this scenario, missed readings occur which will need to be corrected before the data can be utilised in meaningful contexts. The scenario which would most benefit from our concept would be one in which false negative readings occur randomly and rarely. Our methodology has been intended to provide a higher level of intelligence for cases in which missed readings occur consecutively.

Within a stocking warehouse that transfers stocked items via conveyer belt to various locations, tags may not be observed by the readers resulting in a false-negative anomaly. For example, a mock warehouse environment may contain Readers 1 (R1) through to Reader 7 (R7) installed on a linear path that the tag (T1) would take along a conveyor belt between it and the truck it will eventually get loaded onto. However, due to metallic, water or tag interference, the readings observed may only include:

- T1, R1, 2009-11-24 10:30:00.000
- T1, R2, 2009-11-24 10:31:00.000
- T1, R3, 2009-11-24 10:32:00.000
- T1, R7, 2009-11-24 10:26:00.000

Simple false-negative anomalies may be corrected through use of simple logic such as a case in which T1 appears at R1, missing at R2, but present again at R3. This would result in current state-of-the-art techniques, including our own approach, determining that T1 should also be present at R2. However, there may be a situation in which the insertion of the observation may not be straight forward. Within our sample scenario previously mentioned, we can see that T1 is read at readers R1, R2, R3 and R7. From this information, conventional data correction algorithms would either replace all readings between R3 and R7 with the reader location R3 or discard R7 under the assumption that it is a false-positive anomaly. However, our approach would instead derive the shortest path from the available map data, determine if it is feasible, and if so, insert it within the data set thereby increasing the integrity of T1's tracking data.

Methodology

As seen in Figure 1, we have divided our system's architecture into three core components. The first is designed to analyse the data where the missed reading occurred which we have named the Analysis Phase. The data discovered in this Analysis Phase is then passed onto the Intelligence Phase where the correct permutation is selected. After the resulting data set has been chosen, the Loading Phase will complete the program's cycle by inputting the information back into the data warehouse.

The analysis phase consists of our tool locating missed readings and then discovering essential data about the anomaly. The first process is to divide the tags into "Tag Streams" (Definition 1). These tag streams include chronicle information relating only to one individual tag. From these tag streams, certain information is ascertained relating to the nature of the false negative anomaly. This includes finding the reader locations of the observations two readings before and directly before the anomaly (a and b respectively), directly after and two readings after the reader (c and d respectively). Additionally, the shortest path between readings b and c using the map data is found. The total missing readings calculated via the number of missing timestamps (n), and the amount of observations within the shortest path (s), is then calculated. From this information, we ascertain the following analytical data:

- $a == b$
- $b \leftrightarrow c$
- $b == c$
- $d == c$
- $n == (s-2)$

Figure 1. A high level diagram of inner processes used within the false-negative management approach

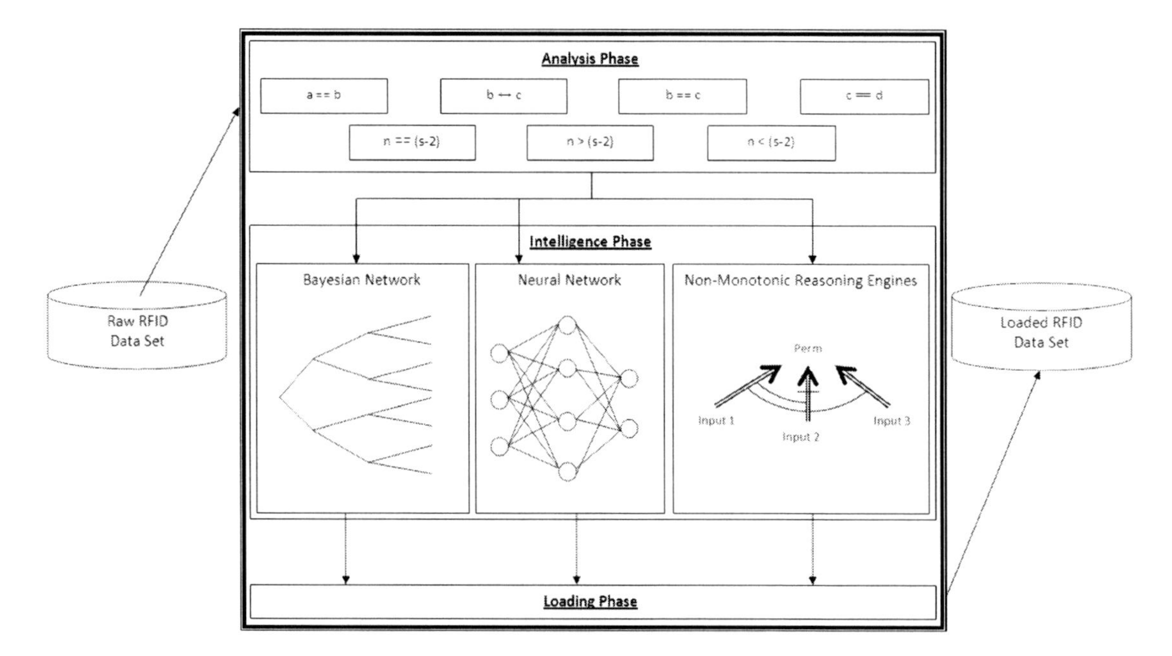

- $n > (s\text{-}2)$
- $n > (s\text{-}2)$

The analytical information obtained from our approach include finding if readers *a* and *b* are equal ($a == b$); determining if readers *b* and *c* are relatively close to each other one according to the map data ($b \leftrightarrow c$); discovering if the readers *b* and *c* are equal ($b == c$); finding if readers *c* and *d* are equal ($c == d$) and discovering if *n* is equal to, less than or greater than *s* minus two ($n == (s\text{-}2)$), ($n < (s\text{-}2)$), ($n > (s\text{-}2)$). The reason as to why we subtract two from value *s* is that the shortest path will include the values of *b* and *c* which are not necessarily a part of the missing gaps of knowledge. All of these analytical Boolean variables are then passed on to the correction phase which utilise it to seek out the most ideal imputed reader values. We utilise four main arithmetic operations to obtain these binary analytical information. These include the equivalent symbol ==, the less < and greater than > symbols, and the ↔ symbol we have elected to represent geographical proximity. The reason as to why *s* is always having two taken away from it is that the shortest path always includes the boundary reader's *b* and *c* which are not included within the *n* calculation.

Definition 1: Tag Stream

The chapter defines Tag Streams as individually analysed streams for one tag from the mass amount of readings.

The intelligence phase is when the various permutations of the missing data are generated as candidates to be restored in the data set. The five different permutations that are been generated are described below:

- **Permutation 1:** All missing values are replaced with the reader location of observation *b*.
- **Permutation 2:** All missing values are replaced with the reader location of observation *c*.

Box 1. Algorithm used to generate the various permutations before passing the information into the various classifiers

Algorithm 2
Input: *Algebraic values of the missing data.* **Output:** *The Permutations* **foreach** *Missing Record* **then** | *Create an Array of size "n".* | *Generate the first permutation by using the value of "b" substituted for each recording.* | *Generate the second permutation by using the value of "c" substituted for each recording.* | *Generate the third permutation by inserting the shortest path in the middle of the array and substitute the values of "b" and "c" on the left and right side of the array to additional missing gaps.* | *Generate the fourth permutation by inserting the shortest path on the right of the array and substitute the value of "b" on for all remaining gaps.* | *Generate the fifth permutation by inserting the shortest path on the left of the array and substitute the value of "c" on for all remaining gaps.* | **end** *Pass the permutations to the various classifiers.*

- **Permutation 3:** The shortest path is slotted into the middle of the missing data gap. Any additional missing gaps on either end of the shortest path are substituted with values *b* for the left side, and *c* for the right.
- **Permutation 4:** The shortest path is slotted into the latter half of the missing data gap. Any additional missing gaps on the former end are substituted with value *b*.
- **Permutation 5:** As the anti-thesis of Permutation 4, the shortest path is slotted into the former half of the missing data gap. Additional missing gaps found at the latter end of the missing data gap are substituted with value *c*.

After the data analysis and permutation formations are completed, all relevant information found in the analytical phase are treated as a feature set definition and passed into the Bayesian Network, Neural Network or the Non-Monotonic Reasoning Engine. The various classifiers will then return the permutation that has been found to best suit the missing gap of data. The various actions included within the Intelligence Phase before the utilisation of the classifiers may be viewed in pseudo-code version in Algorithm 2 in Box 1. With regards to the Bayesian Network, all weights inside the network are first created with small random numbers. After this, we utilise a Genetic Algorithm to train these weights based on a training algorithm of previously correct permutations based on each variation of the feature set definitions. The resulting network configuration from this training will then provide the optimal permutation of readings to be inserted back into the data warehouse.

The ANN accepts seven binary inputs to reflect the analysis data (*a==b, c==d,* etc.) and has five binary outputs to reflect that the permutations that have been found. The ANN also includes nine hidden units found in one hidden layer. We specifically chose this configuration as we have found that there should be a larger number of hidden nodes than inputs and one layer would be sufficient for our network at this moment. If we chose to extend the complexity of our system, we may wish to also either increase the number of hidden units, layers, or both.

We have also applied a momentum term and learning rate whose values are be 0.4 and 0.6 respectively to avoid local minima and network paralysis. Each input and output value will not be 1 and 0 as this may not yield a very high classification rate, instead we will use the values of 0.9 and 0.1 respectively. We will set the stopping criteria as both the RMS error threshold when it

Table 1. The table contains the results of the false-negative management classifiers when attempting to clean 500, 1,000, and 5,000 missing anomalies. The average is also presented at the end and the algorithm performing the lowest, highest, and highest in average have been put in bold typeface.

Cleaning Method	Test Cases	Correct Data Set Percentage
Bayesian Network	**500**	**83.60%**
Bayesian Network	1000	85.00%
Bayesian Network	5000	86.56%
Neural Network	500	85.00%
Neural Network	1000	86.00%
Neural Network	5000	86.46%
Non-Monotonic Reasoning	500	86.40%
Non-Monotonic Reasoning	**1000**	**87.70%**
Non-Monotonic Reasoning	5000	85.40%
	Averages	
Bayesian Network	Average	85.05%
Neural Network	Average	86.46%
Non-Monotonic Reasoning	**Average**	**86.50%**

reaches below 0.1 and 1000 iterations. We have also utilised the sigmoidal activation function to derive our outputs. The pseudo code for both the back-propagation and genetic algorithm training methods can be found Algorithms 3 and 4 (in Boxes 2 and 3) respectively.

In our first permutation, the two rules we first propose have stated both result in the $\sim perm1$ and are specifically $n < (s - 2)$ and $\sim b \leftrightarrow c$. These rules will be beaten by either $a == b \wedge b == c \wedge b \leftrightarrow c \wedge \sim c == d \wedge n > (s - 2)$ or $a == b \wedge b \leftrightarrow c \wedge n > (s - 2)$ to conclude positive $perm1$. The rules $c == d \wedge b \leftrightarrow c$ and $c == d$ have a higher precedence than previous rules and will conclude in $\sim perm1$, however the rules $a == b \wedge b \leftrightarrow c \wedge b == c \wedge \sim c == d \wedge n == (s - 2)$, $a == b \wedge b \leftrightarrow c \wedge n == (s - 2)$ and $a == b \wedge b \leftrightarrow c$ will overtake these to a positive $perm1$. The previously stated rules will be out rendered invalid if any of the $c == d \wedge b \leftrightarrow c \wedge n > (s - 2)$, $c == d \wedge b \leftrightarrow c \wedge n == (s - 2)$, $c == d \wedge n > (s - 2)$ or $c == d \wedge n == (s - 2)$ have been proven correct resulting in a $\sim perm1$ conclusion and can only be beaten by $a == b$ being true turning the returned values into a positive $perm1$. Every single one of these previously proposed rules will be over written resulting in $\sim perm1$ conclusion if the $b == c \wedge n == (s - 2)$, $b == c \wedge n < (s - 2)$ or $b == c$ rules have been proven.

Due to the fact that the data generated in Permutations 1 and 2 are direct opposites, the rules we have created in each of these look up tables are also relatively reversed. The first two rules we have stated both result in the $\sim perm2$ and are specifically $n < (s - 2)$ and $\sim b \leftrightarrow c$. These rules will be beaten by both $\sim a == b \wedge b == c \wedge b \leftrightarrow c \wedge c == d \wedge n > (s - 2)$ and $b \leftrightarrow c \wedge c == d \wedge n > (s - 2)$ resulting in a positive $perm2$ conclusion. The rules $a == b \wedge b \leftrightarrow c$ and $a == b$ will then beat the previous rules concluding $\sim perm2$, however the rules $\sim a == b \wedge b \leftrightarrow c \wedge b == c \wedge c == d \wedge n == (s - 2)$, $b \leftrightarrow c \wedge c == d \wedge n == (s - 2)$ and $b \leftrightarrow c \wedge c == d$ will overwrite this to a positive $perm2$. All previous rules will be beaten if any of $a == b \wedge b \leftrightarrow c \wedge n > (s-2)$, $a == b \wedge b \leftrightarrow c \wedge n == (s - 2)$, $a == b \wedge n > (s - 2)$ or $a == b \wedge n == (s - 2)$ are proven resulting in a $\sim perm2$ conclusion which can only be beaten by $c == d$ being true turning the returned values into a positive $perm2$. Finally, all previous rules can be beaten resulting in $\sim perm2$ conclusion if $b == c \wedge n == (s - 2)$, $b == c \wedge n < (s - 2)$ or $b == c$ are proven.

Permutation 3 has been designed specifically to place the shortest path directly in the middle of the missing data and filling any gaps with reader's b and c, thereby making it the least biased of the five possibilities. The lowest rule $n < (s - 2)$ will conclude a $\sim perm3$, however a positive $perm3$ will be found if $a == b \wedge c == d \wedge n == (s - 2)$ is true. After these rules, $b == c$ is the next high-

Box 2. Algorithm for training the neural network phase utilising the back-propagation algorithm. The goal of this algorithm is to modify the weights using forward and backward passes until the network has been trained to determine the correct outputs

Algorithm 3
Input: *None.* **Output:** *The network configuration* *Initialise the weights to small random values.* *Present the training set to the neural network.* **foreach** *Iteration* **then** | *Determine the actual output of all the neurons (Forward Pass).* | *Check stopping criteria (RMS Error and Iterations).* **if** *Stopping Criteria is True* **then** || *Break the algorithm and store the weights.* | **end** | *Determine the error terms.* | *Adjust the weights using the learning rate, momentum and error terms (Backward Pass).* **end** *Store highest performing network configuration.*

est in precedence which will find ~*perm3*. This will be overwritten by $n == (s - 2)$ which will find a positive *perm3*, unless $\sim a == b \wedge \sim c == d \wedge \sim n > (s - 2)$ is true in which ~*perm3* will be concluded. The next rule we created is $a == b \wedge c == d \wedge n > (s - 2)$ that concludes a positive *perm3*. Finally, the last two rules we have created $\sim a == b \wedge \sim c == d$ and $a == b \wedge c == d$ will prove ~*perm3* and a positive *perm3* respectively.

With regards to Permutation 4, the lowest rules $\sim a==b \wedge \sim n > (s - 2)$ and $\sim n > (s - 2)$ will arrive at the conclusion ~*perm4*. This rule will get overwritten by $n > (s - 2) \wedge a == b$ which will determine positive *perm4* to be the conclusion. The $\sim a == b$ will then take over any previous rule precedence to conclude ~*perm4*; but if $a == b$ has been found to be true, the conclusion will be found to be a positive *perm4*. The final rules we have created include $b == c$ or $c == d$ which will ultimately conclude ~*perm4*.

Similar to the relation between Permutation 2 and Permutation 1, the rules found in Permutation 5 are reversed relatively when compared to Permutation 4. The lowest rules we have created, namely $\sim c == d \wedge \sim n > (s - 2)$ and $\sim n > (s - 2)$ will conclude ~*perm5*. This rule will get overwritten by $n > (s - 2) \wedge c == d$ which will arrive at a positive *perm5* conclusion. The $\sim c == d$ will then outperform any previous rules to conclude ~*perm5*; however, $c == d$ will conclude a positive *perm5*. Finally, in the event that $b == c$ or $a == c$ is proven true, the ultimate conclusion will be found to be ~*perm5*.

Each of the rules specified are combinations found from the analytical data joined by "and" statements $\wedge$ which have been gathered within the Analysis Phase. Within the logic engine build, the precedence of the rules corresponds to the larger number of the rule. In the event that more than one permutation has been found to be ideal in a given situation, we use the following hierarchical weighting: Permutation 3 > Permutation 1 > Permutation 2 > Permutation 4 > Permutation 5. In the unlikely case where no conclusions have been drawn from the Non-Monotonic Reasoning Engine, Permutation 3 will be elected as the default candidate due to it having perfect symmetry within the imputed data. This ordering has been configured to be the most accurate conclusion assuming that the amount of consecutive missed readings are low due to the randomness of the anomalies.

The loading phase consists of the selected permutation being uploaded back into the data storage

Table 2. The table contains the revised results of the false-negative management classifiers when attempting to clean 500, 1,000, and 1,500, 2,000 and 2,500 missing anomalies. The average is also presented at the end, and the algorithm performing the lowest, highest, and highest in average have been put in bold typeface.

Cleaning Method	Test Cases	Correct Data Set Percentage
Bayesian Network	**500**	**78.80%**
Bayesian Network	1000	78.90%
Bayesian Network	1500	84.33%
Bayesian Network	2000	83.90%
Bayesian Network	2500	84.64%
Neural Network	**500**	**88.60%**
Neural Network	1000	83.30%
Neural Network	1500	87.87%
Neural Network	2000	86.65%
Neural Network	2500	86.96%
Non-Monotonic Reasoning	500	85.80%
Non-Monotonic Reasoning	1000	85.90%
Non-Monotonic Reasoning	1500	87.20%
Non-Monotonic Reasoning	2000	86.50%
Non-Monotonic Reasoning	2500	87.48%
Averages		
Bayesian Network	Average	82.11%
Neural Network	**Average**	**86.68%**
Non-Monotonic Reasoning	Average	86.58%

at the completion of the *Intelligence Phase*. The user will have the opportunity to either elect to load the missing data into the current data repository, or copy the entire data set and only modify the copied data warehouse. This option would effectively allow the user to revisit the original data set in the event that the restored data is not completely accurate.

Table 3. The table contains the results of the False-Positive Management classifiers when attempting to clean 1,000, 5,000, and 10,000 missing anomalies. The average is also presented at the end, and the algorithm performing the lowest, highest, and highest in average are put in bold typeface.

Cleaning Method	Test Cases	Correct Data Set Percentage
Bayesian Network	1000	98.70%
Bayesian Network	5000	98.44%
Bayesian Network	10000	98.62%
Neural Network	1000	98.80%
Neural Network	5000	98.54%
Neural Network	**10000**	**98.43%**
Non-Monotonic Reasoning	**1000**	**99.00%**
Non-Monotonic Reasoning	5000	98.78%
Non-Monotonic Reasoning	10000	98.60%
Averages		
Bayesian Network	Average	98.59%
Neural Network	Average	98.59%
Non-Monotonic Reasoning	**Average**	**98.79%**

To store the information recorded from the RFID reader, we utilise portions of the "Data Model for RFID Applications" DMRA database structure found in Siemens Middleware software. Additionally, we have introduced a new table called MapData designed to store the map data crucially needed within our application. Within the MapData table, two Reader IDs are stored in each row to dictate if the two readers are geographically within proximity.

We have made three assumptions that are required for the entire process to be completed. The first assumption is that the data recorded will be gathered periodically. The second assumption we presume within our scenario is that the amount of time elected for the periodic readings is less than the amount of physical time needed to move from

Box 3. Algorithm for training the neural network phase utilising the genetic algorithm. The goal of this algorithm is to set the weights to a point they are able to choose the correct permutation when given various training sets

Algorithm 4
Input: *None.* **Output:** *The network configuration* *Initialise the weights to small random values.* *Present the training set to the neural network.* **foreach** *Generation* **then** | *Evaluate and order the chromosome according to fitness (the classification rate).* | *Check stopping criteria (Generations).* | **if** *Stopping Criteria is True* **then** | | *Break the algorithm and store the weights of the highest performing chromosome.* | **end** | *Delete half the least fit population members and breed two random fitter chromosomes to fill the amount of population.* | **if** *Child chromosome is already present in population* **then** | | *Discard child chromosome.* | | *Cross over two different random parents.* | **end** | *Apply mutation to select members of the chromosome by resetting the weights to random values - 1% the top 10% fittest population members, 5% for other chromosomes.* **end** *Store highest performing network configuration.*

one reader to another. This is important as we base our methodology around the central thought that the different readings will not skip over readers that are geographically connected according to the MapData. The final assumption we make is that all readers and items required to be tracked will be enclosed in a static environment that has readers which cover the tracking area.

Results and Analysis

Within the following section, we have included a thorough description of the setup of the experimentation used in our methodology. First, we discuss the environment used to house the programs. This is followed by a detailed discussion of our experimentation including the four experiments we performed and their respective data sets used. These experiments include the training of the three various classifiers and then taking the highest performing configurations to compare it against each other.

Our methodology has been coded in the C++ language and compiled with Microsoft Visual Studio C++. The code written to derive the lookup table needed for the Non-Monotonic Reasoning data has been written in Haskell and compiled using Cygwin Bash Shell. All programs were written and executed on Dell machine with Windows XP Service Pack 3 operating system installed.

We have conducted four experiments to adequately measure the performance of our methodology. The first experiment we conducted involved finding a Bayesian Network that performs the highest clean on RFID anomalies. After this, we investigated the highest performing Neural Network configuration in our second experiment. The third experiment conducted was to determine which of the Clausal Defeasible Logic formulae performs most successfully when attempting to correct large amounts of scenarios. The training cases used in the each of these experiments consisted of various sets of data consisting of ambiguous false negative anomaly cases.

The fourth experiment we conducted was designed to test the performance of our selected highest performing Bayesian Network, Neural Network and Non-Monotonic Reasoning Logic configurations to determine which classifier yielded the highest and most accurate clean of false-negative RFID anomalies. The reason as to why these techniques were selected as opposed to other related work is that only other state-of-the-art classifying techniques can be compared in respect to seeking the select solution from a highly ambiguous situation.

The fourth experiment testing sets included four data repositories consisting of 500, 1,000, 5,000 and 10,000 ambiguous false negative anomaly cases. We defined our scoring system as if the respective methodologies were able to return the correct permutation of data that had been previously defined. All data within the training and testing set have been simulated to emulate real RFID observational data.

To thoroughly test our application, we devised four different examinations which we have labeled the Bayesian Network, Neural Network, Non-Monotonic Reasoning and False-Negative Comparison Experiments. In the Bayesian Network Experimentation, we attempted to discover the optimal number of chromosomes needed to produce the highest cleaning performance. Similar to the Bayesian Network Experiment, the Neural Network Experimentation included finding the highest training algorithm to configure the network and obtain the highest performing clean. The Non-Monotonic Reasoning experiment compared the cleaning rate of each of the Clausal Defeasible Logic formulae. The highest performing Non-Monotonic reasoning setup was then compared to Bayesian and Neural network approaches to demonstrate the significance of our cleaning algorithm.

The Bayesian Network False-Negative Experiment consisted of the utilisation of the Genetic Algorithm with 250 generation iterations to determine the fittest chromosome. To this end, we trained and compared the genes of chromosomes where 100 through to 1,000 chromosomes were in the population while incrementing by 100 each time. The data set used to determine the fitness of the chromosomes contained every permutation possible with the analysis and its correct permutation answer. From the results obtained, we may have found that the chromosome which was the fittest resulted when 500 chromosomes were introduced into the population. Unlike the other experimentation we have performed in this research, the Bayesian Network has been measured by how many of the inserts were exactly correct as opposed to the measuring how correct the entire data set is. This results in the Bayesian Network achieving a relatively low cleaning rate as permutations that are not an exact match with the training results will be counted as incorrect when the resulting imputed data may actually be correct in the data set.

The Neural Network False-Negative Experiment has the goal of seeking out the training algorithm that yields the highest clean rate. The two training methods used for comparison are the Back-Propagation and Genetic Algorithms. The training set utilised in this experiment is comprised of every possible combination of the inputs and their respective outputs which amount to a total of 128 entries. We have conducted tests upon three different training algorithm setups, the first is the back-propagation algorithm and the other two are genetic algorithms that use 20 and 100 chromosomes to find the optimised solution. Additionally, we conducted each experiment in three trials to further generalise our results. The algorithm that had the hardest time finding the correct configuration was the back-propagation algorithm when it iterated for 50 and 100 times, earning it a 1.56% classification rate. The trainer that performed the best was the Genetic Algorithm both using 20 and 100 chromosomes in every test and iteration number excluding the 20 chromosome configuration which lasted for 5 generations.

From our experimental evaluation, we found that on average the Neural Network Genetic Algorithm performed the best obtaining an 87.5%

cleaning rate. Unfortunately, as discovered before in the results, the back-propagation algorithm performed the weakest within the three algorithms. This is especially present when it is attempting to clean after being trained for 100 iterations. We believe the poor results of the back-propagation algorithm were due directly to over-training the network. We noticed that there was also a particularly low result when attempting to train this algorithm for 50 attempts in trial 2. However, it wasn't until the 100 iteration training that we could clearly see the effects of the training routine on the average cleaning rate.

As a general observation from these averaged results, with the accepting of the 5 iterations experiment, each training result had both genetic algorithms obtaining the same cleaning rate. After examining the weights of both the 20 and 100 chromosome genetic algorithm configurations, we found that although the same result was obtained, the networks that had been generated were different. We believe that this due to there being many certain networks that may be utilised to find an average maximum of classification from the data set which happens to be 87.5% (the highest cleaning rate achieved). Furthermore, we believe that the results obtained from the first experiment were lower for the 20 chromosome network due to there not being enough generations to find this maximum classification network. The reason we believe why the 100 chromosome was able to find this maximum classifying network in such a short amount of time is that it had the advantage of having five times as much weight configurations to evaluate within its population.

We created the Non-Monotonic Reasoning False-Negative Experiment with the goal of determining which of the five CDL formulae would be able to clean the highest rate of highly ambiguous missing RFID observations. We did this by comparing the cleaning results of the μ, α, π, β and δ formulae on various training cases. There were three training sets in all with 100, 500 and 1,000 ambiguous false negative anomalies. Additionally, at the completion of these experiments, the average was determined for all three test cases and was used to ascertain which of the five formulae would be used within the False-Negative Comparison Experiment.

From the results obtained, we have found that the highest average achieving formula is α (Alpha). This is probably due to the fact that it discovers cases in which both the β and π formulae agree upon, increasing the intelligence of the decision. Also of note is that the disjunction of β and π formulae shown within δ achieves a relatively high average cleaning rate also. The lowest performing average cleaning rate has been found to be β, which is probably due to its non-acceptance of ambiguity when drawing its conclusion. We believe it is crucial for the cleaner to have a low level of ambiguity when drawing its conclusions as the problem of missed readings needs a level probability to infer what readings need to be replaced. As stated above, we have chosen the α formula as the highest performing cleaner to be used within the False-Negative Comparison Experiment.

The goal of our fourth experimental evaluation was designed to put three classifiers through a series of test cases with large amounts of ambiguous missing observations. The three different classifications techniques included our Non-Monotonic Reasoning engine with CDL using the α compared against both Bayesian and Neural Networks with the highest performing configurations obtained from previous experimentations. We designed the experiment to have an abnormally high amount of ambiguous false negative anomalies consisting of 500 and 1,000 test cases to thoroughly evaluate each approach. After these experiments had concluded, we derived the average of each technique to find the highest performing classifier.

The results of this experiment, depicted in Table 1, has shown that on average, our Non-Monotonic Reasoning approach achieved the highest cleaning rate. It is important to note that it achieved the highest results when cleaning the data set with 1,000 test cases. We believe that the high results may be due to the deterministic nature of our methodology preventing erroneous permutations being selected. In contrast, the probabilistic methodologies have a higher chance of selecting incorrect imputed values which in turn produces artificial false positive anomalies and not recovering the original missing data.

After completing the False-Negative Comparison Experiment shown in Table 1, we have found that the Neural Network classifier had outperformed the Non-Monotonic Reasoning approach several times. We found that the Non-Monotonic Reasoning classifier obtained an abnormally high result when cleaning 1,000 test cases and therefore, we decided to conduct this experimentation again. Instead of using 500, 1,000 and 5,000 test cases, we instead opted to use 5 test cases that were evenly spaced, specifically 500, 1,000, 1,500, 2,000 and 25,000. The results of this revised experimentation shown in Table 2 have shown that the Neural Network has obtained a higher cleaning average than that of both the Bayesian Network and Non-Monotonic Reasoning classifiers with 86.68% accuracy. Similarly to the previous evaluation, the lowest performing cleaner was the Bayesian Network; however, the Neural Network obtained the highest accuracy in this experimentation.

From our revised findings, we have determined that the Neural Network has provided the highest accuracy when attempting to clean ambiguous false-negative anomalies. It is interesting to note that the probabilistic approach has actually outperformed the deterministic methodology in the case of imputing missing data. We believe that this is due to the fact that there would be a level of ambiguity and probability needed to be introduced when correcting missing data due to the lack of information available to the classifier. In this case, a methodology that attempts to investigate the validity of observations not normally considered by deterministic approaches would yield a higher cleaning rate.

FALSE-POSITIVE MANAGEMENT

To properly illustrate how our methodology performs, we will first review the motivation and architecture of the program. This includes the Feature Set Definition, Bayesian Network, Neural Network, Non-Monotonic Reasoning and Loading phases. Next, we will provide details of the ideal scenario to design our system most effectively. Finally, any assumptions we have made to ensure that the algorithms operates correctly will be listed.

Motivation

We created this system with the intended goal of correcting ambiguous false positive RFID data anomalies. To accomplish this, we created an intelligent analytical feature set definition that ties directly into three classifiers to determine the correct course of action. We chose to integrate classifiers as the current related work (such as rule-based solutions) lack the intelligence needed to handle ambiguous anomalies. This would allow our methodology to be able to counteract the anomalous data in an intelligent and automated manner. This results in our methodology overcoming the current problems that exists in current state-of-the-art techniques and provides a cleaning process that obtains both effective and efficient superiority over existing methodologies. We chose the three classifiers as opposed to other techniques due to the Bayesian Network, Neural Network and Non-Monotonic Reasoning have all been able to effectively correct anomalies in the False-Negative Management we have early experimented upon. We also chose Bayesian and Neural Networks as probabilistic methods to be compared with the deterministic Non-Monotonic approach. For example, the below set of RFID observations actually contain an ambiguous scenario in which a false-positive anomaly exists and may not be easily corrected:

- T1, R3, 2011-11-05 16:20:14.032
- T1, R4, 2011-11-05 16:21:56.012
- **T1, R2, 2011-11-05 16:22:23.005**
- T1, R4, 2011-11-05 16:23:41.043
- T1, R1, 2011-11-05 16:24:01.009
- T1, R1, 2011-11-05 16:25:23.004

The above readings consist of the three vital RFID observational data: the tag's identifier EPC (Electronic Product Code), the reader's identifier and the time of the event. The tag with an EPC of T1 is seen at Location R3, R4, R2, R4, R1 and R1 where the third observation in bold typeface would be flagged as a suspicious reading. In previous methodologies, this flagged observation would automatically be deleted due to the proximity of the readers in the second and fourth observations. However, using "Map Data" our methodology may make a more intelligent cleaning decision.

The map data is used to determine validity of the observation. Specifically for this scenario, the map data would contain the information that all four readers are attached to each other in a linear fashion meaning that R1 would connect to R2, R2 to R3, R3 to R4 and R4 to R5. Knowing this map data, our methodology provides a higher intelligence when it attempts to clean by examining the first and fifth observations where the locations R3 and R1 are read. Knowing that both locations in the first and fifth observation are geographically within proximity to the suspicious reading, it will make a judgment that the reading is actually valid. This will then lead onto the fourth observation which it will discover as a false-positive anomaly due to the second, third, fifth and sixth readings not being within the proximity of the fourth which in this example scenario, is correct. We believe that by enhancing the rules with our concept will apply much needed intelligence in scenarios where the obvious action will not clean the data set.

Methodology

As it may be observed in Figure 2, the design of our system has been broken into three sections: the Feature Set Definition Phase; the Classifier Phase; and the Modification Phase. The Feature Set Definition Phase is the first process that is conducted within our application in which the raw data is searched and sorted to find suspicious readings and the circumstances surrounding each of these observations. The Classification is where the system deviates between three different classifiers, the Bayesian Network, Neural Network or Non-Monotonic Reasoning Engines. Each classifier has one goal which is to determine if the flagged reading should be deleted or kept within the data set. This decision is based solely upon the input gathered from the Feature Set Definition Phase. After each classifier has determined the validity of the observation, it will then pass the decision onto the Modification Phase which will either delete or keep the value being passed to it. This entire process may be viewed in the high level pseudo-code interpretation found in Algorithm 5 in Box 4.

The first stage of the program is the Feature Set Definition Phase whose main goal is to analyse the data to discover suspicious readings and investigate key characteristics surrounding the flagged observation. Initially, this phase breaks the tag readings into Tag Streams designed to analyse the route of one tag. We define Tag Streams as individually analysed streams for one tag from the mass amount of readings. A tag will be flagged suspicious if the difference in timestamps exceed the user-defined duration it should take to reach the location, or if the geographical locations of the readers are not within proximity. To determine the geographic validity of the readers the program utilises a table named "Map Data," which is constructed by the user and reflects the geographic layout of all adjacent readers within the static environment. There are five observational values that are ascertained: *a*, *b*, *x*, *c* and *d*. The values

Figure 2. A high level diagram of inner processes used within the false-positive management approach

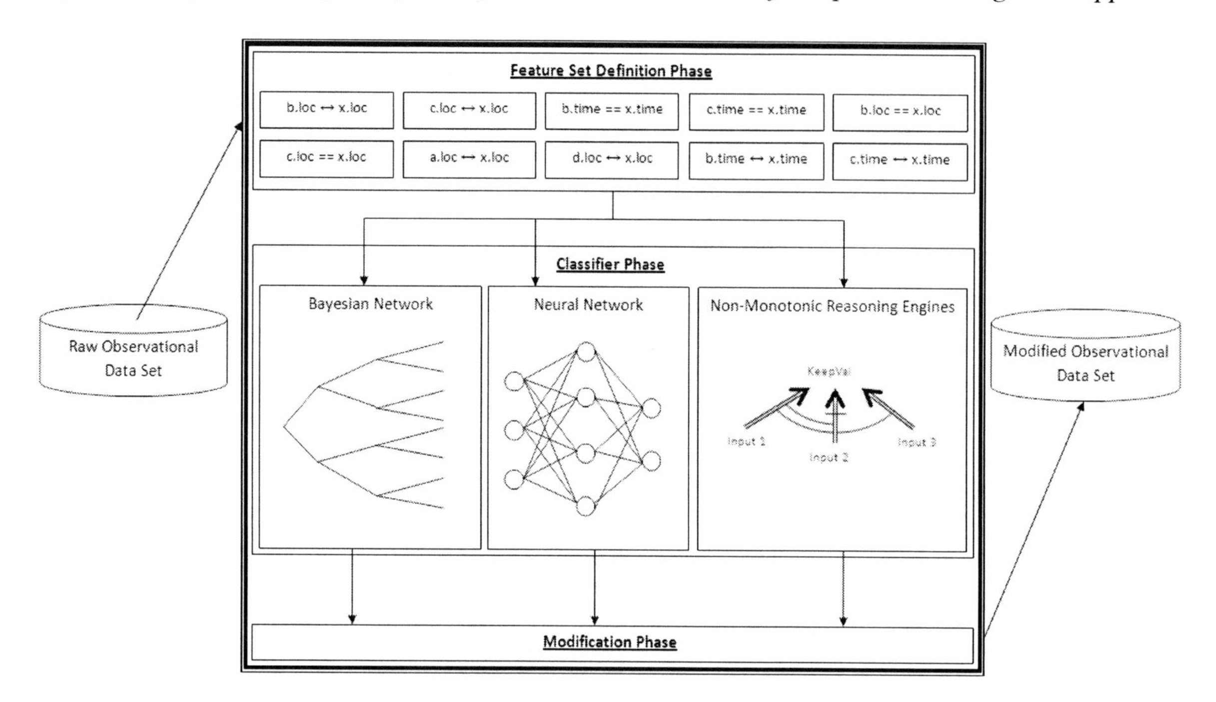

of observations *a* and *b* are the reading taken two readings or one reading respectively before the suspicious reading *x*. The *c* and *d* readings are the observations which been recorded once and twice after the suspicious reading. From all these observations, the timestamp and the location are all recorded and used in further analysis.

After the values of *a*, *b*, *x*, *c* and *d* with their respective timestamps and locations have been found, the Feature Set Definition Phase further investigates key characteristics of the data. The characteristics are comprised of ten different binary mathematical operations, however additional characteristics maybe be added by the user. Each of the characteristics contains spatial and temporal information regarding the observations before and after the suspicious readings. With regards to the proximity of timestamps, we have utilised the value of half a second. This time value may be altered to better suit the application for which it is designed. The characteristics we discover are as follows:

Box 4. A high level pseudo-coded version of the algorithm used to correct false-positive anomalies

Algorithm 5

Input: *Raw Observational Data Set*
Output: *Modified Observational Data Set*
foreach *Reading* **do**
| **if** *The observation is flagged for Suspicion* **then**
| | *Find the readings needed for Feature Set Definition.*
| | *Analyse the readings for key characteristics.*
| | *Pass the characteristics and reading to Classifier.*
| | **if** *The classifier finds the value to be kept* **then**
| | | *Do not modify the data set.*
| | **else**
| | | *Delete the flagged observation.*
| | **end**
| **end**
end
if *The user chooses to replace the data set* **then**
| *Delete the Original Data Set.*
| *Save the Modified Data Set at Original Data Set location.*
else
| *Save the Modified Data Set.*
end

Table 4. The table containing the results from cleaning false-positive anomalies using the integrated classifier when attempting to clean 500, 1,000, 1,500, 2,000, and 2,500 missing anomalies. The average is also presented, at the end and the technique performing the lowest, highest, and highest in average are put in bold typeface.

Cleaning Method	Test Cases	Correct Data Set Percentage
Bayesian Network	**500**	**97.40%**
Bayesian Network	1000	98.60%
Bayesian Network	1500	97.87%
Bayesian Network	2000	98.80%
Bayesian Network	2500	98.12%
Neural Network	500	98.00%
Neural Network	1000	98.30%
Neural Network	1500	98.47%
Neural Network	2000	98.10%
Neural Network	2500	98.76%
Non-Monotonic Reasoning	500	98.40%
Non-Monotonic Reasoning	1000	98.70%
Non-Monotonic Reasoning	1500	98.60%
Non-Monotonic Reasoning	2000	98.55%
Non-Monotonic Reasoning	2500	98.56%
Fused Non-Monotonic Reasoning	**500**	**99.20%**
Fused Non-Monotonic Reasoning	**1000**	**99.20%**
Fused Non-Monotonic Reasoning	1500	98.87%
Fused Non-Monotonic Reasoning	2000	98.80%
Fused Non-Monotonic Reasoning	2500	98.76%
Fused Bayesian Network	500	98.80%
Fused Bayesian Network	1000	98.90%
Fused Bayesian Network	1500	98.53%
Fused Bayesian Network	2000	98.80%
Fused Bayesian Network	2500	98.52%
Fused Majority Rules	500	99.00%
Fused Majority Rules	1000	98.90%
Fused Majority Rules	1500	98.20%
Fused Majority Rules	2000	98.80%
Fused Majority Rules	2500	98.60%
	Averages	
Bayesian Network	Average	98.16%
Neural Network	Average	98.33%
Non-Monotonic Reasoning	Average	98.56%
Fused Non-Monotonic Reasoning	**Average**	**98.97%**
Fused Bayesian Network	Average	98.71%
Fused Majority Rules	Average	98.70%

- $b.loc \leftrightarrow x.loc$
- $c.loc \leftrightarrow x.loc$
- $b.time == x.time$
- $c.time == x.time$
- $b.loc == x.loc$
- $c.loc == x.loc$
- $a.loc \leftrightarrow x.loc$
- $d.loc \leftrightarrow x.loc$
- $b.time \leftrightarrow x.time$
- $c.time \leftrightarrow x.time$

The data used in these characteristics are five different values that have two sub-values each. The five values include the observations of *a*, *b*, *x*, *c*, and *d*, which each have the time (time) and location (loc) for each value stored. The relations our methodology uses in analysis include when values are within certain proximity which is represented as ↔, or are equivalent which is represented as ==. It is important to note that the function that states that the two values are within proximity of each other have different meaning between the location and time. With regards to location, the proximity is determined by the "Map Data" table whereas the proximity with regards to time refers to the time value of the two observations being within the user defined time value of each other. After all these characteristics have been gathered, they are passed onto the various classifying methodologies as inputs to determine whether or not the flagged item should remain within the data set.

The first option that the Classifier Phase can utilise is the Bayesian Network. In this example we have considered the Bayesian Network to have ten inputs that correspond to the analytical characteristics found at the end of the Feature Set Definition Phase. Using these ten inputs the Bayesian Network will then determine, based on the weights it has obtained through training, whether the flagged observational reading should be kept within the database. This will result in one output known as the *keep_value* which will be set to either true or false. This *keep_value* output will be passed to the Modification Phase at the end of this process at which point the entire application will repeat each time a suspicious reading is encountered. We also set all binary input numbers from 0 and 1 to 0.1 and 0.9 respectively to allow for higher mathematical functions to benefit from the avoidance of multiplying by zero.

We have chosen a Genetic Algorithm to train the Bayesian Network weights based on the various test cases that may arise. The Genetic Algorithm will have the population of the chromosomes to determine the ideal number of chromosomes utilised for training purposes. The mutation rate of the Genetic algorithm being utilised will be 1% for the top 10% chromosomes with regards to fitness and 5% for all other chromosomes. After the best weight configuration has been determined, the network will be utilised to compare it against the Neural Network and Non-Monotonic Reasoning approaches.

An Artificial Neural Network (ANN) is the second option we have chosen for the Classifier Phase which utilises weighted neurons to determine the validity of the flagged value. Like the Bayesian Network, this ANN will use the ten inputs gathered from the Feature Set Definition Phase to pass through the network and obtain one output. The network shall be comprised of a single hidden layer with eleven hidden nodes resulting in 121 weights between all the nodes. We specifically wanted to choose more hidden units than inputs and only one layer as we have found that multi-layered networks do not necessarily enhance the performance of the classifier.

We have also set the momentum and learning rates to 0.4 and 0.6 respectively and have utilised a Sigmoidal Activation Function. Additionally, as with the Bayesian Network we shall use the numbers 0.1 and 0.9 rather than the binary numbers of 0 and 1 respectively. Two prominent training algorithms have been utilized to properly configure the Neural Network. The first is the Back-Propagation Algorithm while the second is the Genetic Algorithm which has also been utilised within the Bayesian Network. Both algorithms will use a limited amount of iterations as stopping criteria for the training. Additionally, various amounts of chromosomes will be utilised for the Genetic Algorithm.

The final classifier we have utilised within our implementation is Non-Monotonic Reasoning Logic Engines. The actual algorithm utilises a series of rules that we have created based upon the input analysis variables obtained from the Feature Set Definition Phase. From this, the logic engines determine the correct course of action to either keep the value or not based on the different levels of ambiguity we enforce.

The four symbols that are used to interact with the values within the Non-Monotonic Reasoning rules are the logic AND operator $\wedge$, the negative operator $\sim$, the equal operator $==$ and our use of the double arrow $\leftrightarrow$ to illustrate proximity between the two analysis variables. The rule with the lowest precedence states that if $b.loc == x.loc \wedge c.loc == x.loc \wedge b.time == x.time \wedge c.time == x.time$ is proven true, then conclude $\sim keep\_val$. This will get over taken if $b.loc \leftrightarrow x.loc \wedge c.loc \leftrightarrow x.loc \wedge \sim b.time == x.time \wedge \sim c.time == x.time$ is proven which will result in a positive *keep_val* conclusion. In the event that any of $b.time == x.time \wedge c.time == x.time$ or $b.time == x.time$ or $c.time == x.time$ are proven, all previously stated rules will be ignored and the classifier will conclude $\sim keep\_val$. Similarly, keep_val will be proven if $b.loc \leftrightarrow x.loc \wedge c.loc \leftrightarrow x.loc \wedge \sim b.time == x.time \wedge \sim c.time == x.time$ or $c.loc \leftrightarrow x.loc \wedge c.loc \leftrightarrow x.loc$ or $a.loc \leftrightarrow x.loc \wedge d.loc \leftrightarrow x.loc$ has been proven. The $\sim keep\_val$ will be proven and overwrite these entire conclusions if $\sim b.loc \leftrightarrow x.loc \wedge \sim c.loc \leftrightarrow x.loc$ or $\sim b.loc \leftrightarrow x.loc \wedge \sim c.loc \leftrightarrow x.loc \wedge \sim a.loc \leftrightarrow x.loc \wedge \sim d.loc \leftrightarrow x.loc$ have been proven. Finally, the rules with the highest precedence will prove that *keep_val* is positive if $c.time \leftrightarrow x.time \wedge \sim b.loc == x.loc$ or $c.time \leftrightarrow x.time \wedge \sim c.loc == x.loc$ are true. The process used to find the conclusion from the lookup table created by the logic engine can be seen in pseudocode in Algorithm 6 in Box 5.

Box 5. Algorithm for the non-monotonic reasoning lookup process used in the methodology

Algorithm 6
Input: *DPL Rules, Analytical Values, Formula* **Output:** *Lookup table, Conclusions* *Input DPL Rules into CDL.* *Store lookup table for later use.* *Open up lookup table.* **while** *There are still entries in lookup table* **do** | **if** *Found the entry for given Inputs* **then** | | *Find the conclusion associated with Inputs.* | | *Break the while loop.* | **end** **end** *Output conclusion found.*

As a default case where neither *keep_val* nor *~keep_val* are encountered, the logic engine will keep the flagged reading to avoid artificially introduced false negative observations. Additionally, the order in which they have been written in this document is also the order of priority with regards to finding the conclusion.

After each intelligent classifier has determined whether or not to keep or delete the flagged reading, it will pass it to the Modification Phase. After the decision has been received, the application will then delete the identified value in the original data warehouse. The system will also only flag potential anomalies rather than automatically delete the flagged items if the user would like to examine the observations prior to modification.

Results and Analysis

In order to investigate the applicability of our concepts, we conducted four experiments. The first three were dedicated to finding the optimal configuration of each classifier whereas the last focused on the comparison of the three classifiers. In this section, we describe the scenario, database structure, assumptions and environment in which we conducted these experiments and an analysis

Table 5. The table containing the results from cleaning false-negative anomalies using the Integrated Classifier when attempting to clean 500, 1,000, 1,500, 2,000, and 2,500 missing anomalies. The average is also presented at the end, and the technique performing the lowest, highest, and highest in average are put in bold typeface.

Cleaning Method	Test Cases	Correct Data Set Percentage
Bayesian Network	**500**	**78.80%**
Bayesian Network	**1000**	**78.90%**
Bayesian Network	1500	84.33%
Bayesian Network	2000	83.90%
Bayesian Network	2500	84.64%
Neural Network	500	88.60%
Neural Network	1000	83.30%
Neural Network	1500	87.87%
Neural Network	2000	86.65%
Neural Network	2500	86.96%
Non-Monotonic Reasoning	500	85.80%
Non-Monotonic Reasoning	1000	85.90%
Non-Monotonic Reasoning	1500	87.20%
Non-Monotonic Reasoning	2000	86.50%
Non-Monotonic Reasoning	2500	87.48%
Fused Non-Monotonic Reasoning	500	87.00%
Fused Non-Monotonic Reasoning	1000	85.00%
Fused Non-Monotonic Reasoning	1500	86.93%
Fused Non-Monotonic Reasoning	2000	87.75%
Fused Non-Monotonic Reasoning	2500	87.20%
Fused Bayesian Network	500	85.80%
Fused Bayesian Network	1000	87.40%
Fused Bayesian Network	1500	88.73%
Fused Bayesian Network	2000	87.85%
Fused Bayesian Network	2500	87.88%
Fused Majority Rules	500	87.40%
Fused Majority Rules	1000	87.90%
Fused Majority Rules	**1500**	**88.80%**
Fused Majority Rules	2000	88.60%
Fused Majority Rules	2500	88.68%
	Averages	
Bayesian Network	Average	82.11%
Neural Network	Average	86.68%
Non-Monotonic Reasoning	Average	86.58%
Fused Non-Monotonic Reasoning	Average	86.78%
Fused Bayesian Network	Average	87.53%
Fused Majority Rules	**Average**	**88.28%**

of the experiments. Furthermore, we describe the experimental evaluation and present the results of four experiments which we conducted. The first three are designed to determine the highest achieving configuration of the Bayesian Network, Neural Network and Non-Monotonic Reasoning classifiers. In the fourth experiment, the highest achieving classifiers have been compared against each other to find which one achieves the highest cleaning rate.

The ideal scenario which we envisioned for our application would be an enclosed RFID static environment. It is important that such an environment exists as the program requires knowledge of the geographical landscape with regards to the locations of the readers. The motivation for such an application came from the idea of a building that houses important items which need to be monitored thoroughly such as a hospital or an elderly care unit. Although the scenario has been designed specifically to integrate with an RFID environment for their specific applications, this software may be utilised for an array of domains that utilise spatial and temporal coordinates within the data set. Due to medical facilities already employing RFID systems in this manner (Swedberg, 2005), we believe that this scenario is particularly realistic to be beneficial to similar applications that could be conducted.

To store the observational data within a data warehouse we have designed a database structure modeled after the Data Model for RFID Applications (DMRA) (Liu, 2007). DMRA is currently being used within RFID middleware where it is designed to manage all RFID observational data efficiently. We have chosen only three tables from the many utilised in the application that are relevant to our methodology which houses information regarding the readers (Reader), tags (Object) and interaction between both of them (Observation). The structure of the tables used in this data warehouse is as follows:

READER(ReaderID, Name, Description)
OBJECT(Epc, Name, Description)
OBSERVATION(ReaderID, Value, Timestamp)

Additionally, we have created another table 'Map Data', which stores information vital to our concept. The prime goal of this additional table is to store the readers which are within close proximity of each other. The structure for this additional table is as follows:

MAPDATA(ReaderID1, ReaderID2).

We have two prime assumptions for this application, first, that the environment is static and, secondly, that the locations of the readers are known. This is important as the concept relies heavily on both the mapped data to determine if a reading is suspicious and, subsequently, when it attempts to reason the validity of the observation if it has been flagged. This would require the application to be run in an RFID-enabled environment where the readers are mounted which would prevent it from changing locations. To create this pre-requisite condition, the user would have to first enter the map data into the application before the application has begun to correct anomalies. Additionally, any modifications to the environment or the reader locations would have to be updated within the map data table as well.

The code for this methodology was written and compiled in the C++ language utilising Microsoft Visual Studio C++ 6.0. The experiments were run on a Windows XP operating system running Service Pack 3. As outlined earlier, above, there are four main experiments which were conducted using the methodology. The first experiment was designed to test the highest performing Genetic Algorithm when training the Bayesian Network. For this experiment, the amount of chromosomes in the population was manipulated to find the highest performing number. The second experiment was designed to discover which training algorithm of either the Back-Propagation or Genetic Algorithm obtained the highest cleaning rate. For this experiment, the amount of chromosomes were modified and compared with the back-propagation algorithm to determine the highest achieving algorithm. The third experiment was designed to determine which formulae achieved the highest cleaning rate within the Non-Monotonic Reasoning approach.

We specifically chose only to examine classifier techniques as the related work is not comparable due to either it not being able to clean ambiguous data or not using an automated process. The last experiment which was conducted took the highest achieving configurations of each of the classifiers and compared each methodology against the other. Four data sets were utilised for this experimentation, the first three were training sets in which 500, 1,000 and 5,000 scenarios were used to train the algorithms and find the optimal configuration. Each training set contained different scenarios to avoid the risk of over-fitting the classifiers.

The second data set was three testing sets in which 1,000, 5,000 and 10,000 randomly chosen scenarios were selected and passed to the application to have the anomalies eliminated. Each of these testing sets contained feature set definitions generated within our sample scenario. After each of the training and testing experiments has been conducted, the average of cleaning rate of the experiments has been derived for each technique and used to identify the highest achieving method.

For our first experiment, we conducted an investigation into the optimal amount of chromosomes which are needed to clean the false positive anomalies. To accomplish this, we created three Bayesian Networks that have been configured using a genetic algorithm with 10, 50 and 100 chromosomes. Each network was trained for 10 generations to breed and optimise the configuration. With regards to the set of data being used for training, we used 3 different "Training Cases" comprising 500, 1,000 and 5,000 false-positive anomaly scenarios. After these experiments were completed, the average of the 3 training cases was then extracted and, subsequently, used to determine the amount of chromosomes that are needed to achieve the highest cleaning rate.

From our experimentations, we have found that the configuration that used 10 chromosomes to train the network obtained the highest average cleaning rate. As a result, the Bayesian Network using a Genetic Algorithm with 10 chromosomes will be utilised in the final experiment in which all three classifiers are compared. The lowest achieving configuration using 100 chromosomes tested upon 500 training cases was the Bayesian Network. The highest achieving configuration has been found to be the configurations with 10 and 50 chromosomes against 500 and 1,000 training cases respectively.

The second experiment we conducted was in relation to determining the highest performing network configuration for a Neural Network to clean anomalous RFID data. To do this, we trained the weights of the networks using the back-propagation (BP) and the genetic algorithms with 10 (GA-10), 50 (GA-50) and 100 (GA-100) chromosomes present. The performance of the resulting networks was determined based upon the correctness of the classification from 3 Training Cases using 500, 1,000 and 5,000 false-positive anomaly scenarios. Each configuration had been trained by 10 iterations or generations before the training experiment commenced. The main goal of this experiment was to determine the highest average achieving network trainer, thus the average of the three training cases has also been found.

From our experimental results we have derived a general observation that the performance of the network is vastly improved with 50 and 100 chromosomes using the Genetic Algorithm. The highest performing average of the Neural Network has been found to be the Genetic Algorithms when trained with both 50 and 100 chromosomes. As such, we decided to use the Genetic Algorithm with 100 chromosomes as the attempt to clean 500 training cases performed the highest. The lowest performing cleaning algorithm was the Back-Propagation algorithm when attempting to clean 1,000 training cases.

The main goal of the third experiment was to derive the highest performing Non-Monotonic Reasoning formula from the five different options used in Clausal Defeasible Logic. With this in mind, the μ (Mu), α (Alpha), π (Pi), β (Beta) and δ (Delta) formulae were each trained using 3 training cases containing 500, 1,000 and 5,000 false-positive scenarios each. Like the previous two experiments, the averages of each performing algorithm was ascertained and used to determine which of the five formulae would be utilised to proceed onto the final experiment.

From the results obtained, we have found that the highest performing formulae are μ, α and π. In contrast, the β and δ formulae both performed the least cleaning. With regards to the final experimentation, we have chosen the π formula as it performed the highest and is the most likely to continue to perform highly. The reasons as to why we rejected the μ and α formulae lie in the fact that the μ formula is strict in that it only accepts factual information and the α formula is connected directly to the β. Hence, we determined that the π formula would be superior to the other formulae.

The goal of the final experiment was to determine which of the three highest performing classifier techniques would clean the highest percentage of a large amount of false-positive RFID anomalies. The three classifiers used in this experiment included the Bayesian Network trained by a Genetic Algorithm with 10 chromosomes (BN), the Neural Network trained by Genetic Algorithm with 100 chromosomes (NN) and the π of the Non-Monotonic Reasoning Engine (NMR). The classifiers were all chosen based upon the high performance found within the first three experiments previously discussed. Both of the Bayesian and Neural Networks had both been trained for 10 generations before these tests were conducted. As opposed to the previous experiments, we determined that three "Testing Cases" containing 1,000, 5,000 and 10,000 randomly chosen false-positive scenarios would be utilised to determine the highest performing classifier. To ascertain the highest performance, the average of each of the three test cases has been found from the results.

The results of this experiment is depicted in and Table 3 where the Amount of Test Cases and Classifier has been graphed against the Percentage of Correctness. From these results, it can be seen that the Non-Monotonic Reasoning Engine achieves the highest average cleaning rate among the other classifiers. The highest performing classifier has also been found to be the Non-Monotonic Reasoning when attempting to clean 1,000 test cases; whereas the lowest achieving classifier has been found to be the Neural Network when attempting to clean 10,000 test cases.

The Non-Monotonic Reasoning Engine outperformed the other classifiers in dealing with false-positive data due to the fact that it is a deterministic approach. The Bayesian and Neural Networks, by contrast, rely on a probabilistic nature to train their respective networks. The major drawbacks of this system are that it is specifically tailored for the static RFID cleaning problem; however, we believe that the same concept may be applicable to any static spatial-temporal data enhancement case study. With regards to applying our methodology to other applications where the environment is dynamic, the Feature-Set Definition and Non-Monotonic Reasoning will need greater complexity to accommodate the change in anomalies. Although the test cases utilised in experimentation were small in comparison to the immense amount of RFID readings that get recorded in real world systems, we believe our methodology would behave similarly upon larger data sets.

INTEGRATED CLASSIFIER

After the successful investigation of the False-Negative and False-Positive Management techniques, we have chosen to evolve these approaches through the use of an Integrated Classifier architecture to attempt to obtain an even higher cleaning accuracy. Within this system, we take conclusions drawn from the three classifiers and use fusion techniques, such as a Non-Monotonic Reasoning algorithm, a Bayesian Network or taking the Majority answer, to derive a highly intelligent output. In this section, we identify the motivation behind developing our methodology, the architecture we created, the intended scenario of our approach and the assumptions needed for our concept run as intended.

Motivation

Radio Frequency Identification has been found to have limited functionality due to problems in the system such as data anomalies. If these anomalies were eliminated, the applications that may benefit from RFID would be increased to various other commercial sectors thereby saving cost and effort. Previous approaches have been utilised to eliminate easily found anomalies, such as middleware algorithm used to determine a duplicate observation recorded in the same location in under a second, however these methodologies lack the intelligence needed to properly correct the stored observations to its maximum integrity. Additional past literature has individually stated that there is a need to use both deterministic and probabilistic methodologies to adequately clean the data (Rao, 2006), (Khoussainova, 2008). With this in mind, we propose an approach that took advantage of both probabilistic and deterministic approaches to bring RFID data cleaning to a higher level of integrity. We did this because we fundamentally believed that missing data require a level of probability to find the absent information. In contrast, we believe that both wrong and duplicated data will need to have a deterministic approach due to having the information already present and there is less need to rely on probability. We specifically chose two probabilistic approaches (the Bayesian Network and Neural Network) and one deterministic approach (Non-Monotonic Reasoning) to give a probabilistic advantage to the former methods. This is also the reason as to why we chose the global fusion of the classifiers as opposed to pair wise combinations. To counter this, we chose the novel deterministic Non-Monotonic Reasoning as a fusion technique which permits additional bias to the Non-Monotonic Reasoning conclusion in its logical rules.

Methodology

We have divided our methodology into four core components, the Feature Set Definition, Classification, Classifier Integration and Loader. Due to the vast differences between the False-Positive and False-Negative anomalies, we have different classifier integrations for both the duplicate/wrong data, and the missing data. As seen in Figure 3, the Original Data containing the RFID observations, along with the Geographical Data, is passed into the Feature Set Definition where crucial analytical features of the data are identified. This analytical information is then passed into the Classification component where the Non-Monotonic Reasoning, Bayesian Network and Neural Networks are used to determine if a reading is valid or not. The results of the Classifiers are then passed into the Classifier Integration which uses three fusion methods to intelligently determine the validity of each suspicious reading. Finally, after all the information is gathered, the methodology will finally either delete, keep or insert the correct values into the data set within the loader component.

The first action that the Feature Set Definition takes is to divide up the data into streams that follow the geographical path of each tag using the geographical data passed into the system. Once this is done, suspicious readings are found based

Figure 3. A high level diagram of inner processes used within the integrated classifier approach

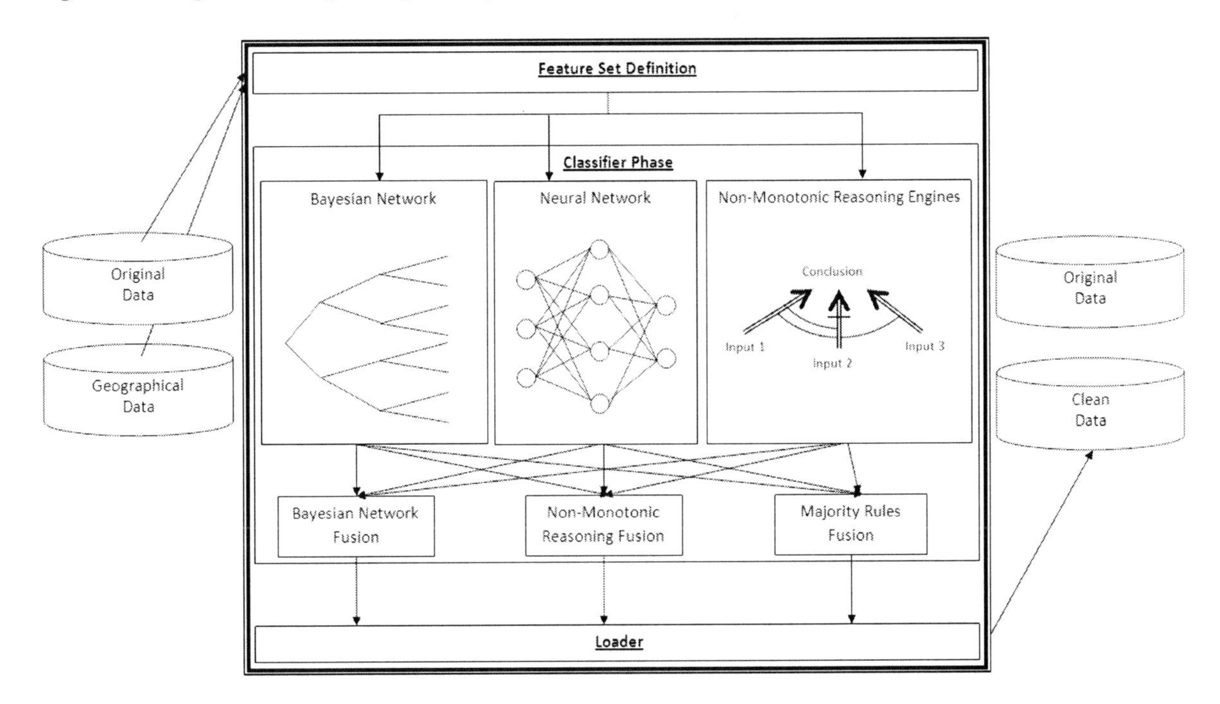

on the geographical data supplied at the beginning. Only suspicious readings will be flagged by the system and all other observations will be ignored by our system. For example, if a reading occurs in two locations not within proximity concurrently, it will be flagged as suspicious as it may be a duplicate anomaly. The information we obtain for both the false-positive and false-negative anomalies are identical to the False-Negative and False-Positive analysis we presented earlier.

After the crucial analytical data has been found, it will be passed on as a feature set to the classification component of the methodology to be determined if the suspicious observations should be deleted, kept, or inserted into the database. For the false-positive anomalies which contain either a suspected duplicate or wrong reading, there are two conclusions that may be drawn from the classifiers: either delete or keep the values. With regards to the false-negative anomalies, there are five possible conclusions that may be made each with different reader values combinations. When handling missing data in various data sets, there is a need to impute the data back into the database (i.e. generate possible answers to be used when lacking factual information). These values to be imputed back into the database are identical to the previously proposed Permutations from our False-Negative Management approach. Additionally, we have retained all the highest performing classifiers and configurations found from our previous False-Negative and False-Positive anomaly experimentation.

In this work, we have proposed various methods of combining the classifiers together to develop a new intelligent means increasing the integrity of the conclusions being made. To this end, we have introduced three main fusion techniques to integrate the classifiers, the Non-Monotonic Reasoning Fusion (NMR Fusion), Bayesian Network Fusion (BN Fusion) and Majority Rules Fusion (MR Fusion). For the False-Positive anomalies, since the returned determination is either to keep or delete a value, we only need to integrate the classifiers once. However, due to the False-Negative anomalies finding five varied conclusions, we need to run the fusion algorithms five different times for each permutation.

With regard to the Non-Monotonic Reasoning Fusion, we took all the conclusions made in the classification component and put them into the logic engine. By default, the result will be negated conclusion, however, if the positive Bayesian Network or positive Neural Network, a positive conclusion will be proven as well. This will be beaten by a negated Non-Monotonic Reasoning; however, a positive Non-Monotonic Reasoning observation will also prove a positive conclusion. However, if a negated Bayesian Network and a negated Neural Network have been found, this will result in a negated conclusion. Finally, the highest rule set is a positive Bayesian Network and positive Neural Network which will result in a positive conclusion and over rule any previous rules stated before. We also have made sure that the Non-Monotonic Reasoning will have a slightly higher bias than the probabilistic techniques as this fusion method is deterministic in nature.

The Bayesian Network contains various weights for both a positive and negative conclusion spanning over the various classifiers. For a positive conclusion, if the Bayesian Network is either 60% (True) or 40% (False), the Neural Network is either 60% (True) or 40% (False), and the Non-Monotonic Reasoning is either 40% (True) or 60% (False). For the negated conclusion, the Bayesian Network values are 40% (True) or 60% (False), the Neural Network is either 40% (True) or 60% (False), and the Non-Monotonic Reasoning classifier is either 60% (True) or 40% (False). Similarly to the NMR Fusion, we wish to give a slightly higher bias to the Bayesian and Neural Networks as the Bayesian Network Fusion is itself a probabilistic technique. The final technique we have employed, the Majority Rules Fusion, is an unbiased approach which will use the conclusion voted most by the three classifiers as its determination. We hope that by having deterministic, probabilistic, and non biased fusion methods, we will observe varying results among the different anomalies. In the unlikely event that none of the permutations have been found to be chosen more than once for the false-negative anomalies, a weighting system is employed based on the scale of most unbiased to biased (3>1>2>4>5).

After the decision has been made by the integrated classifier, our methodology will then proceed to delete, keep, or insert the correct values in the data set. We have made the option to either modify the Original Data set if the user is comfortable with the enhanced data sets or to create a new data set keeping the original data set separate for added integrity. Being that this entire process is at a deferred stage of the capture cycle where all the data has been stored; in this work we did not consider the cost of cleaning. However, in the future, we would like to implement a version of this concept that will run in real-time at the stage of data capture.

We have intended to create our methodology for a scenario in which many readers are mounted around a known environment and tags are passing through the area to be scanned. Applications in which this is already conducted include a hospital, in which surgical patients are monitored, airports which track luggage and the transportation of various items in a supply chain. It is crucial that a known environment is used in the scenario as the geographical locations of each of the readers and their proximity to one another must be recorded in the system.

There are two assumptions we have identified for this scenario relating both to the identification of the false positive and false negative anomalies. The first is that, as stated in the intended scenario, the geographical locations of the readers must be known to the readers so that a tag which is recorded at abnormal locations may be flagged as a suspicious reading. The second assumption we make is that the time used to flag a missed reading is less than the time it takes for the tagged object to move from one readers scan range to another. Both of these assumptions are crucial as they provide the rules that our system follows to identify a suspicious set of observations to be corrected.

Results and Analysis

To properly test our integrated classifier, we have devised four experiments to determine the overall effectiveness and advantages it has over existing techniques. The first two experiments will be carried out to test the effectiveness of each of the fusion techniques for solving both false positive anomalies (duplicate and wrong data), and false negative anomalies (missing data). The second two experiments will take the best performing integrated classifier to compare it to state-of-the-art techniques currently used to enhance the integrity of RFID data. These experiments will be performed on multiple test beds to determine its effectiveness on varying amounts of anomalies.

To properly evaluate the effectiveness of our methodology, we have used simulated test cases of the information obtained from readers. We have created five test beds with 500, 1,000, 1,500, 2,000 and 2,500 test cases each to observe the performance of the approaches where there are various amounts of anomalies. Each of the test cases present within the test bed represents a found anomaly within the data sets. All code used in our methodology was written in the C++ language and executed in Microsoft Visual C++ 6.0. The computer used for this experimentation was a Microsoft Windows XP machine with Service Pack 3 Intel (R) Core 2 Duo CPU E8400 @ 3 GHz 2.99 GHz with 4 GB of RAM.

The first experiment we ran included testing the percentage of clean data for the Bayesian Network, Neural Network, Non-Monotonic Reasoning, Fused Non-Monotonic Reasoning, Fused Bayesian Network and Fused Majority Rules classifiers when attempting to clean 500, 1,000, 1,500, 2,000 and 2,500 False-Positive test cases. From the False-Positive results found in Table 4, the highest performing classifier average has been found to be the Fused Non-Monotonic Reasoning classifier.

The absolute highest performing classifier was also the Fused Non-Monotonic Reasoning classifier when attempting to clean 500 test cases. The least performing classifier for the False-Positive experiment was the Bayesian Network when attempting to clean 500 test cases. We believe that the advantage the Fused Non-Monotonic Reasoning classifier had was due to its deterministic architecture and the nature of False-Positive anomalies.

In the second experimentation we conducted, we took the same classifiers and test case amounts; however we used False-Negative anomalies rather than False-Positive. The results, which may be viewed in Table 5, have shown that the highest performing classifier average has been found to be the Fused Majority Rules classifier. The highest performing clean on the data sets has been found to be the Fused Majority Rules classifier as well when attempting to clean 1,500 test cases. The least achieving classifier for the False-Negative anomalies has been found to be the Bayesian Network classifier when attempting to clean 500 test cases. The results have shown that the unbiased nature of the Fused Majority Rules classifier has given it a clear advantage when cleaning False-Negative anomalies. It is also important to observe that as highlighted in the above results, of the two types of anomalies, it is harder to correct the False-Negative anomalies.

The Non-Monotonic Reasoning fused classifier was able to achieve the highest false-positive clean as its deterministic nature makes it ideal to clean wrong and duplicate data whereas probabilistic techniques would introduce an additional level of ambiguity. With regards to the false-negative anomalies, the Majority Rules fused classifier gained the highest cleaning rate due to it being able to accept all three classifiers without any bias. Additionally, we believe we may have obtained a higher result if we introduced a dynamically trained fused Bayesian Network or created a Fused Neural Network classifier. As this methodology is designed to be applied at a deferred stage of the RFID capture cycle, our experimentation was not concerned with the runtime performance. However, we would like to extend and modify our approach in the future to allow real-time processing in which case we will be taking the processing time into consideration for each classifier and focusing on the amount of time needed to generate rules or train the networks.

FUTURE RESEARCH DIRECTIONS

With regard to future work, we would like to investigate other fusion approaches of additional classifiers such as the Support Vector Machine and other classifier training techniques. We would also like to apply our technique to various databases as we believe that our methodology is not limited to merely cleaning RFID data and may be applied to other spatial-temporal data collections as well. While we have investigate the problems associated with, and tailored our concept towards passive RFID systems, we would like to modify our technique to manage anomalies found in chipless tags. Also, as mentioned earlier, we would like to develop a real time implementation of this concept.

CONCLUSION

In this chapter, we presented several methodologies to clean anomalous Spatial-Temporal data using three various classifiers, and then evolved it into one integrated approach. For this study, we investigated RFID technology as our case study as it continues to generated anomalies within the recorded data sets and has a need to be rectified before it can be employed in various other commercial sectors. Through experimental evaluation, we have found that the highest classifiers for False-Negative and False-Positive Management has been found to be the Neural Network and Non-Monotonic Reasoning respectively. After evolving this technique, we found that the highest performing fusion type for wrong and duplicate data was the Fused Non-Monotonic Reasoning classifier, while the Fused Majority Rules approach was the most effective for cleaning missing readings. From this experimentation, we have solidified our belief that probabilistic approaches are best utilised when cleaning missing observations, in contrast to deterministic methodologies that obtain the highest cleaning rate while cleaning wrong and duplicated readings. While we have focused our investigation and tailored our approach to correct RFID readings, we are confident that our approach can be applied to other data sets that contain any Spatial-Temporal information.

REFERENCES

Antoniou, G., Billington, D., Governatori, G., & Maher, M. (2006). Embedding defeasible logic into logic programming. *Theory and Practice of Logic Programming*, *6*(6), 703–735. doi:10.1017/S1471068406002778

Billington, D. (2007). *An introduction to clausal defeasible logic*. David Billington's Home Page. Retrieved August 12, 2011, from http://www.cit.gu.edu.au/~db/research.pdf

Billington, D. (2008). Propositional clausal defeasible logic. In *European Conference on Logics in Artificial Intelligence* (pp. 34-47).

Billington, D., Estivill-Castro, V., Hexel, R., & Rock, A. (2005). Non-monotonic reasoning for localisation in RoboCup. In *Proceedings of the 2005 Australasian Conference on Robotics and Automation*.

Blumenstein, M., Liu, X., & Verma, B. (2007). An investigation of the modified direction feature for cursive character recognition. *Pattern Recognition*, *40*(2), 376–388. doi:10.1016/j.patcog.2006.05.017

Cha, D., Blumenstein, M., Zhang, H., & Jeng, D. (2008). A neural-genetic technique for coastal engineering: Determining wave-induced seabed liquefaction depth. *In Engineering Evolutionary Intelligent Systems,* (pp. 337-351).

Chawathe, S., Krishnamurthy, V., Ramachandran, S., & Sarma, S. (2004). Managing RFID data. In *Very Large Databases*, (pp. 1189-1195).

Floerkemeier, C., & Lampe, M. (2004). Issues with RFID usage in ubiquitous computing applications. In *Pervasive Computing, No. 3001* (pp. 188–193). Austria: Springer-Verlag. doi:10.1007/978-3-540-24646-6_13

Holland, J. (1975). *Adaption in natural and artificial systems*. University of Michigan Press.

Jeffery, S., Garofalakis, M., & Franklin, M. (2006). Adaptive cleaning for RFID data streams. In *Very Large Databases*, (pp. 163-174).

Khoussainova, N., Balazinska, M., & Suciu, D. (2008). Probabilistic event extraction from RFID data. In *International Conference on Data Engineering,* (pp. 1480-1482).

Liu, S., Wang, F., & Liu, P. (2007). *A temporal RFID data model for querying physical objects.* Technical Report TR-88, TimeCenter.

McCulloch, W., & Pitts, W. (1943). A logical calculus of the ideas immanent in nervous activity. *The Bulletin of Mathematical Biophysics*, *5*, 115–133. doi:10.1007/BF02478259

Rao, J., Doraiswamy, S., Thakkar, H., & Colby, L. (2006). A deferred cleansing method for RFID data analytics. In *Very Large Databases*, (pp. 175-186).

Rooij, A., Johnson, R., & Jain, L. (1996). *Neural network training using genetic algorithms*. River Edge, NJ: World Scientific Publishing Company Incorporated.

Rumelhart, D., Hinton, G., & Williams, R. (1986). Learning representations by back-propagating errors. *Nature*, *323*, 533–536. doi:10.1038/323533a0

Swedberg, C. (2005). Hospital uses RFID for surgical patients. *RFID Journal*. Retrieved August 12, 2011, from http://www.rfidjournal.com/article/articleview/1714/1/1/

Williams, R., & Herrup, K. (1988). The control of neuron number. *Annual Review of Neuroscience*, *11*(1), 423–453. doi:10.1146/annurev.ne.11.030188.002231

ADDITIONAL READING

Darcy, P., Pupunwiwat, P., & Stantic, B. (2011). The challenges and issues facing deployment of RFID technology. In *Deploying RFID - Challenges, solutions and open issues*, (pp. 1-26).

Darcy, P., Stantic, B., & Derakhshan, R. (2007). Correcting stored RFID data with non-monotonic reasoning. *Principles and Applications in Information Systems and Technology*, *1*(1), 65–77.

Darcy, P., Stantic, B., Mitrokotsa, A., & Sattar, A. (2010). Detecting intrusions within RFID systems through non-monotonic reasoning cleaning. In *Intelligent sensors, sensor networks and information processing*, (pp. 257-262).

Darcy, P., Stantic, B., & Sattar, A. (2009a). A fusion of data analysis and non-monotonic reasoning to restore missed RFID readings. In *Intelligent sensors, sensor networks and information processing,* (pp. 313-318).

Darcy, P., Stantic, B., & Sattar, A. (2009b). Augmenting a deferred Bayesian network with a genetic algorithm to correct missed RFID readings. In *Malaysian Joint Conference on Artificial Intelligence*, (pp. 106-115).

Darcy, P., Stantic, B., & Sattar, A. (2009c). Improving the quality of RFID data by utilising a bayesian network cleaning method. In *Proceedings of the IASTED International Conference Artificial Intelligence and Applications*, (pp. 94-99).

Darcy, P., Stantic, B., & Sattar, A. (2010a). Applying a neural network to recover missed RFID readings. In *Australasian Computer Science Conference*, (pp. 133-142).

Darcy, P., Stantic, B., & Sattar, A. (2010b). Correcting missing data anomalies with clausal defeasible logic. In *Advances in Databases and Information Systems*, (pp. 149-163).

Darcy, P., Stantic, B., & Sattar, A. (2010c). Intelligent high-level RFID event transformation utilising non-monotonic reasoning. In *Wireless Communications Networking and Mobile Computing,* (pp. 1-4).

Darcy, P., Stantic, B., & Sattar, A. (2010d). X-CleLo: Intelligent deterministic RFID data transformer. In *International Workshop on RFID Technology,* (pp. 59-68).

Derakhshan, R., Orlowska, M., & Li, X. (2007). *RFID data management: Challenges and opportunities* (pp. 175–182). Institute of Electrical and Electronics Engineers Radio Frequency Identification.

Floerkemeier, C. (2004). *A probabilistic approach to address uncertainty in RFID.* In Auto-ID Labs Research Workshop.

Floerkemeier, C., Lampe, M., & Schoch, T. (2003). The smart box concept for ubiquitous computing environments. In *Proceedings of Smart Objects Conference,* (pp. 118-121).

Gonzalez, H., Han, J., & Shen, X. (2007). Cost-conscious cleaning of massive RFID data sets. In *International Conference on Data Engineering*, (pp. 1268-1272).

Jeffery, S., Franklin, M., & Garofalakis, M. (2008). An adaptive RFID middleware for supporting metaphysical data independence. *Very Large Database Journal, 17*(2), 265–289. doi:10.1007/s00778-007-0084-8

Khoussainova, N., Balazinska, M., & Suciu, D. (2006). Towards correcting input data errors probabilistically using integrity constraints. In *Data Engineering for Wireless and Mobile Access*, (pp. 43-50).

Khoussainova, N., Welbourne, E., Balazinska, M., Borriello, G., Cole, G., Letchner, J., et al. (2008). A demonstration of Cascadia through a digital diary application. In *Proceedings of the Special Interest Group on Management of Data*, (pp. 1319-1322).

Landt, J. (2005). The history of RFID. *Institute of Electrical and Electronics Engineers Potentials, 24*(4), 8–11.

Pupunwiwat, P., & Stantic, B. (2009). Unified Q-ary tree for RFID tag anti-collision resolution. In *Australasian Database Conference*, (pp. 47-56).

Pupunwiwat, P., & Stantic, B. (2010). A RFID explicit tag estimation scheme for dynamic framed-slot ALOHA anti-collision. In *Wireless Communications* (pp. 1–4). Networking and Mobile Computing. doi:10.1109/WICOM.2010.5601080

Pupunwiwat, P., & Stantic, B. (2010a). Dynamic framed-slot ALOHA anti-collision using precise tag estimation scheme. In *Australasian Database Conference*, (pp. 19-28).

Pupunwiwat, P., & Stantic, B. (2010b). Joined Q-ary tree anti-collision for massive tag movement distribution. In *Australasian Computer Science Conference*, (pp. 99-108).

Pupunwiwat, P., & Stantic, B. (2010c). Resolving RFID data stream collisions using set-based approach. In *Intelligent sensors, sensor networks and information processing,* (pp. 61-66).

Stockman, H. (1948). *Communication by means of reected power* (pp. 1196–1204). Institute of Radio Engineers.

Wang, F., Liu, S., & Liu, P. (2010). A temporal RFID data model for querying physical objects. *Pervasive and Mobile Computing*, *6*(3), 382–397. doi:10.1016/j.pmcj.2009.10.001

Welbourne, E., Khoussainova, N., Letchner, J., Li, Y., Balazinska, M., Borriello, G., & Suciu, D. (2008). *Cascadia: A system for specifying, detecting, and managing RFID events*. In Mobile Systems, Applications, and Services.

KEY TERMS AND DEFINITIONS

Bayesian Network: An intelligent classification tool that probabilistically determines conclusions based on the product of all the input validity presented to it.

Classifier Fusion Methods: Three different fusion methods, Bayesian Network, Non-Monotonic Reasoning and Majority Rules, that we have proposed in the latter part of this chapter that combines various classifiers globally to improve the classification rate.

Classifier: An intelligent tool used by computer scientists to categorise fed information into specific information.

Integrated Classifier: The methodology we have proposed in this chapter that combines a Bayesian Network, Neural Network and Non-Monotonic Reasoning to deliver a highly intelligent conclusion.

Neural Network: Another probabilistic classifier that is modeled after the human brain and operates by modifying the input through use of weights to determine what it should output.

Non-Monotonic Reasoning: A deterministic classifier we have used that utilises various logic engines written to classify input into specific output.

Radio Frequency Identification: Wireless technology that uses readers to automatically identify tagged items in certain vicinities.

Chapter 4
Near Field Authentication

Vasileios Lakafosis
Georgia Institute of Technology, USA

Edward Gebara
Georgia Institute of Technology, USA

Manos M. Tentzeris
Georgia Institute of Technology, USA

Gerald DeJean
Microsoft Research, USA

Darko Kirovski
Microsoft Research, USA

ABSTRACT

Counterfeiting affects many different sectors of the world trade, including the pharmaceutical and the aerospace industries, and, therefore, its impact is not only of financial nature but can also have fatal consequences. This chapter introduces a new robust RFID system with enhanced hardware-enabled authentication and anti-counterfeiting capabilities. The system consists of two major components, namely the near-field certificates of authenticity (NF-CoAs), which complement typical RFID tags and serve as authenticity vouchers of the products they are attached to, and a microcontroller-enabled, low-power and low-cost reader. The high entropy and security of this framework stem from the unique, conductive, and dielectric, physical structure of the certificate instances and the highly complex electromagnetic effects that take place when such a certificate is brought in the reactive near-field area of the reader's antenna array. In particular, the reader's main task is to accurately extract the 5 to 6 GHz near-field response (NF fingerprint) of the NF-CoAs. The characterization of the reader's components, with an emphasis on the accuracy achieved, is provided. Rigorous performance analysis and security test results, including uniqueness among different instances, repeatability robustness for same instance and 2D to 3D projection attack resistance, are presented and verify the unique features of this technology. Rendering typical RFID tags physically unique and hard to near-exactly replicate by complementing them with NF-CoAs can prove a valuable tool against counterfeiting.

DOI: 10.4018/978-1-4666-2080-3.ch004

INTRODUCTION

In contrast with piracy, where the buyer is confident that the purchased object is not genuine due to a very low price or some discrepancy with the quality of the product, counterfeiting is the illegal trade in which the adversary fools the buyer into believing that the merchandise is authentic. As a result, the counterfeiter collects substantial revenue with profit margins typically higher than that of the original manufacturer.

Counterfeiting is as old as the human desire to trade and exchange. For example, historians possess evidence of counterfeit coins of the world's first coin, the Lydian Lion (Goldsborough, 2010b). Revealing the interior metal with test cuts, i.e. slashing of the surface of a coin with a hammer, was the first counterfeit detection procedure (Goldsborough, 2010a). To try to prevent detection, some counterfeiters made coins with already engraved fake test cuts; this initiated the cat-and-mouse game of original manufacturers against counterfeiters that lasts to date.

Though it is hard to assess and quantify the market for counterfeit objects of value today, there is no doubt that counterfeiting accounts for a huge economic impact. The World Customs Organization and the International Chamber of Commerce, according to Interpol, estimate that roughly 8% of world trade every year is in counterfeit goods (Robyn, 2008). A 2010 study estimated that the volume of counterfeit U.S. currency in the form of banknotes (paper currency) in circulation worldwide is in the neighborhood of $60 to $80 million (Judson & Porter, 2010). Approximately 8.1 million Americans, or 3.5% of the total U.S. population, experienced fraud in 2010 (Global Card Fraud, 2010). At a global level, the "Global Card Fraud" Nilson Report estimated card fraud losses of $6.89 billion on $14.6 trillion in purchases of goods and services and cash advances in 2009 and projected the amount of fraud losses to rise to $10 billion by 2015 (Global Card Fraud, 2010).

Unfortunately, however, the impact of counterfeiting is not only of financial nature. The numbers get scary when counterfeiters attack industries, such as the pharmaceutical and the aerospace. In particular, Glaxo-Smith-Kline, in a study with the US Food and Drug Administration, estimates that counterfeit drugs account for 10% of the global pharmaceuticals market (Glaxo-Smith-Kline, 2009). U.S. Federal Aviation Authority estimates that each year, 2 percent (520,000 parts) of the 26 million parts installed on airplanes are counterfeit (Stern, 1996). From 1973 to 1993, bogus parts played a role in at least 166 U.S.-based aircraft accidents or less serious mishaps. Four of those were accidents involving commercial carriers that resulted in six deaths (Stern, 1996).

In the battle against counterfeiting, extremely reliable and robust certificates of authenticity, or, in other words, instances of proof of value, that can be used conveniently may prove valuable. This chapter presents the full implementation of a novel near-field (NF) anti-counterfeiting RFID system that aims to address counterfeiting in a hardware-based way (Lakafosis et al., 2011). The fundamental idea is to complement any type of RFID tag with an inexpensive physical object that behaves as a certificate of authenticity (NF-CoA) in the near electromagnetic field so that this "super-tag" is not only digitally but also physically unique and hard to near-exactly replicate. This enables, on one hand, the extraction of the data related to the product in the far field and, on the other hand, the offline verification of its authenticity within its near field with low probability of a false alarm.

Based upon the characterization of the Super High Frequency (SHF) components that comprise the NF-CoA reader and the radiation behavior of the antenna elements of the array, the performance analysis of the reader is presented. Its high efficiency, albeit its simple and low cost implementation, is demonstrated with robustness and repeatability tests using 3D random NF-CoA structures formed by an arbitrary constellation of conductive and dielectric material. The results render the proposed NF-CoA system a unique anti-counterfeiting candidate tool.

RELATED WORK

The majority of the anti-counterfeit methods that have been implemented in the market to this point are of a two-dimensional (2D) style. Black-and-white or colorful barcodes, magnetic strips, watermarking and holograms constitute such examples that have been successfully copied, altered or distorted. Moreover, researchers soon realized that software-only-based solutions could also not be relied upon in the fight against counterfeiting. Traditional RFID tags with encoded digital information can easily be replicated, which means that a complete image of the memory of the RFID tag chip in bit level can be extracted and copied onto other blank RFID tag chips, regardless of if its content is encrypted, signed and/or hashed. On top of that, security groups have demonstrated successful attacks against even enhanced "symmetric-key" RFID tags that are capable of computing symmetric-key functions (Juels, 2006).

As a result, it was soon concluded that hardware-based CoAs that can complement the RFID tags comprise a more effective solution to the problem. Bauder and Simmons were the first to propose CoAs created as a collection of fibers, the positioning of which is fixed on an object using a transparent gluing material, based on the fact that if one end of a fiber is illuminated, the other end will also glow (Bauder, 1983; Simmons, 1991) and using photo detectors. Among only few other efforts, Church and Littman worked on extracting the random optical-fiber patterns in the context of currency anti-counterfeiting (Church & Littman, 1991) and Chen et al. developed the first CoA system based upon fiber-infused paper and public-key cryptography (Chen, Mihcak, & Kirovski, 2005). While efficient and inexpensive, this category of fiber-based CoAs does not provide a "physical one-way function". In other words, given the fingerprint of a particular fiber-based CoA it is not computationally difficult to construct an object of the same fixed dimensions that yields an almost identical fingerprint. This is primarily due to the linearity of the CoA as a system and the lack of interdependence among the responses of distinct fibers; since the positioning of a single fiber is not dependent upon the remaining fibers of a fiber-based CoA, an adversary can launch a simple search process that orients these fibers on paper one by one. Although such a technology is not available, we speculate that its 2_{+}D manufacturing process is substantially easier than creating purely random 3D topologies.

On the contrary, Pappu was able to propose a class of physical one-way functions via speckle scattering (Pappu, Recht, Taylor, & Gershenfeld, 2002). These so-called physical unclonable functions (PUF) of practical cryptography are essentially a mapping between a set of specific challenges applied to a physical, complex structure and their corresponding responses and are used to produce unclonable tokens for identification purposes. Pappu focused on the natural randomness collected from a speckle pattern; a random intensity pattern produced by the mutual interference of coherent wavefronts that are subject to phase differences and/or intensity fluctuations. Skoric was the first to match experimentation and theoretical bounds on the amount of randomness exhibited by keys formed from speckle (Skoric, 2007). Since speckle scattering is sensitive to microscopic changes to the source of scattering, it is difficult to build practical such CoAs that satisfy robustness to environmental changes and ordinary wear and tear; in addition, it is poorly understood how speckle filtering addresses repetitiveness of the extracted fingerprint in different environments and/or at different read-out misalignments and if fingerprint interdependence is effective here, as explained above. Under the same PUF category falls also the proposal of (Tuyls & Batina, 2006) to fabricate RFID-tags, microchips of which are equipped with an additional PUF physical structure that inherently incorporates unpredictable, manufacturing process inaccuracies. Although not supporting off-line scenarios as (Tuyls & Batina, 2006), an actual fabrication of a PUF-enabled RFID chip in 0.18μ technology is presented in (Devadas et al., 2008). The operation of this type of CoAs is not passive as it consumes small, yet perhaps not negligible, power from the tag chip, hence potentially decreasing its read range. Their fabrication is also coupled with the RFID tag chip fabrication process and requires an expensive micro-fabrication technology. Furthermore, all the above PUF-based CoAs require access to non-reusable, one-time challenge-response pairs through online Internet connections with the potential added overhead of recharging the online challenge-response database.

In the far-field communication domain, researchers have also proposed applications that detect the response of the certificate's random structure over the expensive millimeter wave frequency range (Inkode, Inc.). In fact, the cost of the CoA verification here is unnecessarily high compared to currently available semiconductor technologies. Under the same far-field category also fall the chipless RFID tags proposed for authentication applications. However, their common main shortcoming is that the entropy they provide is only limited to a two-digit bit-sequence. For instance, Preradovic et al. (Preradovic & Karmakar, 2009) demonstrate a printable chip-less RFID tag for secure banknote applications in the 5 to 7 GHz frequency band, the anti-counterfeiting robustness of which relies on a bit sequence formed by a multi-resonating circuit and CrossID, Inc (CrossID, Inc.) has tested a chip-less, chemical-material-based RFID tag using 3 to 10 GHz readers with each of the 70 different chemicals being assigned its own position in a 70-digit binary number. In addition to the limitations of the above far-field authentication schemes that are outlined in the next Section, security vulnerabilities of this category include eavesdropping or maliciously jamming the far-field detection.

Finally, regarding the near-field domain, researchers (Romero, Remley, Williams, & Chih-Ming, 2009) have attempted to quantify the electromagnetic characteristics of the near-field coupling nature of ISO 14443 RFID transactions. They actually make use of an expensive real-time oscilloscope with a sampling rate of 20 GHz to measure a fine electromagnetic signature that consists of the fundamental and harmonics up to the ninth harmonic of a 13.56 MHz RF carrier. This method not only incurs a very high "reader" cost but also requires a closed-loop, synchronized control system between the RFID reader and an oscilloscope in addition to potential RFID reader software modifications.

NF-COA TECHNOLOGY

Near Field Electromagnetic Characteristics

The space surrounding an antenna is subdivided into three regions:

- **Reactive Near Field:** This is the area immediately surrounding the antenna and its outer boundary is commonly considered to be at a distance of $0.62\sqrt{D^3/\lambda}$ from the antenna surface, where λ is the wavelength and D is that antenna's largest dimension.
- **Radiating Near Field (Fresnel):** This intermediate area's inner boundary is the same as the outer boundary of the reactive near field and its outer boundary is at the radial distance of $2D^2/\lambda$ far from the antenna. In the Fresnel area, the radiation fields predominate over the reactive ones and the angular field distribution, i.e. the shape of the antenna pattern, is dependent upon the distance from the antenna. The r

variations of the E- and H- field components are not separable from those of θ and φ, there is no TEM wave and the radial field component may be appreciable (Balanis).

- **Far Field (Fraunhofer):** This is the area that extends from $2D^2/\lambda$ to infinity. The field components here are essentially transverse and the angular field distribution is independent of the distance from the antenna.

The amplitude pattern of an antenna, as the observation varies from the reactive near field to the far field, changes in shape. In the reactive near-field region the pattern is more spread out and nearly uniform, with slight variations. As the observation moves to the radiating near-field region, the pattern begins to smooth and form lobes. In the far-field region, the pattern is well formed, usually consisting of few minor lobes and one, or more, major lobes (Balanis; Rahmat-Samii, Williams, & Yaccarino, 1995).

For purposes of both mechanical practicality and higher complexity in theoretical analysis of the signature extraction that yields our system's security, we deliberately bring the CoAs in the near field of the antenna array of our reader; including both the reactive and the radiating areas. In particular, the rationale behind opting for the near-field observation of the incurred electromagnetic effects is manyfold. First of all, the near-field observation enables relatively high variance of the EM field, causing better discriminating characteristics compared to the far-field responses that typically represent certain average characteristics of random discrete scatterers (Tsang, Kong, & Shin, 1985). Additionally, within this near-field region the relationship between the electric field component E and the magnetic field component H becomes often too complex to predict with either E or H field component possibly dominating at any particular point. Moreover, here the waves are spherical and not yet converted to plane waves, so there is no polarization. It is also hard to eavesdrop or maliciously jam near-field communication, compared to the far-field one that is prone to both, potentially devastating, attacks. As an additional advantage of the very short-range observation and the resulting discrimination feature, the reader can operate with low power and use low-efficiency antenna designs. At the same time, the decay of the strength of the fields as a function of the distance from the antenna element is more significant in the near field as opposed to the far field, due to the higher attenuation factor exponent, and as a result eliminates any potential EM compatibility (EMC) issues. Still, since the readout of the CoA does not require a physical contact, NF-CoAs may be built with superior wear-and-tear properties.

Electromagnetic Scattering

Whenever EM radiation encounters an obstacle it may be deflected or absorbed. For large objects the deflection is considered to arise from reflection and refraction. These phenomena can be described with the aid of ray tracing or geometrical optics only if far field is considered and, still, these do not completely describe the interaction (Bayvel, 1981). Diffraction, i.e. spreading out of radiation past a narrow aperture or across an edge as a result of the interference between the wave forms produced, also occurs. When the object is less than or of the order of the wavelength, as is the case with NF-CoA, the interpretation is not so simple as wave optics analysis is required and the deflection process is referred to as scattering. Scattering by obstacles depends upon their size, shape and refractive index. Since the obstacle, our CoA, is not moving or changing in time in any way, there is no need to account for changes in frequency of wave that arise due to the Doppler effect.

Exact solutions to the problem of EM wave scattering must rely on Maxwell equations together with the associated boundary conditions.

First, we consider the example of a wave impinges upon the surface of a wire conducting wire scatterer. Part of the incident electric field $E_i(r)$ impinges on the wire and induces on its surface a current density J_S (Amp/m). For non-perfect conducting wires, as is the case in the real world, some part of the impinging field is absorbed. The induced current density reradiates and produces an electric field that is referred to as the scattered electric field $E_s(r)$. At any point in space the total electric field $E_t(r)$ is the sum of the incident and scattered fields, or $E_t(r) = E_i(r) + E_s(r)$.

Both the incident and scattered fields can be represented by the following radiation integrals:

$$\overline{E} = -j\omega\overline{A} - j\frac{1}{\omega\mu\varepsilon}\nabla\left(\nabla\bullet\overline{A}\right) \tag{1}$$

$$\overline{H} = -j\omega\overline{F} - j\frac{1}{\omega\mu\varepsilon}\nabla\left(\nabla\bullet\overline{F}\right) \tag{2}$$

where the vector potential A for the electric current source J is a solution of the inhomogeneous vector wave equation of

$$\nabla^2\overline{A} + k^2\overline{A} = -\mu\overline{J} \tag{3}$$

and the vector potential F for the magnetic current source M is a solution of the inhomogeneous vector wave equation of

$$\nabla^2\overline{F} + k^2\overline{F} = -\varepsilon\overline{\mathrm{M}} \tag{4}$$

Especially for the scattered field, the electric vector potential stems from the aforementioned induced current J_S. Approximations regarding this current density are not possible as this is not linear across any single Cartesian axis, the wire cannot necessarily be considered 'very thin' so that an independence from the azimuthal angle φ can be assumed; not to mention that in some cases the wires are bent in such a degree that they assume a spheroidal or ellipsoidal form.

For an introduction to basic scattering theory regarding a more generic case, we need to introduce the Green's function. Although this function is primarily the solution of the field equation for a point source, we can use the principle of linear superposition to find the solution of the field due to a general source (Tsang, Kong, & Ding, 2002). The free space dyadic Green's function is given by

$$\overline{\overline{G}}\left(\overline{r},\overline{r}'\right) = \left(\overline{\overline{I}} + \frac{1}{k_o^2}\nabla\nabla\right)\frac{e^{ik_o\left|\overline{r}-\overline{r}'\right|}}{4\pi\left|\overline{r}-\overline{r}'\right|} \tag{5}$$

if the source is placed at a position represented by the primed vector r', as shown in Figure 1. From the vector Green's theorem we derive the Huygen's principle, which is an exact relation that expresses the field in a region of space to the fields on a surface that encloses the region. This means that the fields at any point in space can be calculated if the surface fields are determined. The relation that is known as the Huygen's principle is

$$\overline{E}_s\left(\overline{r}'\right) = \int_{S_1} dS\left[\overline{\overline{G}}\left(\overline{r},\overline{r}'\right)\cdot\hat{n}\times i\omega\mu\overline{H}_i\left(\overline{r}\right) + \nabla'\times\overline{\overline{G}}\left(\overline{r},\overline{r}'\right)\cdot\hat{n}\times\overline{E}_i\left(\overline{r}\right)\right] \tag{6}$$

It should be repeated that with this rigorous approach one needs to first specify $\hat{n}\times\overline{E}$ on the surface S and then solve $\hat{n}\times\overline{H}$ based on the boundary value problem before using the Huygen's principle to calculate the fields everywhere.

In the above introductory analysis, we have only considered single scattering; radiation that is only obstructed by a single scatterer. In other words, we have not considered the fact that the

Figure 1. Graphical representation of the near-field electromagnetic scattering on a random-shaped wire

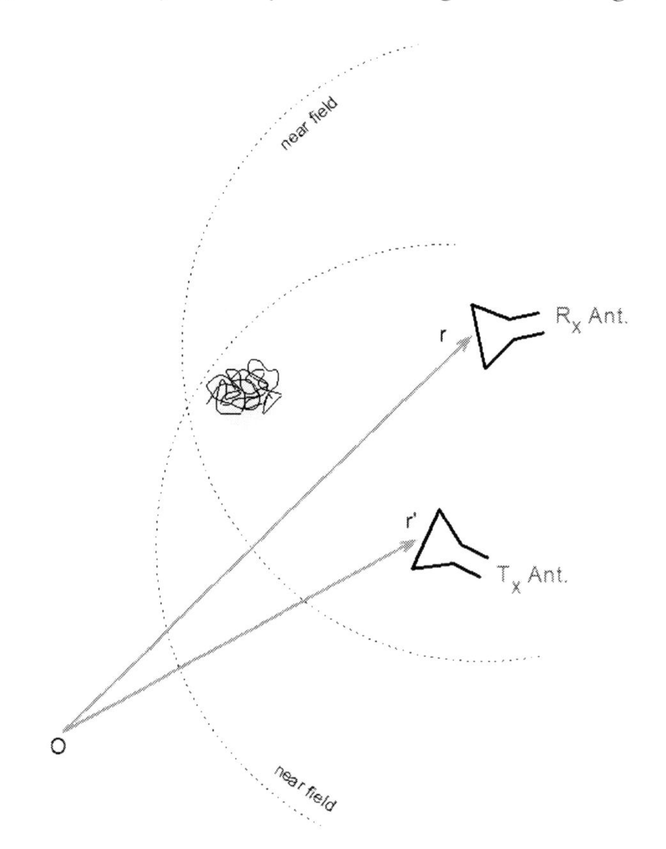

radiation may scatter many times due to spatially grouped together multiple scatterers. For instance, in the example of the conducting wire described above, we could also assume the presence of another wire in close proximity to the first one; in which case, the reradiated wave would have also impinged on the second wire and induce on the surface of the latter a current density J_{S_2}.

Of course, since we have opted for a near-field approach, no far-field approximations can be applied to the above equations. On the contrary, the difficulties in obtaining closed-form solutions that are valid everywhere stem from the inability to perform the integration of the vector potentials $\overline{A}$ (3) and $\overline{F}$ (4).

For an in-depth study of EM scattering by a random collection of scatterers, the reader is referred to concept of fluctuating fields in random media (Foldy, 1945; Lax, 1951). As for non-linear and inelastic scattering processes, useful references are (Anderson, 1971; Herzberg & Crawford, 1946; Zernike & Midwinter, 1973).

THE NEAR-FIELD CERTIFICATE OF AUTHENTICITY

The certificate of authenticity is defined as a physical object that may be attached as a tag, a label or a seal to a physical object or may be integrated as an inseparable part of the surface of the object that can prove its own authenticity and, as a consequence, also prove the authenticity and value of the product it accompanies. In this Chapter, we investigate such physical structures that behave as CoAs in an electromagnetic field; i.e. yield a unique signature in a portion of the electromagnetic radiation spectrum.

The unique physical structure is an arbitrary constellation of small, thin pieces of metallic conductors and/or one or more dielectric materials that are randomly dispersed and spatially immobilized into a 3-dimensional, RF wave permeable dielectric fixative. The conductors may be copper, aluminum or other metal filings and particle scatterers or pieces and plates.

The near-field electromagnetic scattering effects are extracted in the form of reproducible patterns of scattering (S_{11} or S_{21}) parameters or phase information that constitute an electromagnetic signature response from the NF-CoA. We call these sets NF fingerprint or NF signature. In a more rigorous definition, an NF fingerprint of an NF-CoA is a set of S_{21} parameters observed over a specific frequency band and collected for (a subset of or) all possible antenna couplings of a reader's array. A graphical representation of this fingerprint is shown in Figure 2 as extracted from the custom fabricated reader for all its 72 different antenna element permutations using a signal processing method described in detail later in this Chapter.

Not only is the cost of producing these random unique physical structures comparable to the price of typical passive RFID tags in the order of a cent of US dollars but also, because each NF-CoA instance is different, it is almost always infeasible or prohibitively expensive for an adversary to reproduce an NF-CoA with enough accuracy to successfully mimic the electromagnetic fingerprint that certifies authenticity. It is these benefits of physical near-field authentication and the resiliency to potential attacks, as discussed in Section 'Withstanding Potential Attacks', that render the NF-CoAs ideal counterfeit deterrent candidates.

Fabrication Process

The proposed NF-CoA tag is the first, extremely low-cost physical object that enables anti-counterfeiting based completely on its hardware implementation and the resulting NF effects. Although there are numerous ways of creating these certificate instances, the objects that have been tested and used as NF-CoAs can be divided into two main categories, namely the copper-based ones and the inkjet-printed ones.

For the realization of the first generation of certificate objects, silver nano-particle inkjet printing technique on organic substrate (Li, Rida, Vyas, & Tentzeris, 2007), such as regular photo paper, was chosen as a direct-write technique. According to this technique, the design pattern is transferred directly to the substrate in means of multiple inkjet-printed layers, without any requirement of masks, resulting in a very fast, low-cost and in-house process. An example of such a single substrate layer, inkjet-printed NF-CoA that consists of overlapping rhombic loop conductors is shown in Figure 3a. The average loop circumference is 18.6 mm, approximately between half wavelength and one wavelength in electrical length. Another NF-CoA design used consists of a random constellation of 1 mm by 1 mm pixels that trace the form of the final geometry, shown in Figure 3b. A three-dimensional structure, shown in Figure 3c, can be created by tightly stacking multiple 2D CoAs one on top of each other to give the projection of Figure 3a and are used for the 3D attack tests described in the "Performance Evaluation" Section below.

The second category of NF-CoAs includes the copper-based instances; an example of these is shown in Figure 3d. These particular instances have been fabricated in collaboration with an injection molding company (Aero-plastics, Inc.). Essentially the process involves the encapsulation of copper wire of variable gauge into heated plastic mold that is hardened to maintain the wires' position. Randomness in this process is achieved

Figure 2. Graphical representation of an NF fingerprint, as extracted by the custom fabricated reader for all its 72 different antenna element couplings

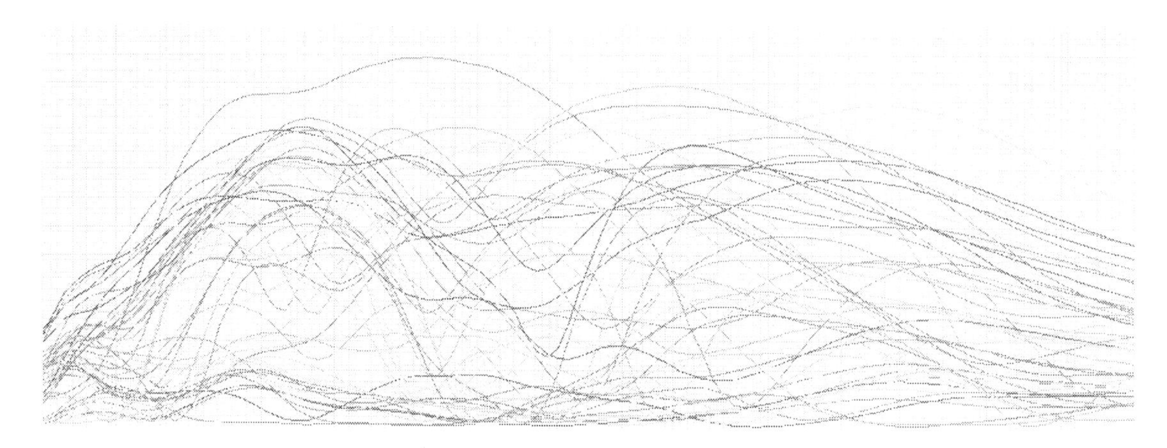

by introducing techniques that are absolutely non-deterministic, such as using large air fans and blending with different speeds and for different amounts of time.

Issuing CoAs

After an NF-CoA has been fabricated, regardless of if it is a copper-based or an inkjet-printed one, its NF response has to be digitally signed by the original issuer using traditional cryptography. This process takes place in the controlled environment of an NF-CoA factory. In particular, first the unique NF fingerprint of the newly fabricated NF-CoA instance is digitized with the use of a reader and compressed into a fixed-length bit string. This information is afterward concatenated to the information associated with the tag, such as product ID, color and expiration date, and the resulting combined bit string is hashed using a cryptographically strong algorithm such as SHA256 (National Institute of Standards and Technology (U.S.), 2008). By adopting a public-key cryptosystem such as RSA (Rivest, Shamir, & Adleman, 1978) or other cryptographic routines based on elliptic curves such as EC-DSA (Johnson, Menezes, & Vanstone, 2001), this hash is signed using the issuer's private key to form, together with the plain initial bit string of the fingerprint, the message that is directly encoded onto the RFID chip. The

Figure 3. a) Inkjet-printed, single substrate layer NF-CoA of rhombic loops; b) NF-CoA as a random trajectory of pixels; c) 3D-stacked NF-CoAs of rhombic loops; d) Copper-based NF-CoA

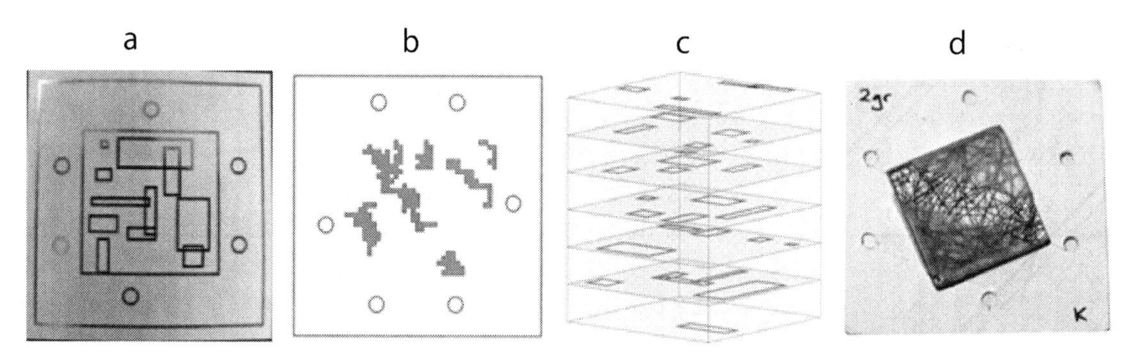

above keys of public key cryptography are related mathematically, but it is virtually impossible to deduce the private key from the public key. So, in our application only the issuer can digitally sign the NF-CoA with the secret private key.

THE NF-COA READER

This Section presents the hardware aspects, as well as the operating details of the fabricated reader that is used to expose the subtle variances of the above described near-field electromagnetic effects resulting from impingement of electromagnetic energy on a certificate instance and derive the fingerprint of an NF-CoA instance. The extraction of the unique signature of an NF-CoA object is briefly performed as follows: electromagnetic power at the neighborhood of 5.5 Ghz is radiated from a particular element of the reader's antenna array, this radiation impinges upon an NF-CoA instance placed at about 2 mm away from the array and the near-field effects, discussed previously, are captured by another antenna element.

The main goal of the reader design is to maximize the entropy, i.e. randomness, of the NF fingerprint as well as guarantee the extraction robustness of the NF signature of the same CoA by different readers, despite the low accuracy provided by the out-of-the-shelf analog and digital circuitry used and the noise due to external factors. However, since the NF-CoA reader is intended to be used, for example, at a large number of customs offices and stores during the check-in or check-out process, its manufacturing cost has to be kept low. In particular, the total cost of the presented prototype, shown in Figure 4b and consisting of an antenna array and a number of out-of-the-shelf components that serve as the analog/digital back-end, is less than 50 US dollars before it is integrated with a microcontroller unit.

Reader Hardware Aspects

The overall reader circuit schematic is shown in Figure 4a. The board's dimensions are 4.3 in by 6.8 in. This board consists of four metallic and three substrate layers, the total thickness of which does not exceed 1.6 mm. The elements of the antenna array are placed on the top and second metal layers at distances of approximately 3 mm between each other. The ground plane is placed on the third metal layer, the SHF part on the bottom layer and the digital lines on the top layer. The substrate for the design is FR-408 with relative dielectric constant $\varepsilon_r = 3.715$ in our frequency band of interest, namely 5 to 6 GHz, relative permeability $\mu_r = 1$ and loss tangent $\tan\delta = 0.01$. The copper conductivity is $1.493 \cdot 10^6$ S / inch.

The reader design consists of a double-stacked layer solution; the Super High Frequency (SHF) plane and the Digital Control plane. As implied by their names, the first layer includes the CoA reading slot, the antenna matrix, as well as all required digital and analog circuitry and the second houses the Micro-controller Unit (MCU) and the data acquisition interface. The circuit part colored with pink lines in the diagram of Figure 4a, corresponds to the control plane. The SHF plane circuit part is shown with green lines.

Super High Frequency Plane

The antenna elements that comprise the array are individual micro-strip patch antennas. Specifically, a folding and meandering minimization technique is applied to the design of these patch antennas, since it is desired to pack as many such elements as possible in a 1 in by 1 in area. The antenna's technical characteristics and design strategy are detailed in (RongLin, DeJean, Tentzeris, & Laskar, 2004). As a result of the exploitation of the folding minimization technique and given that the half wavelength in the 5 to 6 GHz frequency range is around 2.75 cm, it is finally possible to fit 25 elements in a 5 by 5 configuration. During

Figure 4. a) The NF-CoA reader circuit schematic; b) The super high frequency plane (left) and the digital control plane (right) of the fabricated NF-CoA reader

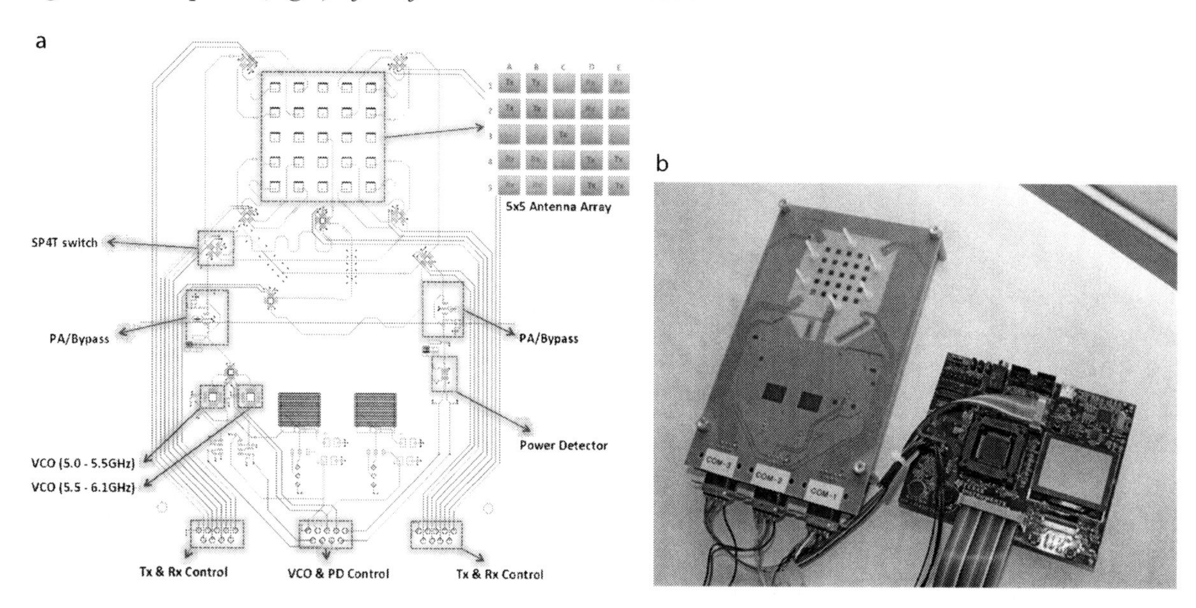

each NF signature read-out, it is ensured that the placement of the NF-CoA instance is fixed and geometrically unique, using short plastic poles, shown in Figure 5, the relative position of which is non-symmetrical on the array's plane.

The 3D antenna element design is shown in the lower right corner of Figure 5. A single such element has been measured by itself, i.e. not in the presence of neighboring elements, and exhibits a return loss of 16.3 dB, measured with a Rohde & Schwarz ZVA 8 Vector Network Analyzer (VNA), at a resonant frequency of 5.149 GHz, as shown in Figure 5. The aforementioned miniaturization technique, however, does not come without cost. The major negative effect is the limited 10 dB bandwidth of this antenna element, which does not exceed 5 MHz. Despite that, the strength of the electric field extracted by these elements, even 500 MHz away from the resonant frequency, is very high (>-10 dB), as shown in performance tests toward the end of this chapter.

Out of all the 25 elements of the 5 by 5 antenna matrix, nine of them are operating as transmit-only and eight of them as receive-only, as shown in the insert pattern of Figure 4a. The transmit- and receive-only elements have been split into four sets of fours, placed as farthest away as possible in the four corners of the array, and the rest eight elements on the cross of the structure, with the exception of the central one, are unconnected. This design helps eliminate the need for single-pole double-throw (SPDT) switches for dual operation of the elements and reduces the number of the required digital input/output control lines. Moreover, the rationale behind this splitting is to minimize the coupling due to proximity and, thus, attribute most of the coupling measured to the presence of the metallic material of the CoA. To quantify this, the S_{21} curves of all possible different antenna element spacing of a subset three by three antenna array, shown in Figure 6a, have been captured over the 4.6 to 6.4 GHz band for 3 MHz step sizes in the absence of any CoA and are presented in Figure 6b. This subset array has been manufactured in the exact same way as the finally used five by five array, so that SMA SHF connectors are available. As expected, all curves have their maximum at the range around the

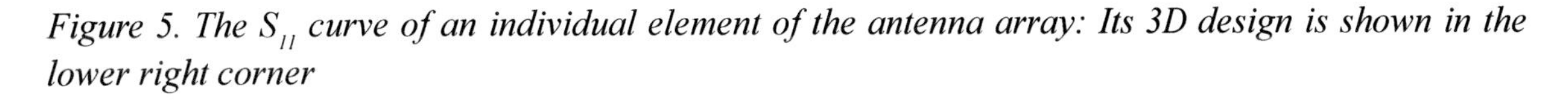

Figure 5. The S_{11} curve of an individual element of the antenna array: Its 3D design is shown in the lower right corner

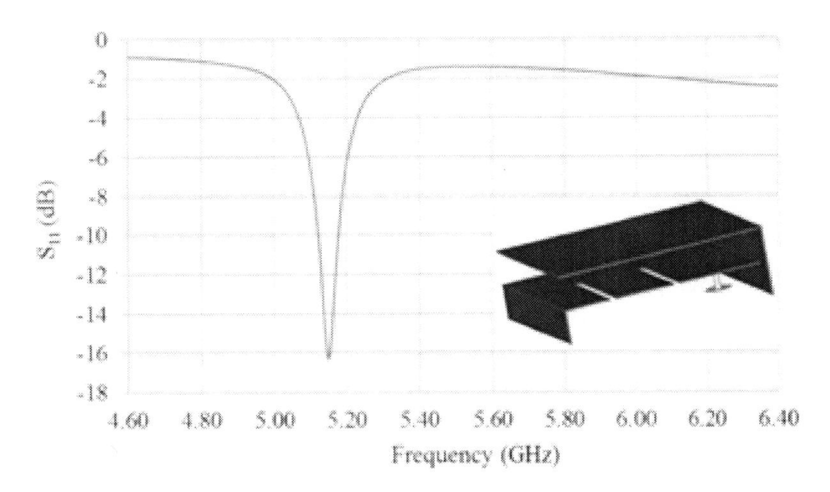

resonant frequency of the antenna elements and the shorter the distance between the transmitting and receiving elements the higher this maximum is (T_x1-R_x4>T_x2-R_x4>T_x2-R_x3>T_x1-R_x2>T_x3-R_x4 > T_x1-R_x3). It should be noted that the maximum coupling for the placement configuration of the antenna elements in the actual five-by-five antenna board, equivalent to T_x1-R_x2 (C3-B4) and T_x2-R_x3 (C3-D2), between any pair of elements of the antenna array never exceeds -20 dB in the absence of a certificate.

Any particular antenna transmit and receive coupling, out of the board's 72 possible permutations shown in Figure 4a, is chosen by digitally controlling eight identical single-pole four-throw (SP4T) switches (GaAs mmIC SP4T Non-reflective positive control switch, DC - 8 GHz), arranged in two hierarchical levels, shown in Figure 7. Based on this arrangement, there are always two switches preceding the transmit-only antenna element and two switches following the receive-only element. The physical location and order of the switches on the board has been optimized so that the coupling between the SHF lines is minimal. All connections are 50 Ohm and it has been ensured that the length of the lines connecting the appropriate switches for any antenna permutation is constant. The average insertion loss introduced by a single SP4T switch across the 4.5 to 6.5 GHz band has been measured to be 1.07 dB using the aforementioned VNA. The S_{21} and S_{11} parameters of the two-port system between the transmit and receive power amplifiers have been shown in (Lakafosis et al., 2010).

The SHF signal radiated toward the NF-CoA instance is generated by two voltage-controlled oscillators (VCO) (mmIC VCO with buffer amplifier, 5.0 - 5.5 GHz; mmIC VCO with buffer amplifier, 5.5 - 6.1 GHz), which both together optimally cover the 5.0 - 6.1 GHz frequency band. The exact VCO mapping of the input control 0 - 10 V voltage range to the 5.1 - 6.3 GHz frequency spectrum is characterized with the Tektronix RSA 3408A Real-Time Spectrum Analyzer (RSA). The peak power of the generated, nearly monochromatic signal, after it is amplified by 11.25 dB by a power amplifier (PA). (General Purpose Amplifier), is measured 4.03 dBm with the same RSA. The scattered, reflected and refracted signal received is amplified by the same PA and the SHF power is monitored by an RMS power detector (PD) (6GHz RMS Power Detector, Linear) that yields a ±1 dB of accuracy. Its voltage output has been accurately mapped to the input

Figure 6. a) Bottom view of a 3 by 3 antenna array fabricated in the exact same way as the reader's 5 by 5 array, b) S_{21} curves of all possible different antenna element spacing of a subset 3 by 3 array in the absence of any CoA

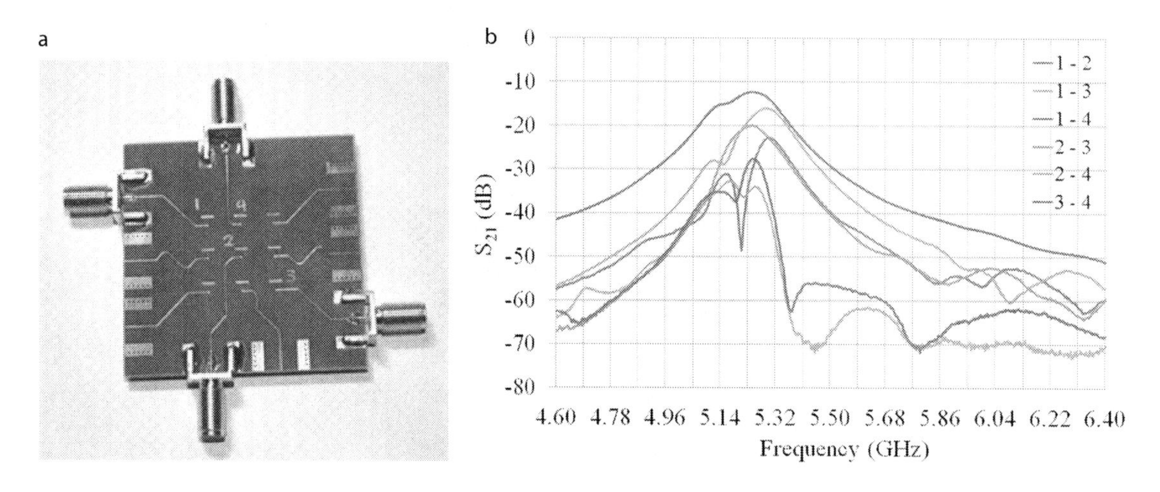

Figure 7. The two-layer SP4T switch hierarchy for enabling any antenna transmit and receive element pair, out of the board's 72 possible permutations

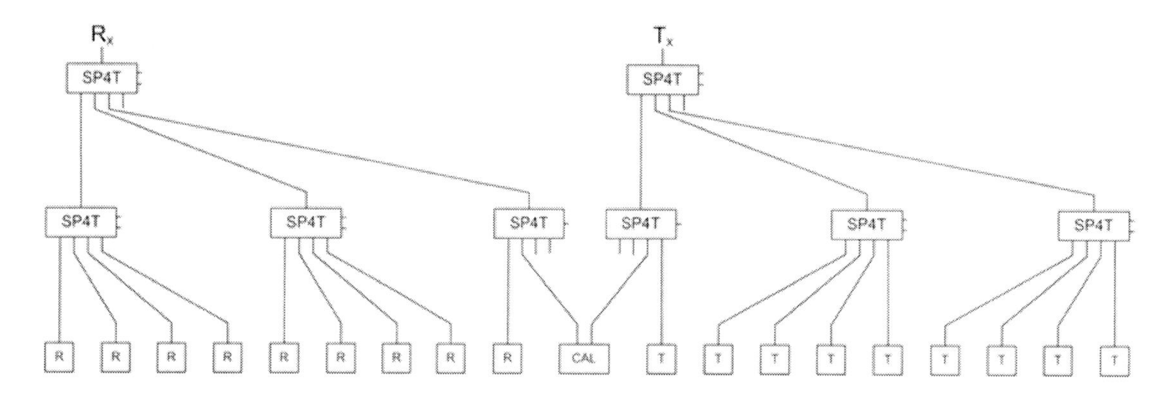

power generated by a signal generator (HP 83622B Swept Signal Generator) before the PA ranging from -55 dBm up to 5 dBm.

The losses due to the reader's components are expected to vary from lot of circuit components to lot and, as a result, from reader to reader. Given that it is very important to achieve consistency and elimination of false negatives among different readers, a calibration technique is deployed. This calibration technique involves measuring all coupling channels (S_{21}) for all possible antenna permutations over the entire supported frequency band in the absence of any CoA, the "No CoA" curves, for each NF-CoA reader and storing these curves into the MCU's non-volatile memory. Additionally, a copper line that is connecting two unused ports of two SP4T switches, out of the overall four lower-level switches, is also used for calibration purposes. With the help of this line, the insertion loss introduced by four consecutive SP4T switches is measured. The average S_{21} value of -16.57 dB as well as the curve across the 4.5 to 6.5 GHz spectrum agree well with the simulated overall loss, which is the sum of the attenuation due to the total line length itself as calculated by ADS (Advanced Design System (ADS) Simulation Environment) with the 4.3 dB attenuation incurred by the four switches. Only after the above two curves are captured right after the reader's fabrication and consequently de-embedded from future measurements extracted by the same reader, i.e. reader is calibrated, can the NF fingerprint be considered valid and checked against the in-chip stored waveform, regardless of the NF-CoA reader used. Similarly, right after the CoA's fabrication described previously, the first set of "No CoA" curves are subtracted from the corresponding S_{21} curves of the CoA's signature and it is this difference that gets stored in the RFID chip. The same "de-embedding," of course, also takes place at the verifier's premises.

Micro-Controller Enabled Digital Control Plane

At the heart of the control plane is a 16-Bit RISC architecture ultra-low-power MCU (MSP-EXP430F5438 Experimenter Board User's Guide (Rev. E)) that features an up to 18-MHz system clock, crystals up to 32 MHz, multiple high-resolution analog-to-digital converters (ADC) and a USB interface for data transfer. The control plane provides not only a very fast means of capturing the NF fingerprint but also the accuracy required toward our effort to maximize the fingerprint's entropy.

For its powering, the control plane does not rely on batteries (although a two AA battery option is available), but on the power supplied by the USB cable, which is anyways used to upload the NF-CoA fingerprint to a desktop or a laptop computer. When the board is initially connected to a computer through the USB cable it finds itself in the low power mode 4 (LPM4) sleep state. This is a deep sleep mode with a current consumption of 1.69 μA at 3.0 V, in which the CPU and all clocks are disabled, the crystal oscillator is stopped, the supply supervisor is operational and full RAM retention and fast wake-up are provided.

Reader Operation

The functionality of the NF-CoA reader is summarized in the following four main tasks:

- Generating the appropriate SHF power and controlling the frequency output of the VCO.
- Dictating the path that the SHF signal follows through the two-layer SP4T switch hierarchy and the coupling, eventually, between the T_x and R_x antenna element.
- Measuring the power captured by monitoring the PD output voltage.
- Uploading the set of measured data that comprise the NF-CoA's fingerprint to a computer.

The 32 kHz auxiliary clock is enabled in short time intervals and used to check if a button on the control plane is pressed. The monitoring time duration of this button lasts for only a few milliseconds, since a human would be pressing the button for at least half a second. As soon as a press on the button is detected, the MCU exits the deep sleep mode and transits to the antenna element pair selection state. In this state a particular antenna element pair is selected by appropriately configuring the two digital logic control pins of all the SP4T switches with a 16-bit sequence generated by the digital output pins of the MCU. The next time the control plane returns to the antenna element pair selection state a check is performed as to whether all 72 antenna permutations have already been selected; in which case the MCU reverts to the sleep mode.

If not all possible antenna element permutations have been enabled, the next step is to generate a sinusoidal power signal at a particular frequency, nearly monochromatic, by controlling the appropriate VCO. The S_{21} NF-CoA fingerprint is captured over the frequency band of 5.1 to 5.9 GHz at 65 steps of 12.3 MHz each. The selection of these steps is achieved by altering the VCO's tune voltage. The MCU provides no DAC functionality and the latter is emulated by a high-frequency pulse width modulation (PWM) signal. The output voltage is configured based on a variable duty cycle that is derived from the ratio between the PWM's emulated voltage and the USB rail of 3.2 V. The next time the control plane returns to the control VCO state a check is performed as to whether the maximum number of frequency steps has already been reached; in which case the MCU returns to the antenna permutation selection state. While the T_x antenna element is radiating power toward the NF-CoA that is placed just 2 mm away from the antenna matrix, the captured reflected and refracted signal is amplified and fed to the PD. For this task, of course, the highest, 12-bit precision mode supported by the ADC of the MCU is chosen.

Based on measurements carried out under a controlled environment of pre-configured signal power amplitude, we have found that a single analog-to-digital conversion is not enough. In particular, although the voltage reference used (3.2V) for the ADC and supplied by the USB input has been measured to remain steady over time, we have recorded AD conversions to be as far as 20% off from the same actual input signal for more than 100 duplicate measurements. In addition, of course, to the inaccuracy introduced by the PD, pointed out above, the two major sources of inaccuracy in the ADC testing of mixed-signal circuits have been identified to be the approximations of IEEE standard for digitizing waveform recorders and IEEE standard for terminology and test methods for analog-to-digital converters (Blair, 1999) and the fact that the DC offset and the amplitude of the input analog signal evaluated on the base of the digital output differ from their true values (Hejn & Pacut, 2003). However, this deviation easily drops to less than 8% by performing 20 consecutive conversions, storing them in the MCU's successive approximation register and afterward simply averaging them. The latency incurred, as a result of the additional 19 conversions, is negligible given that each 12-bit resolution conversion requires only 13 MCU clock cycles, the total time duration of which is less than 0.8 μsec.

Each triplet that consists of antenna permutation, frequency and received power in dB is uploaded to the computer at 57.6 kbps over a UART-to-USB connection. For the currently running software version, the overall time required to extract all 72 curves is around 30 seconds. This time currently includes significantly conservative, long-guard times that are maintained between consecutive operations and will gradually be removed from future software versions.

PERFORMANCE EVALUATION

A number of different types of tests, presented in this Section, have been conducted in order to assess the performance of our proposed NF authenticity certification technology. For all tests presented in this Section, physical objects that are conceptually very close to the final envisioned product have been used as NF-CoAs. These objects, both copper-based and inkjet-printed, are described earlier. Regarding, particularly, the copper-based CoAs, their dimensions were 0.75 in by 0.75 in (with the exception of certificates I and J) and as such could not occupy the whole 1 in by 1 in area of the antenna matrix. As a result, their position has been rotated around the array's central axis so that the radiation of as many antenna elements closer to the circumference of the array as possible is "disturbed" by the NF-CoA. This is shown in the example tag of Figure 3d. As explained previously, the NF-CoA instances are attached to the reader through the plastic poles, shown in Figure 5, against the antenna matrix of the fabricated NF-CoA reader at a distance of 2 mm. For this short spacing, a sturdy dielectric foam material with characteristics close to that of air is used.

In each of the graphs presented hereafter, the y-axis corresponds to the output in dBs of the PD based on its received signal strength of the monochromatic signal, and the x-axis corresponds to the different frequency points sampled by the MCU's ADC. Essential to the evaluation of the results of the following tests is the standard deviation (std); the square root of an estimator of the variance of the received signal strength by the R_x element for different CoAs at the same frequency points and antenna couplings. The std equation used is

$$std \triangleq \left(\frac{1}{N-1} \sum_{i=1}^{N} \left(x_i - \bar{x} \right)^2 \right)^{\frac{1}{2}} \quad (7)$$

Intra-CoA Robustness Tests

This set of tests aims to ensure that always nearly exact replicate NF fingerprints are extracted by the same NF-CoA. According to the test procedure followed, a single certificate instance is placed on the reader, then taken off and then placed back on the reader to indicate any changes in measurement results.

Figure 8a shows the results of an inkjet-printed NF-CoA measured twice against a subset of six different transmitter and receiver couplings, shown in the figure's legend. The figure indicates that there is very small, almost unnoticeable deviation from the two measurement runs. For a quantifiable assessment, the maximum difference between voltage outputs of the power detector of the two runs is 0.025V across the C3-B4 coupling and this corresponds to received signal strength of -79 dB.

A different copper-based certificate has also been tested for repeatability. Figure 8b shows the standard deviation of 10 duplicate measurements of this exact same CoA for eight different transmitter and receiver couplings, shown in the legend, that correspond to all eight R_x elements interchanged for the same A1 T_x element. These std curves exhibit a very low magnitude that does not exceed -86 dB with the exception of two antenna couplings, the maximum of which reaches -83.5 dB in the low part of the spectrum. These results indicate that the resolution captured by the reader for slightly different placements of the CoA in all three directions relative to the antenna array

Figure 8. a) NF responses of a single inkjet-printed NF-CoA measured twice against six different T_x / R_x couplings; b) Standard deviation of 10 duplicate measurements of same copper-based CoA for eight antenna coupling

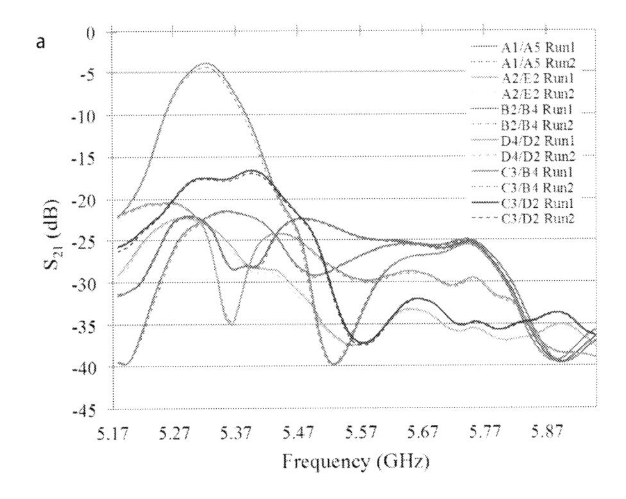

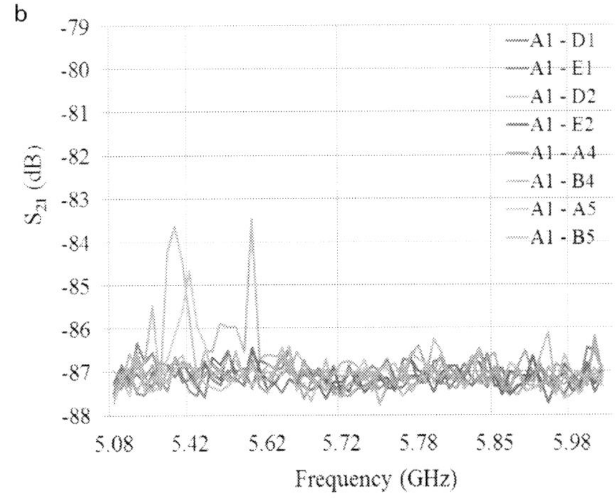

and under different environmental changes, namely NF interference, temperature, etc. is very high. This result is well above the initial precision requirement of 2 dB (Lakafosis, et al., 2011; Lakafosis, et al., 2010) and, consequently, demonstrates the system's repeatability robustness.

Inter-CoA Robustness Tests

In order to demonstrate the uniqueness among different NF-CoA designs, in terms of as high a variability of NF fingerprint extracted as possible, we investigate the entropy of the NF frequency response for different certificate instances across different T_x/R_x couplings.

First, the signature curves of three different inkjet-printed certificate instances across three different T_x/R_x couplings are captured. As can be seen in Figure 9a, the thick, thin, and dotted lines representing a single CoA at different T_x/Rx couplings indicate a significant change in the pattern, which ensures uniqueness. Also, three different CoAs at the same T_x/R_x coupling also show high variation in response. The convergence that can be noted at the very far high end of the frequency range is attributed entirely to the antenna element's low resonant frequency of 5.149 GHz, presented in Figure 5. It should also be noted that this graph only shows three out of the 72 available T_x/R_x couplings. It is our strong belief that, for

Figure 9. a) NF signatures of three different inkjet-printed certificates across three different T_x / R_x couplings; b) Standard deviation of the NF responses of 10 different copper-based CoAs for four different antenna couplings

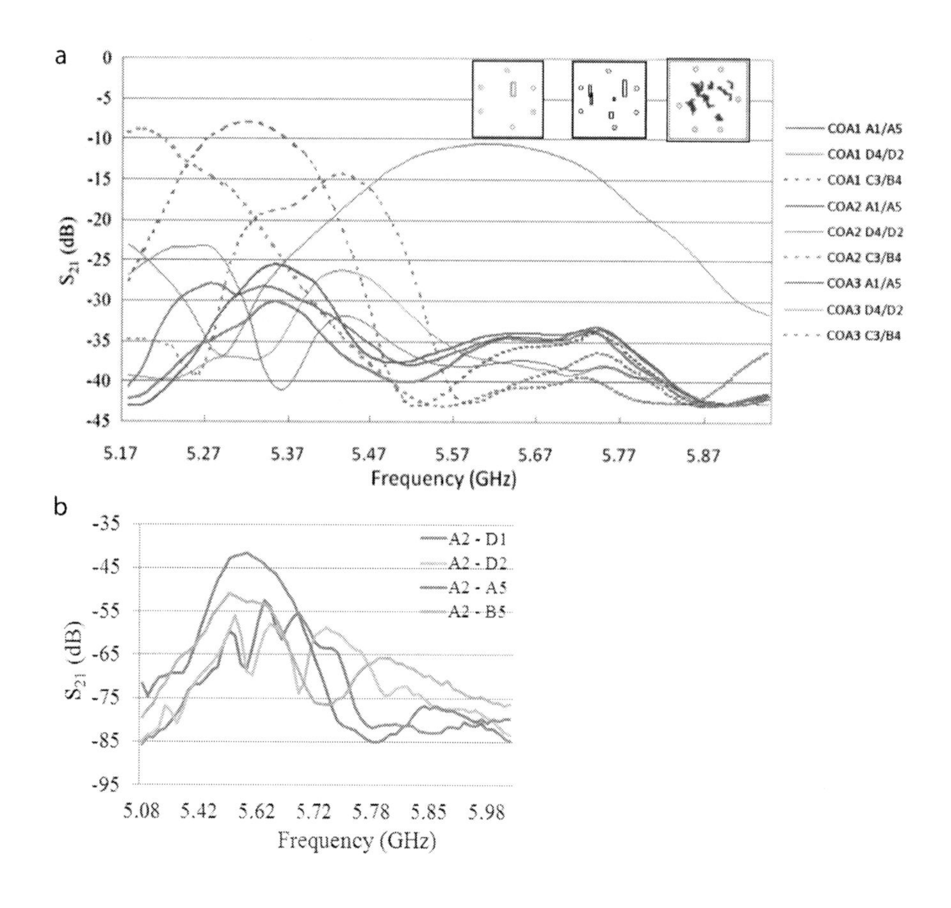

a single scan, 72 couplings give enough data to ensure uniqueness between CoAs.

10 different copper-based certificate objects have also been tested across three different T_x/R_x couplings. An evaluation of the differentiation between the different NF signatures is provided by their std curves shown in Figure 9b. Here, all four couplings provide significant amount of entropy, which is more prominent around the resonant frequency of the antenna elements mentioned above. It is worthy to note that the spatially lengthier antenna couplings A2-A5 and A2-B5 (see Figure 4a) are the ones that yield the highest deviation.

Multiple-2D to 3D CoA Projection Attacks

This experiment aims to test the invincibility of our anti-counterfeiting system against the multiple-2D to 3D CoA projection attacks. This type of attack can be realized by using three-dimensional structures, each created as a different-ordered stack of multiple inkjet-printed 2D CoAs. The separate layers of a 3D stacked certificate are shown in Figure 3c. Of course, all the resulting 3D structures have the same 3D projection when viewed from above.

Figure 10 compares the NF responses of such different stacked sets. In particular, Z1 is a stack of seven NF-CoAs with rhombic loops in a certain order, and Z2 thru Z6 are also stacks of the exact same certificates but in different orders. The remaining clear distinction between the different near-field responses verifies that even slight differentiations across the third dimension, i.e. thickness, of the NF-CoA object yield distinct enough NF fingerprints.

Conductive Material Density Effects

With this set of tests we are investigating the effect of the amount of metal density in the structure of the NF-CoA to the entropy in its NF frequency response. For this purpose, three different sets of copper-based CoAs of different wire gauge, namely 2, 3 and 4 grams per mold, are used. Each of these sets includes 15 certificates of the same

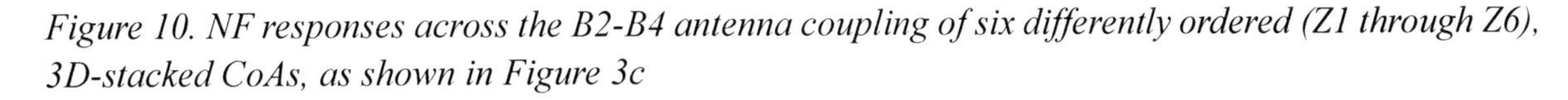

Figure 10. NF responses across the B2-B4 antenna coupling of six differently ordered (Z1 through Z6), 3D-stacked CoAs, as shown in Figure 3c

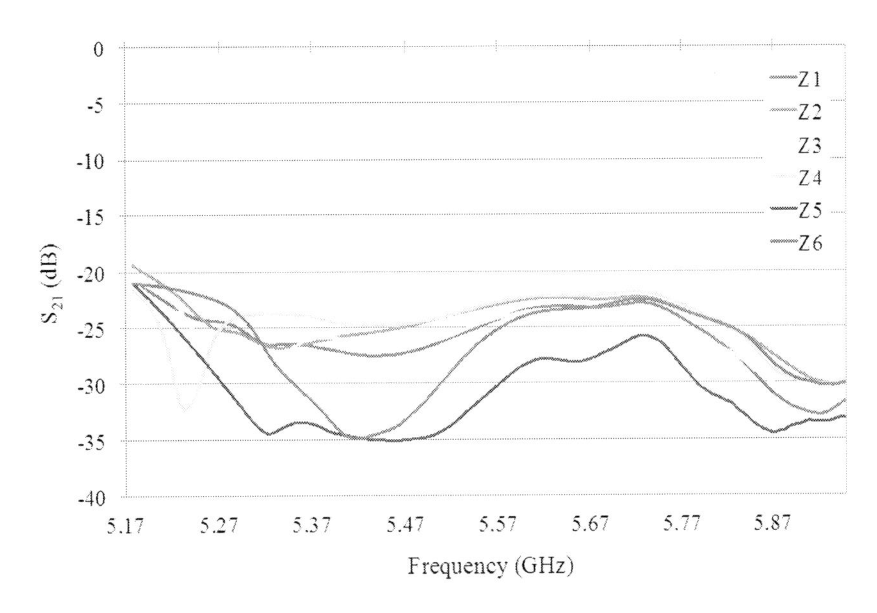

copper weight. The results of the standard deviation of all the certificates for each one of the four antenna permutations are shown in Figure 11.

First of all, from these plots it is verified that the metal density does indeed affect the entropy of the frequency response. For frequencies around the resonant frequency of the antenna elements, this difference in the achieved entropy can be as high as a multiple of 10. An additional conclusion drawn by this particular test is that the lighter copper-based certificates always yield higher response differentiation. On the contrary, although there is not that much of a great difference with the 3 gm ones, the set of heavier, 4 gm, CoAs provides the least entropy.

COA VERIFICATION

The verification process that typically takes place in a store or a warehouse is almost the inverse process of certificate issuing, discussed previously. The digitally encoded message onto the RFID chip is used to validate whether a product is authentic or not. The verifier first reads the aforementioned message and verifies the integrity of the plain bit string of the fingerprint with respect to its hashed and signed version using the corresponding issuer's public key. In case the integrity test is successful, which means that no one else except the possessor of the matching private key has encrypted the message, the original NF fingerprint and associated product data are extracted. This extracted fingerprint is afterward compared to a new reading of the tag's CoA that the verifier

Figure 11. Standard deviation of all the NF fingerprints of the different metal density, copper-based certificates for each of the labeled four antenna permutations

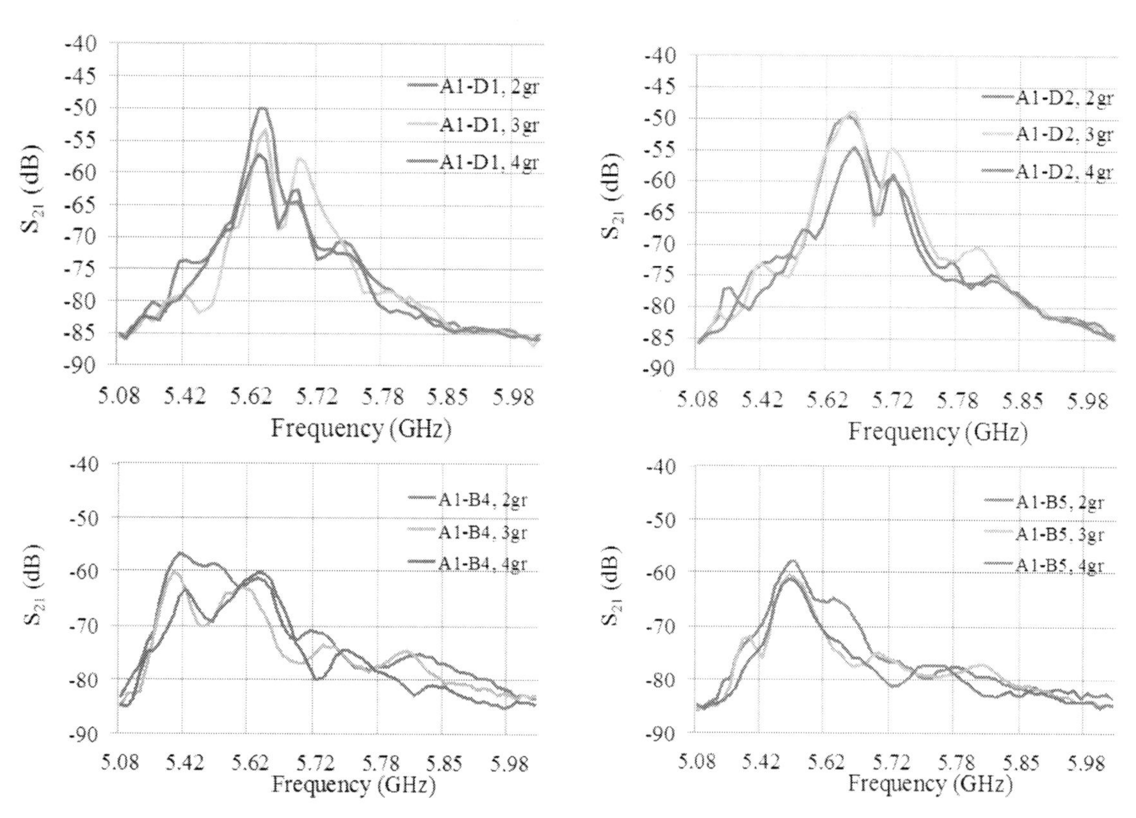

takes with his own reader. Only if the values of the distance metric between these two, read and extracted, fingerprint curves are bounded by a pre-defined and statistically validated threshold all across the frequency range, does the verifier declare that the certificate and, thus, the product is authentic.

The above described verification process is, of course, of offline nature. Nevertheless, the verification in the NF-CoA system can happen both in an offline and in an online fashion. Essentially, the difference between these two types of verification is that the online one is assisted by contacting and accessing a central database that keeps a record of all the queries in the past and is able to tell if one particular serial number has been queried about its authenticity more than once in the past. On the contrary, the offline process makes use of no intranet (corporate) or Internet connection whatsoever and is entirely carried out by the standalone reader itself. In this chapter we have primarily focused on the more attractive offline scenario, as this is applicable to a significantly broader range of applications. One can naively argue that in the online verification process the physical CoA is not required. Of course, this is not true since relying on just monitoring if one serial number has been queried more than once (N times) in the past would only allow the original issuer to know there exist N-1 counterfeits; in fact, the issuer would not be able to track down which exact is the genuine and who N-1 counterfeiters to legally pursue.

WITHSTANDING POTENTIAL ATTACKS

Creating Counterfeit CoAs

Let's assume that an adversary has in his possession an original, physical NF-CoA instance and, as a result, also extracts the exact fingerprint f, where f is a cardinality-N real vector of numbers $\in R^N$, by using a stolen or an own-manufactured reader. An attack against the NF-CoA technology is considered successful when the adversary is able to construct a nearly-exact replica or other three-dimensional object of the same dimensions as the original CoA that is capable of producing a near-field response f' such that $||f'-f|| < \delta$, where the detection threshold δ bounds the proximity of f and f' with respect to a standardized distance metric $|| \cdot ||$, such as the simple Euclidean distance.

Formulating a manufacturing process that can exactly or nearly exactly replicate an already signed CoA instance. Although this task is not infeasible, it requires certain expense by the malicious party. In fact, our NF-CoA designs require by the counterfeiter to accurately scan arbitrary complex 3D structures, with any potential state-of-the-art imaging process, reconstruct them with very high precision in terms of mechanical tolerances and, finally, embed them in a soft or hard encapsulating sealant.

Creating any three-dimensional object of the same dimensions as the original CoA that is producing a nearly-exact same near-field fingerprint. In addition to manufacturing, here an extra layer of difficulty imposed upon the counterfeiter is to numerically compute the extracted NF signature, so that the inverse design methodology can be applied. In general, while all electromagnetic phenomena are analytically explained using the Maxwell equations, the inverse problems for Maxwell equations are ill-defined problems (Romanov & Kabanikhin, 1994) of exceptional computational difficulty with potentially only an approximate or regularized solution. Not to mention that even fundamental problems, such as directly computing responses from simple antennas with regular geometries, may involve intensive computational tasks of arguable accuracy (Europe, 2001). On top of that, this computational complexity exists even despite the fact that for the current generation of NF-CoAs we have only considered isotropic materials. Finding approximations that can accelerate an electromagnetic field solver can be a difficult task if non-linear building elements are used for the fabrication of the CoAs. These elements include ferrites (ceramic compounds with ferromagnetic properties that show various kinds of anomalies in their power absorption at high microwave signal levels (Suhl, 1956)) and metamaterials (artificial materials, the permittivity and permeability of which are both negative resulting in a negative index of refraction and, thus, in a phase velocity that is anti-parallel to the direction of the Poynting vector (Shelby, Smith, & Schultz, 2001; Smith, Pendry, & Wiltshire, 2004)).

Alternatively, the adversary can launch a super-positioning attack, i.e. try to obtain a desired near-field response by combining a number of non-overlapping physical objects with known fingerprints at different highly accurate 3D positions relative to the 2D antenna array. Not only does the slight change in the physical position incur a huge set of different responses for each individual physical object, but also the decision on the number of these objects and the accumulated errors while performing the super-positioning make this task an arduous one. Moreover, we argue that, for a dense population of scatterers in the CoA instance, our system is non-linear in terms of near-field responses produced. One can understand this by simply assuming only two scattering physical objects S_1 and S_2 that, when individually placed in front of an antenna array, yield $f(S_1)$ and $f(S_2)$, respectively. Nevertheless, when both are in proximity to each other and in front of the antenna array, $f(S_1+S_2) \neq f(S_1) + f(S_2)$ because of their mutual interdependence, which is expressed, for example, by additional reflection or refraction on S_1 by the reradiated energy from S_2 (which energy did not exist when S_2 was not in the proximity of S_1, finally yielding $f(S_1)$) or by coupling between S_1 and S_2.

An additional requirement that holds for both aforementioned attack scenarios is that the cost of a massive successful adversarial manufacturing process, including not only the development but also the research, has to be lower than that of the profit that the counterfeit product can fetch on the market. Otherwise, an adversary will never even attempt creating counterfeit CoAs. On the other hand, based on the same perspective, it is safe to conclude that the NF-CoAs can be used to protect objects, the value of which does not exceed the cost of forging a single CoA instance.

It is all the above challenges that inherently enforce the important physical one-way function property and, as a result, the security of the CoA instances.

Computing the Private Key of the Original Issuer

By discovering the private key of the original manufacturer, an adversary can ultimately create his own NF-CoA object, extract its near-field signature with a, possibly, original reader, rather than exact replica, and store and sign the fingerprint bit string into an RFID tag chip.

This task, however, can be rendered arbitrarily difficult by adjusting the key length of the used public-key crypto-system (Menezes, Oorschot, & Vanstone, 1996; Rivest, et al., 1978).

Misappropriating Original Signed CoA Instances

It is possible that one collects CoA tags from already sold products, attaches or embeds them on counterfeit objects that, necessarily, have the same properties, e.g. product model, color and, more importantly, serial number, with the original products, and sells the fake products as authentic merchandise.

This type of attack is eliminated in the case of the on-line authentication process, where the CoA of an already sold item has been invalidated in the central database. Regarding the offline authentication case, one way to address this problem is to assign multiple NF-CoAs for each product; one that vouches for product's genuineness and the rest N, where $N \geq 1$, signaling the first up to

N^{th} owner. Each time the product is sold, it is the retailer's or previous owner's responsibility to devalue a CoA from the latter set. However, this solution requires that the bit strings of all N+1 CoAs are stored into the RFID chip.

APPLICATIONS

Undoubtedly, the CoAs are essentially vouchers of genuineness and as such are, in general, an essential tool for a huge number of transactions occurring on a daily basis. What enhances the proposed and manufactured NF-CoAs' applicability and makes them particularly attractive for several traditional applications as well as for a myriad of new ones is their:

- Ability to withstand wear-and-tear, mainly due the contactless signature readout mechanism as well their firm 3D structure.
- Short physical profile on top of their relatively small 2D projection dimensions and negligible weight. As a result, there are different ways of attaching them to products or documents; fastening them as labels, affixing them onto surfaces or even embedding them inside the products' outer shell.
- Inexpensive issuing and verification by just using a low-cost NF-CoA reader.

Example application categories are those that provide proof of authenticity (credit cards, driver's licenses, titles of ownership, bank notes, checks, money orders, warranties, receipts, endorsements, tickets, coupons), time-stamping and tamper evidence (when NF-CoAs are embedded in seals or product packages and cannot be reassembled if opened or torn).

CONCLUSION

A fabricated, high performing, robust and stand-alone reader for anti-counterfeiting applications has been presented. The SHF characterization of all components comprising the reader, with an emphasis on accuracy and insertion loss introduced, has been presented and the steps followed to extract the NF signatures fast and accurately have been described.

As a means of verifying the reader's performance in extracting the NF frequency response of the NF-CoA objects, a number of tests, namely intra- and inter-CoA robustness, effect of variation in conductive material density of the certificate and 2D to 3D CoA projection attacks, have been conducted yielding very promising results. A strategy to block potential attacks is also discussed. Regarding the effective bit length of entropy (EBLE) achieved with the results presented in this Chapter, this number cannot be cited as simply 56,160 bits, which is the result of 72 (permutations) multiplied by 65 (frequencies) multiplied by 12 (bits of ADC resolution). On the contrary, our achieved EBLE can be considered to be higher, due to the interdependencies of all responses that need to be taken into account, and exceeds what common cryptographic standards demand.

This chapter has demonstrated that uniquely authenticated "super" RFID tags, in the form of badges or product tags of a small dielectric profile, can prove a valuable tool against the ever-increasing activity of counterfeiters.

REFERENCES

Aero-plastics, Inc. (n.d.). *Website*. Retrieved from http://www.aero-plastics.tmcsweb.com/

Agilent. (n.d.). *Advanced design system (ADS) simulation environment*. Retrieved from http://www.home.agilent.com/

Anderson, A. (1971). *The Raman effect*. M. Dekker.

Balanis, C. A. (2005). *Antenna theory - Analysis and design* (3rd ed.). John Wiley & Sons.

Bauder, D. W. (1983). *An anti-counterfeiting concept for currency systems. Research report*. Albuquerque NM Sandia National Labs.

Bayvel, L. P. (1981). *Electromagnetic scattering and its applications* (Bayvel, L. P., & Jones, A. R., Eds.). London, UK: Applied Science. doi:10.1007/978-94-011-6746-8

Blair, J. (1999). Sine-fitting software for IEEE Standards 1057 and 1241. *Proceedings of the 16th IEEE Instrumentation and Measurement Technology Conference, IMTC/99*.

Chen, Y., Mihcak, M. K., & Kirovski, D. (2005). Certifying authenticity via fiber-infused paper. *SIGecom Exchange*, *5*(3), 29–37. doi:10.1145/1120680.1120685

Church, S., & Littman, D. (1991). *Machine reading of visual counterfeit deterrent fea- tures and summary of US research, 1980-90*. Canada: Four Nation Group on Advanced Counterfeit Deterrence.

Cross, I. D. Inc. (n.d.). *Website*. Retrieved from http://innovya.com/CrossID/

Devadas, S., Suh, E., Paral, S., Sowell, R., Ziola, T., & Khandelwal, V. (2008, 16-17 April). *Design and implementation of PUF-based "unclonable" RFID ICs for anti-counterfeiting and security applications*. Paper presented at the 2008 IEEE International Conference on RFID.

Europe, M. E. (2001). *CAD benchmark*. Retrieved from http://i.cmpnet.com/edtn/europe/mwee/pdf/CAD.pdf

Foldy, L. L. (1945). The multiple scattering of waves- I: General theory of isotropic scattering by randomly distributed scatterers. *Physical Review*, *67*(3-4), 107. doi:10.1103/PhysRev.67.107

GaAs mmIC SP4T Non-reflective positive control switch, DC - 8 GHz. (n.d.). Retrieved from http://www.hittite.com/content/documents/data_sheet/hmc345lp3.pdf

Glaxo-Smith-Kline. (2009). *Counterfeiting report*. Retrieved from http://www.gsk.com/responsibility/supply-chain/counterfeiting.htm

Global Card Fraud. (2010). *The Nilson report*.

Goldsborough, R. (2010a). *Ancient Fourree counterfeits*. Retrieved from http://rg.ancients.info/fourees/

Goldsborough, R. (2010b). *A case for the world's first coin: The Lydian lion*. Retrieved from http://rg.ancients.info/lion/article.html

Hejn, K., & Pacut, A. (2003, 20-22 May). Sine-wave parameters estimation - The second source of inaccuracy. *Proceedings of the 20th IEEE Instrumentation and Measurement Technology Conference, IMTC '03*.

Herzberg, G., & Crawford, B. L. (1946). Infrared and Raman spectra of polyatomic molecules. *Journal of Physical Chemistry*, *50*(3), 288–288. doi:10.1021/j150447a021

Johnson, D., Menezes, A., & Vanstone, S. (2001). The elliptic curve digital signature algorithm (ECDSA). *International Journal of Information Security*, *1*(1), 36–63. doi:doi:10.1007/s102070100002

Judson, R., & Porter, R. (2010). *Estimating the volume of counterfeit U.S. currency in circulation worldwide: Data and extrapolation* (F. R. B. o. Chicago & F. M. Group, Trans.). Policy Discussion Paper Series.

Juels, A. (2006). RFID security and privacy: A research survey. *IEEE Journal on Selected Areas in Communications*, *24*(2), 381–394. doi:10.1109/JSAC.2005.861395

Lakafosis, V., Traille, A., Hoseon, L., Gebara, E., Tentzeris, M. M., DeJean, G. R., & Kirovski, D. (2011). RF fingerprinting physical objects for anticounterfeiting applications. *IEEE Transactions on Microwave Theory and Techniques*, *59*(2), 504–514. doi:10.1109/TMTT.2010.2095030

Lakafosis, V., Traille, A., Lee, H., Orecchini, G., Gebara, E., Tentzeris, M. M., et al. (2010, 23-28 May 2010). *An RFID system with enhanced hardware-enabled authentication and anti-counterfeiting capabilities.* 2010 IEEE MTT-S International Microwave Symposium Digest (MTT).

Lax, M. (1951). Multiple scattering of waves. *Reviews of Modern Physics*, *23*(4), 287. doi:10.1103/RevModPhys.23.287

Li, Y., Rida, A., Vyas, R., & Tentzeris, M. M. (2007). RFID tag and RF structures on a paper substrate using inkjet-printing technology. *IEEE Transactions on Microwave Theory and Techniques*, *55*(12), 2894–2901. doi:10.1109/TMTT.2007.909886

Linear. (n.d.). *6GHz RMS power detector*. Retrieved from http://cds.linear.com/docs/Datasheet/5581fa.pdf

Menezes, A. J., Oorschot, P. C. v., & Vanstone, S. A. (1996). *Handbook of applied cryptography*. CRC Press. doi:10.1201/9781439821916

National Institute of Standards and Technology (U.S.). (2008). *Secure hash standard* (SHS). Federal information processing standards publication FIPS PUB 180-3. Retrieved from http://purl.access.gpo.gov/GPO/LPS121031

Pappu, R., Recht, B., Taylor, J., & Gershenfeld, N. (2002). Physical one-way functions. *Science*, *297*(5589), 2026–2030. doi:10.1126/science.1074376

Preradovic, S., & Karmakar, N. C. (2009, 20-22 Aug. 2009). *Design of fully printable chipless RFID tag on flexible substrate for secure banknote applications.* Paper presented at the 3rd International Conference on Anti-counterfeiting, Security, and Identification in Communication, ASID 2009.

Rahmat-Samii, Y., Williams, L. I., & Yaccarino, R. G. (1995). The UCLA bi-polar planar-near-field antenna-measurement and diagnostics range. *Antennas and Propagation Magazine, IEEE*, *37*(6), 16–35. doi:10.1109/74.482029

RFMD. (n.d.). *General purpose amplifier*. Retrieved from http://www.rfmd.com/CS/Documents/3378DS.pdf

Rivest, R. L., Shamir, A., & Adleman, L. (1978). A method for obtaining digital signatures and public-key cryptosystems. *Communications of the ACM*, *21*(2), 120–126. doi:10.1145/359340.359342

Robyn, M. (2008). *Market-driven fraud: The impact and consequences of counterfeit products and intellectual property violations.* ASC Annual Meeting, St. Louis, Missouri

Romanov, V. G., & Kabanikhin, S. I. (1994). *Inverse problems for Maxwell's equations*. Utrecht, The Netherlands: VSP.

Romero, H. P., Remley, K. A., Williams, D. F., & Chih-Ming, W. (2009). Electromagnetic measurements for counterfeit detection of radio frequency identification cards. *IEEE Transactions on Microwave Theory and Techniques*, *57*(5), 1383–1387. doi:10.1109/TMTT.2009.2017318

RongLin, L., DeJean, G., Tentzeris, M. M., & Laskar, J. (2004). Development and analysis of a folded shorted-patch antenna with reduced size. *IEEE Transactions on Antennas and Propagation*, *52*(2), 555–562. doi:10.1109/TAP.2004.823884

Shelby, R. A., Smith, D. R., & Schultz, S. (2001). Experimental verification of a negative index of refraction. *Science*, *292*(5514), 77–79. doi:10.1126/science.1058847

Simmons, G. J. (1991, 1-3 Oct 1991). *Identification of data, devices, documents and individuals.* Paper presented at the 25th Annual 1991 IEEE International Carnahan Conference on Security Technology.

Skoric, B. (2007). *The entropy of keys derived from laser speckle*. ArXiv e-prints.

Smith, D. R., Pendry, J. B., & Wiltshire, M. C. K. (2004). Metamaterials and negative refractive index. *Science*, *305*(5685), 788–792. doi:10.1126/science.1096796

Stern, B. (1996). Warning! Bogus parts have turned up in commercial jets. Where's the FAA? *Business Week*. Retrieved from http://www.businessweek.com/1996/24/b34791.htm

Suhl, H. (1956). The nonlinear behavior of ferrites at high microwave signal levels. *Proceedings of the IRE*, *44*(10), 1270–1284. doi:10.1109/JRPROC.1956.274950

Texas Instruments. (n.d.). *MSP-EXP430F5438 experimenter board user's guide (Rev. E).*

Tsang, L., Kong, J. A., & Ding, K.-H. (2002). *Scattering of electromagnetic waves: Theories and applications*. John Wiley & Sons, Inc. doi:10.1002/0471224286

Tsang, L., Kong, J. A., & Shin, R. T. (1985). *Theory of microwave remote sensing*. New York, NY: Wiley.

Tuyls, P., & Batina, L. (2006). RFID-tags for anti-counterfeiting. In Pointcheval, D. (Ed.), *Topics in Cryptology – CT-RSA 2006* (*Vol. 3860*, pp. 115–131). Berlin, Germany: Springer. doi:10.1007/11605805_8

Zernike, F., & Midwinter, J. E. (1973). *Applied nonlinear optics*. Wiley.

ADDITIONAL READING

Alliance, B. S. (2009). *Seventh annual global software piracy study*. Retrieved from http://portal.bsa.org/globalpiracy2009/studies/globalpiracystudy2009.pdf

Bauder, D. W. (1983). *An anti-counterfeiting concept for currency systems. Research report*. Albuquerque NM Sandia National Labs.

Juels, A. (2006). RFID security and privacy: A research survey. *IEEE Journal on Selected Areas in Communications*, *24*(2), 381–394. doi:10.1109/JSAC.2005.861395

Simmons, G. (1984). A system for verifying user identity and authorization at the point-of sale or access. *Cryptologia*, *8*(1), 1–21. doi:10.1080/0161-118491858737

Staake, T., Thiesse, F., & Fleisch, E. (2009). The emergence of counterfeit trade: A literature review. *European Journal of Marketing*, *43*(3/4), 320–349. doi:10.1108/03090560910935451

KEY TERMS AND DEFINITIONS

Certificate of Authenticity (CoA): A physical object that either vouches its own genuineness and, as a consequence, also proves the genuineness and value of the product it accompanies or can serve as a seal proof.

Near-Field Electromagnetic Scattering: The deflection of an electromagnetic wave as it impinges upon an object (scatterer), when the latter has a size of less than or of the order of the wavelength and is located in the (reactive or radiating) near field of the transmitting antenna element.

NF-CoA: A physical CoA instance that consists of an extremely difficult to replicate, random arrangement of scatterers that produces a unique and repeatable response in the near field for anti-counterfeiting purposes. Because of its slim profile, the NF-CoA may be attached as a tag, a label or a seal to a physical object or may be integrated as an inseparable part of the surface of an object.

NF-CoA Reader: A low-cost reader that can extract an NF-CoA's signature, can verify the issuer of the content of the RFID tag and validate or not the authenticity of an NF-CoA instance.

NF Fingerprint/Signature/Response: A set of S_{21} parameters observed over a specific frequency band and collected for (a subset of or) all possible antenna couplings of a reader's array when the NF-CoA is placed in the array's near field.

Section 2
Middleware

Chapter 5
Edgeware in RFID Systems

Geoffrey Ramadan
Unique Micro Design, Australia

ABSTRACT

An RFID system is an enabling technology encompassing both hardware and software. This chapter presents a software interface that integrates both middleware and enterprise software to automate the RFID system. The developed software is called Edgeware. The chapter presents an industrial perspective of complete implementation of an RFID system to automate the whole process of a big and complex business. In this chapter the definition, evaluation, value proposition, and implementation procedure of Edgeware are presented.

INTRODUCTION

Middleware and Edgeware

In last few decades Enterprise Software had undergone tremendous transformations in terms of both technologies in use and market/user spread. Starting from mainframe computers in few decades ago, enterprise software is now being used across wide verities of computing devices ranging from very powerful supercomputers to everyday objects like mobile phones, PDAs, etc. Software market has widened many folds and includes very complex huge enterprise software like stock market brokerage software, enterprise resource management, etc., to small utility software like calendar, multimedia player, etc. Considerable investment has gone into the development and operation of a number of large enterprise software over last decades. But, the end user perspective, requirement and taste changes more frequently than ever and business software needs to match them in the very competitive market. In most cases it is not practical to rewrite software every time new functionality is required; instead it is easier to extend the core system with the addition of application specific software module, namely middleware, to enhance the functionality (NIST, 2005; Zhang & Jacobsen, 2003).

Until recently, in most enterprise software, data is collected and feed manually. But with the introduction of data capture devices like RFID, sensors, smart cards, etc., the data entry paradigm is shifting from manual to automatic in many busi-

DOI: 10.4018/978-1-4666-2080-3.ch005

nesses. For example, a RFID reader is set at the door of an inventory and each item entered/deport through its door is RFID tagged. Then the reader would read the arrival and departure of stocks and update the inventory management database automatically, thereby no need of manual data entry to the inventory system. Similar to data entry, in many applications, the response/output generated from analyzing input data would drive actuator devices automatically. For example, in a security system, on detecting an intruder from proximity sensor data the central system would activate alarm, flash light, etc. Since recent advances in MEMS technology is developing a wide verity of data capturing and actuator devices, enterprise systems need to incorporate such automatic data capturing and actuation features in their business model. The most quick and cost effective way of doing so is to develop a range of middleware software, which, in one side would interface/drive the physical (data capture or actuator) devices, and on the other side interact with the core enterprise software. Edgeware represents all the hardware and software components that surround business enterprise software (e.g., Enterprise Resource Planning (ERP) and Manufacturing Resource Planning (MRP) systems), and is responsible for the collection of business data from the surrounding physical world using data capturing devices and disseminate automatic responses to the actuator devices (Edgeware, 2011; Wu et.al, 2005; MoreRFID, 2011; InSync, 2011).

The Edgeware Model

Edgeware is represented by Figure 1, and includes the following elements:

- **Physical Activities:** In a typical enterprise this would cover point of service (including fixed and mobile), supply chain, asset management and industrial activities.
- **Data Capture Devices and Sensors:** These are the physical terminal devices that collect data and monitor activities such as RFID, Barcode and mobile devices.
- **RFID/Sensor Network Infrastructure:** Represents the various data transport and

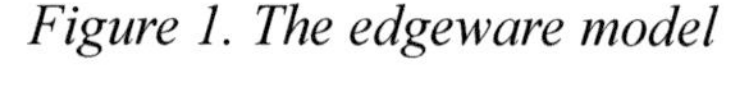
Figure 1. The edgeware model

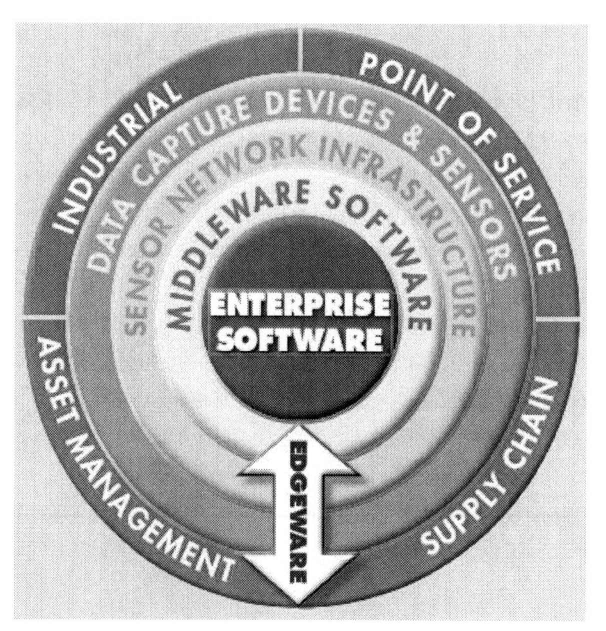

network mechanisms which move information from the data capture devices and sensors to the Enterprise System and may include a mixture of wired and wireless networks.

- **Middleware Software:** Manages the collection of data and integration into enterprise software systems.
- **Enterprise Software:** Typically an ERP/MRP system or asset management system.

The boundary between the data capture layer and physical activities is the Edgeware terminal interface. This in effect represents the "edge" of the enterprise and we define a "terminal" as any electronic data capture / entry device that enables interfacing between the "real world" and "digital world." Typical terminals most people would be familiar with are:

- Computer Terminals
- Portable Data Terminals (PDT)
- Mobile Terminals
- Barcoding Terminals
- EFTPOS Terminals
- POS Terminals
- Warehousing Terminals
- RFID interrogators / terminals

Though Edgeware terminals include human operated data entry devices such as keyboards and touchscreen terminals for example, in their most simple form, the real value of Edgeware terminals is where the data entry is automated via Automatic Data Capture and Identification devices (Auto-ID) which typically include:

- RFID
- Barcoding
- Smart Cards
- Magnetic Strip
- Optical Character Recognition (OCR)
- Sensors

Edgeware for RFID System

As quoted by Wikipaedia (2011), "Services provided by enterprise software are typically business-oriented tools such as online shopping and online payment processing, interactive product catalogue, automated billing systems, security, content management, IT service management, customer relationship management, resource planning, business intelligence, HR management, manufacturing, application integration, and forms automation." Figure 2 shows a complete block diagram of the RFID enterprise system (Cho et. al, 2007). An RFID network is a wireless data capturing technique. The main two components are an RFID tag or transponder which is the tagging device and a RFID reader or interrogator which reads the tag into meaningful information. The interface and control between the core enterprise software and RFID network is established through middleware software. RFID edgeware consists of the RFID network hardware (readers and tags), reader and tag programs, and RFID middleware. The RFID middleware consists of EPC (Electronic Product Code) server, RFID task manager and RFID information management system. EPC server converts the tag information into more meaningful business data such as product code, type, manufacturer, date of manufacturing, etc. RFID information management system receives high level job request from core enterprise system and then convert them into low level RFID network specific instructions understandable to the RFID readers. These low level commands are then passed to the RFID task manager who selects appropriate RFID readers and manages subsequent execution process. On the way back, RFID task manager send the tag information received from EPC server to the RFID information management system which then formulate high level response to the core enterprise system.

Figure 2. Edgeware for RFID system

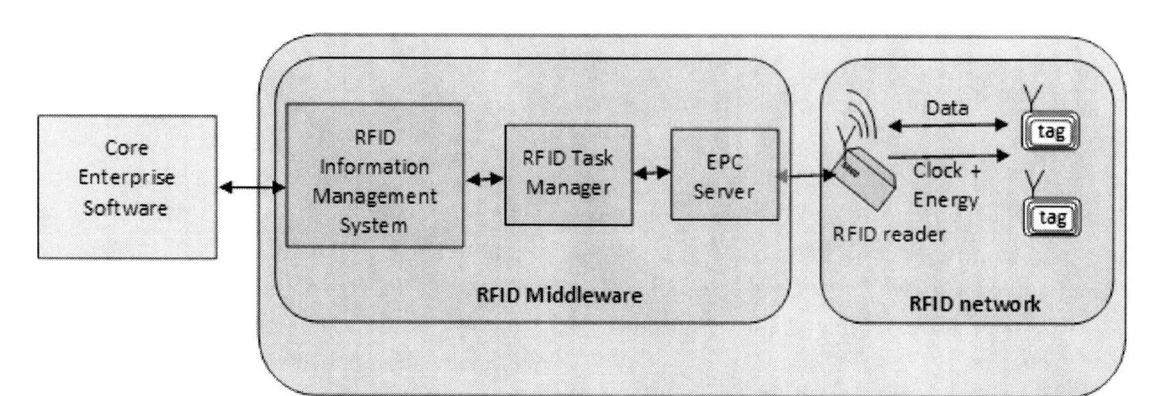

EVOLUTION OF EDGEWARE

Edgeware is the next evolutionary step in Information and Communications Technology (ICT) for computer applications. Edgeware extends the reach of Enterprise Software terminal interfaces to the point of activity and provides real time visibility of logistical and physical events. To understand this, and its implications, we need to take a brief look at the evolutionary nature of computers, particularly from a Business Application or Enterprise Software perspective.

Business (and many other) computer applications rely on synchronizing computer processes and databases with actual events and transactions generated within the Enterprise. These processes include data entry of stock receipts or shipments, customer details, order entry, production data etc. There are delays between these actual physical business events (e.g. Movement of inventory) and when the data is entered into the system and resulting actions require, for example, generation of purchase orders, production schedules, shipping, etc. This delay causes many of the supply chain issues we encounter today.

Originally, the reach of Computers extended to the Terminal interface (such as the original IBM / Wyse "Green Screen" Terminals) and data were entered via their keyboards. To extend the computer's reach, you had to extend the interface cable. With the introduction of Ethernet and Token Ring Networks, the reach of computers was extended within office complexes and campuses. The Internet allowed Enterprise Software to be connected to computer systems anywhere in the world and provided the foundation for which collaboration within the supply chain could be achieved. With the advent of wireless and mobile computers, the reach of Enterprise Software continued beyond the four walls of the enterprise to the point of activity where the data is actually generated.

As considerable investment and time goes into the development and operation of Business systems, especially in software, in many cases it is easier to extend the functionality of the core system with the addition of application specific software, typically called middleware. In Enterprise Software systems, provisions are made for the interchange of data with other applications (including middleware) via Application Program Interfaces (API's). Middleware typically synchronizes other business applications automatically which would have previously been handled via manual data entry. For example, XML enables disparate computers between different Enterprises to electronically transfer business transactions (e.g. purchase orders, shipment details and invoices).

Data capture devices also evolved from manual data entry via keyboards to automatic data entry devices like barcode and magnetic stripe readers.

The speed and accuracy of such devices helps continue to drive business efficiencies. RFID was the next key evolutionary technology. Its ability to read multiple items (RFID tags) at distances, and without line of sight, opened up many new opportunities to add visibility to enterprise systems. This was followed by the addition of sensors such as temperature, proximity and actuators for control.

Similar in concept to a Supervisory Control and Data Acquisition (SCADA) system, which provides real world monitoring and control of production processes, Edgeware adds real time Logistical Control and Data Acquisition (LCADA) to Enterprises Software. In complex Edgeware systems, the Enterprise Software will not have the necessary capability to communicate with, or control, the data capture devices. In these cases, device management may be added to the middleware application functionality.

Middleware may also provide additional business logic functions which the Enterprise Software may not be able to perform. A good example of this is an RFID System in a warehousing environment. Here, a fixed RFID reader may be placed over a door to read all pallets passing through it. Sensors are needed to activate the RFID readers which, once enabled, constantly poll the antenna resulting in multiple readings of the same RFID tag. Middleware would need to control this process and filter this data to generate a list of single valid RFID tag numbers before it is presented to the Enterprise System.

We are now at the evolutionary state where after several decades of development in computer Business Systems, Enterprise Software, communications and Edgeware, the Edgeware Terminal Interface now extends to the point of activity. That is, the system can now sense its own environment. Operators no longer need to key punch in transactional data as the Enterprise Software has "sensed" its movement or activity via Edgeware.

The term "Intelligent Sensor Networks" has been coined to reflect this complexity and the need for additional intelligence in the Edgeware layer, in particular to control and monitor the Edgeware system. For example "mesh sensor networks" rely on the "intelligence" built into the network to ensure data is transferred to where it is needed. This then leads to the next evolutionary step: Intelligence within the Enterprise Software. There are two fundamental requirements to be able to create "intelligent" systems:

1. **Sensor:** So the system can sense its own environment (using Edgeware, of course).
2. **Brain (Computer):** To process data and make decisions based on sensory inputs.

Thus, the foundations are laid to develop intelligent systems based on Edgeware. Emerging intelligent Enterprise Software can make decisions based on sensory data, including communicating with other Intelligent Enterprise Software Systems. The Enterprise system will have the intelligence to sense our activities, watch our consumption, anticipate our needs, schedule our production and manage our supply chain, all without human interaction.

THE VALUE OF EDGEWARE

A major cost to business operations is in the activity of people collecting data and monitoring operations. Connecting people and activities in the physical world to your existing Enterprise System gives you real time visibility. The overwhelming trend in information and communication technologies (ICT) systems today, regardless of industry, is to provide real time visibility of logistical processes and assets. In order to achieve this, sensors and RFID networks need to be added to software, to provide it with "eyes" and "ears" using Edgeware. A key feature of such networks is that they are automated. You do not have to "tell" the enterprise system that you have moved a pallet or asset, the "sensors" can see that you have done this, and can therefore trigger the next process or event automatically. This results in potential savings in:

- **Efficiency:** By collecting data where it is generated.
- **Effectiveness:** Reducing errors and mistakes resulting from poor quality information
- **Empowerment:** Of people to do more with access to accurate real time data
- **Ease:** Edgeware reduces bottlenecks and the burden on IT resources.

METHODOLOGY: HOW TO IMPLEMENT EDGEWARE

Many Enterprise system projects adding Edgeware are based on an "inside-out" approach. That is, they start on the "inside" at the Enterprise Software level, adding application software layers, (middleware), and move towards to the "outside" or the Edgeware layer by adding data capture devices as the last step. The problem with this approach is that may not get results until the end of the implementation process is reached. An "outside-in" approach can be used instead, by implementing the following steps, thus working from the outside of the system's edge (data capture points), to the inside (where enterprise software is interfaced):

- Collecting data from where it is generated by building the physical data capture systems and networks.
- Visualising or displaying this data (independently of the enterprise system) to gain visibility as part of the middleware application features.
- Analysing the data to see what is going on (e.g., identify bottlenecks)
- Automating the analysis where possible by developing optimisation programs, trigger events and process flow monitoring/control.
- Integrating the data into the Enterprise System.

A key outcome of this approach is that results can be achieved quickly by implementing the first 3 to 4 stages above, without any integration into the enterprise system. This allows the enterprise software integration to be done independently without interference, and results in a more immediate return on system investment.

RFID PROJECT IMPLEMENTATION

RFID projects need special consideration. Unlike conventional ICT projects, there is an element of risk in implementing RFID data capture projects, also there is a considerable amount of investigation work required before and during the RFID development process before a finalized solution can be settled on. To minimise this risk, time and cost in implementing RFID projects, the following five-stage process can be used. Each stage builds on the previous stage and learning outcomes leading to a rapid project deployment without excessive consultancy costs.

Stage 1: Proof of Concept

The starting stage of any RFID project is with the feasibility of being able to read an RFID tagged asset. Just being able to read an RFID tagged asset is still not sufficient. Also understanding under what conditions RFID tagged assets can be read is crucial, as this will determine the architectural solution and therefore viability of a project.

The physics of electromagnetic radiation properties (RF), the RFID interrogators (readers), transponders (RFID tags), and the environment determine the readability of a RFID tagged asset. The purpose of the "Proof of Concept" is to provide basic tests with RFID readers and RFID tags in a simulated environment using the actual asset of interest. These basic tests will ascertain under what conditions the RFID tagged asset can be read, and therefore influences the solution.

Factors which effect readability of RFID tagged assets include:

- Frequency choice (Low Frequency (LF), High Frequency (HF) or Ultra-High Frequency (UHF))
- Use of passive or active RFID technology
- RF antenna design (directional properties and antenna gain)
- RFID tag package
- Orientation of RFID tag (tag placement)
- Distance between antenna and tag
- Power output of RFID reader
- Material RFID tag is being attached to
- Material inside of tagged asset (e.g., liquid)
- Surrounding environment
- Surrounding electromagnetic radiation and interference
- Operating temperature of environment of asset.
- Vibration
- Corrosive materials

Stage 2: Workshop

If the Proof of Concept tests are successful and the RFID tagged asset can be read in the required environment then a Workshop is conducted with the customer to:

- Provide an overview of the test results.
- Design a feasible solution based on the outcome of the Proof of Concept in consultation with the customer.

The customers' input into this process is critical, as they will have an understanding of the operational constraints that the system will need to work within. A successful workshop will result in an agreed solution and approach. Consideration is given to:

- RFID Reading system design
- Work flow issues and design
- Host system integration
- Data management (how is the data collected)
- Middleware software requirements and solution
- Pilot stage
- Regulatory and compliance issues
- Implementation plan
- Timelines

Stage 3: Pilot

The Pilot phase usually involves the design and installation of at least one of the RFID reading systems, or portals, in its final environment. It may also include sub-stages:

- **Site-Survey:** Involves location visits to develop an understanding of the Customer's physical environment. Particular consideration is given to the positioning of equipment, and the surrounding environment.
- **RFID Tag Selection:** Also of primary importance is the RFID tag selection as consideration needs to be given to tag mounting options, orientation, RF permeability of material and other environmental issues.
- **Installation:** The physical process of installing equipment into Customer locations as per the systems requirements identified by RFID Site Surveys. It may also include installing wireless and other network infrastructure.

Stage 4: Evaluation and Refinement

Once the pilot RFID reading equipment has been installed, an evaluation process follows which involves testing the RFID reading system and RFID tag assets in the actual environment. Its purpose is to optimize the performance of the system in its real environment and to ensure adequate read rates are achieved and read errors minimized or eliminated. This may involve some additional experimentation with antenna locations, reflectors and RFID tag placements.

Once this process has been completed the project is reviewed and if required, refined based on measured results.

Stage 5: Deployment

Once the pilot RFID tag reading system has been satisfactorily deployed the remainder of the IT infrastructure can be developed, installed and commissioned. This will usually involve:

- Designing, installing and testing the middleware data capture software
- Software integration with host software.
- Designing and installing wired / wireless network
- Equipment procurement and installation, which includes RFID equipment, hand held computers, printers, tags etc.

Though not trivial, once the RFID tag can be read and converted into digital data, conventional IT technologies and consultation processes can be used to develop the desired business process and objectives.

SUMMARY

In summary, by adding Edgeware to the Enterprise system the following trends are being realized:

- Moving from "database" to "process" based computing. Instead of the system collecting data and providing reports, the system manages the process.
- Triggers based on real time sensor inputs are used to drive subsequent actions. Processes no longer need to be linear but can be random by using Edgeware "sensors" to track and monitor.
- Human intervention is not required. The Edgeware sensors "watch" the activities, typically human activities, actioning as required. Thus Operators only need to focus on their task at hand, allowing the "sensors" to watch, control and report.
- Provide real time visibility of inventory and assets, anywhere and anytime, and having this data shared with other stake holders and systems.
- Move from Application Software to Business Intelligence. With the availability of real time data (from many sources) and process monitoring, business intelligence can be used to make automated business decisions such as demand forecasting and process scheduling.

REFERENCES

Cho, J., Shim, Y., Kwon, T., & Choi, Y. (2007). SARIF: A novel framework for integrating wireless sensor and RFID networks. *IEEE Wireless Communications*, *2007*, 50–56. doi:10.1109/MWC.2007.4407227

Edgeware. (2011). *Syspro ERP software for medical device manufacturers*. Retrieved from http://www.edgeware.net/industries/syspro-erp-software-for-medical-device-manufacturers

InSync. (2011). *Auto-ID development platform, 2011*. Retrieved from http://www.insyncinfo.com/INSYNCsensoredgew.asp

MoreRFID. (2011). *FAST Tag™ - RFID edgeware, middleware, tagging and tracking*. Retrieved from http://www.morerfid.com/details.php?subdetail=Product&action=-details&product_id=84&display=RFID

NIST Special Publication. (2005). *PIV middleware and PIV card application conformance test guidelines,* (pp. 800-85). Retrieved from http://csrc.nist.gov/piv-project

Wikipedia. (n.d.). *Enterprise software*. Retrieved September 12, 2011, from http://en.wikipedia.org/wiki/Enterprise_software

Wu, B., Liu, Z., George, R., & Shujaee, K. A. (2005). eWellness: Building a smart hospital by leveraging RFID networks. *IEEE Proceedings of the 27th Annual Conference on Engineering in Medicine and Biology*, Shanghai, China, September 1-4, 2005.

Zhang, C., & Jacobsen, H. A. (2003). Refactoring middleware with aspects. *IEEE Transactions on Parallel and Distributed Systems*, *14*(11). doi:10.1109/TPDS.2003.1247668

KEY TERMS AND DEFINITIONS

Edgeware: Data Capture System that integrates with Enterprise Systems.

Intelligent Sensor Network: A complex Edgeware solution that requires embedded intelligence to function.

Logistical Control and Data Acquisition (LCADA): Typical application of Edgeware that provides real time visibility of inventory and logistical assets.

Middleware: Software add-on to Enterprise Software, to enhance its functionality.

Terminal Interface: Devices use to convert real world activities to digital.

Chapter 6
Design and Implementation of an Event-Based RFID Middleware

Angelo Cucinotta
University of Messina, Italy

Antonino Longo Minnolo
University of Messina, Italy

Antonio Puliafito
University of Messina, Italy

ABSTRACT

The downward trend in the cost of RFID technology is producing a strong impact on the industrial world that is using such powerful technology in order to rethink and optimize most of the existing business processes. In this sense, the chipless technology is playing a key role to facilitate the adoption of RFID in enterprises. All this implies the use of solutions that simplify the adoption of the continuously evolving RFID technology and allow keeping a high-level vision versus the specific technical details. In brief, it is mandatory to abstract the technological level and makes transparent the physical devices to the application level. The widespread use of the RFID technology also produces a large volume of data from many objects scattered everywhere, that have to be managed. In these complex scenarios, the RFID middleware represents an ideal solution that favors the technology integration, reducing costs for application development and introducing real benefits to the business processes. In this chapter, the authors describe the main features of our event-based RFID middleware and its powerful architecture. Their middleware is able to assure an effective process of technological abstraction, switching from a vision linked to the specific issues of interfacing devices (chipless tags, readers, sensor networks, GPS, WiFi, etc.) to the management of the event generated by each device. In brief, "event-based" means to integrate the management logic of different devices.

DOI: 10.4018/978-1-4666-2080-3.ch006

INTRODUCTION

A very important innovation is the possibility for the user to customize the event-response of the system and to adapt its behavior to changing working conditions. An example of the adoption of chipless RFID for document flow management is presented with emphasis on the technological solutions adopted and on the derived benefits. RFID is a very promising wireless technology that will strongly impact on everyday life in several fields such as retail, supply chain management, document management, access control, healthcare and so on. In general, RFID identifies a plethora of heterogeneous location and identification technologies ranging from wireless identification tags to sensors networks and GPS. One of the main opportunities for RFID is the tagging of all things to create a global network of physical objects, the so-called Internet of Things (IoT). To accomplish this result it is necessary to count on very inexpensive tag, and some printed and chipless RFID technologies have demonstrated the potential to achieve this goal. Recent statistics show how the potential of RFID technology is influencing the commercial interests. For example, in the next ten years we will see a rapid gain of the market and in particular of the printed and chipless RFID tags.

The chipless tag is a particular RFID tag that does not contain a silicon chip. It represents a new frontier of RFID technology and at the same time an opportunity that could open a new scenarios as it could be printed directly on the products and packaging for less than 0.1 cents each. Also, chipless tags are smaller and less obtrusive than RF chips and can be easily embedded in products like a paper sheet. In addition, the readers for chipless tags are similar in size and cost to barcode readers. In the near future a progressive replacement of the 2D barcode with chipless tags could be hypothesized because they offer several advantages. Besides the known advantages such as the identification of the object without the requirement to be in line-of-sight or the possibility to hide the RFID tag inside a container (envelope, paper, etc.), the low cost of the chipless tags has to be considered.

The Internet of Things and the broad diffusion of many different RFID technologies for tagging a wide range of objects such as consumer products in supply chains or in other applications (theft deterrence, counterfeit detection, etc.), will produce very complex scenarios and a big amount of data that have to be managed. In order to take advantage from the potential of each technology, to properly manage a big amount of data and then to obtain more effective solutions and services, effort has to be devoted to develop a middleware layer that glues technical hardware aspects and software services. For instance, mobility, identification and localisation are very important issues, as from the position of a user or of an object several implications may arise such as the capability to provide targeted services, optimize paths, increase safety, etc.

In this chapter we intend to present our work in the creation of an "Event-based" RFID Middleware for the automatic identification of objects in several heterogeneous contexts. This middleware provides an effective abstraction of the technological level, facilitating the integration process with company IT systems, reducing costs and improving the effectiveness of business processes. In order to achieve such goals, it was built using a modular and flexible software architecture based on the state-of-the-art technologies such as SOA-Service Oriented Architecture (WS, REST, HTTP), Enterprise Service Bus (XMPP, JMS) and Multichannel Communication (Jabber, VoIP, SMS, MMS).

Our work resulted in the creation of a commercially available product, known as the WhereX© middleware. This product was realized in the Wireless RFID Laboratory of University of Messina and commercialized by Inquadro s.r.l., a spin-off company of the University of Messina. We will present our design choices and implementation details and describe a real case that will enable to understand how WhereX can assure high levels of flexibility, speed up services development and simplify processes integration. In particular, we will describe a software application for workflow management of paper documents through chipless rfid tags is also presented.

BACKGROUND

In an RFID-based application, a main part of the cost of the entire system is related to the cost of tags and the related interaction devices, due to their high number. Assuming a simple scenario, where the application is shortened and the data process is made in a dedicated way, the RFID system could be basically composed of three elements: tags, readers and application system. The tags are placed on the objects; the reader is commonly connected to a host computer, which performs additional signal processing and data tag's elaboration and the application system has to process the data to carry out the tasks for which it was designed.

Almost all the RFID tags include an antenna and an integrated circuit (IC). The IC performs the processing of data when it is powered by the energy that it receives from the RFID reader. This type of tags are called passive because they do not have any form of on-board power supply, to distinguish them from those known as active which have integrated some form of on-board power supply. Passive RFID tags are less expensive and experiences short reading ranges, while active RFID tags are more expensive and offer long reading ranges.

The production cost of a tag is mainly related to the complexity of the on-board IC and this explains why considerable research efforts are constantly carried out to develop RFID tags technology with the minimum possible quantity of IC. These tags, and consequently the developed systems, are known as chipless RFID systems. Most chipless RFID systems use the electromagnetic properties of certain materials and/or the shape/layout of different conductors in order to change the behaviour/properties of electromagnetic waves.

In literature there are three main categories of chipless RFID tags (Preradovic, & Karmakar, 2010):

1. Time domain reflectometry (TDR)-based chipless tags
2. Spectral signature-based chipless tags
3. Amplitude-Phase backscatter modulation-based chipless tags

The time-domain reflectory-based chipless tags are interrogated by detecting the return signal from the tag, obtained by sending a train of pulses from the reader. This signal returned by the tag is used to interpret the data. This type of tag is used in localization/positioning and has the advantage, compared to chipped tags, to be less expensive and ensure a greater reading range. The disadvantages are related to the small number of bits that can be encoded and to the high speed generation and interpretation of pulse trains that RFID readers should have. The TDR-based tags may be distinguished between printable and nonprintable:

- Printable versions of chipless tags are made as thin-film-transistor circuit (TFTC) or microstrip with discontinuities. In particular:
    - The TFTC tags have the disadvantage of requiring more power than other types of chipless tags but offer some extra features such as their small size and low power consumption. The cost of these tags is still not acceptable to justify their wide diffusion.
    - The delay-line-based tags are made of microstrip discontinuity after a section of delay line. The tag is energized by a short EM pulse and reflected at various points of the microstrip. This causes an echo of the electromagnetic

waves with a time delay due to the length of the microstrip. Based on this delay with which these impulses are received from the reader, it is possible to decode the ID of the tag. This technology allows creation of up to 4 bits ID tags, that still represents a limit to the adoption of such technology.
- An example of non printable chipless tags consists of the tag Surface Acoustic Wave (SAW). The SAW tags are turned on by impulses sent by the reader. Impulses, affecting the tags are converted by a Surface Acoustic Wave interdigitated transducer (IDT). The SAW propagate along the piezoelectric crystal and are reflected by the reflectors that create a pulse train with phase shift. These pulses, goes back to the IDT and are converted into electromagnetic waves detected by the reader, which decodes the ID of the tag.

In the spectral signature-based chipless tags, the ID code is encoded using the presence or absence of lines in the spectrum of frequency realized by circuits resonators present in the tags. These peaks, detected by readers, are interpreted as bit. The advantage of these tags is that they are printable, able to store ID codes with a high number of bits, and are also inexpensive. The disadvantages are due to the orientation requirements of the tags, the broad spectrum required to encode the identification code and the bandwidth of the readers.

In the amplitude-phase-backscatter-modulation-based chipless tags, data encoding is achieved by varying the amplitude or phase of signals reflected from the load of the tag antenna. The changes in amplitude or phase can be detected by a reader. The advantage of these tags is due to the narrow bandwidth with which they operate, while the main disadvantages are related to the small number of bits that can be encode on these tags.

Even using only chipless tags, it comes out that different technological solutions are available, that could be even simultaneously adopted. And the situation may complicate even further if other RFID technologies (active, sensors, GPS) are used in with complex scenarios.

Usually, dedicated solutions suffer from technological evolution and hardly offer scalability features. Adapting the system to new needs often involves huge investments and results can still be away from the expected ones. In this case, the adoption of a middleware layer is a good solution to reduce the impact that technological changes/improvements may determine on the whole system.

Similarly, in complex scenarios where a very large area has to be monitored and a mix of different technologies is used, or where there is a need to process and aggregate the information before making them usable by the specific IT application, the middleware adoption becomes a necessary condition. In general, a good definition of middleware could be "a layer of software that resides between the business application and the devices layer of heterogeneous platforms and protocols."

Summarizing, the main role of the middleware is to offer the best solution to easily deal with the integration. The inclusion of a middleware layer allows to decouple the business applications from any dependency form the technological layer (e.g. heterogeneous operating systems, hardware platforms, communication protocols, etc.). Furthermore, the middleware provides the developers several services that can be used to glue all the monolithic applications together (Figure 1).

In other words, the presence of a middleware to build distributed systems is a concrete advantage as developers can focus their attention on the application requirements rather than on the technological details such as concurrency control, transaction management, network communication and so on.

Figure 1. Generic scenario

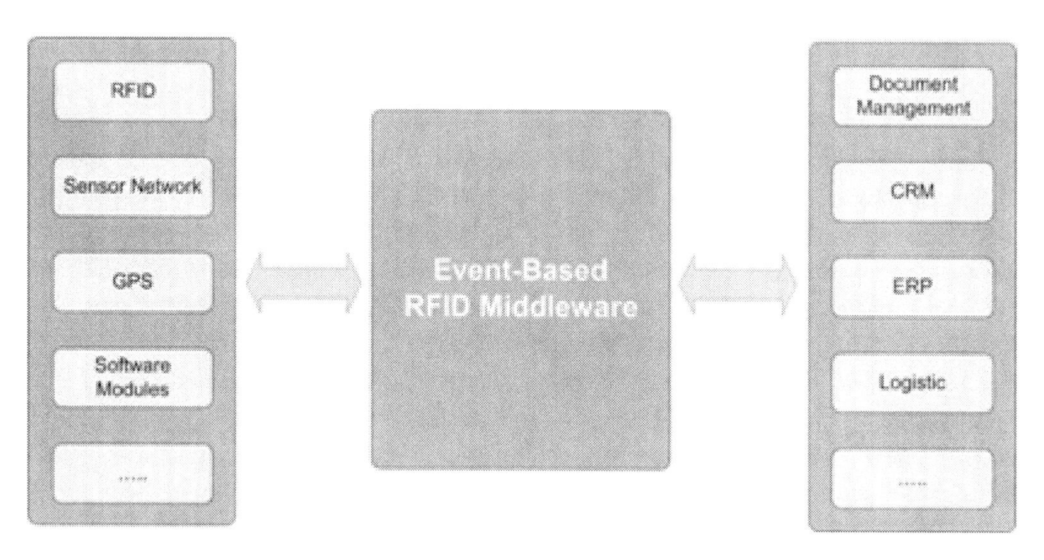

It is therefore evident that the presence of a middleware layer is fundamental and more and more necessary for the connection between the different RFID technologies (RFID, Sensor Network, GPS, WiFi, etc.) (Lòpez, & Kim, 2007) and the application layer. Some examples of middleware successfully used are: CORBA from the Object Management Group (OMG) (Weijie, & Weiping, 2008), Microsoft COM (Aberer, Hauswirth, & Salehi, 2006), SUN Java RMI (Kanoc, 1999), IBM's MQSeries (Pope, 1998), and Remote Procedure Calls (RPCs) introduced by SUN in '80s.

Other researches highlighted how the implementation of a middleware that can integrate some of the technologies mentioned above (Foster, 2005) (MacKenzie, Laskey, McCabe, Brown, Metz, & Hamilton, 2006), or other technologies even in mobility (Pautasso, Zimmermann, & Leymann, 2008), leads to benefits arising from a better use of each potential.

However, the matter is not limited to technological integration. The availability of powerful and reliable mechanisms is crucial for the management of big amounts of information received by the devices that are to be cleaned by errors and filtered according to precise rules. Last but not least, it becomes essential to make data available to company applications (e.g. CRM, ERP, Document Management, etc.) in a simple and reliable way (Clauberg, 2004) (Bruneo, Puliafito, Scarpa, & Zaia, 2005) (Wu, Wang, & Sheng, 2005), by favouring the integration towards the application layer.

What has been above described, allows us to say that: a middleware assures flexibility and scalability, favouring the integration processes, by reducing costs and improving the effectiveness of business strategies. Such need is so strong, that some studies introduce the concept of the index of customer satisfaction (Park, Heo, & Rim, 2008).

So the middleware can affect the system performance. For this reason it is necessary that the middleware architecture has to be able to assure requirements such as scalability and reliability and one possible solution could be the adoption of Cloud Computing principles in the middleware design. A possible evolution of our middleware is then based on the virtualization concept that allows the cloud to execute a software version of a hardware machine into a host system in an isolated way. In this way if the RFID middleware needs new resources, it is possible to ask the cloud for new physical resources and use them transparently.

THE EVENT-BASED RFID MIDDLEWARE

Almost all RFID middleware cover requirements such as scalability, reliability, data management and so on. Furthermore, functional requirements like dissemination support, data filtering and aggregation support, trigger RFID read (Amaral, Hessel, Bezerra, Correa, Longhi, & Dias, 2009) must be assured.

The provisioning of a powerful middleware layer is becoming an always increasing requirement. The number of heterogeneous devices that could be integrated increases day by day and their logic management constantly raises its complexity.

Generally speaking, two fundamental issues have to be solved: the continuous technological evolution on one hand and the integration with company IT systems on the other hand.

In order to address these two open issues, our middleware presents the following specific requirements. The main feature is to be "event based" (Kim, Lee, Lee, & Ryou, 2008) (Ghayal, Khan, & Moona, 2008). This means that it is able to assure an effective process of technological abstraction, switching from a vision linked to the specific issues of device interfacing (Tags/Readers RFID, Sensor Networks, GPS, WiFi, etc.) to the management of the event generated by each device. In brief, "event-based" means to integrate the management logic of different devices. The middleware has a component named "Event Manager" that includes a rule engine able to detect whether a specific situation occurs and respond appropriately. A very important innovation is the possibility for the user to submit new rules; the rule engine is able to interpret these rules and make them immediately operating with no need to recompile the software or reboot the system: they are immediately online. In this way, the system behavior can be adjusted according to the specific needs that may arise. Also, the user can customize the response of the system (e.g. specific data processing, alerts, forwarding of VoIP calls, etc.).

In addition, the events generated by the devices are filtered and cleaned. For instance, it is sometimes sufficient to detect only the time when a tag is detected while entering the range of an antenna, and the time when it is no longer detected, thus discarding all of the detections made between these two times, which would constantly provide the same information about the presence.

Finally, to give the middleware the necessary flexibility and an appropriate response to technological evolution, the core was developed with very flexible and scalable software architecture. In particular, all the components of the middleware were designed with SOA architecture solutions and XMPP communication model.

ARCHITECTURE DESCRIPTION

The features of our middleware allow to extend the interoperability concept and to promote new forms of collaboration. Furthermore, the functionalities ensure a careful management of issues such as context awareness, mobility management and seamless connectivity. Its final goal is to provide also a solution compliant to the use of web applications and able to provide multichannel communications.

Figure 2 shows the main modules of the middleware. The architecture has been created according to the SOA philosophy, paying attention to functional modularity to give the middleware high flexibility and scalability.

Figure 2. Event-based RFID middleware architecture

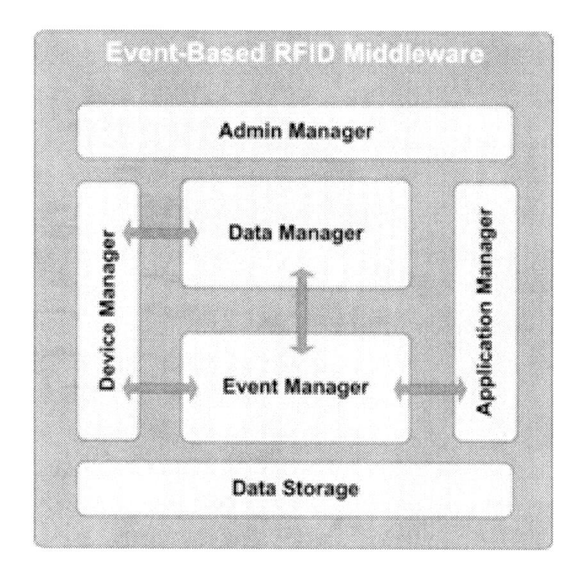

The main functional modules are the following:

- **Device Manager:** Manages the interoperability with physical devices and/or with other software.
- **Data Manager:** Responsible of the data filtering, to remove the data affected by errors, or to remove redundant data. Also, is able to aggregate information.
- **Event Manager:** Its task is to manage customizable events, generated by devices or software.
- **Application Manager:** Has a twofold task. The first one is to manage data exchange to and from the internal components of the middleware. The second one is to dispatch data to the application layer.
- **Data Storage:** The local repository and plays two important roles: storing the configuration parameters and keep the history of the detected data.
- **Admin Manager:** Provides the interactions tools for the middleware administration.

The whole architecture is designed with SOA paradigm permitting to obtain scalability, simplifying the maintenance of the system. Each module exposes its services that are well defined, self contained functions and does not depend on the context or state of other services. The interaction with external system and between internal modules is based on the Enterprise Service Bus (XMPP protocol or JMS) and implements the Multichannel Communication (Jabber, VoIP, SMS, MMS).

Device Manager

The role played by the Device Manager is very important. Its task is solving the interfacing issues with the physical devices from which to receive data. At the same time, it must provide a high level of abstraction of these devices.

In order to provide very flexible solutions, the device manager has two different components: the driver component and the hub one (Figure 3).

The Driver component provides all the basic functionalities to collect data directly from different physical devices directly connected through interfaces such as Ethernet, RS232, USB and Bluetooth. Each driver component is a thread that regularly and independently interrogates the physical devices and takes their data. Also, it can also send commands to the devices.

The Hub component is not too different than the driver one. The difference is that the HUB component implements "logical entities" which are *ports* that allow listening to the data coming from external sources. Thanks to this feature the external source can also be software that generates events. It is evident that the hub component is the one that mostly deals with simplifying integration aspects. For this reason, two design choices exist for its creation: Web Services and XMPP. For example, the hub component could be very useful during the inventory operations made with a PDA equipped with RFID reader. If the PDA operates without communication infrastructures,

Figure 3. Device manager structure

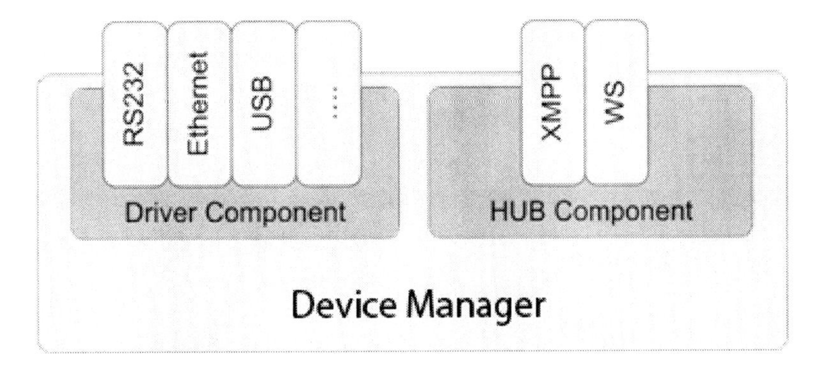

the hub can collect the data coming from the RFID reader in a second time, through a subsequent synchronization.

Data Manager

The Data Manager allows filtering data. In fact, data are generally characterized by large volume, temporality, spatiality, inaccuracy, interferences, and limited active lifespan.

Simplifying the integration process means improving data exchange, but also their quality. In brief, the data must be filtered to remove data affected by errors, to remove redundant information and adapt them for high-level applications.

The filters (Figure 4) are threads that constantly receive data from driver components or hub ones. Each filter has to analyze the acquired data, removing errors or redundant information and finally, sends them to the other modules of the middleware. For example, sometimes is sufficient to detect only the time instant when an event occurs for the first time and the time instant when it is no longer detected, thus discarding all the detections made between these two moments.

Finally, the filters can be applied both to each device and to logic groups of devices. Customized filters can also be developed by programmers with a very limited effort.

Event Manager

The Event Manager is the core of our middleware because it implements a real and efficient abstraction from the technological sublevel. This way, the aspects associated to the event management can be separated from the specific technical issues related to the physical devices that generate the events.

The strength of the component is the rule engine (Figure 5) that allows the user to customize the behaviour of the system and the reaction to the event. The user may submit new rules and the rule engine interprets them without necessity of recompiling or rebooting the system. The new rules are immediately operational.

This component assumes two different data structures to manage the events. The first one,

Figure 4. Data manager structure

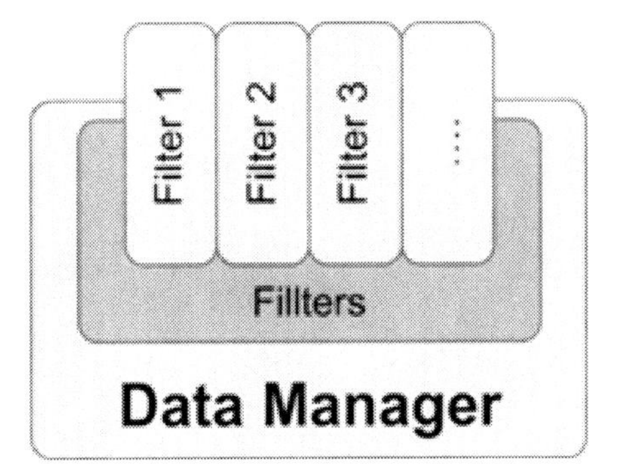

Figure 5. Event manager structure

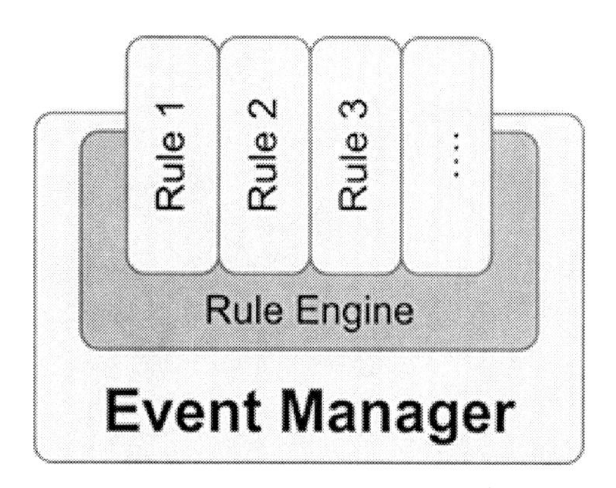

etv_in, manages the event generated by the devices (e.g. RFID Tags, Sensors, GPS, etc.). The second one, *evt_out*, is used to generate the personalized event and forwards data to the other functional components of the middleware (e.g. the Application Manager). Furthermore, the component can use two methods (SET and GET) to acquire/assign information such as *devicename*, *id*, *data*, etc., from/to the data structures it interact with. For example, if you have a surveillance system (Heeseo, Taek, & Hoh, 2006) and you want to track a particular area where a gate (named "Reader1") is installed and also you need to intercept a specific tagged object (tag with ID "123456"), you can use the rule provided in Box 1.

Filters and Rules can be combined. This feature enables the middleware to easily manage safety properties, and to deal with situations where an object must or must not be in a specific area, or a group of specific objects are in a specific area, etc.

The reaction of the system can be customized (e.g. data processing, alerts, SMS, VoIP, e-mails, etc.). Thus, the user can create the most suitable rules for his/her needs, without having to reboot the middleware, and facilitating the integration with legacy systems inside the enterprise. This characteristic makes the middleware more adaptable, and able to increase the effectiveness of the enterprise processes.

Application Manager

The effectiveness of a process of integration may be influenced by the middleware. One of the methods to simplify this process is to have an effectiveness solution to distribute information towards the IT application such as CRM, Document Management, ERP and so on.

The Application Manager is essentially a dispatcher that sends reconstructed information to the different applications. To easily exchange information, the Application Manager can use many protocols (Figure 6) such as HTTP, XMPP, JMS, etc. One of the most important features is the multichannel communication. Thanks this feature, the Application Manager is able to use all the protocols simultaneously thus providing a powerful communication solution.

The Multichannel Communication feature makes the middleware more flexible and scalable, also allows handling multiple contexts (and applications) at the same time.

In order to favour data exchange, the format of information has to be regulated. For this reason

Box 1.

```
if (evt_in.getID() == "123456" && evt_in.getDevicename() = "READER1")
{
evt_out.setID(evt_in.getID());
evt_out.setDevicename(evt_in.getDevicename());
evt_out.setData("OBJECT 12356 IDENTIFIED");
sendMessage();
}
```

Figure 6. Application manager structure

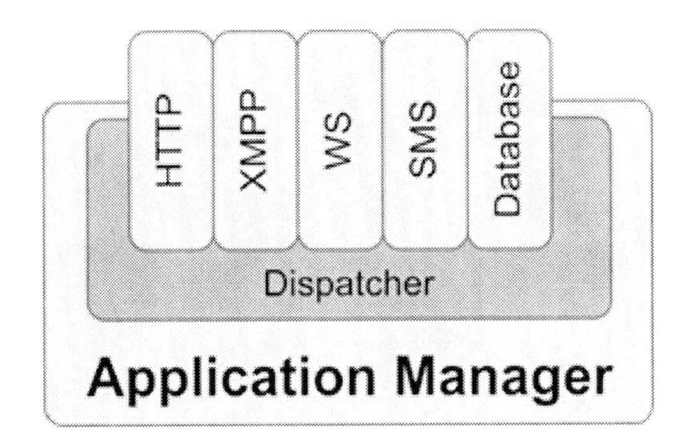

the information has to be reconstructed. This is the main type of reconstructed information:

- **Type:** Specifies the type of operation that has occurred. It depends on the specific technology, such as reading RFID tags, detecting the ZigBee sensor, reading of temperature sensors, etc.
- **Subtype:** Specifies the type of filter applied to the data. Here are some examples: element Entered or Exited from range, transit, operation of object count, etc.
- **Devicename:** Specifies the name of the device that has done the reading.
- **Sitename:** Specifies the name of the site where the sensor is located.
- **Devicegroupname:** Specifies the name of the group the device belongs to.
- **ID:** Indicates the ID code detected.
- **Time:** Date and time when the event has occurred.
- **Data:** Contains additional information detected by the sensor.

The XMPP protocol also provides a further possibility of data exchange. The middleware makes a JavaScript component available that uses the XMPP protocol. This component can be integrated into static HTML pages converting the client-server web solution into a Comet solution. Comet describes a web application model where a long held HTTP request allows a Web server to push data into a browser, without the browser explicitly requesting it. This component integrated into HTML pages, allows the creation of Web-oriented platform able to dynamically update their data. This component is able to respond to events and update the data almost in real time.

The following examples prove the flexibility of the Application Manager and how the integration process can be simplified. The first example, Figure 7 (A), shows the dispatcher configuration equipped with a centralized XMPP. This solution shows how the integration with an enterprise IT system can be simplified by using an external XMPP connector.

Conversely, the second example, Figure 7 (B), shows the configuration of a dispatcher equipped with a distributed XMPP. This solution outlines the high level of flexibility of the middleware that can manage simultaneous and distinct processes at the same time. This solution produces a considerable reduction of costs because the enterprise can improve several services at the same time, even if their types are different (e.g. document management, warehouse, etc.).

However, the dispatcher also provides for a third possible configuration shown in the Figure 7 (C): the DB Dispatcher. Similarly to the previous cases, the DB Dispatcher facilitates the integration processes, directly interacting with an external database. This is certainly a more invasive solution than the previous ones, but it represents another possibility of data exchange, sometime easily to implement. The DB Dispatcher integrates JDBC drivers for Oracle, MySQL, MS SQL Server, etc.

Data Storage

The Data Storage is the local repository whose task is to store the configuration parameters of the system. But another task is keeping the history of the data detected. This aspect is useful if statistical analysis must be done on data even at a later

Figure 7. Examples of application manager configurations

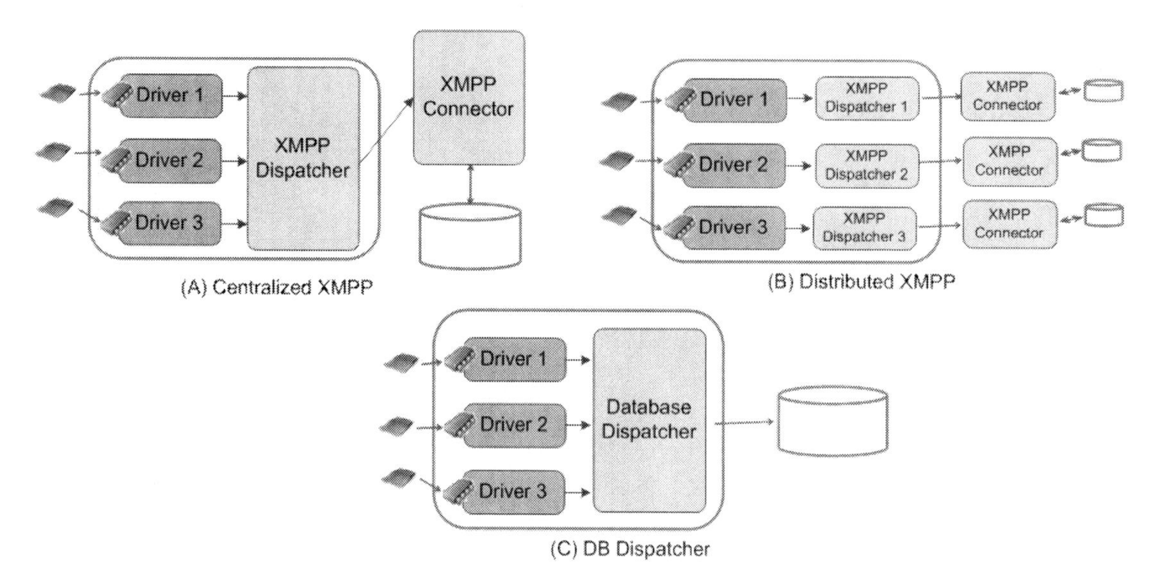

time, for instance to keep track of the movements made by the monitored objects.

However, the local repository can also support processes of data processing that are the basis of business intelligence. The data collected are appropriately processed, and the results can be used to support the company's choices.

Admin Manager

The Admin Manager module provides the tool through which the administrator can configure the middleware and extract reports. Also, through the user-friendly web interface, the administrator can add and remove devices, filters, rules and active dispatchers to exchange data to outside IT applications. It is important that all these functionalities can be executed even from remote stations (from intranet or internet) with no need to be in the place where the middleware was physically installed.

It is also available a graphic module that enables a virtual reconstruction of the context: the administrator can load a map and arrange the objects-devices managed by the middleware, into the map. This virtual representation enables to locate and trace objects more easily, and to select areas where some rules can be fixed (Figure 8).

In summary, the functionalities of the Admin Manager module allow you to configure the middleware in any new possible scenario. In fact, the middleware is able to manage different problems at the same time simply by configuring new devices, filters, rules and other necessary components.

The WhereX® Middleware

Our work resulted in the creation of a commercially available product, known as the WhereX© middleware. This product was realized in the Wireless RFID Laboratory of University of Messina and commercialized by Inquadro s.r.l., a spin-off company of the University of Messina.

As we have outlined several times, the main purposes that must be achieved by the middleware are flexibility, scalability, integration, and reduction of costs. All this turns out into an improvement in the effectiveness of business processes. The description of some real cases (WhereDoc, WhereObj, WhereArt and WhereSafe) will prove

Figure 8. Graphic tool

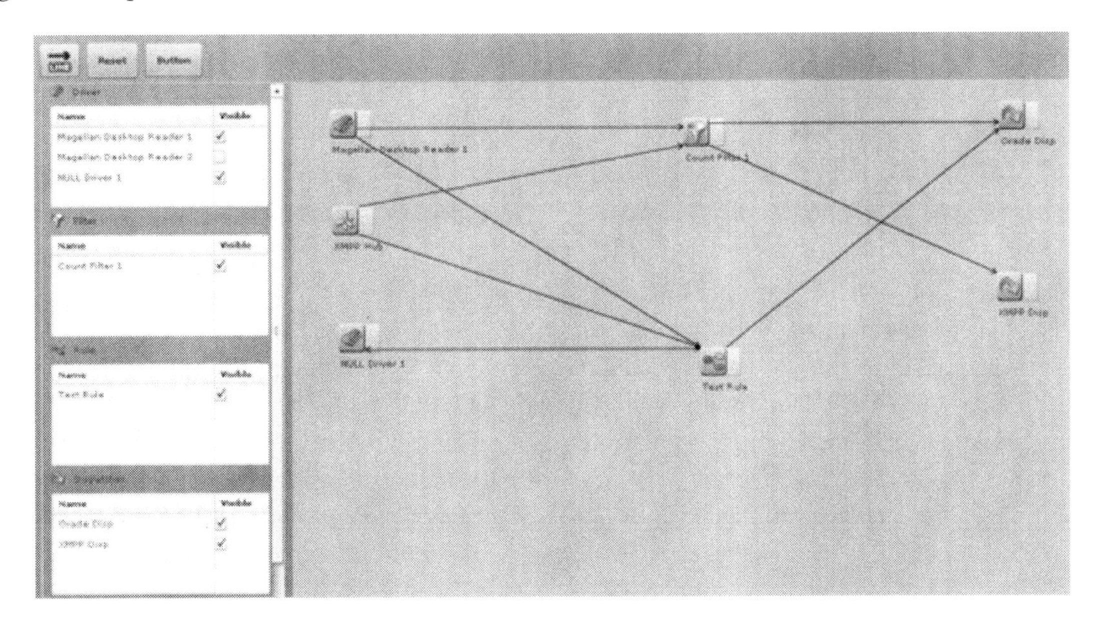

how different issues can be managed – even simultaneously – with the same middleware, providing companies with real and competitive benefits. Such applications are clear examples of the great potentialities of the WhereX® middleware, that allows the programmer to focus on the specific problem to be solved, taking care of all the aspects related to the integration with legacy systems and enterprise systems in general, providing very powerful features to manage user and components mobility, thus reducing the time to market of the final solution, as well as its quality and performance.

WhereDoc® Solution

WhereDoc starts a new way for the workflow management of paper documents, which allows you to identify and keep track of each document equipped with a RFID tag (also chipless tags). Even if the recent regulations are going towards a digital management of documents, the need to manage paper documents is still fundamental in many contexts, and will certainly remain fundamental for many years.

WhereDoc originates from these thoughts. It is an application for the identification, tracing and management of paper archives. This developed solution is going to use a mixed strategy, by combining the "virtual life" of the electronic document with its real life, through the use of the RFID technology.

The system allows mapping out office structures (desks, cabinets, shelves, etc.) on the related digital correspondent through RFID radio-frequency sensors and readers. Once the mapping is done, the system enables to define a hierarchy of people allowed to access document management at various levels of permits, visibility and possibility. The system architecture envisages a protocol office dealing with the acquisition of the document, recording it on the digital system together with its related general information (sender, recipient, date and time of arrival, etc.), providing it with an RFID label and dispatching it to the proper office or area. After the protocol phase, the paper document reaches the proper office or area. At this stage, the document is classified more specifically, and the area manager defines and assigns a specific workflow to the file. The person responsible for the document is also designated and is duly informed through the system's internal message service of all movements during the file's itinerary (Figure 9).

According to the workflow information input (itinerary, priorities, etc.), the system provides services such as traceability up to desk or shelf level, viewing of document processing status, integrated messaging system for users involved in file management, multichannel information (email, messages, sms, voice) of events such as error or anomaly during the itinerary or unauthorized removal and management of processing and file priority schedules.

The final step concluding the document life cycle is its filing. The document management system rapidly identifies filed documents, also by using maps and allows managing unauthorized removals from or wrong locations of documents on shelves and in cabinets.

On the whole, what is especially important is the monitoring function, aimed both at retrieving the document in real time and – above all – at analysing the information flow process. By studying the real itinerary of each file, processing times can be calculated and process bottlenecks identified, which, if resolved, simplify the work and speed up administration time in every office. Analysis functions show how correct the information flow is and facilitate an on-the-spot analysis of personnel equipment and other resources.

The WhereDoc uses the WhereX middleware previously described to interface with RFID devices, and allows using different equipments easily. In fact, the middleware allow creating and changing customised environments. For example, such environments will use PDAs equipped with mobile readers, document trays, cabinets, barriers, etc. The middleware is in charge of the abstraction issues regarding the interfacing with fixed devices, and the synchronisation with mobile devices. The use of customised rules allows setting document workflows in the offices, and the use of filters allows you to create alert mechanisms

Figure 9. User interface: real-time monitor of documents

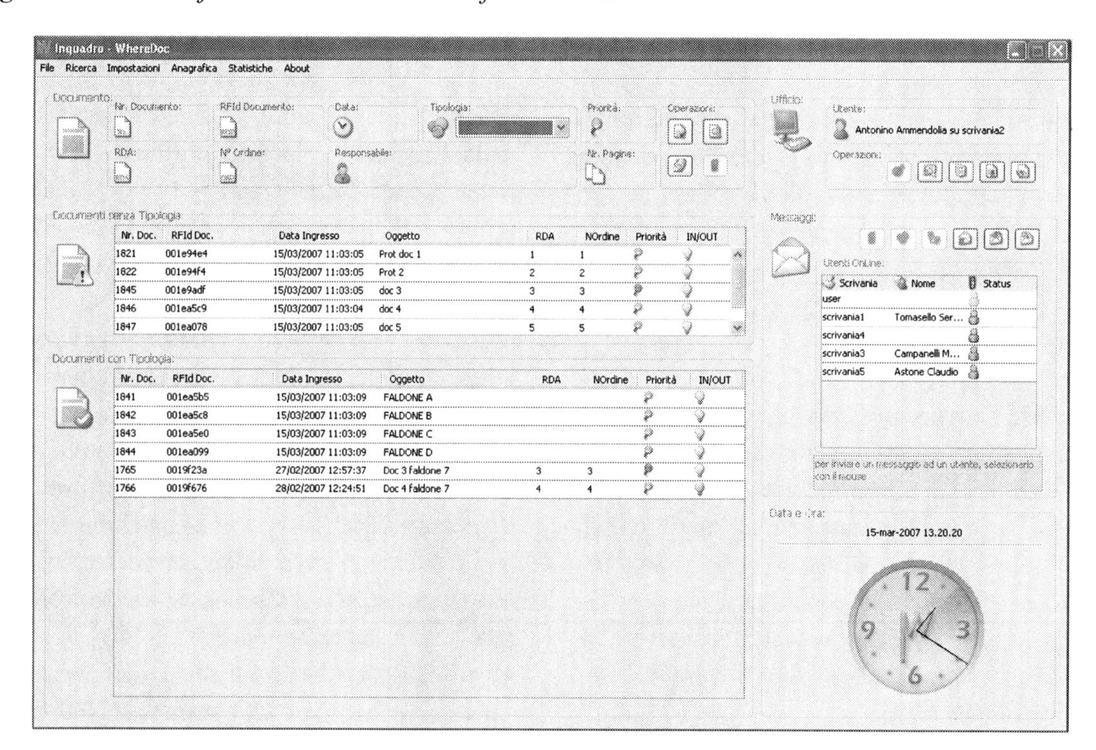

if the pre-set workflows are breached. Finally, the components such as Device Manager and Application Manager are widely used to facilitate the integration process to enterprise IT systems. The use of XMPP mechanisms allows to connect remote devices, and to overcome the barriers of enterprise firewalls effectively and safely.

The WhereObj® Solution

WhereObj is an application for asset management that makes extensive use of RFID technology and sensor networks. Special attention is paid to the usability of the system and to the possibility of integrating it with the procedures and the systems used by the customer. This application is used in wide and dynamic realities, where the management of assets is a complex work, with long processes and management times. If this management is made partly automatic, this can bring to better accuracy of data, as well as to cost saving.

Inventory operations are made by an operator who is equipped with a RFID reader, and regularly checks the buildings, detecting the tags placed on each asset. Thanks to the RFID technology, WhereObj allows you to:

- Locate the tools, and to have a graphical map containing the information about the presence of each asset in each room.
- Analyze the movements of the assets, keeping track of the previous locations of each asset.
- Optimize the resources, by making statistics about the assets that have been subject to maintenance more than other assets. In fact, the integration of WhereObj® with the most common systems for the management of technical actions on IT assets, allows to integrate and to manage the information about the identity and the location of the asset with a single user interface, together with the information about the life cycle of the asset.
- Automate the registration of events concerning the management of assets, which are generally registered manually.
- Increase safety, by introducing a tool that reduces the risk of theft. Placing a RFID barrier consisting of two antennas on each entrance/exit turnstile in the buildings involved can make the inventory operations more complete, by collecting data about the presence/absence of assets in the building, and detecting any theft immediately.

This application is used in wide and dynamic realities, where the management of assets is a complex work, with long processes and management times. If this management is automated, this can bring to better accuracy of data, as well as to cost saving (Figure 10).

WhereObj application uses the WhereX middleware to interface transparently with several devices such as RFID readers, GPS and Wireless Sensors. Also in this case, by using WhereX drivers, mixed configurations can be created, where a whole production chain using different technologies can be traced. WhereObj uses the functions of WhereX drivers for the synchronisation with PDAs in inventory operations, and uses the functions of rules to create workflows about objects and goods.

The Hub and Dispatcher components, and XMPP in particular, are widely used to integrate WhereObj with enterprise information systems such as Asset Management Systems, ERP, Access Control, and Warehouse Management. The Hub component allows you to manage objects geographically spread on large areas (big enterprises eventually distributed on several distinct sites).

The WhereArt® Solution

The exploitation of natural, historical and artistic sites and the preservation of cultural heritage is an important indicator of economic and cultural development. WhereArt® intends to provide a

Figure 10. WhereObj operating mode

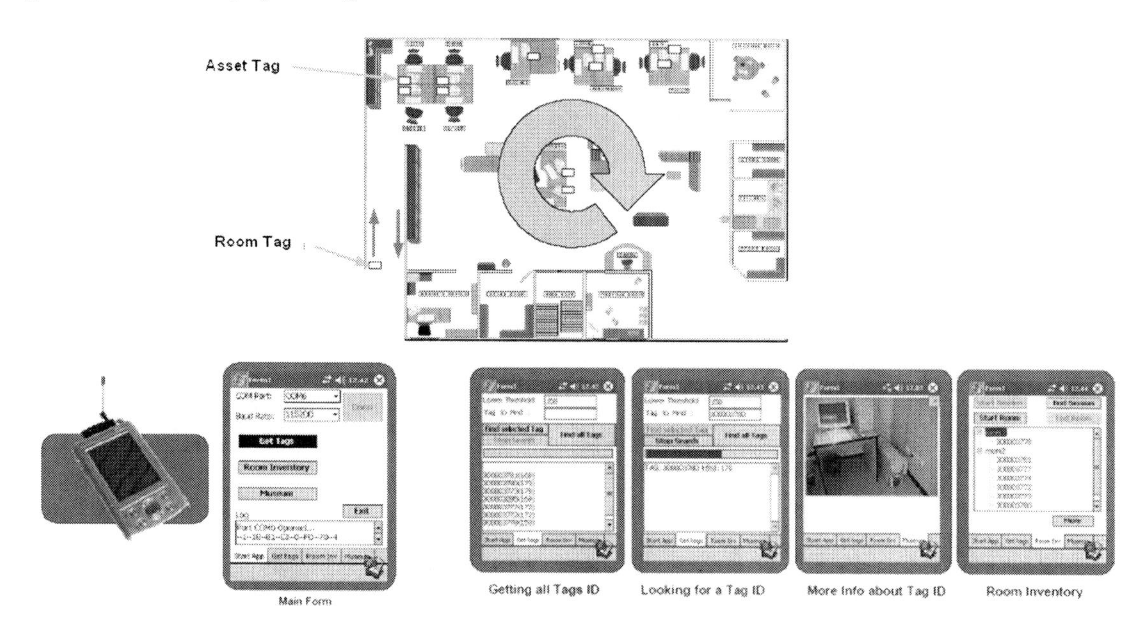

valuable solution to enrich the cultural offering of public and private realities (museums, parks, cultural events) introducing a new way to access and distribute information to the users.

The proposed solution is an audio-video virtual guide executed on handheld devices or tablet PCs to be provided to visitors of a cultural event in general, able to suggest personalized pathways according to age, preferences or cultural level, and to dynamically adapt the information on the basis of the position of the user. WhereArt expoits the services offered by the middleware WhereX and makes wide use of RFID and GPS technologies.

The main features of the developed solution are:

- Multimedia content accessed via handheld computers (PDA) or tablet PC.
- Personalized user profiles (adult, children, students, archaeologist, etc.) which allow information filtering and customization.
- Generation and combination of multimedia contents through a web-based back-end panel.
- Automatic conversion of text into synthesized audio.
- Location of users nearby a point of interest and automatic delivery of multimedia contents.

Figure 11 shows a diagram of the system, highlighting the main modules and their functionalities.

The WhereSafe® Solution

Building and industrial construction is one of the productive sectors with the greatest risk for health and safety of workers. Constructions, in fact, significantly contribute to injury complains, both for the high frequency with which accidents occur and for the even higher severity of injury they produce.

In order to improve the effectiveness of the prevention measures and health protection of workers, we developed the WhereSafe solution that, through the use of sensors to collect data to be managed by WhereX, allows you to monitor the safety devices traditionally used for workers protection.

The monitoring system receives information from sensors applied to helmet, shoes, etc., pro-

Figure 11. WhereArt operating mode

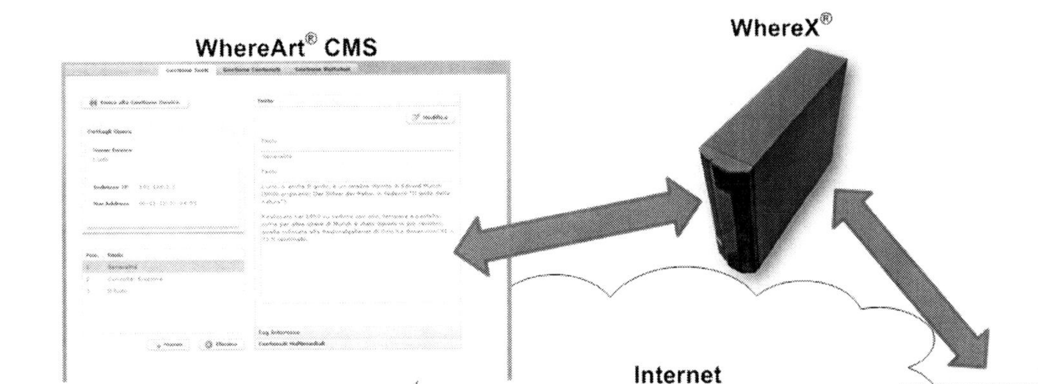

cesses them and reports any dangerous situations to the responsible of safety management. The system is organized as follows:

- **Sensor Network:** It makes use of ZigBee sensors organized in the form of a mesh network. The end-nodes are associated to different objects such as helmets, shoes, etc., and transmit data on the proper usage of the specific safety equipment to a router-node or the coordinator-node. Using the mesh network, the coordinator periodically transmits such data to the middleware WhereX for further processing.
- **Middleware WhereX:** It implements the integration between the hardware and the application layers. Through the interpreter of rules, it allows to customize the system behavior and adapt the monitoring logic to several management strategies. Moreover, WhereX transmits data to the applications layer.
- **Application Layer:** It implements the front end through which it is possible to monitor the construction plant. In particular, alarms are activated in relation to the occurrence of specific events signaled by the middleware to the application layer (not properly dressed safety clothes, worker entering a forbidden area, etc).

In Figure 12 the Sensor nodes (generator nodes) are associated with safety equipment of workers. The routing-protocol ensures forwarding alarm data (i.e. "Helmet is not worn") toward sink nodes that, finally, send them to the application layer to manage critical situation that could happen in the working-area.

Figure 12. WhereSafe operating mode

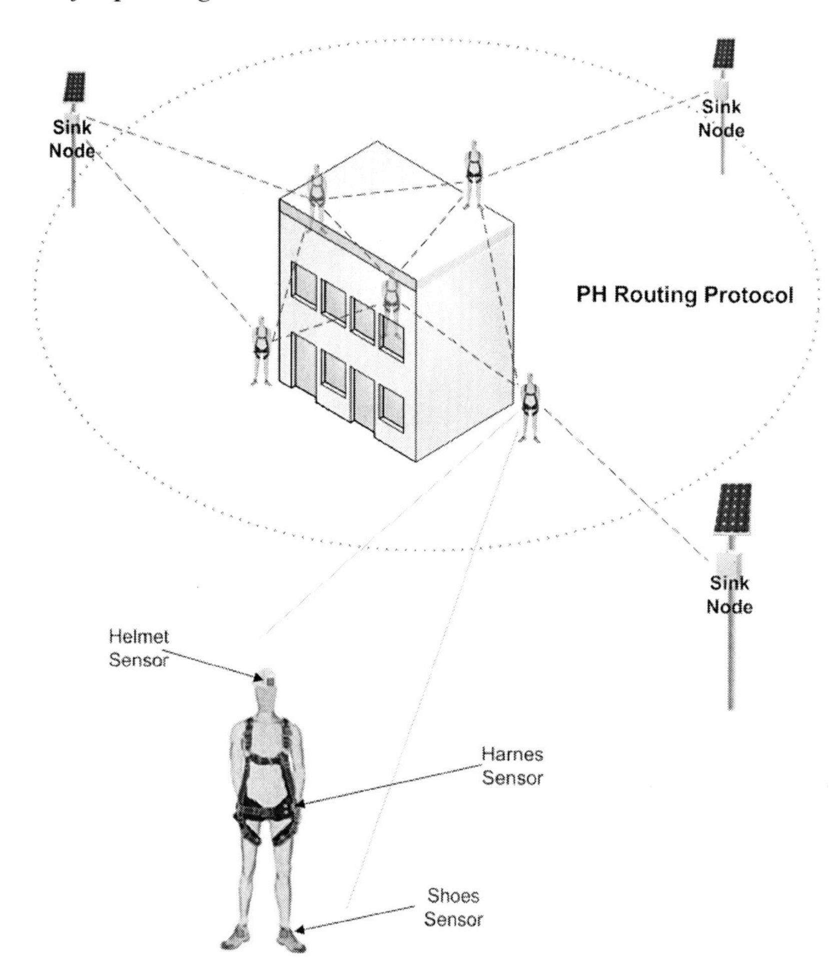

FUTURE RESEARCH DIRECTIONS

The new paradigm of Internet of Things represents a very important evolution as simple physical objects become smart objects able to automatically distribute behavioural information enriched with other data such as location, surrounding environment, etc. As a result the IT system has to process a huge amount of data generated by objects (RFID, Sensors, GPSs, etc.) and that can cause different problems: overloading, data errors, low performance of the system, problem of scalability, low reliability and so on. The middleware represents the crucial element and it is necessary that its architecture has to be able to meet needs like high scalability, high reliability, high performance.

To obtain these features the future research is oriented towards the Cloud Computing and in particular we are working on a new Cloud-based RFID Middleware that allows solving two fundamental and critical problems. The first one is the scalability property that measures how well the middleware can maintain its performance under increasing load. The second one is fail-over capacity in the event of the server failure. Successful fail-over makes outages transparent to the users, while maintaining their state within the application.

Cloud computing is generally considered as one of the more interesting topic in the IT world. There are many definitions of Cloud but in general it could be considered like a paradigm that brings together various distributed technologies for accessing hardware resources such as CPUs and storage, and logical resources such as the software. It is an extremely complex infrastructure that includes aspects of GRID Computing (Foster, 2002) (reliable architecture, consistent, pervasive and inexpensive access to high-end services), Utility Computing (Ross, & Westerman, 2004), Autonomic Computing (Kephart, & Chess, 2003) and Green Computing (Murugesan, 2008). Companies like Amazon (Amazon Elastic Compute Cloud (Amazon EC2): http://aws.amazon.com/ec2.2011), IBM, Sun Microsystems, Microsoft (Windows Azure: http://www.microsoft.com/azure/default.mspx.2011), etc. have expressed great interest and also numerous scientific and academic communities have developed projects such as Reservoir (Resources and Services Virtualization without Barriers: http://www.reservoir-fp7.eu/.2007) or OpenNEbula (The Open Source Toolkit for Cloud Computing: http://www.opennebula.org/.2002). Service-centric perspective is the main issue of the Cloud computing paradigm: all capabilities and resources also geographically distributed are provided, by internet, to the user as a service. To use the resources is not necessary to have knowledge about the technological infrastructure. In order to achieve such goals, a Cloud implements a level of abstraction of physical resources (computational, storage, devices, etc.), standardizes their interfaces and provides mechanism for managing them, adaptively to user requirements. All these are done through virtualization, service mashup and service oriented architecture.

In general, clustering addresses scalability by distributing the load among several physical servers or virtual servers on a Cloud Computing infrastructure.

Figure 13 shows two main components of the cloud-enabled middleware solution: the physical servers and the virtual ones. Each server hosts

Figure 13. Cloud-based RFID middleware

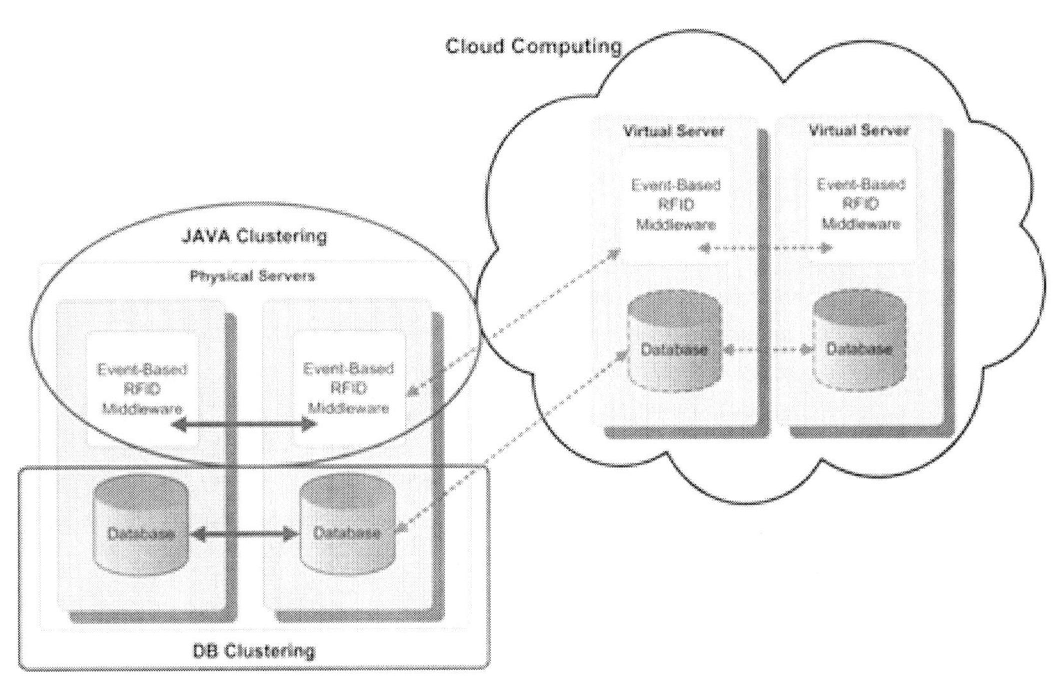

a specific instance of the Event-Based RFID Middleware. Thanks to virtualization, each virtual server has the same software configuration of the physical node. If the workload increases, the physical cluster asks news resources to the Cloud that can allocate one or more nodes.

Theoretically perfect scalability is linear. As new servers are added to a cluster, each server adds support for a constant number of users. For example, if one server can support 100 users, then two servers can support 200 users and so on.

In this specific situation, each server hosts the Middleware. When a new server is added (physical or virtual server) to the cluster, it results in adding extra capacity to handle an increased number of RFID readers, Sensors, GPSs, etc. In order to properly identify the operating limits of each single node, it is mandatory to count on a very flexible configuration and, at the same time, on a very robust solution.

The second issue of the cluster solution is the robustness of the middleware. If only one single machine is available to host the RFID middleware, all the services are concentrated in a unique possible point of failure. The cluster implements the strategy for replicating a user's state to a single or several secondary servers and then, if the first server goes down, to redirect all the subsequent requests to the secondary server(s).

CONCLUSION

This chapter introduced the concept of middleware and the advantages derived from its adoption in an RFID environment. Having in mind the plethora of different RFID technologies (considering tags with and without on-board chip), the presence of a reliable and powerful middleware layer allows you to quickly develop RFID based solutions, without being an expert of technological details and solutions. Details on the organization of such middleware layer were provided and specifically the WhereX solution was presented. An application related to the adoption of chipless tags was also described, which provides an interesting example of the adoption of such technology in the field of document workflow management. Future directions related to the virtualization of the middleware layer and its execution in a cloud-based infrastructure were also presented.

ACKNOWLEDGMENT

The authors would like to thank Inquadro srl, and specifically Dr. A. Zaia, A. Ammendolia and S. Tomasello, for the fruitful discussions and support in preparing this chapter.

This research work was partially supported by POR Sicilia 2007-2013 measure 4.1.1.1 under grant number 179.

REFERENCES

Aberer, K., Hauswirth, M., & Salehi, A. (2006). A middleware for fast and flexible sensor network deployment. *VLDB*, *06*(September), 1215.

Amaral, L. A., Hessel, F. P., Bezerra, E. A., Correa, J. C., Longhi, O. B., & Dias, T. F. O. (2009). An adaptative framework architecture for RFID applications. 33rd Annual IEEE Software Engineering Workshop (SEW), 13-14 October 2009, (pp. 15-24).

Amazon. (2011). *Elastic compute cloud.* Retrieved from http://aws.amazon.com/ec2

Bruneo, D., Puliafito, A., Scarpa, M., & Zaia, A. (2005). *Mobile middleware in enterprise systems.*

Clauberg, R. (2004). *RFID and sensor networks.* RFID Workshop, University of St. Gallen, Switzerland, Sept. 27, 2004.

FP7. (2007). *Resources and services virtualization without barriers*. Retrieved from http://www.reservoir-fp7.eu/

Foster, I. (2002). What is the grid? - A three point checklist. *GRID Today, 1*(6).

Foster, I. (2005). Service-oriented science. *Science*, *308*(5723), 814–817. doi:10.1126/science.1110411

Ghayal, A., Khan, M. Z., & Moona, R. (2008). *SmartRF: A flexible and light-weight RFID middleware*. IEEE International Conference on e-Business Engineering. 978-0-7695-3395-7/08

Heeseo, C., Taek, L., & Hoh, P. (2006). Situation aware RFID system: Evaluating abnormal behavior detecting approach. *Proceeding of the Fourth Workshop on Software Technologies for Future Embedded and Ubiquitous Systems and Second International Workshop on Collaborative Computing, Integration, and Assurance* (SEUS-WCCIA'06), 2006 IEEE.

Kanoc, T. (1999). *Mobile middleware: The next frontier in enterprise application integration*. White paper Nettech Systems Inc.

Kephart, J. O., & Chess, D. M. (2003). The vision of autonomic computing. *Computer*, *36*(1), 41–50. doi:10.1109/MC.2003.1160055

Kim, M., Lee, J. W., Lee, Y. J., & Ryou, J. (2008). COSMOS: A middleware for integrated data processing over heterogeneous sensor networks. *ETRI Journal, 30*(5).

Lòpez, T. S., & Kim, D. (2007). A context middleware based on sensor and RFID information. *Proceedings of the Fifth Annual IEEE International Conference on Pervasive Computing and Communications Workshops* (PerComW'07) IEEE.

MacKenzie, C. M., Laskey, K., McCabe, F., Brown, P. F., Metz, R., & Hamilton, B. A. (2006). *Reference model for service oriented architecture 1.0*. OASIS SOA Reference Model Technical Committee, 2006.

Murugesan, S. (2008). Harnessing green it: Principles and practices. *IT Professional*, *10*(1), 24–33. doi:10.1109/MITP.2008.10

OpenNebula. (2002). *The open source toolkit for cloud computing*. Retrieved from http://www.opennebula.org/

Park, Y., Heo, P., & Rim, M. (2008). Measurement of a customer satisfaction index for improvement of mobile RFID services in Korea. *ETRI Journal, 30*(5).

Pautasso, C., Zimmermann, O., & Leymann, F. (2008). RESTful web services vs. big web services: Making the right architectural decision. *Proceedings of the 17th International World Wide Web Conference* (WWW2008) Beijing, China.

Pope, A. (1998). *The Corba reference guide: Understanding the common object request broker architecture*. Addison-Wesley, 1998.

Preradovic, S., & Karmakar, N. C. (2010). Chipless RFID: Bar code of the future. *Microwave Magazine*, *11*(7), 87–97. doi:10.1109/MMM.2010.938571

Rhino. (2011). *JavaScript for Java*. Retrieved from http://www.mozilla.org/rhino/

Ross, J. W., & Westerman, G. (2004). Preparing for utility computing: The role of it architecture and relationship management. *IBM Systems Journal*, *43*(1), 5–19. doi:10.1147/sj.431.0005

Weijie, C., & Weiping, L. (2008). *Study of integrating RFID middleware with enterprise applications based on SOA*. IEEE.

Windows Azure. (2011). *Website*. Retrieved from http://www.microsoft.com/azure/default.mspx

Wu, J., Wang, D., & Sheng, H. (2005). *Design an OSGi extension service for mobile RFID applications*. IEEE International Conference on e-Business Engineering 2007.

ADDITIONAL READING

Bertolini, M., Bottani, E., Rizzi, A., & Volpi, A. (2009). *The benefits of RFID and EPC in the supply chain: Lessons from an Italian pilot study*. The Internet of Things: 20th Tyrrhenian Workshop on Digital Communications.

Giusto, D., Iera, A., Morabito, G., Atzori, L., Pirri, F., & Pettenati, M. C. … Ciofi, L. (2009). *InterDataNet: A scalable middleware infrastructure for smart data integration*. TIWDC 2009 20th Tyrrhenian Workshop on Digital Communications "The Internet of Things", Pula, Sardinia, Italy, September 2-4, 2009

Guojin, Z., Kewen, Y., Xiaoli, Z., & Dong, W. (2010). Complex event process infrastructure in mobile distributed system. *2010 IEEE International Conference on RFID-Technology and Applications* (RFID-TA), (pp. 57-60).

Koswatta, R., & Karmakar, N. C. (2010). Development of digital control section of RFID reader for multi-bit chipless RFID tag reading. *2010 International Conference on Electrical and Computer Engineering* (ICECE), (pp. 554-557).

Li, Y., Rongwei, Z., Staiculescu, D., Wong, C. P., & Tentzeris, M. M. (2009). A novel conformal RFID-enabled module utilizing inkjet-printed antennas and carbon nanotubes for gas-detection applications. *Antennas and Wireless Propagation Letters*, *8*, 653–656. doi:10.1109/LAWP.2009.2024104

Parlanti, D., Paganelli, F., & Giuli, D. (2009). A scalable grid and service-oriented middleware for distributed heterogeneous data and system integration in context-awareness-oriented domains. *Proceeding of 2009 Tyrrhenian International Workshop on The Internet of Things.*

Preradovic, S., Balbin, I., Karmakar, N. C., & Swiegers, G. F. (2009). Multiresonator-based chipless RFID system for low-cost item tracking. *IEEE Transactions on Microwave Theory and Techniques*, *57*(5), 1411–1419. doi:10.1109/TMTT.2009.2017323

Preradovic, S., Karmakar, N., & Zenere, M. (2010). UWB chipless tag RFID reader design. 2010 IEEE International Conference on RFID-Technology and Applications (RFID-TA), (pp. 257-262).

Rosales, P., Kyuhyup, O., Kyuri, K., & Jae-Yoon, J. (2010). Leveraging business process management through complex event processing for RFID and sensor networks. 2010 40th International Conference on Computers and Industrial Engineering (CIE), (pp. 1-6).

Shrestha, S., Balachandran, M., Agarwal, M., Phoha, V. V., & Varahramyan, K. (2009). A chipless RFID sensor system for cyber centric monitoring applications. *IEEE Transactions on Microwave Theory and Techniques*, *57*(5), 1303–1309. doi:10.1109/TMTT.2009.2017298

Tedjini, S., Perret, E., Deepu, V., & Bernier, M. (2009). Chipless, the next RFID frontier. *Proceeding of 2009 Tyrrhenian International Workshop on The Internet of Things.*

KEY TERMS AND DEFINITIONS

Chipless RFID Tag: Encodes data into the spectral signature in both magnitude and phase of the spectrum. It represents a new frontier of RFID technology and at the same time an opportunity that could open a new scenarios as it could be printed directly on the products and packaging.

Cloud Computing: Could be considered like a paradigm that brings together various distributed technologies for accessing hardware resources such as CPUs and storage, and logical resources such as the software.

Internet of Things: A new paradigm which makes the thing's information be shared on a global scale. It represents a very important evolution as simple physical objects become smart objects able to automatically distribute behavioural information enriched with others data such as location, surrounding environment, etc.

Middleware: Represents the solution that favors the technology integration, reducing costs for application development and introducing real benefits to the business processes.

Radio Frequency Identification (RFID): An automatic identification and data capture technology.

Service Oriented Architecture (SOA): An architectural pattern providing agility to align technical solutions to modular business services that are decoupled from service consumers.

XMPP Protocol: Provides a push based notification functionality that is appropriate for the event-driven paradigm. Also it provides the authentication and encryption required for data protection.

Section 3
Anti-Collision Protocol

Chapter 7
RFID Tag Anti-Collision Protocols

Ching-Nung Yang
National Dong Hwa University, Taiwan

Jyun-Yan He
National Dong Hwa University, Taiwan

Yu-Ching Kun
National Dong Hwa University, Taiwan

ABSTRACT

A tag collision problem (or missed reads) in Radio Frequency Identification (RFID) system happens when multiple tags respond to a reader simultaneously. At this time, the reader cannot differentiate these tags correctly. This problem is often seen whenever a large volume of RFID tags are read together in the same radio frequency field. Tag collisions will degrade identification efficiency, and this unreliable identification will compromise the usefulness of RFID system. This chapter introduces tag collision problem and discusses tag anti-collision protocols, including ALOHA-based protocol, Binary Tree (BT) protocol, and Query Tree (QT) protocol. To date, most tag anti-collision protocols are QT protocols. Thus, in this chapter, the authors briefly describe some elegant researches on QT protocols, and also introduce their recent research results on QT protocols.

1. INTRODUCTION

Radio frequency identification (RFID) system consists of radio tags, readers, and the backend database system that associates RFID tag data collected by readers for verification or special need. A Reader communicates with tags at distance through wireless transmission. Its main tasks are to activate tags, communicate tags, and identify the tags uniquely. After the successful identification, a reader then sends the collected data from tags to a data processing system, which can be an application or database, according to the special application need (Shih, Sun, Yen, & Huang, 2006). Many manufacturers see the intended applications of RFID technology, and deploy RFID in

DOI: 10.4018/978-1-4666-2080-3.ch007

inventory control, distribution industry and supply chain management. As far, there are already many concrete applications using RFID, e.g., inventory control, distribution industry and supply chain management (Bardaki, Karagiannaki, & Pramatari, 2008; Bo, Yehua, & Caijiang, 2008; Chalasani & Boppana, 2007; Chuang & Shaw, 2007; Jea & Wang, 2008; Kapoor, Li & Ding, 2007; Liang & Li, 2007; Kapoor, Wei, & Piramuthu, 2008; Yan, Chen, & Meng, 2008). However, there are two types of collision problems in RFID system, the tag collision (see Figure 1(a)) and the reader collision (see Figure 1(b)), which will compromise the usefulness of RFID system. Tag collision happens when multiple tags respond to a reader simultaneously and the reader cannot differentiate these tags correctly (Floerkemeier & Lampe, 2004). On the other hand, when multiple readers are operating in proximity of one another, several readers interrogate one tag at the same time. This is so-called the reader collision (Sarma, Weis, & Engels, 2003). In this chapter, we discuss the tag collision problem in RFID system.

Several technologies on resolving tag collision had been proposed. There are two major types of anti-collision protocols. One is ALOHA-based protocol (Lee, Joo, & Lee, 2005) and the other is tree-based protocol. ALOHA-based protocol reduces the tag collisions, while it has the starvation problem (a tag cannot be identified for a long time). To address such problem, two tree-based protocols- binary tree (BT) protocol and the query tree (QT) protocol- were accordingly proposed. In the BT protocol (Cui, & Zhao, 2008; Feng, Li, Guo, & Ding, 2006; Lai, & Lin, 2009), every tag generates a binary random number (0 or 1). Afterwards, the tags having "0" transmit their EPCs to the reader first, and then the tags having "1" transmit later. This procedure is repeated until all tags are successfully identified. QT protocol does not need the additional memory and thus is referred to as the memoryless protocol. A reader sends a prefix of EPC to query tags, and the tags matching the prefix respond. By extending the prefixes until only on tag's ID matches, the tag is identified successfully.

Most researches on RFID anti-collision protocol are QT-based protocols. Some QT-like protocols are briefly described in the following. Chiang et al. (2006) proposed a prefix-randomized QT protocol. A reader first scans the neighboring tags to determine the *M*-ary tree for querying tags. After finishing queries by *M*-ary tree, a reader then uses binary tree for inquiries. Adaptive query splitting protocol (Myung, Lee, Srivastava, & Shih, 2007) used extra candidate queue to store the prefix bits of responded tags to speed up the identification process. A hybrid QT (Ryu, Lee, Seok, Kwon,

Figure 1. Collision problems in RFID system: (a) tag collision (b) reader collision

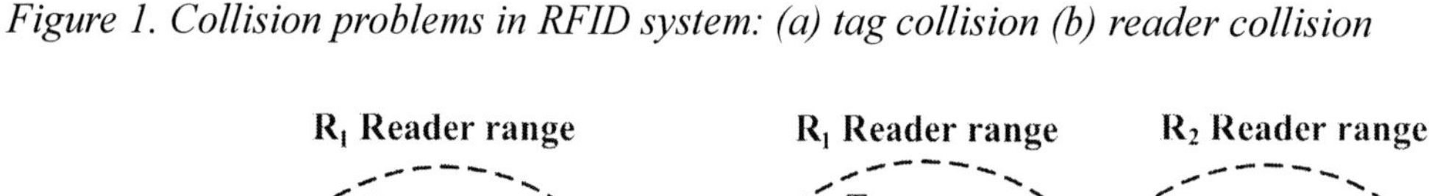

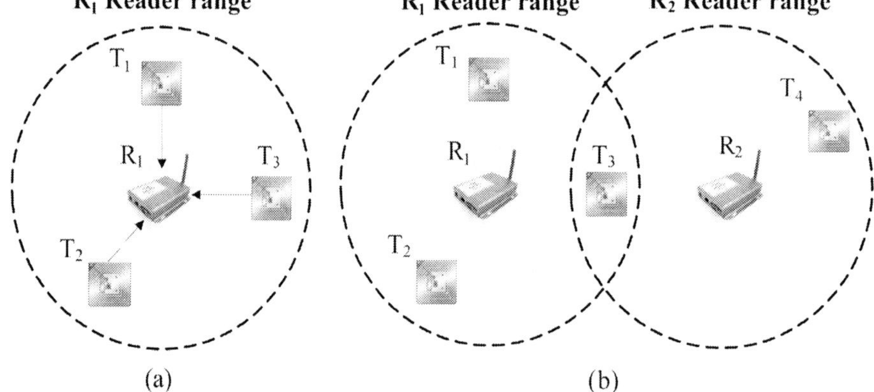

& Chio, 2007) adopted 4-ary QT and the slotted back-off tag response mechanism to avoid collision. Recently, Choi, Lee, and Lee (2007) proposed a 16-bit random number QT protocol for RFID tag anti-collision by using a RN16 as the tags temporary ID. Finally, this protocol successfully reduces the time consumption for tag identification than the present identification implemented in EPC Class 1 Gen. 2 (EPC-C1G2). In an enhanced QT (Hsu, Yu, Huang, & Ha, 2008), the length of prefix code is adjusted dynamically according to the length of tag's ID. Cho, Shin, and Kim (2008) addressed the problem of identifying tags with the consecutive ID number. They used the reversed ID to avoid the consecutive serial number. For example, the books from a same publisher may have the same company ID and the consecutive serial numbers. In (Pupunwiwat & Stantic, 2009), two *M*-ary trees were combined as the unified QT to enhance the identification efficiency. In (Yang, Kun, Chiu, & Chu, 2010a), the authors proposed adaptive QT protocol combining multiple *M*-ary trees, which successfully reduce the collisions and the idle time simultaneously. Collision tree (Jia, Feng, & Ma, 2010) protocol is based on QT and Manchester code. It uses the characteristic of Manchester code to find the location of different bit in responded strings. Thus, the reader can skip the unnecessary queries, and avoid idle time and collisions. A ternary tree is impossible in QT protocol, but it brings about the optimum performance for tag identification. Therefore, a practical ternary QT protocol was proposed for identifying a binary EPC to achieve the better identification performance (Yang, Kun, He, & Wu, 2010b). An enhanced protocol on Choi et al.'s 16-bit random number QT protocol (Choi et al., 2007) was proposed in (Yang & He, 2011) to solve the tag collision caused by the short length of RN16.

All the above QT-like protocols solved some specific problems in tag anti-collision, or enhanced other protocols. They considerably mitigate the tag collision in tags identification, and achieve the reliable and efficient identification. In this chapter, we will briefly describe ALOHA-based protocol, BT-based protocol, and QT-based protocol. Also, we introduce our recent research results on QT protocol published in (Yang et al., 2010b) and (Yang & He, 2011).

2. ANTI-COLLISION PROTOCOL

2.1 ALOHA-Based Protocols

The chapter is intended to cover a wide range of QT protocols on solving the tag collision in tags identification. Therefore, we only briefly describe four basic ALOHA-based protocols: Pure ALOHO (PA) protocol, Slotted ALOHO (SA) protocol, Framed Slotted ALOHO (FSA) protocol, and Dynamic Framed Slotted ALOHO (DFSA) protocol, which are used in RFID standard and products, e.g., ISO 18000-3, ISO 14443-3, ISO 18000-6A, EPCGlobal Class 1, and Philips I-Code (Finkenzeller, 2010; Bolic, Simplot-Ryl, & Stojmenovic, 2010).

2.1.1 PA Protocol

PA protocol is a TDMA scheme. When receiving a querying request from the reader, each tag randomly responds. If only one tag responds, the reader transmits an acknowledgment (ACK) command and the tag send back its tag ID. On the other hand, the reader sends a negative ACK (NACK) when tags collide. All tags will then retransmit after a random delay. A large volume of tags may cause such retransmissions frequently. This compromises the reading rate, and thus PA protocol is unsuitable for RFID system. Some approaches, muting, slow down and fast mode, are accordingly proposed to enhance the throughput of PA protocol.

2.1.2 SA Protocol

To enhance the reading rate of PA protocol, we should solve the collision problem. The timeline is divided into different time slots. Tags can only transmit at the beginning in time slots, and the transmitted data should be less than a time slot length. If there is a collision, tags retransmit data after a random number of time slots. The difference between PA protocol and SA protocol is that the collision of SA protocol only happens at the start point in time slots, while PA protocol may occur at any time. Finally, the collision is reduced and the reading rate is enhanced. Generally, the throughput of SA protocol doubles that of PA protocol. Some variants on SA protocol are also proposed to enhance the efficiency.

2.1.3 FSA Protocol

In PA and SA protocols, each tag needs to respond at least one time in a reading cycle. A Frame including several time slots is defined in FSA protocol. Besides, tags only respond in next frame if the collision occurs. Thus, tags only reply once in each frame. Obviously, the frame size (the number of time slots in a frame) affects the collision performance in FSA protocol. For example, if there are only a few tags and a large frame size, it will have many idle slots. On the contrary, many tags and the small frame size cause many collisions. Therefore, a reader should carefully determine the frame size to reduce collisions and idle slots according to the number of tags.

2.1.4 DFSA Protocol

To achieve the better optimization of collisions and idle slots, a dynamic frame size rather than the fixed frame size is adopted in DFSA protocol. The frame size is dynamically adjusted according to the estimationof the number of tags. One approach of tag estimation is described as follows. After the last cycle, reader could estimate the tag population from the collision status to determine the frame size for the next cycle. Actually, EPC-C1G2 protocol adopts the DFSA as its tag anti-collision protocol.

2.2 BT Protocol

In BT protocol, each tag needs a counter. All counters in tags are initially set to zero. If the collision occurs, a random binary value in a tag is generated such that the counter increments the count when the binary value is 1. Afterwards, when the counter value in a tag is "0" the tag replies to a reader. Meantime, the counters in other tags decrement the count. We repeat this procedure until all tags are successfully identified. Box 1 shows the operation procedure of BT protocol by using a pseudocode, where c_i is the counter value of tag T_i, $i \in [1, n]$.

Example 1. Suppose that there are 9 tags T_1–T_9. At first nine tags' counters are initiated as 0, i.e., $\{c_1, c_2, c_3, c_4, c_5, c_6, c_7, c_8, c_9\}=\{0, 0, 0, 0, 0, 0, 0, 0, 0\}$. At this time, all tags will reply and there is a collision. Afterwards, a random binary value in a tag is generated such that the counter increments and decrements the count when the binary value is 1 and 0, respectively. As shown in Table 1, in cycle 2, suppose that we have T_6–T_9 increment the count, i.e., $\{c_1, c_2, c_3, c_4, c_5, c_6, c_7, c_8, c_9\}=\{0, 0, 0, 0, 0, 1, 1, 1, 1\}$. T_1–T_5 will respond and collide, so that these 5 tags randomly increment the counter. Suppose that, in cycle 3, for T_1–T_5, we have T_2–T_2 increment the count. Also, at this time, other tags, T_6–T_9, should increment the count. So, we have $\{c_1, c_2, c_3, c_4, c_5, c_6, c_7, c_8, c_9\}=\{0, 1, 1, 1, 1, 2, 2, 2, 2\}$. Since only T_1 has the counter value "0", T_1 transmits its ID to reader. Other tags without competition collision decrease their counter by 1, i.e., $\{-, c_2, c_3, c_4, c_5, c_6, c_7, c_8, c_9\}=\{-, 0, 0, 0, 0, 1, 1, 1, 1\}$ (see cycle 4). By repeating this procedure, we will successfully identify the tags. The detail procedures are listed in Table 1.

Box 1. Pseudocode of BT protocol

```
/* initialize counter of all tags */
        { c1, c2, ..., cn}={0, 0, ..., 0}; /* n tags for identification */
    begin
    while(the number of remaining tags!=NULL)
    If (ci=0) Ti responds; /* tag action */
     switch (the number of responded tags: rn) /* reader action */
        case (rn=1): /*success cycle*/
            Ti be successful identified;
            other tags' counters decrement the count; /* tag action */
         end
        break;
        case (rn>1): /* collision cycle */
            if(ci=0) /* tag action */
               then {Ti randomly generate a binary value;
                    ci adds the binary value; }
               else ci +1
    end
        break;
        case(rn=0): /* idle cycle */
    every tag' counter decrements the count; /* tag action */
        break;
    end while
    end
```

Table 1. Identification of nine tags in example 1 using BT

cycle	c_1	c_2	c_3	c_4	c_5	c_6	c_7	c_8	c_9	Reader
1	0	0	0	0	0	0	0	0	0	collision
2	0	0	0	0	0	1	1	1	1	collision
3	0	1	1	1	1	2	2	2	2	T_1 is identified
4	–	0	0	0	0	1	1	1	1	collision
5	–	0	0	1	1	2	2	2	2	collision
6	–	0	1	2	2	3	3	3	3	T_2 is identified
7	–	–	0	1	1	2	2	2	2	T_3 is identified
8	–	–	–	0	0	1	1	1	1	collision
9	–	–	–	0	1	2	2	2	2	T_4 is identified
10	–	–	–	–	0	1	1	1	1	T_5 is identified
11	–	–	–	–	–	0	0	0	0	collision
12	–	–	–	–	–	0	1	1	1	T_6 is identified
13	–	–	–	–	–	–	0	0	0	collision
14	–	–	–	–	–	–	1	1	1	idle
15	–	–	–	–	–	–	0	0	0	collision
16	–	–	–	–	–	–	0	1	1	T_7 is identified
17	–	–	–	–	–	–	–	0	0	collision
18	–	–	–	–	–	–	–	0	1	T_8 is identified
19	–	–	–	–	–	–	–	–	0	T_9 is identified
20	–	–	–	–	–	–	–	–	–	FINISH

There are total 20 interrogation cycles, which includes 9 collision cycles, 1 idle cycle and 9 successful cycles.

2.3 QT Protocol

In QT protocol, a reader sends a prefix of EPC to query the tags. Tags matching the prefix will then respond to the reader. The reader extends the prefixes until only one tag's ID matches, i.e., a tag can be identified successfully. First, the reader sends a binary string q to ask tags whether their IDs contain a prefix $(t_1, t_2, \ldots, t_{|q|})$ same to q. If multiple tags answer, these tags have the same prefix and a collision is detected at communication channel. The reader then appends bit 0 and 1 to form a longer prefixes ($q0$) and ($q1$) to query tags. The reader repeats the query procedure until all tags are uniquely identified. Box 2 illustrates QT protocol. Some notations used in QT protocol and subsequent anti-collision protocols are first defined in Table 2.

Box 2. Pseudocode of QT protocol

```
/* initialize Q */
if (Q = NULL)
  Push(Q, 0);
  Push(Q, 1);
end
begin
 while (Q!=NULL)
  q=Pop(Q);
  reader broadcast q to tags; /* reader query tags*/
  switch(M(q, t)) /*tag compare its ID with the query q*/
   case (M(q, t)=1):
    if (collision) /* multiple tags respond */
     Push(Q, q0);
     Push(Q, q1);
    else /* only one tag responds */
     tag transmit its ID t to reader;
    end
   break;
   case (M(q, t)=0):
    no tag responds; /* idle cycle */
  break;
 end while
end
```

An instance of QT protocol for the identification of nine tags using QT protocol is shown as follows.

Example 2. Suppose the length of tag's ID is 6 bits (i.e., y=6), and all $t=(t_1, t_2, t_3, t_4, t_5, t_6)$ of 9 tags are {(000000), (010011), (010010)(010001), (011000), (011110), (110010), (110100), (110111)}. A reader first broadcasts 1-bit query string q=0, six tags will respond since there are 6 tags having $t_1=0$. A collision is detected. By using Push (Q, 00) and Push (Q, 01), a reader adds two query strings into queue Q={1, 00, 01}. Also, there is a collision when a reader sends q=1 (since 3 tags have $t_1=1$), and then the queue Q is updated as={00, 01, 10, 11}. The detail procedures of identification are listed Table 3. Finally, there are total 23 interrogation cycles including 10 collision cycles, 4 idle cycles and 9 successful cycles.

3. SOME QT-BASED PROTOCOLS

To solve tag collision problem, many improvements on base QT protocols are accordingly proposed. Three QT-like protocols are briefly described in the following. A collision tree (CT) protocol (Jia et al., 2010) using Manchester code successfully reduced the required iterations for identification. An adaptive QT (AQT) protocol (Yang et al., 2010a) combined multiple *M*-ary trees to reduce the collisions and the idle time simultaneously. A RN16-based QT (RN16QT) protocol (Choi et al., 2007) took RN16 as the tag's temporary ID (*TID*), and then applied QT to identify this *TID*. Finally, RN16QT protocol successfully reduced the time consumption for tag identification than the identification implemented in EPC-C1G2 tag.

Table 2. Notations used in QT protocol

Notations	
t	the tag's ID (e.g. EPC), a y-tuple $t=(t_1, t_2, \ldots, t_y)$, where $t_i \in \{0, 1\}$, $1 \le i \le y$, and y is the number of bits of tag's ID; for example EPC has y=96
q	the query string sent from a reader, the x-tuple $q=(q_1, q_2, \ldots, q_x)$, where $q_i \in \{0, 1\}$, $1 \le i \le x$, and $x(\le y)$ is the length of a query string
Q	a queue maintained by a reader, which stores the query strings
Push (Q, s)	push a query string s into the queue Q
Pop (Q)	pop a query string q=Pop(Q) from the queue Q
$M(q, t)$	verify the query string q whether matches the prefix of tag's ID t, if $q = (t_1, t_2, \ldots, t_{\lvert q \rvert})$, then $M(q, t)$=1 else $M(q, t)$=0

Table 3. Identification of nine tags in example 2 tags using QT

cycle	query string q	response	queue Q
1	NULL	collision	0,1
2	0	collision	1,00,01
3	1	collision	00,01,10,11
4	00	**000000**	01,10,11
5	01	collision	10,11,010,011
6	10	idle	11,010,011
7	11	collision	010,011,110,111
8	010	collision	011,110,111,0100,0101
9	011	collision	110,111,0100,0101,0110,0111
10	110	collision	111,0100,0101,0110,0111,1100,1101
11	111	idle	0100,0101,0110,0111,1100,1101
12	0100	collision	0101,0110,0111,1100,1101,01000,01001
13	0101	idle	0110,0111,1100,1101,01000,01001
14	0110	**011000**	0111,1100,1101,01000,01001
15	0111	**011110**	1100,1101,01000,01001
16	1100	**110010**	1101,01000,01001
17	1101	collision	01000,01001,11010,11011
18	01000	**010001**	01001,11010,11011
19	01001	collision	11010,11011,010010,010011
20	11010	**110100**	11011,010010,010011
21	11011	**110111**	010010,010011
22	010010	**010010**	010011
23	010011	**010011**	NULL

3.1 CT Protocol

CT protocol is based on QT and Manchester code to enhance the performance of QT protocol. It can reduce the required iterations for identification. By using Manchester code, we can detect the location of collided bits due to the characteristic of Manchester code. Therefore, we can find the location of different bits from all responded tags and the reader can skip some prefixes to reduce the collisions and idle time. Manchester code was also used in binary search protocol (Finkenzeller, 2010, pp.204-209).

In Manchester coding, the bit "0" is encoded by a positive transition, while bit "1" is encoded by a negative transition. Such coding let reader can detect the location where there are more than two different bits "1" and "0" simultaneously received. We define the collided string *cs* as the same prefix for these tags. A reader can thus skip the *cs* without unnecessary queries to improve the reading efficiency. Box 3 illustrates CT protocol.

Suppose that there are three tags have the same prefix 0010. When a reader queries "0", all three tags will reply. We will have the same waveforms for the previous four bits from Manchester code. So, a reader can send the query strings (00100) and (00101) (note: skip *cs*=(0010)), and avoid three collisions when sending (00), (001), (0010), and three idle steps (01), (000), (0011) in QT protocol. CT protocol improves the identification efficiency than QT protocol at the cost of analyzing the received signals to find the location of different bits every time when receiving the responded tags.

Box 3. Pseudocode of CT protocol

```
/* initialize Q */
if (Q = NULL)
  Push(Q, 0);
  Push(Q, 1);
end
begin
  while (Q!=NULL)
    q=Pop(Q);
    reader broadcasts q to tags; /* reader query tags*/
    switch(M(q, t)) /*tag compare its ID with the query q*/
      case (M(q, t)=1):
        if (collision) /* multiple tags respond */
          find cs using Manchester code;
          Push(Q, q, cs,0);
          Push(Q, q, cs,1);
        else /* only one tag responds */
          tag transmit its ID t to reader;
        end
      break;
      case (M(q, t)=0):
        no tag responds; /* idle cycle */
    break;
  end while
end
```

3.2 AQT Protocol

QT protocol is based on binary tree. A so-called *M*-ary QT protocol is to adopt *M*-ary ($M=2^i$, $i\in[1, b]$) for the identification of tags. It is observed that using the large *M*-ary tree implies the longer prefix, so that the tags may have the same prefix with the small probability. This small matching probability reduces the tag's collisions. However, the less matching of the query string and the tag's ID increases the invalid inquiries. On the contrary, the small *M*-ary tree has the more valid inquiries but increases the collision. Therefore, there exists a dilemma of using the large *M*-ary tree and the small *M*-ary tree.

AQT protocol is a combination of different *M*-ary QTs. It uses the suitable *M*-ary tree according to the number of responded tags. AQT protocol is shown in Box 4. In the initial phase, we first set the threshold ranges, $R_1, R_2, \ldots, R_b$, and add *M* ($M=2^i$, $i\in[1, b]$) query string with the *i*-tuples if the number of total tags for identification *n* belongs the R_i region ($n\in R_i$). When the collision occurs, AQT uses 2^i-ary tree for identification by checking the number of collided tags n_i belongs the R_i ($n_i\in R_i$). Also, we skip the remainder leaves of this collided node to avoid the unnecessary inquiries if we had queried all the tags in a collided node.

Example 3. Consider the identification of nine tags in Example 2 by using AQT protocol with

Box 4. Pseudocode of AQT protocol

```
/* initialize Q */
if (Q = NULL)
 Set R_i, i ∈ [1, b];

 if (n ∈ R_i) {Push (Q, 0···00 (i bits)); ...; Push (Q, 1···11 (i bits));};   /* add M query strings, where M=2^i */
 1) end
 2) begin
 3) while (Q!=NULL)
 4) if (the remainder number of tags in a collided node =0) skip the remainder leaves of this collided node;
  5) q=Pop(Q);
  6) reader broadcast q to tags; /* reader query tags*/
  7) switch (M(q, t)) /*tag compare its ID with the query q*/
   8) case (M(q, t)=1):
    9) if (collision) /* two or more tags respond */

    10) if (n_t ∈ R_i) {Push (Q, q0···00 (i bits)); ...; Push (Q, q1···11 (i bits));};   /* add M query strings, where M=2^i */
   a) else /* only one tag responds */
    b) tag transmit its ID t to reader;
   c) end
  d) break;
  e) case (M(q, t)=0):
   f) no tag responds; /* idle cycle */
  g) break;
 h) end while
  i) end
```

M=2, 4, 8. Also, the threshold ranges are R_1=[1, 2], R_2=[3, 4], R_3=[5, 9]. At first, a reader detects n=9 tags for identification; since n∈R_3 therefore an 8-ary tree is used. When broadcast the query string (000), a tag with ID (000000) responds. At the collided nodes, 010, 011, 110, there are, respectively n_t=3, 2 and 3 collided tags. The remainder number of collided nodes for the original root is n_1=9−1−3−2−3=0, and thus the query string (111) can be discarded. Since the numbers of collided tags at the nodes (010) and (110) are n_t=3∈R_2, both collided nodes use 4-ary trees. So, the query strings (01000), (01001), (01010), (01011), and (11000), (11001), (11010), (11011) are put into the queue Q. On the other hand, the value of n_t in (011) node is 2∈R_1, so the query strings (0110), (0111) are added into the queue Q. At the collided node (010), a reader finishes the querying (01000) by receiving the unique response from the tag with ID (010001). When querying (01001), a reader detects two collided tags from (010010) and (010011). Therefore the remainder number of collided nodes for the (010) root is n_2=3-1-2=0, so the querying strings (01010) and (01011) can be discarded. Repeat checking the remainder number of collided tags in a collided node. We could avoid the unnecessary inquiries. Table 4 lists the detail of identifying procedure. There are 5 collision cycles, and 4 idle cycles. The overall interrogation cycles are 5(collision)+4(idle)+9(successful)=18 is better than QT (23 interrogation cycles).

3.3 RN16-Based Query Tree (RN16QT)

RN16 is a 16-bit random number generated by a 16-bit pseudo-random number generator (PRNG) on EPC-C1G2 tag. This RN16 is chiefly used to preload the tag's slot counter for avoiding the tag

Table 4. Communication between reader and tags in example 3 using AQT

cycle	q	response	collided tags	rem. collided tags	queue Q
1	NULL	collision	n=9		**000,001,010,011,100,101,110,111**
2	000	**000000**	–	n_1=8	001,010,011,100,101,110,111
3	001	idle	–	–	010,011,100,101,110,111
4	010	collision	n_t=3	n_1=5	011,100,101,110,111,**01000,01001,01010,01011**
5	011	collision	n_t=2	n_1=3	100,101,110,111,01000,01001,01010,01011,**0110,0111**
6	100	idle	–	–	101,110,111,01000,01001,01010,01011,0110,0111
7	101	idle	–	–	110,111,01000,01001,01010,01011,0110,0111
8	110	collision	n_t=3	n_1=0	01000,01001,01010,01011,0110,0111,**11000,11001,11010,11011**
9	01000	**010001**	–	n_2=2	01001,01010,01011,0110,0111,11000,11001,11010,11011
10	01001	collision	n_t=2	n_2=0	0110,0111,11000,11001,11010,11011,**010010,010011**
11	0110	**011000**	–	–	0111,11000,11001,11010,11011,010010,010011
12	0111	**011110**	–	–	11000,11001,11010,11011,010010,010011
13	11000	idle	–	n_3=3	11001,11010,11011,010010,010011
14	11001	**110010**	–	n_3=2	11010,11011,010010,010011
15	11010	**110100**	–	n_3=1	11011,010010,010011
16	11011	**110111**	–	n_3=0	010010,010011
17	010010	**010010**	–	–	010011
18	010011	**010011**	–	–	NULL

collision. Recently, Choi et al. (2007) proposed a RN16-based QT (RN16QT) protocol for anti-collision. Box 5 illustrates RN16QT protocol. It took RN16 as the tag's temporary ID (*TID*), and then applied QT to identify this *TID*. Because a tag responds reader with *TID* which the length is smaller than EPC length, tags will save the average responded bits. After the successful identification, the reader sends an ACK signal to ask the tag to send its whole EPC.

RN16QT protocol successfully reduces the time consumption for tag identification than the identification implemented in EPC-C1G2 tag. Also, it can solve the similar EPC problem that many tags have the same EPC prefix. The same EPC prefix often occurs in real application environment. As we know, EPC has 96 bits embracing four sections- the header (H: 8 bits), the GMN (G: 28 bits), the object class (O: 24 bits), and the serial number (S: 36 bits). Suppose that items are manufactured by the same company, and are stacked together in a large warehouse. These items with the similar IDs may have the same encoding scheme (H), the same company prefixes (G), the same object class (O), and the different serial Number (S). Because RN16QT protocol uses the *TID* (a random RN16) for identification, it can avoid the frequent collisions due to the same prefix as described above.

4. THE PROPOSED TAG ANTI-COLLISION PROTOCOLS

In this section, we introduce our two recent research results on QT protocols. The first is effective RN16QT (ERN16QT) protocol (Yang & He, 2011) to address the problem that the length

Box 5. Pseudocode of RN16QT protocol

```
/* initialize Q */
if (Q = NULL)
  Push(Q, 0);
  Push(Q, 1);
end
begin
  while (Q!=NULL)
    q=Pop(Q);
    reader broadcast q to tags;
    switch(M(q, TID)) /* tag uses RN16 as its TID;
                        tag compare its TID with the queried q */
      case (M(q, TID)=1):
        if (collision) /* multiple tags respond */
          Push(Q, q0);
          Push(Q, q1);
        else /* only one tag responds */
          Reader send the ACK to ask the tag sends its tag's ID
        end
      break;
      case (M(q, TID)=0):
        no tag responds; /* idle cycle */
    break;
  end while
end
```

of RN16 is not enough in real environments to avoid collision. Our ERN16QT protocol really solves the tag collision caused by the short RN16. Moreover, our ERN16QT protocol returns the lesser responded bits than RN16QT protocol. The second is Ternary QT (TQT) protocol. Mathys and Flajolet (1985) showed that ternary tree will bring about the optimum performance for tag identification. However, a ternary tree cannot be really implemented in the QT protocol. We propose a practical implementation of a ternary query tree for identifying a binary EPC (Yang et al., 2010b).

4.1 ERN16QT Protocol

For ERN16QT protocol, as defined in EPC-C1G2, the probability that any two or more tags have the same sequence of RN16, for 10,000 tags, should be less than 0.1%. However, a well known approximation for the collision is

$$N \approx \sqrt{2R \times \ln\left(1/(1-P)\right)}$$

where n is the number of tags, R is the size of the range of random number, and P is the probability that at least two random numbers have the same value. This approximation can be easily derived from the birthday attack. Suppose two RN16s have the same value with half chance (P=0.5), then we have $N \approx 302$. Through this theoretical estimation, RN16QT protocol is effective only for less than about 300 tags. Some tags cannot be identified since they have the same RN16. The length of RN16 is not enough for the usage of ID in real environments. Even though one can trivially generate another RN16 as *TID*, but tags need to perform PRNG generator one more time. We call RN16QT protocol using two or more RN16s a modified RN16QT (MRN16QT) protocol. In the MRN16QT, if the identification of all tags cannot be successfully finished, i.e., some tags have the same RN16. We need to regenerate new RN16 for the remaining tags to proceed with the identification. Our new ERN-16QT protocol still uses one RN16 but can thoroughly solve the collision problem for any numbers of tags. When compared with MRN16QT, our ERN16QT needs fewer responded bits and meantime do not need the regeneration of RN16.

As shown in Figure 2, a 96-bit EPC is divided into 6 sub EPCs EPC_i, $1 \leq i \leq 6$. By RN16 and EPC, we could construct 6 16-bit random numbers $RXE_i = RXE_{i-1} \oplus EPC_i$, $1 \leq i \leq 6$, where RXE_0=RN16. Let TID_i be RXE_i. Then, apply QT protocol to identify TID_i, $1 \leq i \leq 6$, in order. Six TIDs in our ERN16Q protocol work like RN-96, and thus could solve the collision problem in tag identification.

Pupunwiwat and Stantic (2009) show a possible application scenario of RFID, a so-called warehouse distribution. It is reasonable to assume that the EPC data of most items from the same warehouse will be very similar since the items are manufactured by the same company, and are stacked together in a large warehouse. These items with the similar IDs may have the same prefix, and the different serial Number. In the above ap-

Figure 2. Generation of RXE16$_i$, $1 \leq i \leq 6$

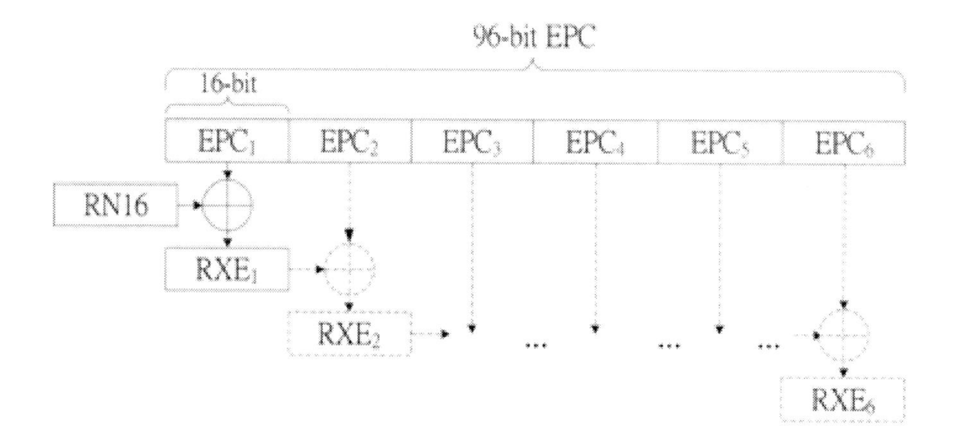

plication scenario, all tags may have the same EPC_1, EPC_2, and EPC_3. Since RXE_1= ($RN16 \oplus EPC_1$), RXE_2=($RN16 \oplus EPC_1 \oplus EPC_2$), and RXE_2= ($RN16 \oplus EPC_1 \oplus EPC_2 \oplus EPC_3$), so tags also have the same RXE_1, RXE_2, and RXE_3. A QT protocol with reversed IDs (Cho et al., 2008) can be adopted in our ERN16QT protocol to address this problem. At this time, the generation of RXE RXE_i= ($RXE_{i-1} \oplus EPC_{7-i}$), $1 \leq i \leq 6$, using reverse EPC is shown in Figure 3.

Figure 4 is the ERN16QT protocol. EPC are fist divided into six sub EPCs. Tags then use one RN16 to generate six RXE_i as TID_i for identification in the i-th stage. Every stage has the same function like RN16QT. The major difference is that the response is different. After the successful identification, in ERN16QT, a reader will send an ACK to tag. Tag returns its remaining sub EPCs. For example, when the identification is now on TID_i, tag should return its (RN16, EPC_{i+1}, ..., EPC_6). Our ERN16QT could calculate (EPC_1, ..., EPC_i) from EPC_1=$RN16 \oplus TID_1$, ..., EPC_i=$TID_{i-1} \oplus TID_i$; also, the reader receives (EPC_{i+1}, ..., EPC_6). Finally, the reader can recover the whole EPC=($EPC_1 || EPC_2 || \ldots || EPC_6$).

One experiment is conducted to evaluate RN16QT protocol and the proposed ERN16QT protocol. We use LAMED (a PRNG for EPC-

Figure 3. Generation of RXE16$_i$, $1 \leq i \leq 6$, using reverse EPC

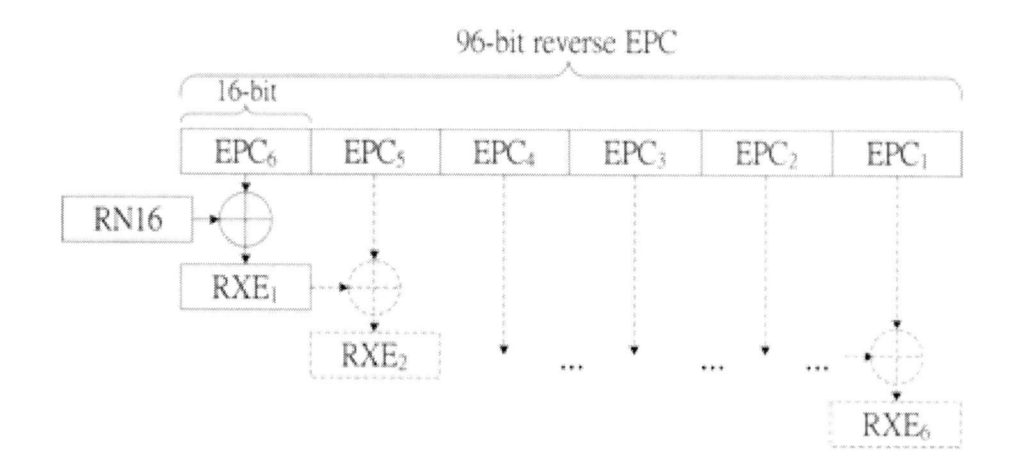

Figure 4. Flow chart of ERN16QT protocol

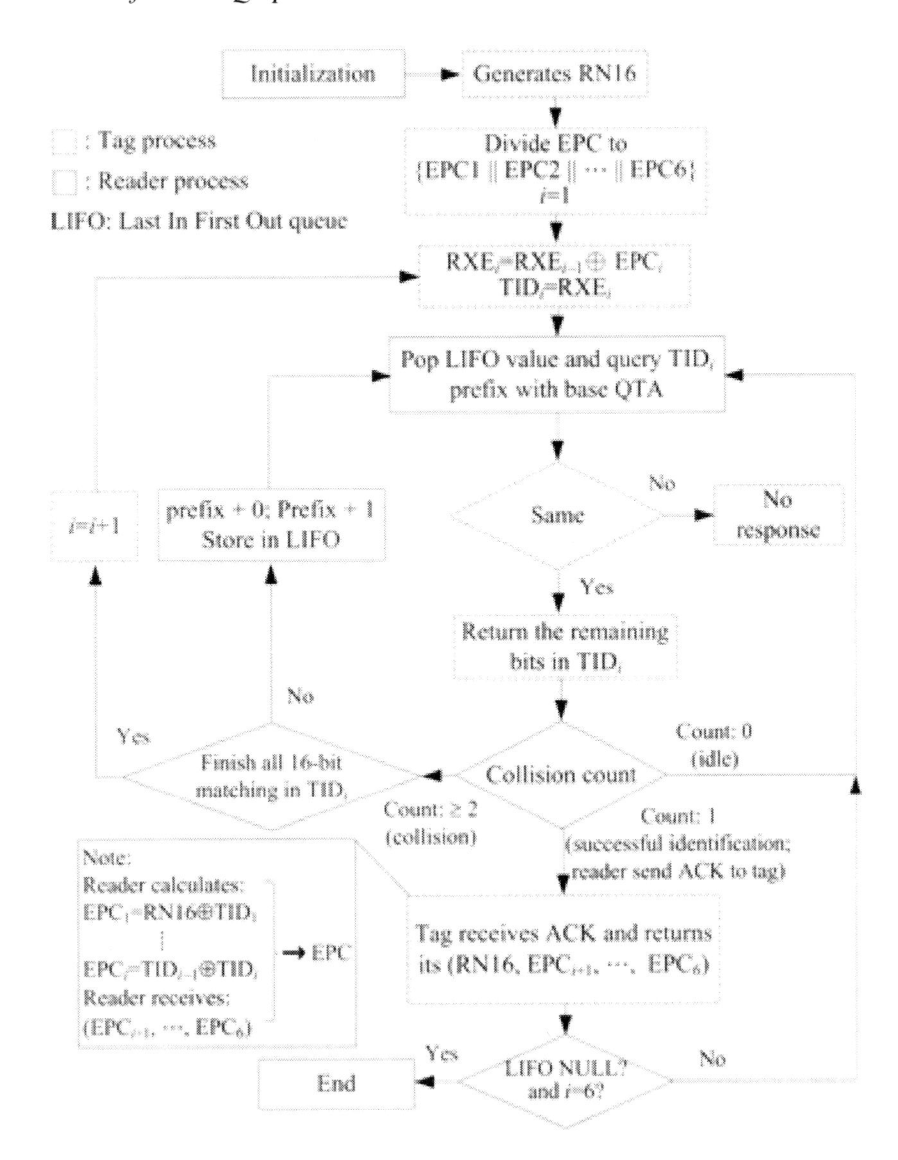

C1G2 specification) in (Peris-Lopez, Hernandez-Castro, Estevez-Tapiador, & Ribagorda, 2009) to generate the required RN16s. LAMED not only fulfils the requirements of EPC-C1G2 standard, but also passes some very strict randomness test batteries, e.g, ENT, DIEHARD, NIST, and SEXTON. Test tags in experiment have random EPCs. Table 5 shows the average number responded bits (N_R), the remaining number of unidentified tags (N_U), the prefix length of query string for finishing all identifications (L) for some values of N. RN16QT protocol is only effective for $L \leq 16$ (note: using RN16 as TID). According to the theoretical estimation, two tags may have the same RN16s about N=302, simulation results demonstrate this situation. From Table 5, it is observed that RN16QT protocol cannot identify all tags for more than 400 tags. MRN16QT protocol needs the second RN16 for N=400, and the third RN16 for N=5000. In Table 5, the asterisk denotes using MRN16QT

Table 5. The values of N_R, N_U, and L for different N

N	RN16QT (MRN16QT*)			ERN16QT		
	N_R	N_U	L	N_R	N_U	L
50[#1]	180	0	11	178	0	10
100[#1]	186	0	11	186	0	14
400[#2]	199*	4*	18*	198	0	20
1000[#2]	207*	18*	25*	206	0	22
5000[#3]	223*	411*	36*	216	0	29
10000[#3]	233*	1451*	41*	220	0	29

#1: use the first RN16 (RN16QT Protocol)
#2: use the second RN16 (MRN16QT Protocol)
#3: use the third RN16 (MRN16QT Protocol)

protocol. If the identification is finished within $L \leq 16$, the responses for RN16QT and ERN16QT protocols are, respectively, EPC and (RN16, EPC_2, ..., EPC_6). The numbers of returned bits for both algorithms are exactly same (96 bits). When the identification is finished within the i-th stage, i.e., MRN16QT protocol responds EPC, while our ERN16QT protocol responds (RN16, EPC_{i+1}, ..., EPC_6), and thus saves $16 \times (i-1)$ for one response.

The detail comparison of RN16-like protocols (RN16QT, MRN16QT, and ERN16QT protocols) with other tree based protocols including QT, BS, and CT can be found in (Yang & He, 2011). In RN16-like protocols, tags respond TID in the identification process. So, they have the less responded bits. CT uses Manchester code to avoid the idle cases and reduces the collision cases, and thus it requires the less number of iterations. When RN16-like protocols successfully identify tags by TID, they still need one iteration to send EPC information. Our ERN16QT protocol works for any numbers of tags; however, from experimental results, RN16QT works only for about N=400 (this consists with the approximate value $N \approx 302$). MRN16QT protocol can solve the collision caused by the short length of RN16, but it needs regeneration of RN16.

4.2 TQT Protocol

As we described in the section of AQT protocol, the choice of M in a M-ary QT protocol is critical. The different values of M have different effectiveness. The large M-ary tree reduces the collisions but increases the unnecessary inquiries. Contrarily, the small M-ary tree has the less idle time but increases the collisions.

Mathys & Flajolet (1985) showed that ternary tree has the optimum performance (i.e., the best identification time). However, a ternary tree cannot be directly adopted in QT protocol to identify a binary EPC. So using TQT delivers a problem how to efficiently convert 96-bit $(EPC)_2$ to a ternary $(EPC)_3$. We construct a practical implementation of TQT protocol for identifying a binary EPC by a workable 3 bits to 2 ternary digits (3B2T) conversion. In TQT protocol, we also design an approach to delete the unnecessary inquiries to further enhance the efficiency of identification. If we had queried all the nodes (the successful identified nodes and the collided nodes), we may skip the remainder leaves of this collided node to avoid the unnecessary inquiries. The TQT protocol combines 3B2T conversion and the approach discarding the unnecessary inquiries to enhance the identification efficiency.

Since $2^{96} < 2^{61}$, we can convert a binary 96-bit $(EPC)_2$ to 61 ternary digits to $(EPC)_3$. However this conversion cannot be implemented efficiently by a light-weight tag. Our 3B2T coding method converts 3 bits to 2 ternary digits one time. As shown in Figure 5, we need total 2×(96/3)=64 ternary digits, which is slightly greater than 61. Actually, we can identify tags successfully without requiring all 64 ternary digits for most cases. Therefore, applying TQT protocol on 61-digit $(EPC)_3$ and 64-digit $(EPC)_3$ will have the almost same performance.

Let 3B2T(·) be a conversion function to derive $(EPC)_3$=3B2T($(EPC)_2$). The string t and q used in TQT protocol are ternary digits, which t_i and $q_i \in \{0, 1, 2\}$. A tag's ID is first converted to a $(EPC)_3$ code $t=(t_1, t_2, \ldots, t_{64})$. Notice that we only process 3B2T conversion in need of t_i, $1 \leq i \leq 64$, and thus we do not have to covert 96-bit $(EPC)_2$ one time. When a tag is uniquely identified, it responds its $(EPC)_2$. TQT protocol is shown in Box 6.

Example 4. Suppose that there are nine tags of tag's ID 9 bits are (000000000), (001010000), (0010111 10), (001110101), (011000001), (100111010), (101010100), (110000111), and (110111010) for identification by using TQT protocol. Three bits of nine tags {(000), (001), (001), (001), (011), (100), (101), (110), (110)} are first converted to $\{(t_1t_2)\}$={(00), (01), (01), (01), (10), (11), (12), (20), (20)}, respectively by 3B2T. A reader initially pushes (Q, 0), (Q, 1) and (Q, 2) in a null queue Q, i.e., the initial queue Q={0, 1, 2}. When a reader broadcasts the query string q=0, four tags with ID t_1=0 responds. A reader then applies Push (Q, 00), Push (Q, 01) and Push (Q, 02) to add three query strings into queue, and obtain Q={1,2,00,01,02}. Afterwards, the reader queries q=1, and three tags with the prefixes (10), (11) and (12) respond. By using TQT protocol and checking the remainder number of collided tags in a collided node to avoid the unnecessary inquiries, we finally finish the identification. Let n, r, c_i and r_i be the number of total tags, the number of remainder tags at root, the number of collided tags at the node i, and the number of remainder tags at the node i, respectively. At the collided nodes 0, 1, and 2, there are, respectively c_0=4, c_1=2 and c_2=3, where $n=c_0+c_1+c_2$=9. When finishing the inquiry at node "01", the number of remainder tags is $r_0=c_0-1$(a successful node "00")$-c_{01}$=4−1−3= 0, and thus the query string (02) can be discarded. After using t_2 for match the query string, the next 3 bits of tags are then converted to (t_3t_4). Finally, the overall interrogation cycles are 6(collision)+1(idle)+9 (successful)=16. Table 6 lists the detail of identifying procedure, and Figure 6 is its corresponding tree plot.

Figure 5. 3B2T conversion

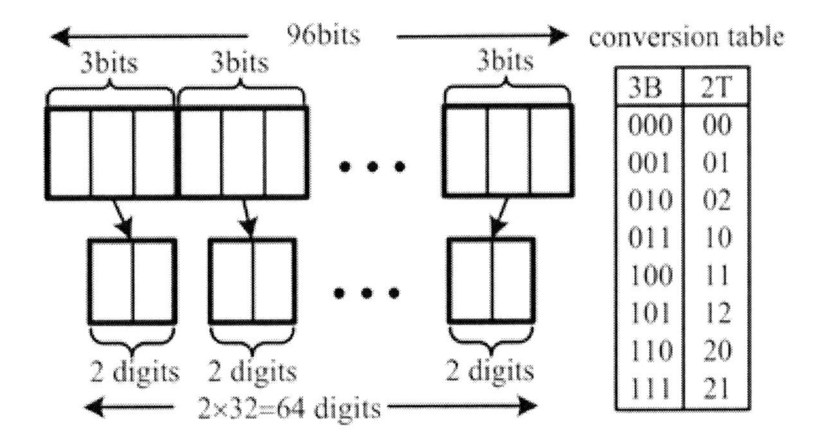

Box 6. Pseudocode of TQT protocol

```
/* initialize Q */
if (Q = NULL)
   Push(Q, 0); Push(Q, 1); Push(Q, 2);
end
begin
  While (Q!=NULL)
 if (the remainder number of tags in a collided node=0)
skip the remainder leaves of this collided node;
/* apply discarding idle nodes at the tail leaves */
 end
q=Pop(Q);
reader broadcasts q to tags; /* reader queries tags*/
 switch(M(q, t))
/* tag compares its ID with the query Q;
(EPC)_2 is converted to gain the required t_i by 3B2T */
       case (V(q, t)=1):
 if (collision) /* multiple tags respond */
Push(Q, q0);
Push(Q, q1);
Push(Q, q2); /* note: if the front digit is 2 then we do not put q2 in
the queue (see the conversion table in Figure 5) */
 else /* only one tag responds */
tag transmit its (EPC)_2 to reader;
 end
 break;
 case (V(q, t)=0):
no tag responds; /* idle cycle */
      break;
  end while
end
```

Table 6. Communication between reader and tags using TQT

cycle	q	response	collided tags	rem. tags	queue Q
1	NULL	collision	n=9	-	**0,1,2**
2	0	collision	c_0=4	r=5	1,2,**00,01,02**
3	1	collision	c_1=3	r=2	2,00,01,02,**10,11,12**
4	2	collision	c_2=2	r=0	00,01,02,10,11,12,**20,21**
5	00	**00 (000000000*)**	-	r_0=3	01,02,10,11,12,20,21
6	01	collision	c_{01}=3	r_0=0	02,10,11,12,20,21,**010,011,012**
7	10	**10 (011000001*)**	-	-	11,12,20,21,010,011,012
8	11	**11 (100111010*)**	-	-	12,20,21,010,011,012
9	12	**12 (101010100*)**	-	-	20,21,010,011,012
10	20	collision	c_{20}=2	r_2=0	21,010,011,012,**200,201,202**
11	010	**010 (001010000*)**	-	-	011,012,200,201,202
12	011	**011 (001011110*)**	-	-	012,200,201,202
13	012	**012 (001110101*)**	-	-	200,201,202
14	200	**200 (110000111*)**	-	-	201,202
15	201	idle	-	-	202
16	202	**202 (110111010*)**	-	-	NULL

* The $(EPC)_2$ code; 02 in cycle 6 and 21 in cycle 10 are discarded idle nodes

Figure 6. Identification using TQT

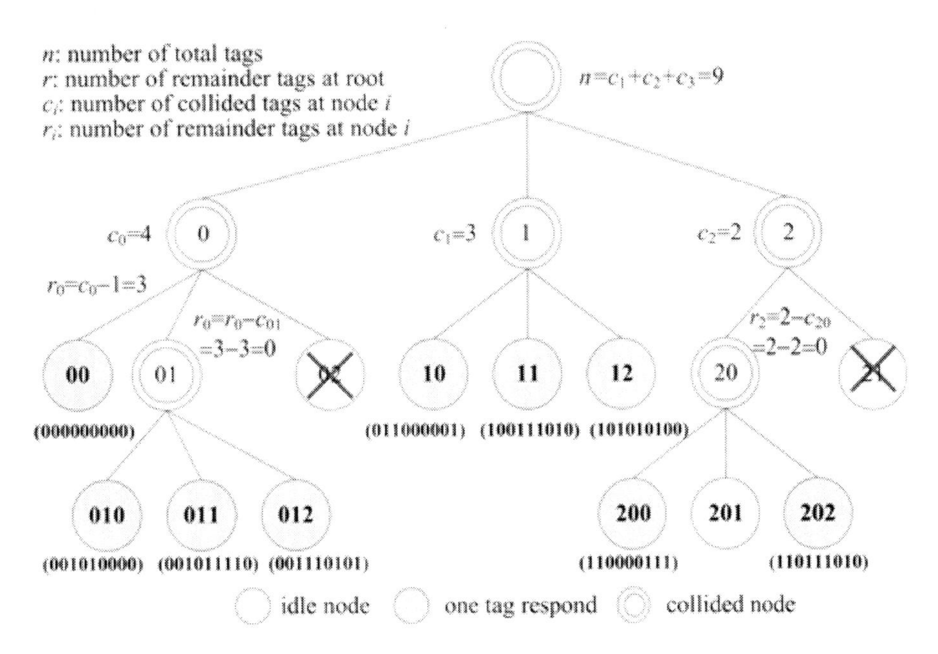

Performance evaluation for QT protocol, M-ary (M=16) QT protocol, and the proposed TQT protocol are described as follows. The 96-bit EPC data of test tags are randomly generated. The number of total tags for test are n=100, 200, 300, 400, 500, 1000 and 2000. We randomly generate n tags' IDs (96-bit EPC). Apply the above three protocols, respectively, to identify n tags. For each n, we repeat the same experiment five times to calculate the average number of collision cycles N_C, the average number of idle cycles N_I, and the average number of total interrogation cycles N_T, where $N_T= n+N_C+N_I$.

Table 7 summarizes test results, where the values in parentheses are the values by discarding the tail leaves. It is observed that MQT protocol has the least collisions of all three protocols but significantly increases the idle cycles; finally, it has the most interrogation cycles. For n=2000, the values of N_T are 5498, 9824, 5286 for of QT, MQT and TQT protocols. By discarding tail leaves, the numbers of idle cycles are reduced. Finally, the numbers of total interrogation cycles are 5135, 7945, 4610. The approach of discarding tail leaves really works, and can reduce the number of idle cycles for all MQT protocols. No matter whether we discard tail leaves or not, our TQT protocol outperforms QT protocol and MQT protocol. Experiments reveal that the proposed TQT has better performance than other M-ary QT ($M \neq 3$).

Redo the above experiment, but use test tags with the very similar EPC data, which all tags have the same first 60 bits (H+G+O) and the random last 36 bits (S). For example, the items are manufactured by the same company, and are stacked together in a large warehouse. Experimental results of N_C, N_I and N_T are shown in Table 8. It is observed that all N_T of QT, MQT and TQT protocols increase due to the same prefixes. For n=2000, our TQT needs 5571 total interrogation cycles. When discarding tail leaves, TQT save 828 idle cycles and has 4762 total interrogation cycles. However, BOT and MQT with (without) discarding tail leaves needs 5834 (5366) and 10534 (8257) total interrogation cycles, respectively. Our TQT protocol still has the better performance than other two protocols for very similar EPC.

Table 7. The values of N_C, N_I and N_T for tags with random EPC

	QT			16-QT			TQT		
n	N_C	N_I	N_T	N_C	N_I	N_T	N_C	N_I	N_T
100	141	43 (21)	284 (262)	31	387 (283)	518 (414)	93	77 (43)	270 (236)
200	287	89 (40)	576 (527)	68	834 (576)	1102 (844)	189	158 (77)	547 (466)
300	418	120 (59)	838 (777)	112	1399 (972)	1811 (1384)	279	225 (115)	804 (694)
400	554	156 (79)	1110 (1033)	148	1842 (1302)	2390 (1850)	371	296 (159)	1067 (930)
500	705	207 (96)	1412 (1301)	195	2447 (1713)	3142 (2408)	489	382 (199)	1371 (1168)
1000	1396	398 (202)	2794 (2598)	348	4242 (3145)	5590 (4493)	936	761 (400)	2697 (2336)
2000	2748	750 (387)	5498 (5135)	613	7211 (5332)	9824 (7945)	1838	1448 (772)	5286 (4610)

Table 8. The values of N_C, N_I and N_T for tags with similar EPC

	QT			16-QT			TQT		
n	N_C	N_I	N_T	N_C	N_I	N_T	N_C	N_I	N_T
100	206	108 (45)	414 (351)	49	651 (382)	800 (531)	139	167 (74)	406 (313)
200	348	150 (68)	698 (616)	83	1067 (673)	1350 (956)	230	238 (113)	668 (543)
300	487	189 (87)	976 (874)	126	1609 (1044)	2035 (1470)	321	306 (151)	927 (772)
400	641	243 (117)	1284 (1158)	168	2139 (1429)	2707 (1997)	427	405 (207)	1232 (1034)
500	781	283 (134)	1564 (1415)	215	2747 (1858)	3462 (2573)	524	483 (244)	1507 (1268)
1000	1494	496 (238)	2990 (2732)	375	4635 (3271)	6010 (4646)	1000	884 (446)	2884 (2446)
2000	2916	918 (450)	5834 (5366)	657	7877 (5600)	10534 (8257)	1934	1637 (828)	5571 (4762)

5. FUTURE RESEARCH DIRECTIONS

Since using the prefixes of EPC for identification, the performance is very sensitive to the distribution of tags' EPCs. Therefore, when addressing RFID tag collision in real application, we should find more effective anti-collision protocols for the specific pattern of tag's ID, e.g., the same company code (used in a warehouse management), consecutive serial number, …, etc. Hybrid protocol combing different anti-collision protocols is also a good approach to achieve reliable identification, and deserves further studying.

Actually, there are two types of collisions- the reader collision and the tag collision (Sarma, Weis, & Engels, 2003). Due to the fast and successful deployment of large scale RFID system, this results in a situation that multiple readers are located in a small area. This multi-reader RFID environment is referred to as dense reader environment (DRE). In DRE, multiple readers are operating in proximity of one another. DRE makes the RF channel noisy such that readers cannot successfully identify tags. So, the reader collision is that several readers interrogate at the same time in the vicinity, and two or more readers scan a tag simultaneously. Therefore, to assure RFID system of a reliable identification, we need to solve not only the tag collision but also the reader collision. There are some approaches to solve reader collision in DRE, such as TDMA, FDMA, CDMA and SDMA. This chapter does not discuss the problem of reader collision. Here, we give some references about reader collision in additional reading section.

6. CONCLUSION

In this chapter, we described some well-known tag anti-collision protocols. Some QT-like protocols are introduced. We also show our recent researches on QT protocols. One is ERN16QT protocol and the other is TQT protocol. ERN16QT protocol enhances RN16QT protocol. Theoretical estimation and simulations demonstrate that RN16QT protocol is ineffective for large number of tags. By XOR-ing RN16 with 6 sub EPCs, we derive RXE_i, $1 \leq i \leq 6$, for the identification of tags. Finally, our ERN16QT protocol solves the collision caused by the short length of RN16, and works for any numbers of tags. In TQT protocol, we design a 3B2T conversion to convert $(EPC)_2$ to $(EPC)_3$, and then apply TQT protocol for identification.

ACKNOWLEDGMENT

This work was supported in part by Testbed@ TWISC, National Science Council under the Grants NSC 100-2219- E-006-001.

REFERENCES

Bardaki, C., Karagiannaki, A., & Pramatari, K. (2008). A systematic approach for the design of RFID implementations in the supply chain. *Panhellenic Conference on Informatics* (pp. 244-248). IEEE Press.

Bo, Y., Yehua, H., & Caijiang, Z. (2008). Apparel supply chain management based on RFID. *27th Chinese Control Conference* (pp. 419-423). IEEE Press.

Bolic, M., Simplot-Ryl, D., & Stojmenovic, I. (Eds.). (2010). *RFID systems: Research trends and challenges*. Wiley Press. doi:10.1002/9780470665251

Chalasani, S., & Boppana, B. (2007). Data architectures for RFID transaction. *IEEE Transaction on Industrial Informatics*, *3*(3), 246–257. doi:10.1109/TII.2007.904147

Chiang, K. W., Hua, C., & Yum, T. S. (2006). Prefix-randomized query-tree protocol for RFID system. *IEEE International Conference on Communication* (pp. 1653-1657). IEEE Press.

Cho, J. S., Shin, J. D., & Kim, S. K. (2008). RFID Tag anti-collision protocol: Query tree with reversed IDs. *10th International Conference on* Advanced Communication Technology (pp. 225-230). IEEE Press.

Choi, J. H., Lee, D., & Lee, H. (2007). Query tree-based reservation for efficient RFID tag anti-collision.. *IEEE Communications Letters*, *11*(1), 85–87. doi:10.1109/LCOMM.2007.061471

Chuang, M. L., & Shaw, M. H. (2007). RFID: Integration stages in supply chain management. *IEEE Engineering Management Review*, *35*(2), 80–87. doi:10.1109/EMR.2007.899757

Cui, Y., & Zhao, Y. (2008). Mathematical analysis for binary tree algorithm in RFID. *IEEE Vehicular Technology Conference* (pp. 2725-2729). IEEE Press.

EPCglobal Inc. (2005). *EPC™ radio-frequency identification protocols class-1 generation-2 UHF RFID protocol for communications at 860MHz-960MHz Version 1*.0.9. Specification for RFID air interface. Retrieved July 15, 2010, from http://www.epcglobalinc.org

Feng, B., Li, J. T., Guo, J. B., & Ding, Z. H. (2006). Id-binary tree stack anti-collision algorithm for RFID. *11th IEEE Symposium on Computers and Communications* (pp. 207-212). IEEE Press.

Finkenzeller, K. (2010). *RFID handbook: Fundamentals and applications in contactless smart cards and identification*. Munich, Germany: John Wiley & Sons, Ltd.

Floerkemeier, C., & Lampe, M. (2004). Issues with RFID usage in ubiquitous computing applications. In F. Alois & M. Friedemann (Eds.), *PERVASIVE 2004, Lecture Notes in Computer Science: Vol. 3001,* (pp. 188-193). Berlin, Germany: Springer-Verlag press.

Hsu, C. H., Yu, C. H., Huang, Y. P., & Ha, K. J. (2008). An enhanced query tree (EQT) protocol for memoryless tag anti-collision in RFID systems. *2008 Second International Conference on Future Generation Communication and Networking* (pp. 427 - 432). IEEE Press.

Jea, K. F., & Wang, J. Y. (2008). An RFID encoding method for supply chain management. *Asia-Pacific Services Computing Conference* (pp. 601-606). IEEE Press.

Jia, X. L., Feng, Q. Y., & Ma, C. Z. (2010). An efficient anti-collision protocol for RFID tag identification. *IEEE Communications Letters*, *14*(11), 1014–1016. doi:10.1109/LCOMM.2010.091710.100793

Kapoor, G., Wei, Z., & Piramuthu, S. (2008). RFID and information security in supply chains. *4th International Conference on Mobile Ad-hoc and Sensor Networks* (pp. 59-62). IEEE Press.

Lai, Y. C., & Lin, C. C. (2009). Two blocking algorithms on adaptive binary splitting: Single and pair resolutions for RFID tag identification. *IEEE/ACM Transactions on Networking*, *17*(3), 962–975. doi:10.1109/TNET.2008.2002558

Lee, S. R., Joo, S. D., & Lee, C. W. (2005). An enhanced dynamic framed slotted ALOHA algorithm for RFID tag identification. *2nd Annual International Conference on Mobile and Ubiquitous Systems: Networking and Services* (pp. 166-172). IEEE Press.

Li, Y., & Ding, X. (2007). *Protecting RFID communications in supply chains. 2nd ACM syMposium on Information, Computer and Communications Security* (pp. 234–241). ACM Press.

Liang, Y., & Li, L. (2007). Integration of intelligent supply chain management (SCM) system. *International Conference on Service Systems and Service Management* (pp. 1-4). IEEE Press.

Mathys, P., & Flajolet, P. (1985). Q-ary collision resolution algorithms in random-access systems with free or blocked channel access. *IEEE Transactions on Information Theory*, *31*(2), 217–243. doi:10.1109/TIT.1985.1057013

Myung, J., Lee, W., Srivastava, J., & Shih, T. K. (2007). Tag-splitting: adaptive collision arbitration protocols for RFID tag identification.. *IEEE Transactions on Parallel and Distributed Systems*, *18*(6), 763–775. doi:10.1109/TPDS.2007.1098

Peris-Lopez, P., Hernandez-Castro, J. C., Estevez-Tapiador, J. M., & Ribagorda, A. (2009). LAMED – A PRNG for EPC class-1 generation-2 RFID specification. *Computer Standards & Interfaces*, *31*(1), 88–97. doi:10.1016/j.csi.2007.11.013

Pupunwiwat, P., & Stantic, B. (2009). Unified q-ary tree for RFID tag anti-collision resolution. In B. Athman & L. Xuemin (Eds.), *20th Australasian Database Conference: Vol. 92, Conferences in Research Research and Practice in Information Technology* (pp. 49-58). Australian Computer Society Press.

Ryu, J., Lee, H., Seok, Y., Kwon, T., & Chio, Y. (2007). A hybrid query tree protocol for tag collision arbitration in RFID system. *IEEE International Conference on Communications* (pp. 5981-5986). IEEE Press.

Sarma, S., Weis, S., & Engels, D. (2003). RFID systems and security and privacy implications. *Workshop on Cryptographic Hardware and Embedded Systems: Vol. 2523, Lecture Notes in Computer Science* (pp. 454-470). Springer-Verlag Press.

Shih, D. H., Sun, P. L., Yen, D. C., & Huang, S. M. (2006). Taxonomy and survey of RFID anti-collision protocols. *Computer Communications*, *29*(11), 2150–2166. doi:10.1016/j.comcom.2005.12.011

Yan, B., Chen, Y., & Meng, X. (2008). RFID technology applied in warehouse management system. *ISECS International Colloquium on Computing, Communication, Control, and Management* (pp. 362-367). IEEE Press.

Yang, C. N., & He, J. Y. (2011). An effective 16-bit random number aided query tree algorithm for RFID tag anti-collision. *IEEE Communications Letters*, *15*(5). doi:10.1109/LCOMM.2011.031411.110213

Yang, C. N., Kun, Y. C., Chiu, C. Y., & Chu, Y. Y. (2010). A new adaptive query tree on resolving RFID tag collision. *IEEE International Conference on RFID-Technology and Applications* (pp. 153-158). IEEE Press.

Yang, C. N., Kun, Y. C., He, J. Y., & Wu, C. C. (2010). A practical implementation of ternary query tree for RFID tag anti-collision. *IEEE International Conference on Information Theory and Information Security* (pp. 283-286). IEEE Press.

ADDITIONAL READING

Chen, Y. H., Horng, S. J., Run, R. S., Lai, J. L., Chen, R. J., & Chen, W. C. (2010). A novel anti-collision algorithm in RFID systems for identifying passive tags. *IEEE Transactions on Industrial Informatics*, *6*(1), 105–121. doi:10.1109/TII.2009.2033050

Choi, J. H., Lee, D., & Lee, H. (2006). Bi-slotted tree based anti-collision protocols for fast tag identification in RFID systems. *IEEE Communications Letters*, *10*(12), 861–863. doi:10.1109/LCOMM.2006.061348

Eom, J. B., Lee, T. J., Rietman, R., & Yener, A. (2008). An efficient framed-slotted ALOHA algorithm with pilot frame and binary selection for anti-collision of RFID tags. *IEEE Communications Letters*, *12*(11), 861–863. doi:10.1109/LCOMM.2008.081157

Karmaker, N. C., Roy, S. M., & Ikram, M. S. (2008). Development of smart antenna for RFID reader. *IEEE International Conference in RFID* (pp. 16-17). IEEE Press.

Kim, J. G. (2008). A divide-and-conquer technique for throughput enhancement of RFID anti-collision protocol.. *IEEE Communications Letters*, *12*(6), 474–476. doi:10.1109/LCOMM.2008.080277

Kim, S. H., & Park, P. G. (2007). An efficient tree-based tag anti-collision protocol for RFID systems.. *IEEE Communications Letters*, *11*(5), 449–451. doi:10.1109/LCOMM.2007.070027

Klair, D. K., Chin, K. W., & Raad, R. (2010). A survey and tutorial of RFID anti-collision protocols.. *IEEE Communications Surveys & Tutorials*, *12*(3), 400–421. doi:10.1109/SURV.2010.031810.00037

Kun, Y. C., & Yang, C. N. (2010). *A new adaptive query tree on resolving RFID tag collision. Unpublished Master's theses*. Taiwan: National Dong Hwa University.

Myung, J., Lee, W., & Shih, T. K. (2006). An adaptive memoryless protocol for RFID tag collision arbitration. *IEEE Transactions on Multimedia*, *8*(5), 1096–1101. doi:10.1109/TMM.2006.879817

Myung, J., Lee, W., & Srivastava, J. (2006). Adaptive binary splitting for efficient RFID tag anti-collision. *IEEE Communications Letters*, *10*(3), 144–146. doi:10.1109/LCOMM.2006.1603365

Shin, W. J., & Kim, J. G. (2009). A capture-aware access control method for enhanced RFID anti-collision performance. *IEEE Communications Letters*, *13*(5), 354–356. doi:10.1109/LCOMM.2009.081970

Waldrop, C. J., Engels, D. W., & Sarma, S. E. (2003). Colorwave: An anti-collision algorithm for the reader collision problem. *IEEE International Conference on Communications* (pp. 1206-1210). IEEE Press.

Waldrop, C. J., Engels, D. W., & Sarma, S. E. (2003). Colorwave: A MAC for RFID reader networks. *IEEE Wireless Communications and Networking Conference* (pp. 1701-1704). IEEE Press.

Wang, T. P. (2006). Enhanced binary search with cut-through operation for anti-collision in RFID systems. *IEEE Communications Letters*, *10*(4), 236–238. doi:10.1109/LCOMM.2006.1613732

Yu, J., Liu, K. H., Huang, X. D., & Yan, G. (2008). An anti-collision algorithm based on smart antenna in RFID system. *International Conference on Microwave and Millimeter Wave Technology* (pp. 21-24). IEEE Press.

Zhou, S. J., Luo, Z. W., Wong, E., & Tan, C. J. (2007). Interconnected RFID reader collision model and its application in reader anti-collision. IEEE International Conference on *RFID* (pp. 26-28). IEEE Press.

KEY TERMS AND DEFINITIONS

Anti-Collision Protocol: Using for prevent or reduce the collisions of the protocol.

EPC-C1G2: EPCglobal Inc. published important standards for RFID, clearly defined relevant regulations of RFID reader and RFID tag communication, operating procedures, instructions and labels.

Query Tree: According the inquiry results, dividing to the child nodes of the tree.

Tag Collision: Multiple tags respond to a reader simultaneously and the reader cannot differentiate these tags correctly.

Tag Identification: Reader uses communication via radio waves to exchange data between a reader and an electronic tag.

Ternary Tree: A tree data structure, each node has at most three child nodes.

Chapter 8
Managing Tag Collision in RFID Data Streams using Smart Tag Anti-Collision Techniques

Prapassara Pupunwiwat
Griffith University, Australia

Bela Stantic
Griffith University, Australia

ABSTRACT

Radio Frequency Identification (RFID) is considered an emerging technology for advancing a wide range of applications, such as supply chain management and distribution. However, despite the extensive development of the RFID technology in many areas, the RFID tags collision problems remain a serious issue. Collision problems occur due to the simultaneous presence of multiple numbers of tags within the reader zone. To solve collision problems, different anti-collision methods have been mentioned in literature. These methods are either insufficient or too complex, with a high overhead cost of implementation. In this chapter, the authors propose a novel deterministic anti-collision algorithm using combinations of Q-ary trees with the intended goal to minimise memory usage queried by the RFID reader. By reducing the size of queries, the RFID reader can preserve memories, and the identification time can be improved. In addition, the chapter introduces a novel probabilistic group-based anti-collision method to improve the overall performance of the tag recognition process.

INTRODUCTION

Radio Frequency Identification (RFID) technology uses radio frequency waves to automatically identify people or objects. The main RFID systems consist of fast capturing radio frequency tags and networked electromagnetic readers. RFID technology is currently emerging as an important technology for advancing a wide range of applications. It has the potential to improve the efficiency of business processes by providing automatic identification and data capture. The current interest in RFID technology has grown rapidly and can now be certified by CompTIA RFID+ certification

DOI: 10.4018/978-1-4666-2080-3.ch008

in order to validate the knowledge and skills of professionals who work with RFID technology. In the modern world, RFID technology is used in different applications such as distribution and retail packaging, security, library system, defence and military, health care, and baggage and passenger tracing at the airport.

Chip-based RFID systems are mainly comprised of the following components:

- Chipped-Tag, which has a microchip attached to an antenna that transmits and responds to radio signals of a particular frequency. Chipped-Tag types are separated into three categories known as Passive Tag, Semi-Passive Tag, and Active Tag. In this chapter, we focus on passive tag, which does not have its own power source, and has no battery on-board. The tag obtains power from radio waves received from the reader. Passive Tags are small and light weight, and their functionalities are limited due to power source. Due to a lack of enough power, it cannot support an active transmitter to communicate with the reader. Passive tags are well suited in applications for which tags are not reusable, because of their low cost.
- Reader, which sends and receives RFID data to and from tags via antennas. Readers come in multiple formats, which can be separated into three main categories: Fixed readers, Handheld readers, and Vehicle-mount readers.
- Middleware, which pre-processes the RFID data and converts it into a meaningful Data.
- Application software, which is a specific component that resides on host computer.

In traditional RFID systems where chipped-tags are presented, there are several methods of identification. The most common is to store a serial number that uniquely identifies a person or object such as Electronic Product Code (EPC). All EPC numbers contain strings of binary numbers, which provide a unique identity for every physical object. All data captured by RFID readers before any further process are known as dirty data. In order to improve efficiency of database, dirty data must be filtered at the earlier stage soon after they were captured. The filtering of RFID data streams is known as filtering at the edge, where data are still meaningless and easier to eliminate. The main issue that usually arises in RFID data streams is the data stream errors. There are four typical errors, which include unreliable reads, noises, missed reads, and duplications/redundancies.

Several techniques for filtering RFID data have been proposed in literatures. However, these techniques only filter specific kind of errors generated. Therefore, the amount of wrong data is still recorded into the database. The most common errors are missed reads, which usually happen in a situation of low-cost and low power hardware that lead to a frequently dropped reads (Derakhshan et al., 2007). Another cause of missed reads is simultaneous transmissions in RFID systems, which lead to collisions as the readers and tags typically operate on the same channel. Tag collisions in RFID systems happen when multiple tags simultaneously reflect their respective signals back to the reader at the same time, preventing the reader from identifying all tags. Filling in dropped reads is one way to alter missed reads but it is sufficient to prevent missing data from the beginning. RFID collision problem can be solved by using anti-collision techniques, to prevent two or more tags from responding to a reader at the same time, and to re-identify them again when collisions occurred.

The current deterministic anti-collision methods suffer from identification delay and high memories usage during the identification process, while the probabilistic anti-collision methods suffer from tag starvation problems due to inaccurate frame-size estimation and low performance efficiency. In this research, a "Joined Q-ary tree" anti-collision schemes are proposed, based on deterministic Q-ary Tree. The motivation of this work is the improvement of data quality obtained, and the minimal use of memories required per complete identification. Methodologies for Joined Q-ary Tree are first derived and experimental evaluations are then conducted, in order to prove the efficiency of the proposed technique. The results and analysis of the experiments have indicated that Joined Q-ary Tree can effectively reduce total memories required, compared with current state-of-the-art techniques, which then results in the minimal identification delay.

Additionally, we propose a group-selection approach called a Probabilistic Cluster-Based Technique (PCT) method, to improve identification time and minimise number of frames and slots used during an identification process. The experiment results have indicated that, to achieve the best performance and solve the tag starvation problem, tags should be grouped into specific size according to PCT threshold and the parameters for frame-size prediction should be dynamically adjusted over the identification process.

COLLISION HANDLING IN RFID DATA STREAMS

RFID collision handling is one of the most heavily researched topics because it is a very important step to determine a quality of captured data. The better quality of data at the earlier stage of data processing means less complex algorithms are needed for RFID event process and database management. This section explains the type of each collision, taxonomy of RFID tag anti-collision protocols, and literature surveys on existing deterministic and probabilistic anti-collision methods.

RFID Collision Types

Simultaneous transmissions in RFID systems lead to collisions as the readers and tags typically operate on the same channel. Three types of collisions are possible: Reader-Reader collision, Reader-Tag collision, and Tag-Tag collision.

Reader-to-Reader Collisions

Interference occurs when one reader transmits a signal that interferes with the operation of another reader, and prevents the second reader from communicating with tags in its interrogation zone (Jain & Das, 2006). Reader-to-reader collision can be easily avoided by determining the appropriate reader's deployment that prevents direct signal interference between two or more readers.

Reader-to-Tag Collisions

Interference occurs when one tag is simultaneously located in the interrogation zone of two or more readers, where more than one reader attempts to communicate with that tag at the same time (Jain & Das, 2006).

Tag Collisions

Tag collision in RFID systems, sometimes known as Multi-Access, happens when multiple tags are energised by the RFID reader simultaneously, and reflect their respective signals back to the reader at the same time. This problem is often seen whenever a large volume of tags must be read together in the same reader zone. The reader is unable to differentiate these signals.

Taxonomy of RFID Tag Anti-Collision Approaches

The various types of anti-collision methods for multi-access/tag collision can be reduced to two basic types: probabilistic method and deterministic method (Klair et al., 2007; Choi & Lee, 2007; Bang et al., 2009; Alotaibi et al., 2009; Li et al., 2009; Klair et al., 2010; Zhu & Yum, 2011).

The deterministic method begins an identification process by issuing a prefix until it gets matching tags. Then it continues to ask for additional prefixes until all tags within the region are found. This method is slow but leads to fewer collisions and have high successful identification rate.

In probabilistic methods, tags respond at randomly generated times. If a collision occurs, colliding tags will have to identify themselves again after waiting for a random period of time. This technique is faster than deterministic but suffers from tag starvation problem where not all tags can be identified due to the random nature of chosen time.

There are also some hybrid anti-collision protocols that combine the advantages of tree-based and ALOHA-based approaches. From literature, it is clear that most hybrid protocols combine the Query Tree protocol with ALOHA variant (Klair et al., 2010).

Tree-Based Anti-Collision Methods

Deterministic methods can be classified into a Memory tree-based algorithm and a Memoryless tree-based algorithm. In the Memory algorithm, which can be grouped into Tree Splitting, Binary Search, and Bit Arbitration, the reader's inquiries and the responses of the tags are stored and managed in the tag memory. This results in an equipment cost increase especially for RFID tags. In contrast, in the Memoryless algorithm, the responses of the tags are not determined by the reader's previous inquiries. The tags' responses are determined only by the present reader's inquiries so that the cost for the tags can be minimised. Query Tree is classified as a Memoryless algorithm.

Depending on the number of tags that respond to the interrogator, there are three cycles of communication between tag and reader in deterministic approaches.

1. **Collision Cycle:** Collision cycle occurs when the number of tags that respond to the reader is more than one. The reader cannot identify the ID of tags.
2. **Idle Cycle:** Idle cycle occurs when there is no response from any tag to the reader. This type of cycle is unnecessary and should be minimised.
3. **Successful Cycle:** Successful cycle happens when exactly one tag responds to the reader and the reader can identify the ID of that tag.

Binary Search

Binary Search (BS) algorithm (Finkenzeller, 2003) involves the reader transmitting a string of EPC to tags, which the tag then compares against its ID. Those tags respond, with ID equal to or lower than the requested string. The reader then monitors tags' responses bit by bit using Manchester coding, where the value of bit is defined by the change in level (negative or positive transition) within a bit window. A logic 0 is coded by a positive transition, while a logic 1 is coded by a negative transition. The no transition state is not permissible during data transmission and is recognised as an error. Once a collision occurs in BS, the reader splits tags into subsets, based on collided bits.

The enhanced version of the BS protocol is called the Dynamic Binary Search Algorithm (DBSA) (Finkenzeller, 2003). In DBSA, the reader and tags do not use the entire length of EPC and tags ID during the identification process. For example, if a reader receives the response 01X, tags only need to transmit the remaining part of their ID since the reader has already identified the prefix 01. This enhancement effectively reduces the amount of data sent by the reader to tags.

Bit Arbitration

Bit Arbitration (BA) algorithms are memory-based anti-collision and are less robust than those within memoryless category. The key feature of

BA algorithms is that bit replies are synchronised, meaning that multiple tags' responses of the same bit value will result in no collision. A collision is observed only if two tags respond with different bit values. Moreover, the reader has to specify the bit position it wants to read. There are several algorithms in this category including ID Binary Tree Stack (Feng et al., 2006), Bit-by-bit Binary Tree (Jacomet et al., 1999), Modified Bit-by-Bit Binary Tree (Choi et al., 2004), and Enhanced Bit-by-Bit Binary Tree (Choi et al., 2004).

Tree Splitting

Tree Splitting (TS) protocols operate by splitting responding tags into multiple subsets, using a random number generator. In this category of anti-collision, the reader needs less preserving memory than those within Binary Search and Bit Arbitration categories because TS only needs to store information of random binary numbers. We present two algorithms in this category: "Binary Tree Splitting" and "Adaptive Binary Splitting".

The Binary Tree Splitting (BTS) uses random binary numbers generated for the splitting procedure (Myung & Lee, 2006a). The tag has a counter initialised to 0 at the beginning of the process. The tag transmits ID when the counter value is 0. The reader transmits a response to inform tags of the event of tag collision. The tag randomly generates a binary number when its transmission causes collision. By adding the selected binary number to the counter, a set is split into two subsets, '0' or '1'.

Tag identification in BTS protocols starts from one tag set including all tags, which cause more tag collisions for splitting tag sets. The colliding tag needs to re-transmit ID whenever tag collision occurs. The Adaptive Binary Splitting (ABS) uses information on the last frame of the tree and makes a new tag identification start from multiple tags sets. Hence, the reader can recognise tags with less collision. ABS begins tag identification from only readable cycles of the last frame and uses random numbers for splitting tag sets. This technique is an improvement on the BTS protocol. However, it requires extra memory to store information from previous frame. ABS requires tags to support both the transmission and reception at the same time.

Query Tree

In TS variants, tags require a random number generator and a counter to track their tree position, thus making them costly and computationally complex. Query Tree (QT) algorithms overcome these problems by storing tree construction information at the reader, and tags only need to have a prefix matching circuit. Numerous variants of query tree algorithms exist. Among all tree protocols, QT protocols promise the simplest tag design (Klair et al., 2010).

For the tree-based anti-collision, we focus on QT- based protocols because it is the most acceptable and is an effective anti-collision technique for passive Ultra-High Frequency (UHF) tags (Klair et al., 2010). There are several improved anti-collision methods based on QT, such as Adaptive Query Splitting (AQS) proposed by Myung & Lee (2006b), and a "Hybrid Query Tree" (HQT) proposed by Ryu et al. (2007). The AQS requires tags to support both the transmission and reception at the same time, thereby making it difficult to apply to low-cost passive RFID systems. On the other hand, the HQT managed to reduce collision cycles but at the same time introduce too many idle cycles. Accordingly, the QT Algorithm, which is currently adopted as the anti-collision protocol in EPC Class 1, may be limited to the tree based anti-collision protocol that can be implemented effectively (Choi et al., 2008).

The QT (Law et al., 2000) is a data structure for representing prefixes that is sent by the RFID reader. The QT algorithm consists of loops, and in each loop, the reader issues a query with specific prefixes, and the matching tags respond with their information. If only one tag replies, the reader successfully recognises the tag. If more than one tag tries to respond to reader's query, tag collision occurs and the reader cannot get any information about the tags. The reader, however, can recognise the existence of tags to have ID that matches the query. To further identify collided tags, the QT algorithm tries to query with 1-bit longer prefixes in next round of identification. By extending the prefixes, the reader can recognise all the tags.

The Adaptive Query Splitting (AQS) uses information on the last frame of the tree for tag identification, so that the reader can recognise tags with less collision. The AQS recognises tags with query that is sent by a reader, which includes a bit string. The basic idea of AQS is based on QT where tag responds with its ID when its first bits of ID are equal to the bit string of the query. The reader has queued Q, which maintains bit strings for queries. At the beginning of the frame, Q is initialised with queries of all the leaf nodes in the tree of the last frame. AQS keeps information that is acquired during the last identification process, in order to shorten the collision period. This technique also requires tags to support both the transmission and reception at the same time as in ABS. In addition, according to Bhatt & Glover (2006), the adaptive splitting protocol is only compatible with EPC class 0 and class 1 generation 1; and is more complex than basic BTS and QT.

Hybrid Query Tree (HQT) utilises a 4-ary query tree instead of a binary query tree (Ryu et al., 2007). This technique increases too many idle cycles despite reducing collision cycles, while extra memory needed also increases, as an identification process gets longer. This is because each query increases the prefixes by 2-bits instead of 1-bit. There is a basic Idle cycles elimination (slotted back-off tag response mechanism) for HQT, but this requires more time and memory. The extended version of HQT also requires extra memory, since it mimics the AQS for the last identification information to be kept. Nevertheless, HQT is better than QT in reducing collision between tags, especially at higher number of tags.

There are other improved version of QT, which enhances the performance but increases implementation cost, due to the more complex execution algorithms. These techniques include the Improved QT (ImpQT) algorithm (Zhou et al., 2004), the QT-based Reservation (QTR) algorithm (Choi et al., 2007), and the Intelligent Query Tree (IntQT) (Bhandari et al., 2006). For the deterministic anti-collision approaches, it is preferred that the algorithms are simple, since the adoption of the tree-based techniques is in the older RFID system. The recent technology uses ALOHA-based anti-collision algorithms rather than the tree-based. From the observation and literature survey, we discover that for tree-based approaches, the number of identification cycles, the total memory bits required, and the similarity of IDs, mostly affects the delay of tags' identification. Therefore, we make the assumption that by taking advantage of EPC pattern and bulky movement of items, the identification ability of the reader can be improved without the need for complex anti-collision algorithm.

ALOHA-Based Anti-Collision Methods

In a probabilistic approach, tags respond to readers at randomly generated times. If a collision occurs, colliding tags will have to identify themselves again after waiting a random period of time (Choi & Lee, 2007; Li et al., 2009; Bang et al., 2009; Klair et al., 2010). When we mentioned the probabilistic anti-collision approach in RFID, we usually refer to the ALOHA-based approach, which is the most widely used type of anti-collision. Slotted ALOHA (Quan et al., 2006), which initiates discrete time-slots for tags to be identified by reader at the specific time, was first employed as an anti-collision method in an early days of RFID technology. The principle of Slotted ALOHA techniques is based on the Pure ALOHA introduced in early 1970s (Abramson, 1970), where each tag is identified randomly. To improve the performance and throughput rate, different anti-collision schemes were suggested in the past literature. Framed-Slotted ALOHA technique is the most improved ALOHA-based technique currently applied in many applications. The three most accepted Framed-Slotted ALOHA techniques are Basic Framed-Slotted ALOHA, Dynamic Framed-Slotted ALOHA, and Enhanced Dynamic Framed-Slotted ALOHA. Several researchers (Wang et al., 2007; Lee et al., 2005, 2008a; Cho et al., 2007) have also attempted to improve the throughput rates by implementing a more accurate Frame-size Estimation algorithm.

BFSA Method

The Basic Framed-Slotted ALOHA (BFSA) is the most basic ALOHA-based algorithms that use a fixed frame-size throughout the identification round. The reader offers information to the tags, including the frame-size specification and the random number selected by each slot within the frame. Each tag selects a slot using the random number and then sends its ID back to the reader (Ding & Liu, 2009; Lee & Lee, 2006; Lee et al., 2008b). Since the frame-size of the BFSA is fixed, its implementation is simplistic. However, the system's efficiency drops significantly in the event of there being too large or too small tag counts. For instance, no tag may be identified in a read cycle if there are too many tags within the interrogation zone. On the other hand, under small tag counts where large frame-size is used, lots of empty slots are produced resulting in decreased system efficiency.

DFSA Method

The Dynamic Framed-Slotted ALOHA (DFSA) overcomes the problems associated with BFSA, by dynamically changing the frame-size according to estimated number of Backlog, which is a number of tags that have not been read. In DFSA, each tag in an interrogation zone selects one of the given N slots to transmit its identifier; and all tags will be recognised after a few frames. Each frame is formed of specific number of slots that is used for communication between the readers and the tags. To determine the frame-size, it gathers and uses information such as number of successful slots, empty slots, and collision slots from previous round, to predict the appropriate frame-size for the next identification round (Ding & Liu, 2009; Lee & Lee, 2006; Devarapalli et al., 2007). DFSA can identify the tag efficiently because the reader adjusts the frame-size according to the estimated number of tags. However, the frame-size change alone cannot sufficiently reduce the tag collision when there are a number of tags because it cannot increase the frame-size indefinitely. DFSA has various versions depending on different tag estimation methods used. There have been several researches to improve the accuracy of frame-size by implementing frame-size estimation techniques (Lee et al., 2005, 2008a; Cho et al., 2007).

According to the DFSA protocol, the reader picks tag within an interrogation zone by the command "Select", then issues "Query", which contains a 'Q' parameter to specify the frame-size (frame-size $F = 2Q - 1$). Each selected tag will pick a random number between 0 to $2Q - 1$ and put it into its slot counter. The tag, which picks zero as its slot number, will respond and backscatter its EPC to reader. Then reader issues "QueryRep" or "QueryAdjust" command to initiate another slot (Wang et al., 2007; Zhu & Yum, 2009).

Similar to the Tree-based anti-collision, there are three kinds of slot in ALOHA-based anti-collision: 1) Empty slot where there is no tag reply; 2) Successful slot where there is only one tag reply; and 3) Collision slot where there is more than one tag reply. The term initial Q refers to the first 'Q' or frame-size, which applies to a specific identification cycle. The reader first initiates a "Query" and broadcasts the signal to nearby tags. Since there is no tag that picks zero as its slot counter, the slot is counted as an empty slot. After the first "Query" was sent, each tag deducted its slot counter by one. The reader then sends "QueryRep" to tags in close proximity; and any tag that has zero as its slot counter replies. If there is only one tag that responds, a successful slot occurs and the tag replies to the reader with its RN16. When two tags respond to the reader at the same time, a collision slot occurs and in this case, no information is transmitted.

EDFSA Method

The DFSA algorithms change the frame-size to increase the performance efficiency of the tag identification. However, as the number of tags becomes larger than the frame-size, the probability of collision increases rapidly. If the number of unread tags can be estimated accurately, frame-size can be determined to maximise the system efficiency or minimise the tag collision probability. For instance, when the number of tags is large, the probability of tag collision can be reduced by increasing the frame-size. However, the frame-size cannot be increased indefinitely. When the number of unread tags is too large to achieve high system efficiency, the number of responding tags somehow must be restricted so that the optimal number of tags responds to the given frame-size (Lee & Lee, 2006; Lee et al., 2005).

The Enhanced Dynamic Framed-Slotted ALOHA (EDFSA) first estimates the number of unread tags. If the number of tags within the interrogation zone is larger than the maximum frame-size, the EDFSA algorithm splits the number of Backlog into number of groups and allows only one group of tags to respond. When the reader limits the number of responding tags, it transmits the number of tag sets and a random number to the tags, when it issues the query. Only the tag that picks zero as its slot counter responds to the request. If the number of estimated Backlog is below the threshold, the reader adjusts the frame-size without grouping the unread tags. After each read cycle, the reader estimates the number of unread tags and adjusts its frame-size. This procedure repeats until all the tags are read (Lee & Lee, 2006; Lee et al., 2005).

In EDFSA method, if the number of estimated unread tags is equal to or less than 354 tags, the EDFSA algorithm will not split tag into group. However, according to the rule, if there are more than 354 tags remaining in the interrogation zone, the EDFSA algorithm will split unread tags into groups. For instance, if there are 1245 estimated remaining tags, the EDFSA algorithm will divide tag into four groups (Lee & Lee, 2006). The problem with EDFSA method is that it assumes that 256 is the optimal frame-size and splits tags into group by using the power of two (2,4,8,...). This results in decreased system efficiency when the number of tags is just above the threshold and the number of group doubled.

Other ALOHA-Based Methods

There have been a number of methodologies proposed to improve the performance efficiency of ALOHA-based anti-collision methods. This includes partitioning algorithms (Shin & Kim, 2007; Kim, 2008), which have claimed to have had higher efficiency than the EDFSA approach, but lacks signalling robustness. Despite the wide array of approaches, only the BFSA, DFSA and EDFSA methods (Klair et al., 2010) are commonly used for comparative analysis in past literature. Additionally, Backlog estimation approaches have also been a popular research topic in this domain.

Backlog Estimation Techniques

In order to predict accurate number of unread tags and to determine the new frame-size for the next identification round, BFSA, DFSA, and EDFSA algorithms gather and use information such as number of successful slots, empty slots, and collision slots from previous round. There have been several other methods mentioned in literature related to Backlog estimation, including Schoute method (Schoute, 1983), Lowerbound method, Chen1 and Chen2 methods (Chen, 2006), Vogt method (Vogt, 2002), and Bayesian method

(Floerkemeier, 2007). Some of these methods are either having worse performances than simple Schoute and Lowerbound methods, or are too complicated to be implemented for RFID system.

- **Schoute Backlog Estimation Technique:** Schoute (1983) developed a *Backlog* estimation technique for Dynamic Framed-Slotted ALOHA using Poisson distribution. The *Backlog*, after the current frame Bt, is given by equation:

  $$\mathrm{Bt} = 2.29 \times \mathrm{c}$$

  where c represents the number of collided slot in the current frame, and Bt represents the remaining *Backlog*. This technique has the best performance where fewest frames were used, compared with other algorithms. Schoute method is the simplest, easy to implement with low overhead computation, and provides accurate tag estimation.
- **Lowerbound Backlog Estimation Technique:** The Lowerbound estimation function is obtained under the assumption that a collision involves at least two different tags. Therefore, Backlog after the current frame Bt is defined by equation:

  $$\mathrm{Bt} = 2 \times \mathrm{c}$$

  where c is the number of collided slot in the current frame, and Bt represents the remaining Backlog. Lowerbound method is also simple, easy to implement with low overhead computation, and provides accurate tag estimation.
- **Chen1 and Chen2 Estimation Techniques:** Most of the static algorithms estimate the *Backlog* with the number of collided slot. However, Chen1 method (Chen, 2006) estimates the *Backlog*, based on the empty slot information, through the probability of finding h empty slots after completing a frame. Chen2 method (Chen, 2006) is a simpler way to estimate the number of tags, which is illustrated by the following equation:

  $$\mathrm{n} = \left(\mathrm{L} - 1\right) \times \frac{s}{h}$$

  where n is the number of *Backlog*, L is frame length, s is the number of successful slots, and h is the number of empty slots. If h = 0, n is set to a certain upper bound for the tag's estimate. According to Wang et al. (2007), Chen1 and Chen2 methods have worse performances than simple Schoute method. Chen1 method also requires complex computation, which leads to high overhead and delays the tag identification process.
- **Vogt Estimation Techniques:** In (Vogt, 2002), a procedure to estimate *Backlog* is presented by minimising the difference between the observed value, including number of empty slot h, successful slot s, collision slot c, and the expected value E(H), E(S), E(C). In order to find the comparative precise *Backlog*, the reader needs to resolve the equation below:

  $$\min \left| \begin{pmatrix} h \\ s \\ c \end{pmatrix} - \begin{pmatrix} En(H) \\ En(S) \\ En(C) \end{pmatrix} \right|$$

  Vogt method presents the most accurate tag estimation. However, the complexity of the algorithm resulted in high overhead and therefore cannot be applied to EPC Gen2 protocol (Lee et al., 2008b).
- **Bayesian Estimation Techniques:** Bayesian method (Floerkemeier, 2007; Wu & Zeng, 2010) first computes the frame size L, based on the current probability

distribution of the random variable N that represents the number of tags transmitting. Then it starts frame with L slots, waits for tag replies, and updates probability distribution of N, based on evidence from the reader at the end of the frame. The evidence comprises the number of empty slots, successful slots, and collision slots in the last frame. The method then adjusts probability distribution N by considering newly arrived tags and departing tags, including the ones that successfully replied and did not transmit in subsequent slots. Bayesian method requires the most complex computation and implementation of algorithm. This results in high overhead and therefore delays the identification process.

Limitations of Existing Methods and Research Problem

Many challenges remained for missed reads, which is the most crucial issue in RFID applications, and is the hardest to identify and filtered. Filling in dropped readings is one way to alter missed reads, but it is easier to fix the error from the source where data is missing in the first place. The cause of these missed reads is the radio-frequency collision, which occurs when two or more tags attempt to respond to a reader at the same time. To solve collision problem, several anti-collision protocols are proposed in the literature. However, these approaches still suffer from performance inefficiency, high delay in identification time, and overhead computation of algorithms.

Limitation on Tree-Based Anti-Collision Techniques

There are several tree-based anti-collision techniques that can effectively prevent tag collisions. Most memory anti-collision algorithms including Binary Search, Bit Arbitration, and Tree Splitting, require higher computational complexity compared with the memoryless Query Tree. Nevertheless, some techniques from QT category still have drawbacks and limitations, as described below:

- Query Tree protocols suffer from a long identification delay in the case where there are a large number of tags within an interrogation zone. The delay is also caused by similarity of ID and mobility of tags, where tags are not static (stay at the same spot at all time).
- Adaptive Query Splitting technique introduces more complexity than QT because information on last identification must be kept, in order to accelerate identification process. This technique also requires tags to support both the transmission and reception at the same time, thereby making it difficult to apply to low-cost passive RFID systems.
- Hybrid Query Tree reduces collision cycles by querying 2-bits of prefixes for each loop instead of 1-bit as in QT. However, it produces even more idle cycles than QT because at one level it generates 4-leaf nodes, especially if there are not many tags in the interrogation zone.
- There is a basic Idle cycles elimination (slotted back-off tag response mechanism) for HQT, but this requires more time and memory. The extended version of HQT also requires extra memory since it mimics the AQS based for last identification information to be kept. However, HQT is better than QT in reducing collision between tags, especially at higher number of tags.

Limitation on ALOHA-based Anti-Collision Techniques

ALOHA-based anti-collision technique is the most widely used type of anti-collision within the probabilistic category. The earlier type of ALOHA anti-collision such as Pure ALOHA and Slotted

ALOHA perform poorly, while the more advanced Framed-Slotted ALOHA has better performance. However, some techniques from Framed-Slotted ALOHA category still have drawbacks and limitations, as described below:

- Basic Framed-Slotted ALOHA has the worst performance compared with other Framed-Slotted ALOHA methods. This approach suffers from inaccurate frame-size for each round of identification because it uses fixed frame-size. Therefore, the system's efficiency drops significantly in the event of there being too large or too small tag counts.
- Dynamic Framed-Slotted ALOHA suffers from different level of insufficiency, depending on frame-size prediction technique applied. If the number of unread tags are not estimated accurately, correct frame-size cannot be determined to maximise the system efficiency or minimise the tag collision probability. Thus, the performance of DFSA depends highly on the selection of frame-size estimation technique.
- Enhanced Dynamic Framed-Slotted ALOHA assumes that the optimal frame-size is fixed to 256. The number of group in EDFSA increases, using the power of two (2,4,8...), which results in decreased system efficiency when the number of tags is just above the threshold, and the number of group doubled.
- Current Backlog Estimation methods suffer from low performances or are too complicated to be implemented for RFID system. Schoute's method is the simplest, easy to implement with low overhead computation, and provides accurate tag estimation. Other Backlog Estimation methods such as Chen1 and Chen2 methods, Vogt method, and Bayesian method, have good simulated performance but cannot be realistically applied to the actual passive RFID system. Specifically, the Bayesian method requires the most complex computation and implementation of algorithm.

Overall, the literature review on current state-of-the-art techniques demonstrates that some of the existing techniques are inefficient, while other methods are too complex with high overhead cost of implementation. Some approaches cannot be further improved but we can take advantage of other constraints, to improve their capability. For instance, basic tree-based methods such as QT (2-ary) and HQT (4-ary) are the best naive tree-based methods but cannot be improved any further, in terms of simplicity, without the need for complex algorithm. Thus, we need to take advantage of other constraints such as EPC pattern, and a possible use of a combination of two trees, in order to improve memory and power efficiency. Additionally, for probabilistic anti-collision, the DFSA method is the simplest and most accurate method. However, in this case, to keep the simplicity of the DFSA algorithm, only *frame-size* prediction scheme can be further improved.

It remains an open problem to find optimal solutions, to improve performance of the current RFID anti-collision techniques. Two main goals for both tree-based deterministic and ALOHA-based probabilistic anti-collision methods are to achieve the maximum efficiency and to minimise identification time and resource wasted during the identification process. Structuring anti-collision methods in RFID system is extremely important because it is a step that determines the effectiveness and the overall quality of data captured.

In this chapter, the research problem is to investigate a suitable structure of tree-based deterministic and ALOHA-based probabilistic anti-collision approaches such that new efficient methods can be developed to improve performance of anti-collision technique in RFID system. Given the limited resources of RFID components including the readers and the tags, it is important to develop the anti-collision method that minimises power and memory usage in the RFID reader, and to simplify the structure of algorithm so that identification time can be minimised.

There are two main constraints in developing effective anti-collision algorithms. These include limited power source from RFID reader and limited memory in both readers and tags. By constructing complex anti-collision algorithms, high memory capacity and power sources are needed, which is impractical in RFID system. Therefore, our aim is to develop anti-collision schemes that are simple, with low overhead computation, and perform effectively, compared with existing techniques. To address our research problem, we compare our newly proposed methods to specific existing approaches, which have simple algorithm structure, high robustness, and have accomplished high performance with minimum time requirement.

DETERMINISTIC ANTI-COLLISION TECHNIQUES

In this section, we tackle problems on existing deterministic tree-based anti-collision schemes including the amount of identification cycles produced and total memories used during the identification process. We introduce a Joined Q-ary Tree with the intended goal to minimise memory usage queried by the RFID reader. As indicated in literature (Choi et al., 2008), most implementation of Tree-based algorithms are deployed with older type of EPC class 1, which has limited memory and capability. Although recent technology uses ALOHA-based anti-collision algorithms rather than the Tree-based, it is necessary to improve the Tree-based approach with simple implementation in order to suit the backward compatibility for older RFID systems. The remaining of this section comprises the explanation of GID-96 bits EPC encoding schemes, the typical scenarios discussion, the Splitting Fitness justification, the foundation of Joined Q-ary Tree, and the experimental evaluation.

General Identifier 96 Bits

The General Identifier (GID) is defined for a 96-bit EPC, and is independent of any existing identity specification or convention. In addition to the Header which guarantees uniqueness of the encoding type, the General Identifier is composed of three fields; the General Manager Number (GMN), Object Class (OC) and Serial Number (SN), as shown below:

- **Header (H):** 8 bits; Binary 0011 0101
- **General Manager Number (GMN):** 28 bits; Max Decimal of 268,435,455
- **Object Class (OC):** 24 bits; Max Decimal of 16,777,215
- **Serial Number (SN):** 36 bits; Max Decimal of 68,719,476,735

The GID-96 includes three fields in addition to the Header, with a total of 96-bits binary value. Only 'H' is shown in Binary, while the rest are shown in Decimal. The general structure of EPC tag encodings is a string of bits, consisting of a fixed length (8-bit) Header followed by a series of numeric fields whose overall length, structure, and function are completely determined by the Header value. There are four major fields in the GID-96 bits.

Warehouse Distribution Scenarios

We examine specific scenarios based on the assumption that items tend to move and stay together through different locations especially in a large warehouse. We focus on Crystal warehouse scenario using GID-96 bits encoding scheme, which can be classified into four different scenarios: 1) Unique Item-Level, 2) Unique Container-Level, 3) Unique Company-Level, and 4) Unique Warehouse-Level.

Unique Item-Level Scenario

This scenario occurs when two collided tags (GID-96 encoding) are captured and they have the same Encoding Scheme (Header), same GMN, same OC, but different SN. We can assume that all items are from the same warehouse that uses the same encoding scheme throughout the warehouse, and the warehouse also keeps different kind of products from different companies. The Crystal Warehouse Scenario follows:

1. **Unique Item-Level:** Two containers of crystal red-wine have the same Header (=), GMN (=), and OC (=), but different SN (≠).
2. **Unique Container-Level:** Crystal white-wine and crystal red-wine containers have the same Header (=) and GMN (=), but different OC (≠) and SN (≠).
3. **Unique Company-Level:** Crystal white-wine and crystal plate containers have the same Header (=), but different GMN (≠), OC (≠), and SN (≠).
4. **Unique Warehouse-Level:** Crystal plate and plastic plate containers have different Header (≠), GMN (≠), OC (≠), and SN (≠).

For Unique Item-Level circumstance, by using the above Crystal Warehouse Scenario example 1), it can be seen that two collided tags are captured with the same Encoding Scheme, General Manager Number, and Object Class. We believe that both tags are each attached to two different cases of crystal red-wine.

Unique Container-Level Scenario

The Unique Container-Level Scenario takes place when two collided tags are captured and they have the same Header, same GMN, different OC, and different SN. Crystal Warehouse Scenario 2) shows that crystal red-wine glasses and crystal white-wine glasses are packed in different case and pallet because they are different type of wine glasses. Within this scenario, each case of wine glasses will have a unique SN attached to it, with different OC for each pallet of white-wine or red-wine.

Unique Company-Level Scenario

The Unique Company-Level Scenario is shown in Crystal Warehouse Scenario 3). Two collided tags are captured and they have the same Header, and unique GMN, OC, and SN. We believe that one tag is attached to crystal plate case, while the other tag is attached to crystal white-wine case. We can assume that there are two different companies producing separate crystal ware; and the wine glasses and plates are from different companies but share the same warehouse because they are both crystal.

Unique Warehouse-Level Scenario

Unique Warehouse-Level Scenario occurs when two collided tags are captured and they have different Header, GMN, OC, and SN. We can assume that all items are from different companies that use different encoding schemes. For example, Crystal Warehouse Scenario 4) shows that two wine glasses with different sculpture, one made from crystal and the other from plastic, are allocated in the same warehouse. This Unique Warehouse-Level scenario will not be discussed any further in this chapter because we are only looking at a large warehouse distribution where most items move together as a group. Therefore, most items from the same type of manufacturing will stick together until they are deployed to smaller retailer.

Splitting Fitness

Splitting Fitness is the measurement level for the performance of our proposed tree-based anti-collision methods. Splitting Fitness can be classified into Worst-Case splitting, Perfect splitting, and Random splitting.

Worst-Case Splitting

Worst-Case splitting is when tags spliced into an unbalanced tree, where one child node has no further node in a binary tree case. Figure 1a) shows that there are 16 tags at Level 0 tree; then at Level 1, tags spliced into 16 tags on the left-hand node and no tag on the right-hand node. As there is no tag left, no further splitting is necessary on the right-hand node. This case of splitting will likely happen for the first few bits of EPC identification in real world warehouse environment because most items have Massive tag movement and usually belong to the same EPC pattern with similar ID. The Worst-Case splitting caused more Idle cycles because all tags will be travelling down to only one side of the tree, which results in further collision.

Perfect Splitting

Perfect splitting happens when a set of tags spliced to the left and right child node equally. Figure 1b) shows that there are 16 tags at Level 0 tree; then at Level 1, tags spliced equally into 8 nodes. Further splitting is required for both left-hand and right-hand nodes until only one tag is left. This case of splitting is almost impossible in real world scenario but will be the closest case to the latter stages (bits) of EPC identification within warehouse environment because most items belong to the same group of EPC pattern. For example, one pallet of white-wine glasses containing 20 cases move into one interrogation zone. All items from the pallet will have the same OC and will travel along the same side of child node at earlier levels; resulting in Worst-Case Splitting. However, the remaining few bits will be unique for each EPC because they belong to SN. These remaining bits encoded within the same EPC pattern will split almost equally to the left and right child nodes. Both child nodes of left-hand side and right-hand side of binary tree will not be exactly equal since data captured are not always even. Therefore, we call this situation Partial-Perfect splitting.

Random Splitting

Random splitting happens when a set of tags spliced to the left and right child node randomly and splitting pattern cannot be found. Figure 1c) shows that there are 16 tags at Level 0 tree; then at Level 1, tags spliced into 5 and 11 tags. Further at Level 2, tags spliced into 2, 3, 4, and 7 where no specific splitting pattern exists. Thus, this situation is called Random Splitting, which will likely happen in retail distribution environments (belong to Unique Warehouse-Level Scenario) because all items usually come from different locations. Therefore, this splitting case will not be

Figure 1. Splitting fitness: a) Worst-case splitting, b) Perfect splitting, and c) random splitting

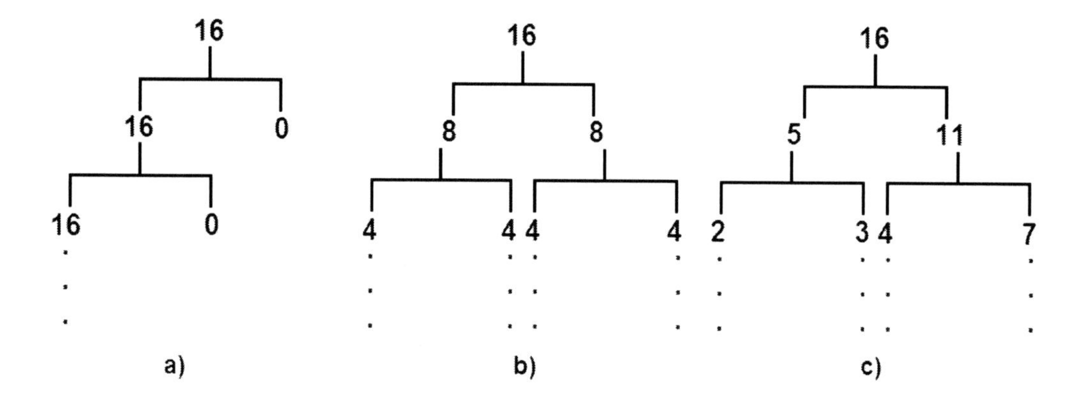

further discussed as this chapter will only focus on warehouse environment.

Joined Q-ary Tree

The Joined Q-ary Tree employs the right combination of Q-ary trees for each specific scenario. The joined Q-ary Tree adaptively adjusts its tree branches to suit EPC pattern. This procedure will further reduce accumulative bits from the reader's queries and improve the robustness of the overall identification process. The Joined approach is a combined Q-ary trees, specifically 2-ary tree and 4-ary tree, which have been identified to be the best Q-ary trees in literatures. The Joined approach will be applied on each collided tags EPC, which will be split using every 1 or 2-bits of tag ID for the first few queries; and then at one point, every 1 or 2-bits will be queried. In order to optimise the performance of Joined Q-ary Tree, the right Separating Point (SP) between the two Q-ary trees needs to be configured. The objective of Joined Q-ary Tree is to reduce the Bits Length queried by a reader so that identification time can be minimised. In this section, we will investigate and compare the "Naive Q-ary Tree" approach and our newly proposed "Joined Q-ary Tree".

Figure 2 shows the example of a) Naive 2-ary, b) Naive 4-ary, and c) Joined Q-ary Tree. Joined Q-ary Tree bonded both 2-ary and 4-ary trees together and applied to specific bits of EPC, depending on how Identical or Unique they are.

EPC Bits Prediction and Classification

In warehouse distribution environment according to Unique Item-Level and Unique Container-Level Scenarios, it is known that the first 36-bits of EPC (Header and GMN) are definitely identical. However, 24-bits of OC can be both Identical and Unique for all tags, depending on how many pallets existed within one interrogation zone. For example, if there are five pallets of 12 cases each in the interrogation zone, there will be five different OC and sixty unique SN for all sixty items (cases). Since OC involved 24-bits of EPC (allow 16,777,215 unique tags) but only five unique OC is needed, we must calculate a certain number of Unique bits needed in order to apply the right Q-ary tree. This also applies to SN that contains 36-bits of string. Assuming that EPC pattern is used, not all 36-bits of these strings will be Unique.

Our method is executed based on the assumption that the approximate number of tags (pallets, cases) is known, prior to the identification process. This information is needed for Unique bits calculation: UOC and USN, as shown below. However, in most circumstances, number of tags is usually unknown until the first query is issued by the reader. Therefore, UOC and USN of Joined

Figure 2. A sample of: a) a Naive 4-ary Tree, b) a Naive 2-ary Tree, and c) a Joined Q-ary Tree

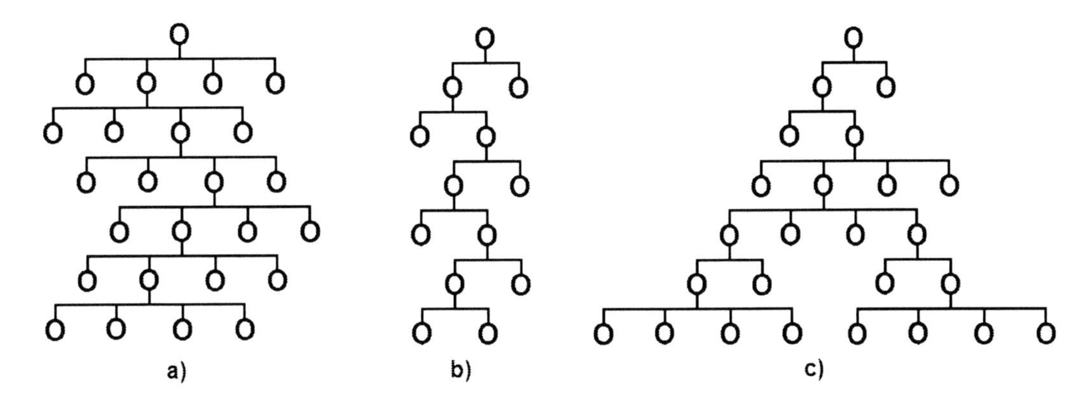

Q-ary Tree can be initially set to zero and after the first round of identification, these two parameters can be computed.

- **Header:** 8 bits Total; 8 bits Identical; 0 bit Unique
- **General Manager Number:** 28 bits Total; 28 bits Identical; 0 bit Unique
- **Object Class:** 24 bits Total; 24-UOC bits Identical; UOC* bit Unique
- **Serial Number:** 36 bits Total; 36-USN bits Identical; USN** bit Unique

The above shows formal structure of bits classification of EPC GID-96 bits. *UOC is number of Unique bits within Object Class and **USN is number of Unique bits within Serial Number

Joined Q-ary Tree adaptively adjusts their tree branches at specific SP. These SP is configured according to Identical bits and Unique bits within an EPC data. In order to calculate the estimated number of Unique bits within an EPC, we need the average number of tags within an interrogation zone, and then to apply the equation below:

$$B = \log_2 n \quad (1)$$

where N = Number of tags, B = Unique Bits of EPC.

Unique Bits Computation

To demonstrate a calculation of Unique bits of EPC, we examine a Massive tag movement of 720 tags within 12 pallets (OC). By using the following equations, UOC and USN can be calculated:

$$UOC = \frac{\log_{10}(N)}{\log_{10}(2)} = \frac{\log_{10}(12)}{\log_{10}(2)} = 4$$

$$USN = \frac{\log_{10}(N)}{\log_{10}(2)} = \frac{\log_{10}(60)}{\log_{10}(2)} = 6$$

Therefore, number of Unique bits required to cover all unique OC is approximately 4-bits and approximately 6-bits for SN.

Experimental Evaluation

In order to show the significance of our proposed Joined Q-ary Tree methods, we conducted an experimental evaluation and compared our methods with existing techniques. There are three major data sets in the experiment. We performed ten runs on each test case and presented the average results.

Experiment Data Sets

In this experiment, we conducted an experiment using three different tag sets: 288 tags, 576 tags, and 864 tags. The impact of different number of tags in an interrogation zone and performances of Joined Q-ary Tree approach is to be evaluated.

There are three test cases used in this experiment:

- **Test Case A:** 12 pallets, 24 cases each, total 288 tags
- **Test Case B:** 24 pallets, 24 cases each, total 576 tags
- **Test Case C:** 36 pallets, 24 cases each, total 864 tags

For the joined Q-ary Tree approach, the SP of each test case must be calculated using Equation 1.

Theoretical Bits Prediction

Assuming that the existence of tags are known before identification process, by using Equation 1, UOC and USN of Test case A, B, and C can be calculated as follows:

- **Test Case A:** UOC = $\log_2(12) = 4$;USN = $\log_2(24) = 5$
- **Test Case B:** UOC = $\log_2(24) = 5$;USN = $\log_2(24) = 5$

- **Test Case C:** UOC = $\log_2(36) = 6$;USN = $\log_2(24) = 5$

Results

Based on the experiment results shown in Figure 3a), the Joined Q-ary Tree always performed the best out of the three approaches considered, while the 4-ary tree has the worst performance, regardless of number of tags within an interrogation zone. This corresponds with our methodology that if the Separating Point and the Q-ary trees are applied correctly to the EPC data, the optimal results can be achieved by the Joined Q-ary Tree.

Figure 3b) demonstrates that 2-ary tree has the better performance than 4-ary tree by about 1 percent, while Joined Q-ary Tree's performance is approximately 12 percent better than the 2-ary tree. The 4-ary tree has the worst performance out of the three approaches considered. The percentage of improvement increases more slowly once the number of tags within the interrogation zone gets higher.

PROBABILISTIC ANTI-COLLISION TECHNIQUE

In this section, we tackle issues of existing probabilistic anti-collision schemes, such as the amount of slots and frames produced during each identification process, and the performance efficiency. We introduce the Probabilistic Cluster-Based Technique (PCT) anti-collision method to improve the performance of tag recognition process and provide a sufficient performance over existing methodologies. The remaining of this section comprises the mathematic fundamental for probabilistic anti-collision schemes, the foundations of the proposed PCT methods, and the experimental evaluation.

Mathematic Fundamental for ALOHA-Based Tag Estimation

In the Framed-Slotted ALOHA based probabilistic scheme, to estimate the number of present tags, Binomial distribution is a good fundamental

Figure 3. Performances comparison (a) and Percentage of improvement (b) between Naive approaches and Joined Q-ary approach

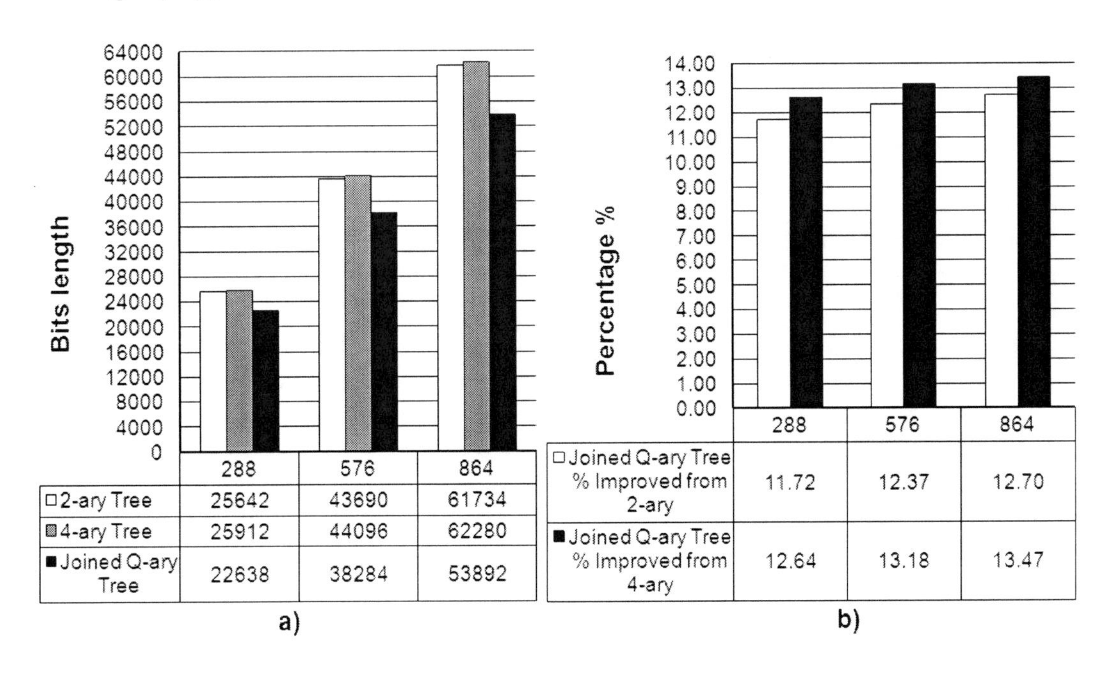

method (Wang et al., 2007; Li et al., 2009; Fan et al., 2008a, b). For a given initial Q in a frame with F slots and n tags, the expected value of the number of slots with occupancy number x is as follows:

$$a_x = n \times C_x^n \left(\frac{1}{F}\right)^x \left(1 - \frac{1}{F}\right)^{n-x}$$

Therefore, the expected number of Empty slot e, Successful slot s, and Collision slot c is given by the following equations:

$$\begin{cases} e = a_0 = F\left(1 - \frac{1}{F}\right)^n \\ s = a_1 = n\left(1 - \frac{1}{F}\right)^{n-1} \\ c = a_k = F - a_0 - a_1 \end{cases}$$

Thus, the system efficiency (E) is defined as the ratio between the number of Successful slot and the frame-size, as per the following equations:

$$E = \frac{s}{F} = \frac{n\left(1 - \frac{1}{F}\right)^{n-1}}{F} = n\frac{1}{F}\left(1 - \frac{1}{F}\right)^{n-1}$$

It has been proven that the highest efficiency can be obtained if the frame-size F is equal to the number of tags n, provided that all slots have the same fixed length (Wang et al., 2007):

$$\mathrm{F(optimal)} = \mathrm{n}$$

Therefore, we make the assumption that by keeping the number of tags close to the available *frame-size*, the optimal performance efficiency can be obtained. Moreover, from past literatures (Cheng & Jin, 2007), it is possible to achieve the theoretically optimal efficiency of 36.8 percent in ALOHA-based systems.

Probabilistic Cluster-Based Technique

The PCT method employs a dynamic probabilistic algorithm concept, and uses group-splitting rule, to split Backlog into group if the number of unread tags is higher than the maximum frame-size. The PCT approach first estimates the number of Backlog, or the remaining tags, within the interrogation zone. If the number of Backlog is larger than the specific frame-size, it splits the number of Backlog into a number of groups and allows only one group of tags to respond. The reader then issues a "Query", which contains a 'Q' parameter to specify the frame-size (frame-size F(min) = 0; F(max) = 2^Q - 1). Each selected tag in the group will pick a random number between 0 to 2^Q - 1 and put it into its slot counter. Only the tag that picks zero as its slot counter responds to the request. When the number of estimated Backlog is below the threshold, the reader adjusts the frame-size without grouping the unread tags. After each read cycle, the reader estimates the number of Backlog and adjusts its frame-size.

PCT Algorithm

PCT approach first estimates the number of unread tags, then it decides if the number of tags needs to be spliced or not. The probabilistic anti-collision algorithm with frame-size prediction is then applied to each selected group of tag.

Algorithm 1 contained in Figure 4 demonstrates the probabilistic anti-collision algorithm applied to each selected group of tags, where only one group of tags responds to the reader. There are three kinds of slot:

1. **Successful Slot:** Where there is only one tag reply, the reader sends ACK(RN16) to

a tag. The tag then backscatters its EPC to the reader and the reader issues QueryRep for the next slot.

2. **Empty Slot**: Where there is no tag reply, the reader then issues QueryRep for the next slot.
3. **Collision Slot:** Where there is more than one tag reply, the reader then issues QueryRep for the next slot.

After "QueryRep" command is received, each tag decreases its slot counter by 1. At the end of each frame, the reader checks if all tags have been identified. Then, the reader estimates the number of Backlog using frame-size estimation algorithm, and adjust its frame-size.

PCT Preliminary

Instead of splitting tags into group randomly, the PCT approach derived new rules using particular equations, according to the optimal system efficiency obtained for specific number of tags. To the best of our knowledge, the optimal system efficiency achieved by the probabilistic ALOHA method is approximately 38 percent and the optimal number of tags is close to the maximum frame-size. Efficiency is calculated as shown in Equation 2:

$$\text{Efficiency} = \frac{S}{(S + C + E)} \quad (2)$$

Figure 4. Probabilistic anti-collision algorithm with frame-size prediction

Algorithm 1

```
Reader sends Query
for (Identification procedure) do
|   Every tags generate RN16 and slot counter;
|   for (Current frame) do
|   |   if (Slot counter == 0) then
|   |   |   Tag replies its RN16;
|   |   |   if (A single tag replies) then
|   |   |   |   Reader sends ACK(RN16) to a tag;
|   |   |   |   if (RN16 received by tag == RN16 tag saved data) then
|   |   |   |   |   Tag sends (EPC+PC+CRC) to reader;
|   |   |   |   end
|   |   |   |   Reader sends QueryRep;
|   |   |   end
|   |   |   else if (Multiple tags reply) then
|   |   |   |   Reader sends QueryRep;
|   |   |   end
|   |   |   else if (No tag replies) then
|   |   |   |   Reader sends QueryRep;
|   |   |   end
|   |   end
|   |   if (Tag receives QueryRep) then
|   |   |   slot counter = slot counter - 1;
|   |   end
|   end
|   Reader uses frame-size estimation algorithm to adjust the size of the new frame;
|   Reader sends QueryAdjust;
end
```

where S is the number of Successful slots, C is the number of Collision slots, and E is the number of Empty slots.

From the results acquired for performance efficiency evaluation, we have developed Equations 3, 4, 5, 6 and 7 to find a minimum and maximum number of tags suitable for particular frame-size. These minimum and maximum numbers of tags are derived to acquire the optimal performance efficiencies. Each equation is then used to exploit rules for PCT.

$$\begin{aligned} \max &= 2^{Q} + 2^{(Q-1)} - 2^{(Q-2)} + 2^{(Q-3)} \\ \min &= \left(2^{(Q-1)} + 2^{(Q-2)} - 2^{(Q-3)} + 2^{(Q-4)}\right) + 1 \end{aligned} \tag{3}$$

$$\begin{aligned} \max &= \left(2^{Q} + 2^{(Q-1)} - 2^{(Q-2)} + 2^{(Q-3)}\right) \\ &\quad + \left(2^{(Q-2)} + 2^{(Q-3)} - 2^{(Q-4)} + 2^{(Q-5)}\right) \\ \min &= \left(2^{Q} + 2^{(Q-1)} - 2^{(Q-2)} + 2^{(Q-3)}\right) + 1 \end{aligned} \tag{4}$$

$$\begin{aligned} \max &= \left(2^{Q} + 2^{(Q-1)} - 2^{(Q-2)} + 2^{(Q-3)}\right) \\ &\quad + \left(2^{(Q-1)} + 2^{(Q-2)} - 2^{(Q-3)} + 2^{(Q-4)}\right) \\ \min &= \left(2^{Q} + 2^{(Q-1)} - 2^{(Q-2)} + 2^{(Q-3)}\right) + 1 \end{aligned} \tag{5}$$

$$\begin{aligned} \max &= \left(2^{Q} + 2^{(Q-1)} - 2^{(Q-2)} + 2^{(Q-3)}\right) \\ &\quad + \left(2^{(Q-1)} + 2^{(Q-2)} - 2^{(Q-3)} + 2^{(Q-4)}\right) \\ \min &= \left[\begin{matrix} \left(2^{Q} + 2^{(Q-1)} - 2^{(Q-2)} + 2^{(Q-3)}\right) + \\ \left(2^{(Q-2)} + 2^{(Q-3)} - 2^{(Q-4)} + 2^{(Q-5)}\right) \end{matrix}\right] + 1 \end{aligned} \tag{6}$$

$$\begin{aligned} \max &= \left(2^{Q} + 2^{(Q-1)} - 2^{(Q-2)} + 2^{(Q-3)}\right) \\ &\quad + \left(2^{(Q-1)} + 2^{(Q-2)} - 2^{(Q-3)} + 2^{(Q-4)}\right) \\ &\quad + \left(2^{(Q-2)} + 2^{(Q-3)} - 2^{(Q-4)} + 2^{(Q-5)}\right) \\ \min &= \left[\begin{matrix} \left(2^{Q} + 2^{(Q-1)} - 2^{(Q-2)} + 2^{(Q-3)}\right) + \\ \left(2^{(Q-1)} + 2^{(Q-2)} - 2^{(Q-3)} + 2^{(Q-4)}\right) \end{matrix}\right] + 1 \end{aligned} \tag{7}$$

In order to simplify the derived equations, we employ the use of β(Beta), κ (Kappa), and μ (Mu), and assigned these three icons to express each rule. In this research, we proposed three rules for PCT: PCT256, PCT128, and PCT-E (PCT-Extended). All rules split the number of Backlog into groups then used one of initial Q8 (frame-size 256), Q7 (frame-size 128), or Q6 (frame-size 64), to identify a current set of tags. Equation 8 shows the conversion of all three key sets, from Equations 3 to 7, into β, κ, and μ.

$$\begin{cases} \beta = 2^{Q} + 2^{Q-1} - 2^{Q-2} + 2^{Q-3} \\ \kappa = 2^{Q-1} + 2^{Q-2} - 2^{Q-3} + 2^{Q-4} \\ \mu = 2^{Q-2} + 2^{Q-3} - 2^{Q-4} + 2^{Q-5} \end{cases} \tag{8}$$

From Equation 3, 4, 5, 6 and 7, we derived three key sets within these equations. These key sets are converted into β, κ, and μ and are applied into each PCT rule, as shown in Table 1. Table 1 displays the conversion of Equations 3 to 7, with the minimum and maximum boundaries for each rule. For instance, Equation 3 is applied to all three rules: PCT256, PCT128, and PCT-E. However, Equation 4 only applies to PCT-E.

PCT Rules

PCT approach derived new rules using particular equations expressed by β Beta, κ Kappa, and μ Mu. All rules split the number of Backlog into groups then used one of Q8 (frame-size 256), Q7 (frame-size 128), or Q6 (frame-size 64), to identify a current set of tags. We make the assumption that the performance efficiency can be improved by dividing tags into accurate number of groups, and then performing the tag identification separately for each group. In this research, we have chosen the *frame-size* of 256, 128, and 64 for our PCT rules since the initial Q of 8, 7 and 6 provide the most appropriate range for the current RFID reader and passive tags specification. Generally, the UHF reader is capable of capturing variety numbers of

Table 1. The conversion of PCT rules to β Beta, κ Kappa, and μ Mu

	PCT256	PCT128	PCT-E
Max Min	β κ + 1 (3)		
Max Min	β + κ β + 1 (5)		β + μ β + 1 (4)
			β + κ [β + μ] + 1 (6)
			β + κ + μ [β + κ] + 1 (7)

passive tags, depending on the reader type and tag class (e.g. Class 0: Read-only tag). Thus, selected initial Qs are the most suitable for our proposed rules. Each PCT rule, with the minimum and maximum boundaries, is explained as follows:

PCT256

PCT256 uses either frame-size of 256 (Q = 8) or frame-size of 128 (Q = 7) for tag identification. We assume that the identification time and performance efficiency of our proposed PCT256 will advance from the existing probabilistic approaches. From the preliminary for all PCT rules, we obtained specific equations to calculate minimum and maximum boundaries for the PCT256 rule. For example, the minimum boundary for PCT256 is calculated by 3β + 1 when the number of group division comprises three groups of 256 and one group of 128, and the maximum boundary is calculated by 3β + κ. Following the computation, the minimum and maximum boundaries are 1057 and 1232 respectively, as shown in Table 2. The detailed calculation is present as follows:

$$\begin{aligned} \max &= 3\beta + \kappa \\ &= 3\left(2^{8} + 2^{(8-1)} - 2^{(8-2)} + 2^{(8-3)}\right) \\ &\quad + \left(2^{(8-1)} + 2^{(8-2)} - 2^{(8-3)} + 2^{(8-4)}\right) \\ &= 1232 \\ \min &= 3\beta + 1 \\ &= 3\left(2^{8} + 2^{(8-1)} - 2^{(8-2)} + 2^{(8-3)}\right) + 1 \\ &= 1057 \end{aligned}$$

After applying specific equations for each group division, Table 2 shows the final PCT rule for PCT256. For instance, if the number of Backlog equals to 900 tags, the PCT256 algorithm will split the unread tags into three groups of Q8 (256).

Algorithm 2 in Figure 5 demonstrates the group splitting algorithm using PCT256 rule, and either keep tag in a single group or split tag into number of groups according to PCT256 rule.

PCT128

PCT128 uses either frame-size of 128 (Q = 7) or frame-size of 64 (Q = 6) for tag identification. The PCT128 contains higher number of groups in some cases, compared with the PCT256 method, which may result in worse performance efficiency for specific number of tags. We calculate minimum and maximum boundaries for the PCT128 rule according to specific equations. For example, the minimum boundary is calculated by 5β + 1 when the number of group division comprises five groups of 128 and one group of 64, and the maximum boundary is calculated by 5β + κ. Following the computation, the minimum and maximum boundaries are 881 and 968 respectively, as shown in Table 3.

Table 3 shows the PCT rule for PCT128. For instance, if the number of *Backlog* equals to 900 tags, the PCT128 algorithm will split the unread tags into five groups of Q7 (128) and one group of Q6 (64).

Algorithm 3 in Figure 6 demonstrates the group splitting algorithm using PCT128 rule, and either keep tag in a single group or split tag into number of groups according to PCT128 rule.

PCT-Extended

The rules of PCT-Extended (PCT-E) are more complex than the PCT256 and PCT128. This is because the PCT-E identifies tags using three different frame-size of 256 (Q = 8), 128 (Q = 7), and 64 (Q = 6) instead of two. We assume

Table 2. PCT256 rule - The number of unread tags, optimal frame-size (A and B), and number of group (A and B)

PCT256 rule				
Backlogs	**FS A**	**G A**	**FS B**	**G B**
...	...	...	...	...
1233 to 1408	256	4	-	-
1057 to 1232	256	3	128	1
881 to 1056	256	3	-	-
705 to 880	256	2	128	1
529 to 704	256	2	-	-
353 to 528	256	1	128	1
177 to 352	256	1	-	-
89 to 176	128	1	-	-
45 to 88	64	1	-	-
23 to 44	32	1	-	-
12 to 22	16	1	-	-
6 to 11	8	1	-	-
...	...	...	...	...

that the performance efficiency of PCT-E can improve further from the PCT256. However, the identification time may increase due to the higher number of group applied in each identification round. From the preliminary for all PCT rules, we obtained specific equations to calculate minimum and maximum boundaries for the PCT-E rule.

For instance, the minimum boundary is calculated by $[2\,\beta + \kappa] + 1$ when the number of group division comprises two groups of 256, one group of 128, and one group of 64; and the maximum boundary is calculated by $2\,\beta + \kappa + \mu$. Following the computation, the maximum and minimum boundaries are 881 and 968 respectively, as shown in Table 4.

Table 4 displays the PCT-E rule. For instance, if the number of *Backlog* equals to 900 tags, the PCT-E algorithm will split the unread tags into two groups of Q8 (256), one group of Q7 (128), and one group of Q6 (64).

Algorithm 4 in Figure 7 demonstrates the group splitting algorithm using PCT-E rule, and either keep tag in a single group or split tag into number of groups according to PCT-E rule.

Experimental Evaluation

In order to show the significance of our proposed PCT method, we conducted an experimental evaluation and compared our methods to existing techniques.

Figure 5. Group splitting algorithm using PCT256 rule

Algorithm 2

```
Input: Tagcount
Output: Number of Group
for (Group Splitting procedure) do
    if Tagcount less than 353 tags then
        Keep tag into a single group;
    end
    else
        while Looking up PCT256 Rule Table do
            if Found Matched rule for specific Backlog then
                Split tags into groups;
            end
        end
    end
    Output number of groups;
end
```

Table 3. PCT128 rule - the number of unread tags, optimal frame-size (A and B), and number of group (A and B)

PCT128 rule				
Backlogs	**FS A**	**G A**	**FS B**	**G B**
...	...	...	...	...
1321 to 1408	128	8	-	-
1233 to 1320	128	7	64	1
1145 to 1232	128	7	-	-
1057 to 1144	128	6	64	1
969 to 1056	128	6	-	-
881 to 968	128	5	64	1
793 to 880	128	5	-	-
705 to 792	128	4	64	1
617 to 704	128	4	-	-
529 to 616	128	3	64	1
441 to 528	128	3	-	-
353 to 440	128	2	64	1
265 to 352	128	2	-	-
177 to 264	128	1	64	1
89 to 176	128	1	-	-
45 to 88	64	1	-	-
23 to 44	32	1	-	-
12 to 22	16	1	-	-
6 to 11	8	1	-	-
...	...	...	...	...

The aim of the experiment is to compare the performance of our proposed PCT method to the existing probabilistic DFSA and EDFSA anti-collision approaches. In this experiment, we considered different number of tags, from 100 to 1400, within the interrogation zone. The numbers of simulated tags are assumed to be no more than 1400 tags, due to maximum range of UHF reader and passive tags. All methods are applied separately to different randomly generated data sets.

Results

Our experiment evaluates the performance of our proposed PCT method to existing DFSA and EDFSA approaches. From Figure 8, it can be seen that both PCT256 and PCT-E produced minimal number of slots during identification process, compared with other methods. Specifically, PCT256 and PCT-E technique minimised the number of slots from EDFSA approach when the number of tags is between 400 and 500, and between 800 and 1200 tags. This is because the number of group sets for EDFSA will be doubled when the number of Backlog reached the specific threshold; while PCT increased number of group slowly, according to the estimated number

Figure 6. Group splitting algorithm using PCT128 rule

Algorithm 3

```
Input: Tagcount
Output: Number of Group
for (Group Splitting procedure) do
    if Tagcount less than 177 tags then
        Keep tag into a single group;
    end
    else
        while Looking up PCT128 Rule Table do
            if Found Matched rule for specific Backlog then
                Split tags into groups;
            end
        end
    end
    Output number of groups;
end
```

Table 4. PCT-E rule - the number of unread tags, optimal frame-size (A, B, C), and number of group (A, B, C)

PCT-E rule						
Backlogs	FS A	G A	FS B	G B	FS C	G C
...	...	...	...	...	...	...
1321 to 1408	256	4	-	-	-	-
1233 to 1320	256	3	128	1	64	1
1145 to 1232	256	3	128	1	-	-
1057 to 1144	256	3	-	-	64	1
969 to 1056	256	3	-	-	-	-
881 to 968	256	2	128	1	64	1
793 to 880	256	2	128	1	-	-
705 to 792	256	2	-	-	64	1
617 to 704	256	2	-	-	-	-
529 to 616	256	1	128	1	64	1
441 to 528	256	1	128	1	-	-
353 to 440	256	1	-	-	64	1
265 to 352	256	1	-	-	-	-
177 to 264	256	1	-	-	-	-
89 to 176	128	1	-	-	-	-
45 to 88	64	1	-	-	-	-
23 to 44	32	1	-	-	-	-
12 to 22	16	1	-	-	-	-
6 to 11	8	1	-	-	-	-
...	...	...	...	...	...	...

of unread tags. As a result, the number of slots are minimised for PCT256. On the other hand, PCT128 performed better than DFSA but did not outperform the EDFSA. According to optimal efficiency, the initial Q of 8 (frame-size = 256) has a wider range of optimal efficiency compared with the initial Q of 7. Therefore, PCT256 with initial frame-size of 256 has a better performance than the PCT128 with initial frame-size of 128.

Figure 8a shows that there is no improvement to our proposed methods compared with existing methods when the number of tags are low (up to around 300 tags). This is because PCT methods start dividing tags into groups only when the number of tags reaches the specific threshold. As a result, for certain tag sizes, the number of slots and performance efficiency remained unchanged due to the same identification procedure, compared with DFSA and EDFSA methods. Moreover, Figure 8b) also demonstrates that the PCT128 is the only method that has different results when the number of tags are 100 and 200 tags. This is due to the fact that PCT128 is the only method that uses initial frame-size of 128 to predict Backlog. Therefore, even when the number of tags is still low, the PCT128 starts splitting tags into group, resulting in different tag outcomes.

Figure 8b) shows that both PCT256 and PCT-E maintained their system efficiency above other methods and has the most stable performance. Nevertheless, the PCT-E required additional number of group sets from the PCT256 method throughout the identification process (see Table 2 versus Table 4). As a result, the PCT-E required extra time to initiate a new group compared with the PCT256 method. On the other hand, the DFSA's efficiency dropped dramatically when the number of tags increase, while the EDFSA's efficiency become unstable during the time when number of groups doubled-up from 1 to 2 and from 2 to 4. The PCT128 has steady performance but does not perform as good as PCT256.

The PCT256 has a better performance than the EDFSA by about 4 percent on average, while it is approximately 11 percent better than the DFSA approach as demonstrated in Figure 8c). The optimal percentage of improvement of PCT256 method can achieve up to 14 percent and 21 percent compared with the EDFSA and DFSA respectively, depending on the number of tags within the interrogation zone. Nevertheless, the PCT-E method required additional number of groups from the PCT method, and acquired slightly lower percentage of improvement, compared with the PCT method. On the other hand, the PCT128

Figure 7. Group splitting algorithm using PCT-E rule

Algorithm 4

```
Input: Tagcount
Output: Number of Group
for (Group Splitting procedure) do
    if Tagcount less than 353 tags then
        Keep tag into a single group;
    end
    else
        while Looking up PCT-E Rule Table do
            if Found Matched rule for specific Backlog then
                Split tags into groups;
            end
        end
    end
    Output number of groups;
end
```

has a better performance than the DFSA method by around 6 percent on average, but does not show any improvement from the EDFSA technique, as displayed in Figure 8c). However, the PCT128 still shows some improvement in some cases and is able to achieve up to 16 percent compared with the EDFSA and DFSA methods. Therefore, we conclude that our proposed PCT256 method is the most effective method, in terms of system efficiency and number of slots minimisation.

FUTURE RESEARCH DIRECTIONS

In this chapter, we have addressed issues directly related to anti-collision, which is a part of major data streams filtering in RFID. However, the implementation and integration of efficient anti-collision techniques is only the first step to improve data quality, and further process of data is required in order to successfully complete all stages of RFID data stream management. Although anti-collision is the most important step to determine the quality of RFID data, it is also important to consider other potential issues in data management. There are several open research questions regarding RFID data streams filtering and management, which should be addressed in the future. These open research questions are described as follows:

- In the context of RFID anti-collision, we have not looked into evaluating Reader-to-Reader or Reader-to-Tag anti-collision techniques. Despite the fact that Tag-to-Tag collision is the most critical problem for RFID data management, it is necessary to look into other perspective collisions in order to optimise the quality of captured data.
- It is also feasible to perform a comparative cost analysis for both Tree-based and ALOHA-based anti-collision techniques. While performance studies have been done on identification time, memory usage, and performance efficiency, there is no evidence confirming the cost analysis for RFID anti-collision.
- For both deterministic and probabilistic anti-collision methods, there is a need for a new measurement model that will determine the capability of each anti-collision approach, based on every constraint including identification time, memory usage, performance efficiency, and cost consump-

Figure 8. Number of slots comparison (a), Performance efficiency (b) for DFSA(D), EDFSA(ED), PCT128(P128), PCT256(P256), and PCT-E(P-E) methods on different number of tags, and Percentage of improvement of PCT compared with DFSA and EDFSA methods (c)

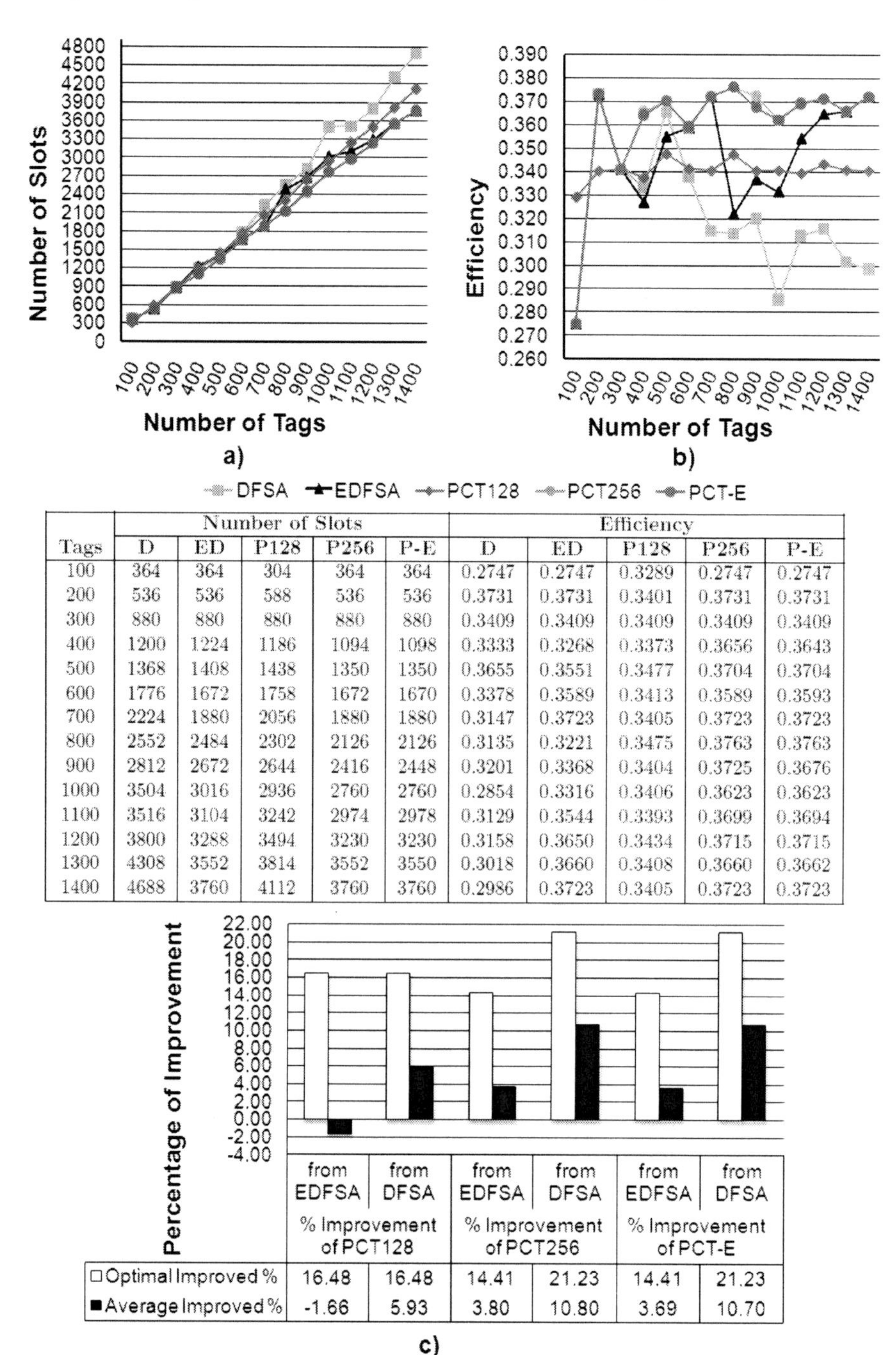

Tags	Number of Slots					Efficiency				
	D	ED	P128	P256	P-E	D	ED	P128	P256	P-E
100	364	364	304	364	364	0.2747	0.2747	0.3289	0.2747	0.2747
200	536	536	588	536	536	0.3731	0.3731	0.3401	0.3731	0.3731
300	880	880	880	880	880	0.3409	0.3409	0.3409	0.3409	0.3409
400	1200	1224	1186	1094	1098	0.3333	0.3268	0.3373	0.3656	0.3643
500	1368	1408	1438	1350	1350	0.3655	0.3551	0.3477	0.3704	0.3704
600	1776	1672	1758	1672	1670	0.3378	0.3589	0.3413	0.3589	0.3593
700	2224	1880	2056	1880	1880	0.3147	0.3723	0.3405	0.3723	0.3723
800	2552	2484	2302	2126	2126	0.3135	0.3221	0.3475	0.3763	0.3763
900	2812	2672	2644	2416	2448	0.3201	0.3368	0.3404	0.3725	0.3676
1000	3504	3016	2936	2760	2760	0.2854	0.3316	0.3406	0.3623	0.3623
1100	3516	3104	3242	2974	2978	0.3129	0.3544	0.3393	0.3699	0.3694
1200	3800	3288	3494	3230	3230	0.3158	0.3650	0.3434	0.3715	0.3715
1300	4308	3552	3814	3552	3550	0.3018	0.3660	0.3408	0.3660	0.3662
1400	4688	3760	4112	3760	3760	0.2986	0.3723	0.3405	0.3723	0.3723

	% Improvement of PCT128		% Improvement of PCT256		% Improvement of PCT-E	
	from EDFSA	from DFSA	from EDFSA	from DFSA	from EDFSA	from DFSA
□ Optimal Improved %	16.48	16.48	14.41	21.23	14.41	21.23
■ Average Improved %	-1.66	5.93	3.80	10.80	3.69	10.70

tion. Currently, different types of analysis have been done separately, thus results may be bias.

- As our main research goal concentrates on the implementation of simple anti-collision algorithms, we have not looked into the Hybrid anti-collision techniques, which combine both deterministic and probabilistic anti-collision together. Hybrid approaches may be a good solution for specific real-world applications.
- Data management, which is the most crucial project in RFID system, involves many procedures beside the data filtering process. These include data transformation, data aggregation, event management, data warehousing, and data mining. Each stage of data management must be integrated together before they can be deployed to the actual organisation. Thus, more research must be developed case-by-case to verify that all necessary stages of data management are compatible with each other and that they can be integrated and deployed to the chosen organisation.

CONCLUSION

In this chapter, we attempt to address a question on making anti-collision methods for RFID data streams collision more efficient than current available approaches. The chapter discusses the background information on RFID, why it is important, and motivation for investigating suitable anti-collision techniques toward different scenarios. We assess relevant literature and provide information on existing state-of-the-art anti-collision methods.

The chapter focuses on two main research problems: one is designing and developing a new deterministic anti-collision technique; and the other is introducing a new group based technique for probabilistic anti-collision. The contributions of the chapter are summarised as follows:

- We reviewed the literature, and identified and analysed the benefit and detriment of each existing method. We discovered that for deterministic anti-collision method, major limitations are due to memories and computational complexity requirements in memory anti-collision algorithms. The memoryless QT algorithms also perform poorly in the case where there are a large number of tags within an interrogation zone. Several improved version of QT enhance the performance and reduce identification delay but comes with huge implementation cost. For probabilistic anti-collision method, the earlier type of ALOHA anti-collision performs poorly, while the more advanced Framed-Slotted ALOHA has better performance. However, literature review demonstrated that existing methods from Framed-Slotted ALOHA category are either inefficient or too complex, with high overhead cost of implementation.
- We defined the research questions in relation to RFID anti-collision. Given that both deterministic and probabilistic anti-collision algorithms have different advantages and disadvantages toward development of real world applications, it is important to construct anti-collision methods that are simple, with low overhead computation, and perform effectively, compared with existing techniques. We concluded that the basic Query Tree is the most effective deterministic anti-collision approach, and the Framed-Slotted ALOHA is arguably the best approach in probabilistic anti-collision. Thus, our research focused on the development of new methods that perform better than these existing techniques.
- By constructing complex anti-collision algorithms, high memory capacity and

power sources are needed, which is impractical in RFID system. No study has been undertaken to construct a basic deterministic anti-collision, under the assumption that we have a limited power source from RFID reader and limited memory in both readers and tags. Thus, we proposed a Joined Q-ary Tree, which adaptively adjusted its breaches between the two Q-ary Trees, according to the EPC pattern. From experimental evaluation, the Joined Q-ary Tree outperformed the existing approaches in terms of memory usage, which resulted in minimal identification time.

- We also proposed a Probabilistic Cluster-Based Technique, in order to improve the performance efficiency from current probabilistic anti-collision methods. The Probabilistic Cluster-Based Technique utilised group splitting rules derived by using particular equations, according to the optimal system efficiency obtained for specific number of tags. We have developed several equations to calculate boundaries and exploit rules. The experimental evaluation demonstrated that our proposed method is the most effective method, in terms of system efficiency and number of slots minimisation.

REFERENCES

Abramson, N. (1970). The ALOHA system - Another alternative for computer communications. *In Proceedings of Fall Joint Computer Conference, AFIPS Conference* (pp. 281–285). Houston, Texas.

Alotaibi, M., Postula, A., & Portmann, M. (2009). Tag anti-collision algorithms in RFID systems: A new trend. *WSEAS Transactions on Communications* (WTOC). *WSEAS Transactions on Communications*, *8*, 1216–1232.

Bang, O., Choi, J. H., Lee, D., & Lee, H. (2009). *Efficient novel anti-collision protocols for passive RFID tags*. Auto-ID Labs White Paper. Retrieved August 11, 2011, from http://www.autoidlabs.org/single- view/dir/article/1/323/page.html

Bhandari, N., Sahoo, A., & Iyer, S. (2006). Intelligent query tree (IQT) protocol to improve RFID tag read efficiency. In *The 9th International Conference on Information Technology (ICIT'06)* (pp. 46–51).

Bhatt, H., & Glover, B. (2006). *RFID essentials*. Sebastopol, CA: O'Reilly.

Chen, W. T. (2006). An efficient anti-collision method for tag identification in a RFID system. *IEICE Transactions*, *89-B*(12), 3386–3392. doi:10.1093/ietcom/e89-b.12.3386

Cheng, T., & Jin, L. (2007). Analysis and simulation of RFID anti-collision algorithms. In *The 9th International Conference on Advanced Communication Technology (ICACT), Vol. 1*, (pp. 697–701). Phoenix Park, Korea. IEEE Computer Society.

Cho, H., Lee, W., & Baek, Y. (2007). LDFSA: A learning-based dynamic framed slotted ALOHA for collision arbitration in active RFID systems. In *Advances in Grid and Pervasive Computing Second International Conference, Vol. 4459*, (pp. 655–665). Paris, France. Berlin, Germany: Springer.

Choi, H., Cha, J. R., & Kim, J.-H. (2004). Fast wireless anti-collision algorithm in ubiquitous ID system. In *Vehicular Technology Conference (VTC2004), IEEE* (pp. 4589–4592).

Choi, J., & Lee, W. (2007). Comparative evaluation of probabilistic and deterministic tag anti-collision protocols for RFID networks. In *EUC Workshops '07* (pp. 538–549).

Choi, J. H., Lee, D., & Lee, H. (2007). Query tree-based reservation for efficient RFID tag anti-collision. *IEEE Communications Letters*, *11*(1), 85–87. doi:10.1109/LCOMM.2007.061471

Choi, J. H., Lee, H. J., Lee, D., Lee, H. S., Youn, Y., & Kim, J. (2008). *Query tree based tag identification method in RFID systems*. Retrieved August 11, 2011, from www.freshpatents.com/Query-tree-based-tag-identification-method-in-rfid-systems-dt20080508ptan20080106383.php

Derakhshan, R., Orlowska, M. E., & Li, X. (2007). RFID data management: Challenges and opportunities. In *IEEE International Conference on RFID 2007* (pp. 175–182). Texas, USA.

Devarapalli, M. R., Sarangan, V., & Radhakrishnan, S. (2007). AFSA: An efficient framework for fast RFID tag reading in dense environments. In *QSHINE '07: The Fourth International Conference on Heterogeneous Networking for Quality, Reliability, Security and Robustness Workshops*, (pp. 1–7). New York, NY: ACM.

Ding, J., & Liu, F. (2009). Novel tag anti-collision algorithm with adaptive grouping. Wireless Sensor Network (WSN). *Wireless Sensor Network*, *1*(5), 475–481. doi:10.4236/wsn.2009.15057

EPCGlobal. (2008). *EPCGlobal tag data standards version 1.4: Ratified specification*. Retrieved August 11, 2011, from http://www.epcglobalinc.org/standards/tds/

Fan, X., Song, I., & Chang, K. (2008a). Gen2-based hybrid tag anti-collision Q algorithm using Chebyshev's inequality for passive RFID systems. In *19th International Symposium on Personal, Indoor and Mobile Radio Communications, IEEE* (pp. 1–5). Cannes, France.

Fan, X., Song, I., Chang, K., Shin, D. B., Lee, H. S., Pyo, C. S., & Chae, J. S. (2008b). Gen2-based tag anti-collision algorithms using Chebyshev's inequality and adjustable frame size. *ETRI Journal*, *30*(5), 653–662. doi:10.4218/etrij.08.1308.0098

Feng, B., Li, J. T., Guo, J. B., & Ding, Z. H. (2006). ID-binary tree stack anti-collision algorithm for RFID. In *Proceedings of the 11th IEEE Symposium on Computers and Communications, IEEE Computer Society* (pp. 207–212). Washington, DC, USA.

Finkenzeller, K. (2003). *RFID handbook - Fundamentals and applications in contactless smart cards and identification* (2nd ed.). Munich, Germany: John Wiley and Sons.

Floerkemeier, C. (2007). Bayesian transmission strategy for framed ALOHA based RFID protocols. In *RFID, 2007 - IEEE International Conference on RFID Gaylord Texan Resort* (pp. 228–235). Grapevine, TX, USA.

Jacomet, M., Ehrsam, A., & Gehrig, U. (1999). *Contactless identification device with anti-collision algorithm*. In IEEE Computer Society, Conference on Circuits, Systems, Computers and Communications. Athens, Greece.

Jain, S., & Das, S. R. (2006). Collision avoidance in a dense RFID network. *In WiNTECH'06: Proceedings of the 1st International Workshop on Wireless Network Testbeds, Experimental Evaluation & Characterization*, (pp. 49–56). New York, NY: ACM.

Kim, J. G. (2008). A divide-and-conquer technique for throughput enhancement of RFID anti-collision protocol. *IEEE Communications Letters*, *12*(6), 474–476. doi:10.1109/LCOMM.2008.080277

Klair, D. K., Chin, K. W., & Raad, R. (2007). On the suitability of framed slotted Aloha based RFID anti-collision protocols for use in RFID-enhanced WSNs. In *Proceedings of 16th International Conference on Computer Communications and Networks 2007 (ICCCN 2007)* (pp. 583–590).

Klair, D. K., Chin, K. W., & Raad, R. (2010). A survey and tutorial of RFID anti- collision protocols. *IEEE Communications Surveys Tutorials*, *12*(3), 400–421. doi:10.1109/SURV.2010.031810.00037

Law, C., Lee, K., & Siu, K. Y. (2000). Efficient memoryless protocol for tag identification. In *Proceedings of the 4th International Workshop on Discrete Algorithms and Methods for Mobile Computing and Communications, DIALM '00, ACM* (pp. 75–84). New York, NY, USA.

Lee, C. W., Cho, H., & Kim, S. W. (2008a). An adaptive RFID anti-collision algorithm based on dynamic framed ALOHA. *IEICE Transactions, 91-B*(2), 641–645. doi:10.1093/ietcom/e91-b.2.641

Lee, J. G., Hwang, S. J., & Kim, S. W. (2008b). Performance study of *anti-collision* algorithms for EPC-C1 Gen2 RFID protocol. In *Information Networking. Towards Ubiquitous Networking and Services, Vol. 5200,* (pp. 523–532). Estoril, Portugal. Berlin, Germany: Springer.

Lee, S. R., Joo, S. D., & Lee, C. W. (2005). An enhanced dynamic framed slotted aloha algorithm for RFID tag identification. In *MOBIQUITOUS '05: Proceedings of The Second Annual International Conference on Mobile and Ubiquitous Systems: Networking and Services,* (pp. 166–174). Washington, DC: IEEE Computer Society.

Lee, S. R., & Lee, C. W. (2006). An enhanced dynamic framed slotted ALOHA anti- collision algorithm. In *Emerging Directions in Embedded and Ubiquitous Computing, Vol. 4097,* (pp. 403–412). Seoul, Korea. Berlin, Germany: Springer.

Li, B., Yang, Y., & Wang, J. (2009). *Anti-collision issue analysis in Gen2 protocol - Anti-collision issue analysis considering capture effect.* Auto-ID Labs White Paper. Retrieved August 11, 2011, from http://www.autoidlabs.org/single-view/dir/article/6/320/page.html

Myung, J., & Lee, W. (2006a). Adaptive binary splitting: A RFID tag collision arbitration protocol for tag identification. *Mobile Networks and Applications, 11*(5), 711–722. doi:10.1007/s11036-006-7797-6

Myung, J., & Lee, W. (2006b). Adaptive splitting protocols for RFID tag collision arbitration. In *MobiHoc '06: Proceedings of the 7th ACM International Symposium on Mobile Ad Hoc Networking and Computing,* (pp. 202–213). New York, NY: ACM.

Pupunwiwat, P., & Stantic, B. (2009a). Performance analysis of enhanced Q-ary tree anti-collision protocols. In *The First Malaysian Joint Conference on Artificial Intelligence (MJCAI), Vol. 1* (pp. 229–238). Kuala Lumpur, Malaysia.

Pupunwiwat, P., & Stantic, B. (2009b). Unified Q-ary tree for RFID tag anti-collision resolution. In A. Bouguettaya & X. Lin (Eds.), *The Twentieth Australasian Database Conference (ADC), Vol. 92 of CRPIT, ACS* (pp. 47–56). Wellington, New Zealand.

Pupunwiwat, P., & Stantic, B. (2010a). A RFID explicit tag estimation scheme for dynamic framed-slot ALOHA anti-collision. In *The Sixth Wireless Communications, Networking and Mobile Computing (WiCOM)* (pp. 1–4). Chengdu, China: IEEE. doi:10.1109/WICOM.2010.5601080

Pupunwiwat, P., & Stantic, B. (2010b). Dynamic framed-slot ALOHA anti-collision using precise tag estimation scheme. In H. T. Shen & A. Bouguettaya (Eds.), *The Twenty- First Australasian Database Conference (ADC), Vol. 104 of CRPIT, ACS* (pp. 19–28). Brisbane, Australia.

Pupunwiwat, P., & Stantic, B. (2010c). Joined Q-ary Tree anti-collision for massive tag movement distribution. In B. Mans & M. Reynolds (Eds.), *The Thirty-Third Australasian Computer Science Conference (ACSC), Vol. 102 of CRPIT, ACS* (pp. 99–108). Brisbane, Australia.

Pupunwiwat, P., & Stantic, B. (2010d). Resolving RFID data stream collisions using set- based approach. In *The Sixth International Conference on Intelligent Sensors, Sensor Networks and Information Processing (ISSNIP), IEEE* (pp. 61–66). Brisbane, Australia.

Quan, C. H., Hong, W. K., & Kim, H. C. (2006). Performance analysis of tag anti-collision algorithms for RFID systems. In *Emerging Directions in Embedded and Ubiquitous Computing, Vol. 4097,* (pp. 382–391). Seoul, Korea. Berlin, Germany: Springer.

Ryu, J., Lee, H., Seok, Y., Kwon, T., & Choi, Y. (2007). A hybrid query tree protocol for tag collision arbitration in RFID systems. In *MobiHoc '06: Proceedings of the 7th ACM International Symposium on Mobile Ad Hoc Networking and Computing,* (pp. 5981–5986). Glasgow, UK. IEEE Computer Society.

Schoute, F. C. (1983). Dynamic frame length ALOHA. *IEEE Transactions on Communications, 31*(4), 565–568. doi:10.1109/TCOM.1983.1095854

Shin, W. J., & Kim, J. G. (2007). Partitioning of tags for near-optimum RFID anti-collision performance. In *IEEE Wireless Communications and Networking Conference, WCNC 2007,* (pp. 1673–1678).

Vogt, H. (2002). Efficient object identification with passive RFID tags. *In Pervasive '02: Proceedings of the First International Conference on Pervasive Computing,* (pp. 98–113). London, UK: Springer.

Wang, Z., Liu, D., Zhou, X., Tan, X., Wang, J., & Min, H. (2007). *Anti-collision scheme analysis of RFID system.* Auto-ID Labs White Paper. Retrieved August 11, 2011, from http://www.autoidlabs.org/single-view/dir/article/6/281/page.html

Wu, H., & Zeng, Y. (2010). Bayesian tag estimate and optimal frame length for anti- collision Aloha RFID system. *IEEE Transactions on Automation Science and Engineering,* 7(4), 963–969. doi:10.1109/TASE.2010.2042957

Zhou, F., Chen, C., Jin, D., Huang, C., & Min, H. (2004). Evaluating and optimizing power consumption of anti-collision protocols for applications in RFID systems. In *Proceedings of the 2004 International Symposium on Low Power Electronics and Design (ISLPED04),* (pp. 357–362). New York, NY: ACM.

Zhu, L., & Yum, P. T. (2009). The optimization of framed Aloha based RFID algorithms. In *MSWiM '09: Proceedings of the 12th ACM International Conference on Modeling, Analysis and Simulation of Wireless and Mobile Systems,* (pp. 221–228). New York, NY: ACM.

Zhu, L., & Yum, P. T. (2011). A critical survey and analysis of RFID anti-collision mechanisms. *IEEE Communications Magazine, 49,* 214–221. doi:10.1109/MCOM.2011.5762820

KEY TERMS AND DEFINITIONS

Backlog: A remaining number of tags that have not been read. The number of Backlog can be estimated using specific frame-size estimation techniques.

EPC Pattern: In order to manage and monitor the traffic of RFID data effectively, the EPC pattern is usually used to keep the unique identifier on each of the items arranged within a specific range. The EPC pattern does not represent a single tag encoding, but rather refers to a set of tag encodings. 25.1545.[3456-3478].[778-795] is a sample of the EPC pattern in decimal, which later will be encoded to binary and embedded onto tags.

Frame-Size: In ALOHA-based anti-collision approaches, each tag in an interrogation zone selects one of the given N slots to transmit its identifier; and all tags will be recognised after a few frames. Each frame is formed of specific number of slots that is used for communication between the readers and the tags. To determine the frame-size, it gathers and uses information such as number of successful slots, empty slots, and collision slots from previous round, to predict the appropriate frame-size for the next identification round. Frame-size is calculated by $2^Q - 1$, where Q is an integer ranged between 0 and 15.

Number of Bit: The measurement of memory usage in deterministic anti-collision approaches. The minimal Number of bit means the higher robustness of the anti-collision procedure.

Number of Cycle: Measures how many cycles have been produced by the deterministic anti-collision methods during the identification process. The Number of cycle can, at one point, clarify the performance of all tree-based anti-collision methods. However, in order to measure the robustness of each tree-based approach, the calculation of the Number of bit is required.

Number of Frame: Used to measure the performance efficiency in probabilistic anti-collision methods. Each frame contains specific number of slots. The lower Number of frames along with minimal Number of slots signifies the higher efficiency of the anti-collision procedure.

Number of Slot: Used to measure the performance efficiency in probabilistic anti-collision methods. The lower Number of slots along with minimal Number of frames indicates the higher efficiency of the anti-collision process. Although the number of frames and slots appear to be relative, the minimal number of slots does not specify the minimal number of frames and vice versa. This is because each frame may contain different number of slots. For example, three frames of 2, 16, and 64 slots form the total number of 82 slots; while four frames of 4, 6, 8, and 32 slots form the total number of 50 slots.

Section 4
Applications

Chapter 9
Passive UHF RFID Technology Applied to Automatic Vehicle Identification:
Antennas, Propagation Models and Some Problems Relative to Electromagnetic Compatibility

Salvador Ricardo Meneses González
ESIME Zacatenco, México

Roberto Linares y Miranda
ESIME Zacatenco, México

ABSTRACT

In this chapter, propagation channel aspects in current passive UHF RFID systems applied for automatic vehicular identification (AVI) are presented, considering the antennas design for passive UHF RFID tag and some problems relative to the electromagnetic compatibility. These issues are focused on RFID link, reader-tag-reader, and the channel modelling that is supported with measurements, and reader-reader interference problems are analysed.

1. INTRODUCTION

Automatic identification is the broad term given to a number of technologies used to help identify objects. This process is coupled with automatic data capture. That is, the capture of information should carry out without the human being involved in the process, only in the interpretation. The goal of most automatic identification systems is to increase efficiency and reduce errors in data capture.

There are number of technologies that can be included in the process of automatic identification. These include bar codes, smart cards, voice recognition, some biometric technologies (retinal scans, for example), optical character recognition, radio frequency identification (RFID) and others.

DOI: 10.4018/978-1-4666-2080-3.ch009

Currently, passive RFID technology at UHF band (860~960 MHz) is being widely used for identification of different objects, retail inventory management, and tracking applications. This is due to low production cost of tag with reasonable readable range. Recently, passive UHF RFID systems have been also applied to wireless sensor. Respect to the automatic identification of vehicles with passive RFID technology, the applications are for the payment of tolls road and something about the traffic control, but this issue is briefly mentioned (Blythe, 1999).

Passive UHF RFID technology is basically a wireless communication. Communication takes place between a reader and a tag. The reader is an interrogator and tag is a transponder. Tags can be presented in different ways depends on the objects to identify, and they must be coupled electromagnetically with the material of the object where they are sticked, for the case of the AVI, the tag is strapped on the vehicle windshield.

Passive UHF RFID technology presents a better solution for AVI with respect to other near-field technologies currently available, such as: barcode, magnetic cart, passive LF (Low Frequency) RFID tag, and passive HF (high frequency) RFID tag. The passive UHF RFID systems can carry out the automatic vehicle identification in far field to several meters, depending of tag characteristics (sensitivity-chip and antenna-gain). This technology can be used in heavy rain, snow, cold, or in subzero temperatures. Readers of passive RFID UHF technology recognize tags placed on the windshield of moving vehicles, so that drivers do not even need to stop for identification, or to open the vehicle window to enter access codes or push buttons to get a ticket.

Most wireless communication systems operate in high electromagnetic pollution environments. These environments in addition to multipath effect are problems for the RF signal propagation that also affects to passive UHF RFID systems. Even though passive UHF RFID systems normally operate in line-of-sight (LOS), they are also affected by typical electromagnetic environment and for AVI application is required to consider other propagation problems, such as: if the vehicle identification takes place on a highway, the tags detection probability decreases with increasing vehicle speed; the multipath effect (reflections diffraction and scattering) if the vehicle moves at slow speed (city traffic); the interference reader-reader when they operate at the same time (toll booths) and each one is close to the other.

The communication protocol used normally for the identification of vehicles is based on the standard ISO / IEC 18000-6 C, this is a standard that has a good performance. However, the vehicle detection at speed greater than 90 km/H, even the communication protocols proprietary are used, these ones require some modifications. The AVI in roads using passive UHF RFID technology has not yet been discussed extensively, in spite of being applied. Typical problems of propagation of the RF signal in passive UHF RFID systems, such as: coupling and alignment of the tag for a high performance in LOS are described perfectly by (Nikitin, 2008). However, the issue of the propagation problems of RF signals for AVI application has not been identified widely, because the scenarios where the vehicles pass at different speeds are complex, for example, with heavy traffic (slow speed) or/and light traffic (high speed).

The purpose of this chapter is to describe the main problems of Automatic Vehicular Identification with Passive UHF RFID Technology on roads, with three issues: antennas, propagation channel and electromagnetic compatibility, which is organized as follow: Section 2 both Active and Passive UHF RFID Tag Technologies are described. Section 3 antennas for passive UHF RFID tag are presented. Propagation model for passive UHF RFID system applied to the vehicular identification in the roads is presented in Section 4. Electromagnetic Compatibility aspects are described in Section 5; finally conclusions and remarks are given in Section 6.

2. ACTIVE AND PASSIVE UHF RFID TAG TECHNOLOGIES

Automatic Vehicle Identification (AVI) with RFID technology at UHF frequency bands can be active or passive. The classification is by type of tag: The active tags are powered by batteries always transmit RF electromagnetic energy (ID) at constant intervals, in order to be located by a reader. The passive tags does not incorporate a battery, they are powered with the energy provided by a reader/writer. The passive tag's chip receives power from the antenna and responds by varying its input impedance. The impedance variation switches between two states, matched and mismatched, of the antenna. This generates a signal modulated of amplitude shift keying (ASK) that transmits the tag, which is backscattered. The two type of tags have advantages and disadvantages (Nikitin, 2008), and for the AVI application is a discussion subject.

The active tags can be installed anywhere of the vehicle, the line of sight (LOS) is not necessary for her detection. These tags can be carried several tasks and processing information, since the tag's chip can include some sensors, as such: temperature sensor, pressure sensor and other. This chip also has memory of KB (Pais, 2011). Some typical functions that are carried out with the active tags for control and vehicular identification are: open or close doors, alarm system activation. The detection range of these tags is up 500 m, depends on the power and for the detection of circulating vehicles they have a good performance at speeds of 160 km/H.

For cars, active tags currently are used as a remote electronic key that has different security functions. These tags have a price of 20 to 300 U.S. dollars and depend on the functions that have. The performance of tags is a function of its power emitted and of the sensitivity of readers. In these systems RFID, the tags constantly emit electromagnetic energy to the environment, so they are high electromagnetic pollution sources. On the other hand, the active tags also are vulnerable to electromagnetic interference, so that they can be easily cancelled (jammer). This is the major technique disadvantage of RFID systems with active tags.

The passive tags applied to the vehicular identification requires line of sight (LOS) and they are strapped to the vehicle windshield in order to be detected by the reader/writer. Normally, the reader/writer is placed in a portal through which the vehicle is traveling. The tag-reader distance must be in accordance with the concept of far field. With passive RFID tags current a good performance is obtained with a distance of 10m between the vehicle and reader at speeds of up to 160 km/H. The typically price of passive RFID tag is 0.2 U.S. Dollar and they can be manufactured as holograms in order to be a visual identification, this attractive for security systems of the governments.

When one analyzes the characteristics of passive RFID tags in the UHF frequency band, about its detection range and ability to process information, active tags are better. However, passive tags are attractive for implementing automatic toll payment, identification and access vehicles. This is because the passive RFID tag for AVI in many application, it is enough just to ID, in addition to being robust to electromagnetic interference and much cheaper than active tags.

In Mexico, several performance analyses for passive and active RFID tags in the UHF frequency band have been carried out, in order to be applied in the vehicular identification. ISO/IEC 18000-7 standard was considered for active RFID technology. For passive RFID systems the Mexican Standard NOM 121 and ISO/IEC 180006-C standard was considered. ISO/IEC TR 18047-6 standard was considered for test methods. The ISO/IEC standards define air interface requirements, the operation frequency and Effective Isotropic Radiated Power (EIRP) for the region. The passive RFID systems in Mexico operate at nominal frequency of 915MHz between 902-928 MHz with 4W EIPR. For AVI some results of RFID systems tests are the following:

2.1. Active RFID Technology Applied to AVI

The Active RFID Technology is a typical radio communication system. This technology is made up by tags that act as transmitter/receptor and a state base. The tag emits a constant RF signal at short intervals time, to be located by a base station. The chip of tags keeps in her memory the of ID information, which is transmitted and when it is received by the base station is processed and transferred to a local database or remote to verify the information. Tag and base state can be fixed or at continuous movement, also someone of them can be fixed and the other at continuous movement.

It is made up by the tag acting as transmitter/receptor, installed anywhere of the vehicle under test, keeping the ID information in an electronic memory, and the receptor/transmitter acting as a base station installed outside it, processing and transferring the data to the database in a local or remote computer in order to verify the information. Both of them can be fixed, at continuous movement or one of them fixed and the other at continuous movement. A basic system of Active RFID Technology applied to AVI in Figure 1 is shown.

Evaluation of the active RFID technology applied to vehicular identification was carried out under a typical environment of Mexico City with some basic characteristics such as:

- **Temperature Range:** 15°C - 30°C
- **Relative Humidity Range:** 57% - 82%
- **Operation Frequency:** 433 MHz and 915 MHz
- **Modulation:** ASK and PSK
- **Bit Rate Range:** 0.1- 600kbps
- **Sensitivity:** < -122 dBm

The speed test of active RFID technology was carried out up to 160 km/H and was successful.

Figure 1. Basic system of active RFID technology applied to AVI

Because of the high sensitivity of active RFID technology was necessary to carry out tests of electromagnetic interference. These tests were conducted in an anechoic chamber according to the diagram in Figure 2. The jammer power level for active RFID tags at 915 MHz nominal frequency was 1.6mW Effective Isotropic Radiated Power (EIRP)

2.2. Passive RFID Technology Applied to AVI

The passive RFID technology with frequency hopping can be detected up to 20 m. For the AVI application the performance can be secured at 10 m with 4W EIRP and -18dBm of tag sensitivity at 902-928MHz range. This technology is readable for a very long time, and the useful life is approximately of twenty years, resistant to corrosion and physical damage. Typical problems of this technology are on the tag, which are: the sensitivity of the chip, antenna, antenna-chip matching and antennas alignment tag reader. The best performance is achieved at line of sight and a same polarization of reader-tag antennas. Some aspects regarding of tag sensitivity and its alignment are described below, aspects of tag antenna are presented in the next section.

The tag sensitivity is determined by the number of readings (backscatter power) of a tag for an incident power in a controlled environment, such as a chamber anechoic. In this regard, knowing the EIRP in the reader and using the Friis transmission (Equation 1), the tag sensitivity is determined calculating the uplink and downlink power for power threshold of detection.

The normalized response of reading a passive RFID tag Avery - Dennison Mod 224, respect to the attenuation of EIRP of the reader with 4W of nominal power is shown in Figure 3. The sensitivity is determined for the downlink with 10% of the response of the tag. This is a good approximation for AVI application, because with once reading of tag ID, it is sufficient to identify the vehicle. In

Figure 2. RFID system under jamming environment

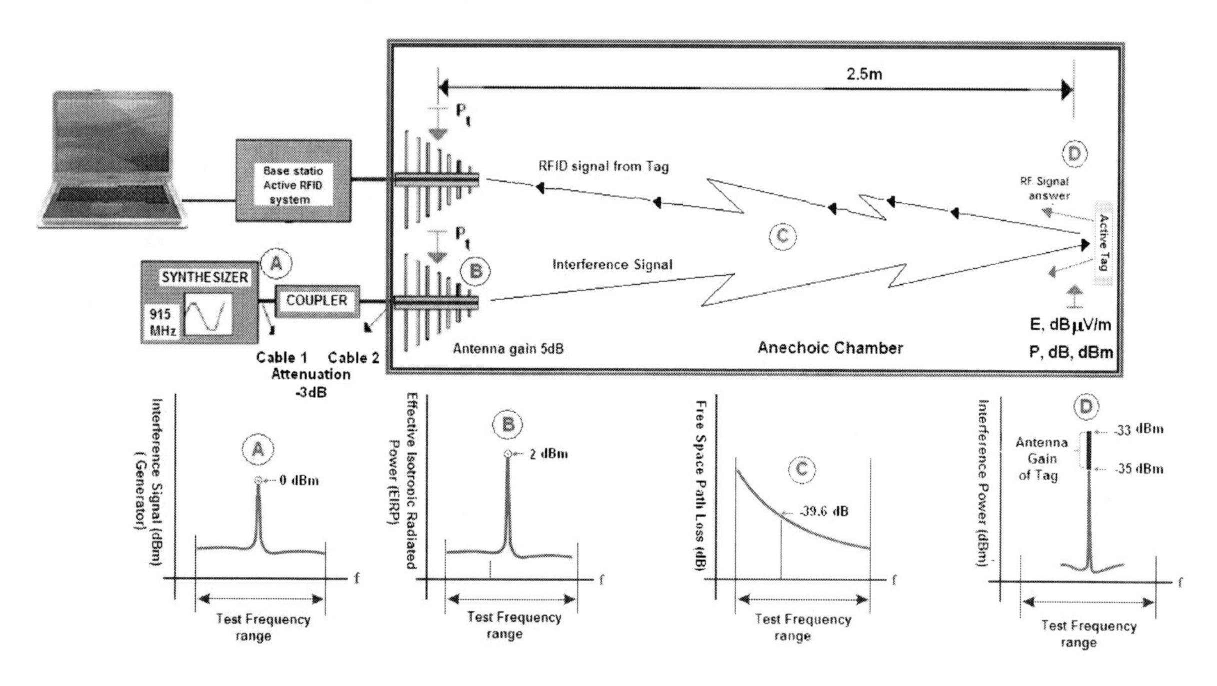

Figure 3. Normalized response of reading a passive RFID tag Avery - Dennison Mod 224

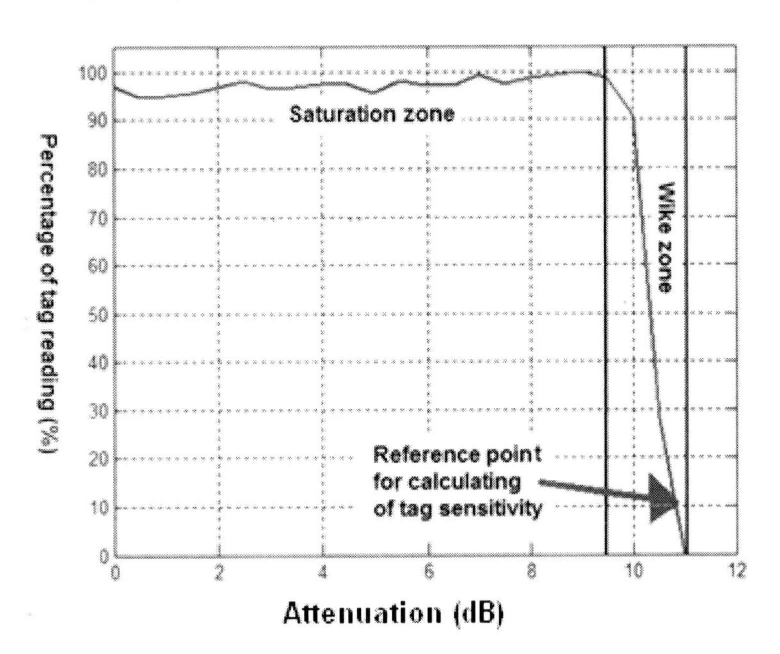

this case, an efficiency of 75% tag and an antenna gain of tag of 2dBi is considered for obtaining a sensitivity of -18 dB.

In the identification of vehicles, the passive RFID tag is located on the car's windshield and the reader's antenna is located in a portal on the road where vehicles pass. The antennas (tag-reader) should have the same polarization, i.e. line of sight aligned. Typically, the tag's antennas are symmetrical dipoles meandering, easy to become misaligned, which causing decrease system performance

The alignment characteristic of the passive RFID tag antenna with respect to its rotation and tilt, considering a horizontal antenna of reader is analyzed. This analysis is carried out within an anechoic chamber by turning the RFID tag at the azimuth, as well as tilting angle. The results are shown in Figure 4. It shows (a) the rotation angle and (b) the angle of inclination with respect to the percentage of reading and the attenuation level for a nominal EIRP of 4 W. Also, the null and maximum is marked.

According to Figure 4, tags detection can be carried out only up to 45° with respect to the reference alignment for both tilt and azimuth. Figure 5 shows the passive tag strapped to the car windshield. Based on the successful readings of tag in vehicles moving to high speed (100 km/H), the tag placed in a horizontal position is the best polarization, which must be the same of transmitting antenna. However, as we´ll see on section 4, the horizontal polarization is more sensible to multipath environment than the vertical polarization.

3. ANTENNAS FOR PASSIVE RFID TAGS AT UHF FREQUENCY BAND

Passive UHF RFID tag antennas include a lot of structures, and they provide different practical aspects, and one of them is the RFID tag (antenna-Chip) design and analysis, specifically the sensitivity and noise analysis, that is an important aspect in chip AC-DC converter, which

Figure 4. (a) The rotation angle and (b) the angle of inclination of passive RFID tag

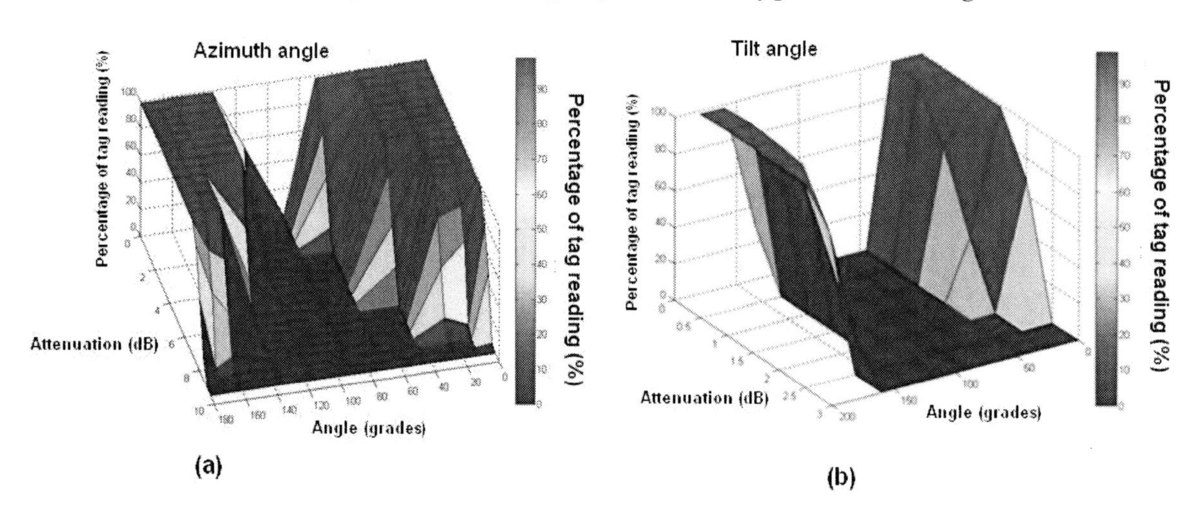

define the read range performance, that is, the maximum distance at which RFID reader can detect the backscattered signal from the tag. Because of the reader sensitivity is typically higher than the tag sensitivity the read range is defined by the tag response threshold, similarly, defined by gain antenna and equivalent noise input of chip. Read range is also sensitive to the tag orientation, the kind of material of the object where the tag is placed on, and to the propagation environment.

The passive tag antenna is constructed for an antenna attached to a chip that has a rectifier circuit, which one provides the voltage and current to power on the chip. For UHF passive RFID systems, the relation between the input impedance of the rectifier circuit and the antenna imped-

Figure 5. Alignment of the passive RFID tag (polarization)

ance determines the maximum read range of the tag (Iizuka et al., 2007), moreover the antenna defines the size tag that is a major parameter for the application. Then, it must have a compromise between antenna gain and impedance matching antenna-chip to attain a good performance of the tag (Marrocco, 2008).

The impedance of the chips are not standardized like other communications systems, and then one of the main challenges in designing antennas for passive RFID tags is to have the maximum matching to achieve maximum efficiency and minimal effects of the electromagnetic environment. The chips are made to cover three UHF RFID frequency ranges for different regions: Europe (866.5 MHz), North America (915 MHz), and Asia (953 MHz). These chips have typically input impedances with a real part of one order of magnitude, smaller than imaginary part. Due to the problems of the chip impedance, it is first necessary to knowing or measure impedance to design antennas that appropriates, which also presents a challenge because it is not easy to measure these impedances. Many methodologies for measuring impedance of RFID chip has been reported, some values are shown in Table 1.

To understand the basic principles of operation of passive UHF RFID Tags, first a basic scheme of RFID system is analyzed with respect to link budget, which is shown in Figure 6. The polarization losses in this system are not taking in account, and Line of Sight (LOS) communication is assumed, their performance is defined by the matching quality at the interface between of RFID tag antenna and RFID chip input impedance, which can be expressed by Standing Wave Ratio (SWR) or reflection coefficient ($\Gamma_{in,TAG}$) at the interface. Then for backscattering power (P_b), it is obtained by:

$$P_b = P_i \left(\frac{SWR - 1}{SWR + 1} \right)^2 = P_i \left| \Gamma_{in,TAG} \right|^2 \qquad (1)$$

The received power on the reader is:

Table 1. Input impedance of various RFID chips

Model	Input-Impedance (Ohms)	Frequency (MHz)	References
ATA5590 Atmel	6.7 –j210	915	Atmel Co.;Pascal, 2005; Jari-p, 2007
XRA00 ST	6.7 -j197	915	Jari-p, 2007; ST Microelectronics, 2005
MM9647 NSC	73 –j113	915	Popovic, 2011
ALL-9238 Alien Technology	20 –j127	900	Popovic, 2011; Alien, 2011
ALN-9338.R Alien Technology	6.2 –j127	915	Popovic, 2011; Alien, 2011
EPC1.19G2 Phillips	16 –j315	914	Popovic, 2011
NXP Uncode Gen2	22 –j404	867	Kumar, et. al, 2010; Phillips Semiconductor, 2006
NXP Uncode Gen2	16 –j380	915	Kumar, et al, 2010; Phillips Semiconductor, 2006
AD220 Avery Dennison	8 –j98	866.5	Loo, et al, 2008; Avery Dennison, 2011
AD220 Avery Dennison	8 –j91	915	Loo, et. al, 2008; Avery Dennison, 2011
AD220 Avery Dennison	8 –j85	953	Loo, 2008; Avery Dennison, 2011

Figure 6. Passive UHF RFID system

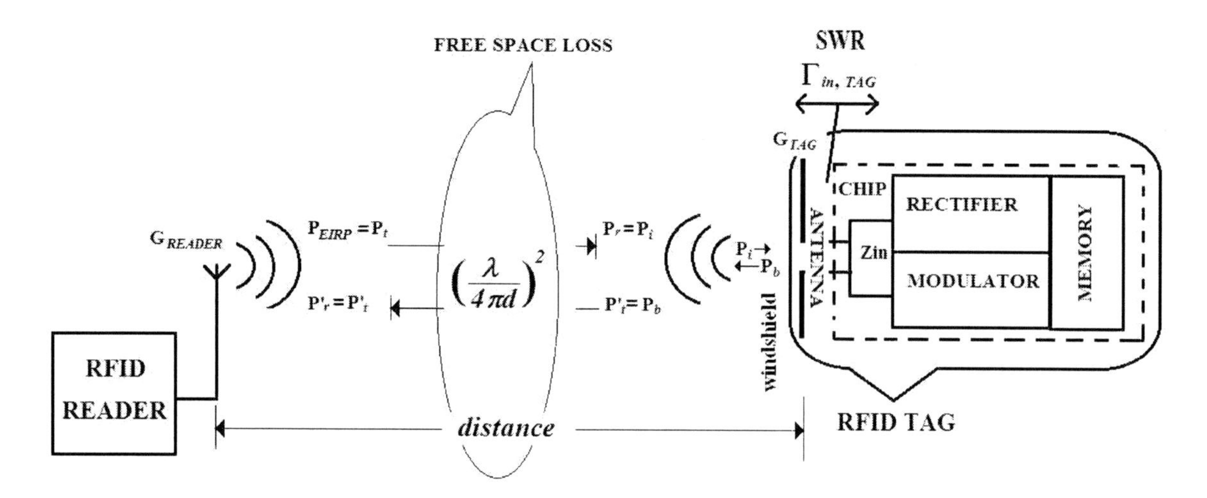

$$P_r' = G_{READER}\left(G_{TAG}\right)^2 P_{EIRP}\left(\frac{\lambda}{4\pi d}\right)^4 \left|\Gamma_{in,TAG}\right|^2 \quad (2)$$

In general, the matching problems are analyzed by the power reflection coefficient for complex impedances, for which there are many analysis tools, the most widely known is the Smith chart, which is a graphical method that normalize the real part of impedance and it can be used to determine the matching between two complex impedances (Kurukkawa, 1965). Another widely used method which is based on the Smith chart to determine the power reflection coefficient for when both generator and load impedances are complex, which is described in (Nikitin, 2005), where the objective is the power wave reflection coefficient too. Then, in accordance to scheme of Figure 7 adapted for passive RFID tag at receiving mode we have the following expressions:

Figure 7. Equivalent circuit of passive RFID tag at receiving mode

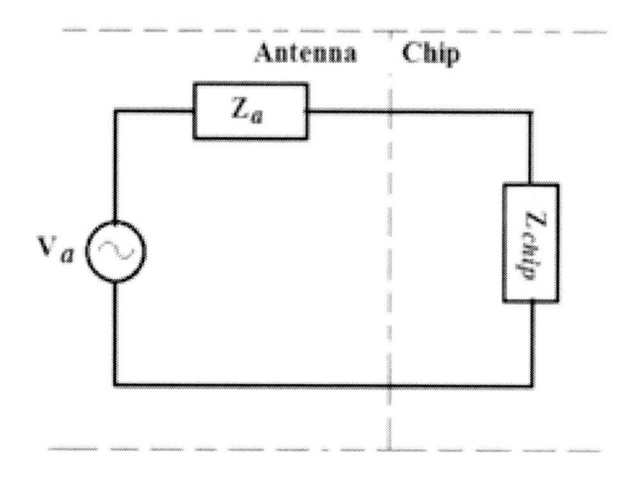

The voltage wave reflection coefficient:

$$\Gamma_{in,TAG} = \frac{Z_{chip} - Z_a^*}{Z_{chip} + Z_a} \quad (3)$$

The power reflection coefficient is given by:

$$\left|\Gamma_{in,TAG}\right|^2 = \left|\frac{Z_{chip} - Z_a^*}{Z_{chip} + Z_z}\right|^2 \quad (4)$$

The power reflection coefficient shows the fraction of the maximum power available from the generator which is not delivered to the load, for this to happen it must have a perfect complex conjugate match, namely: $Z_{chip} = Z_a^*$.

3.1. Matching Analysis for Passive UHF RFID Tag

As specified in the Equations 1 and 2, the performance of a passive RFID system at UHF frequency band depends that the RFID tag has a good matching between the antenna-chip. On this issue there is a lot of information (Monti et al., 2010), but a simple way to analyze antenna-chip RFID tags is using two equivalent circuits: one at receiving mode and other at transmitting mode, as shown in Figure 8 (Balanis, 2005).

For the case of equivalent circuit at transmitting mode, the antenna is represented by its input impedance which is frequency dependent, as well as the generator impedance also, where jX_a represents energy stored in electric $\left(E_E\right)$ and magnetic $\left(E_M\right)$ near-field components. For an approximation of a resonant antenna (dipole), it is possible to say that if $\left|E_E\right| = \left|E_M\right|$, then $X_a = 0$ and the real part is represented by R_r, which correspond to energy radiated into space of the far-field components, and R_l, which correspond to energy lost, that can manifest as heat in the structure of the antenna. For the analysis of impedance matching, one can assume that V_G, R_G, X_G are constant and $R_a = R_r + R_l$; R_a, X_a vary according to ambient conditions, then for determine the maximum energy transfer, we start from the power absorbed by antenna given by $P = I^2 R_a$.

Determining the current power:

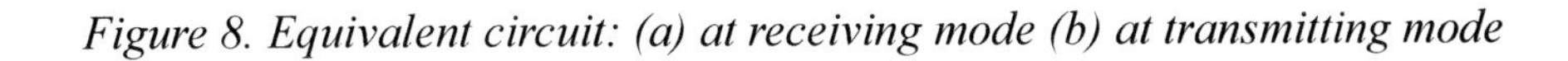
Figure 8. Equivalent circuit: (a) at receiving mode (b) at transmitting mode

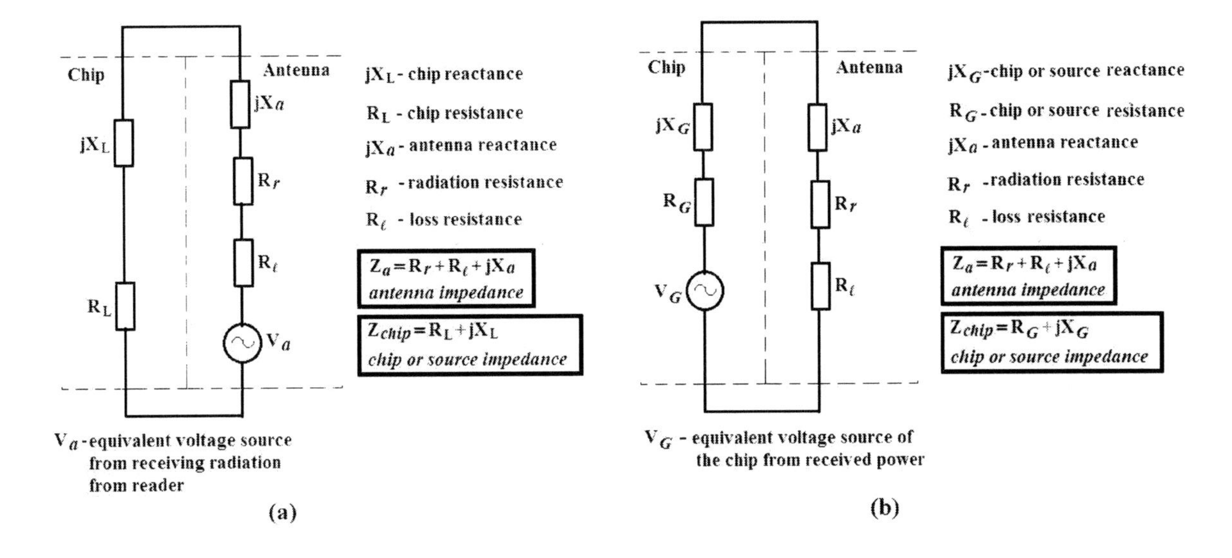

$$P = V_G^2\left(\frac{R_a}{\left(R_G + R_a\right)^2 + X_G^2 + 2X_GX_a + X_a^2}\right) \quad (5)$$

For the optimal value of X_a, we obtain

$$\frac{\partial P}{\partial X_a}$$

and when $X_a = -X_G$ we have

$$\frac{\partial P}{\partial X_a} = 0$$

Now, for the optimal value of R_a we have that $X_a + X_G = 0$ and then

$$P = V_G^2\frac{R_a}{\left(R_a + R_G\right)^2}$$

so,

$$\frac{\partial P}{\partial R_a} = 0$$

when $R_a = R_G$.

The conditions of maximum energy transfer have to:

$$\frac{\partial P}{\partial R_a} + \frac{\partial P}{\partial X_a} = 0$$

implying that $X_a = -X_G$ and $R_a = R_G$ and the power is

$$P = \frac{V_G^2}{4R_G}$$

Finally, we can say for the case of passive RFID tags, the maximum power is delivered to the antenna or from antenna, when the antenna impedance and the impedance of the generator (chip) or load (chip) are matched.

3.2. UHF RFID Tag Antenna Structures

The main challenge in designing antennas for RFID tags is to achieve better matching between the antenna-chip, both cases have complex impedance and most of the chip impedance cannot be changed, so the fit must be to the structure of the antenna.

Different methodologies have been presented and analyzed for the optimization of RFID antenna impedance. The RFID antenna usually can be treated as an array, as it is formed by a folded dipole and a reflector in order to optimize the matching. The basic principles to optimize the antenna impedance are presented in (Balanis, 2005) and a review of the design of RFID antennas with this methodology is presented by (Marrocco, 2008).

As was mentioned above, it is clear that the maximum power transfer to RFID tag is achieved when the impedance of the antenna $\left(Z_a\right)$ is equal to the conjugate complex of the chip impedance $\left(Z_c\right)$ i.e. is a conjugate pair. Maximum power transfer is one of the basic problems on antennas design. An example is showed in Figure 9, with the parameter S11 of a dipole tuning at 915 MHz (resonance frequency), when it is matched to different chips. From the graphic, it is possible to observe that the antenna impedance variation modifies the resonance frequency, obtaining a high mismatching.

Figure 9. Dipole operating at 915 MHz coupled to different impedances

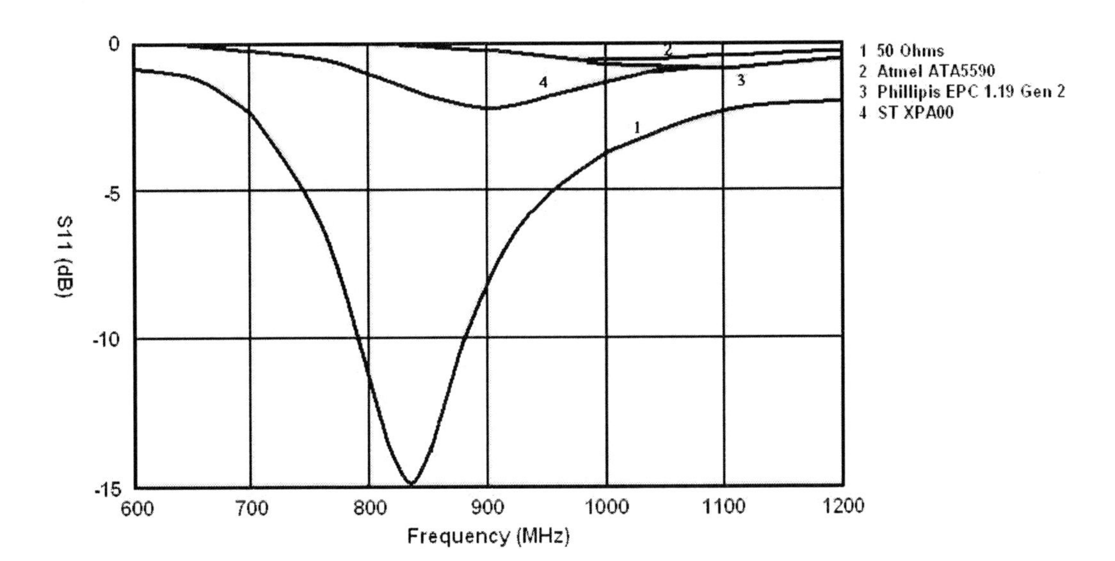

The matching problem in the antenna design has been solved using different techniques; the most-used of them follow (Balanis, 2005; Marrocco, 2008).

T-Match

T-match is based on dipoles; its structure and equivalent circuit is shown in Figure 10, where input impedance is given by:

$$Z_{in} = \frac{2Z_t\left(1+\alpha\right)^2 Z_a}{2Z_t + \left(1+\alpha\right)^2 Z_a} \tag{6}$$

where:

$Z_t = jZ_0 \tan\left(k\frac{a}{2}\right)$, transmission-line impedance,

Figure 10. T-Match a) Structure, b). Equivalent circuit

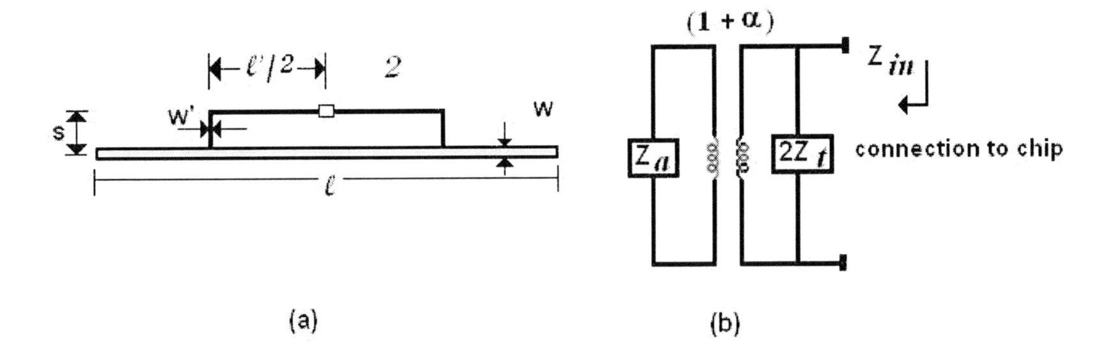

Figure 11. Inductive match feed loop (a) Structure, (b) Equivalent circuit

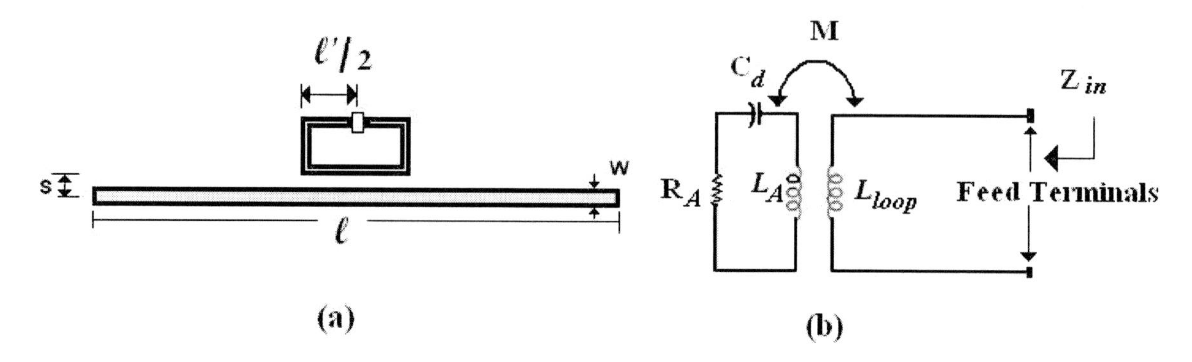

$Z_a = R_a + jX_a$, the antenna impedance,

$\alpha = \frac{w}{w'}$, strip line wide relation

Inductive-Match Feed Loop

Inductive-match feed structure for wideband impedance between antenna-chip of the RFID tag without additional matching networks was proposed by (Son, 2005). This structure matching is used to compensate the large capacitive reactance of tag chip. Figure 11 shows the structure and its equivalent circuit.

Input impedance is given by:

$$Z_{in} = R_{in} + jX_{in} = Z_{loop} + \frac{(\omega M)^2}{Z_A} \quad (7)$$

where:

Z_{loop}, loop impedance

Z_A, radiating impedance

M, mutual inductance

Z_{loop} and Z_A can be expressed as a function of frequency in terms of resonance frequency f_0, radiation resistance $R_{A,0}$ and quality factor Q_A, then:

$$Z_{loop} = j\omega L_{loop}$$

$$Z_A = R_{A,0} + R_{A,0} Q_A \left(\frac{\omega}{\omega_0} - \frac{\omega_0}{\omega} \right)$$

Referred dipole $\lambda / 2$ to the antenna impedance operating at 915 MHz, in accordance to with Figure 10, Figure 11 and Figure 12, where the variable parameters are the length ℓ and the separation s, the T-match structure results, with a strip line wide W equal to 3mm and 1mm, are shown in Figure 13 and relative to the inductive

Figure 12. Loop structure

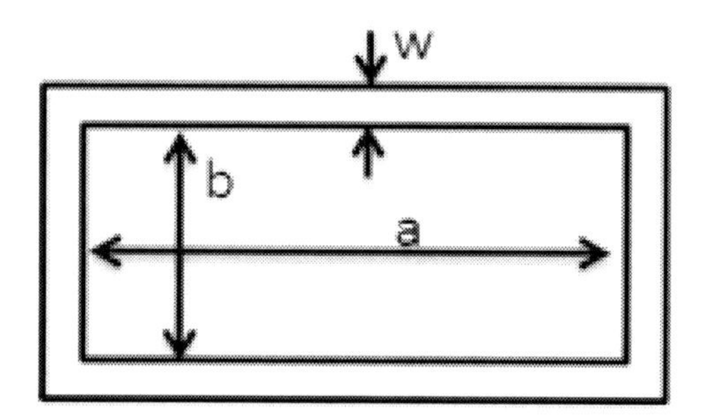

Figure 13. Real and imaginary part of the antenna impedance considering T-Match

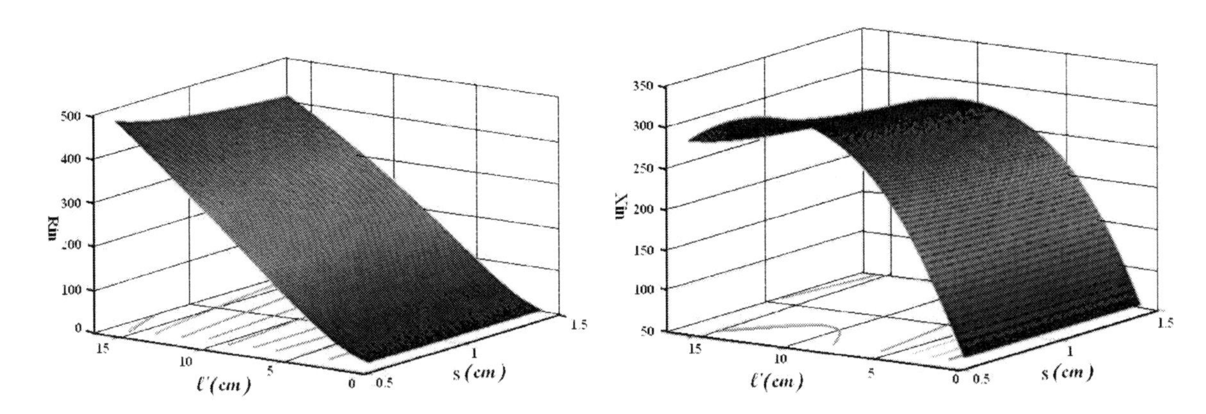

Figure 14. Real and imaginary part of the antenna impedance considering inductive match

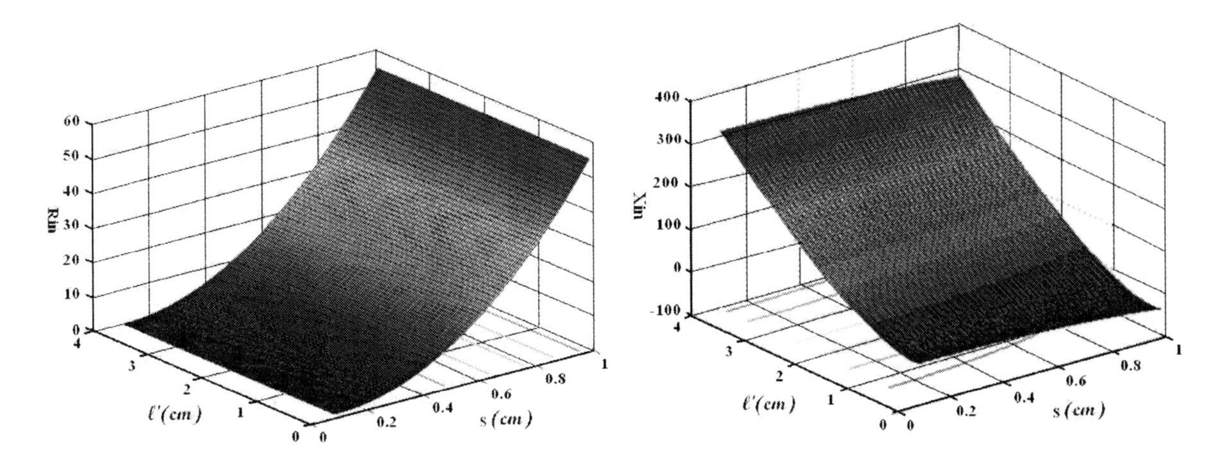

match, with a strip line wide equal to 3mm the results are shown in Figure 14.

From the graphics, it is evident that the antenna impedance is affected by geometry. Relative to the inductive match, the real part value is closer to the impedance real part of the chip. On the other hand, the real part impedance T match has higher values than the real part impedance chip. Respect to the reactive part of the inductive match, this one shows a better lineal performance than the T Match. From the results obtained, it is possible to observe, that, it is easier to take the control at the inductive match.

UHF RFID Tag Antenna

The most important challenge in the RFID tag antenna design which must be strapped to the vehicle windshield is the small size, and the meandering lineal antenna (MLA) is the suitable method to be applied, and so, based in the design method of (Marrocco, 2008), where the design parameters are the length of the vertical *(h)* and horizontal *(w)* segments and the round number *(N)*, the tag antenna proposal for the vehicle identification showed in the Figure 15 must satisfy with the following requirements:

Figure 15. The tag antenna

Figure 16. Parameter vs. frequency

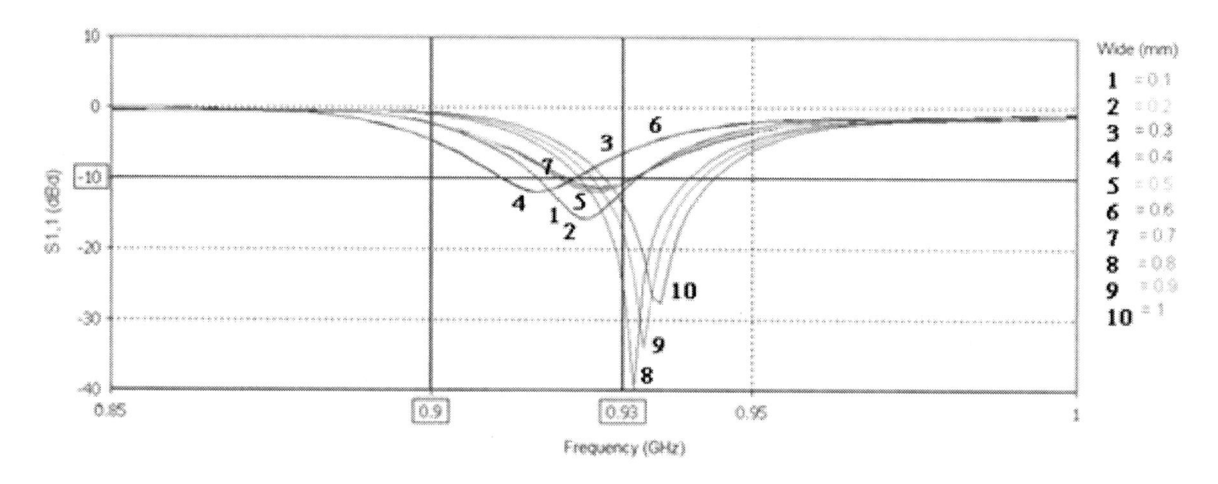

Figure 17. Wideband antenna

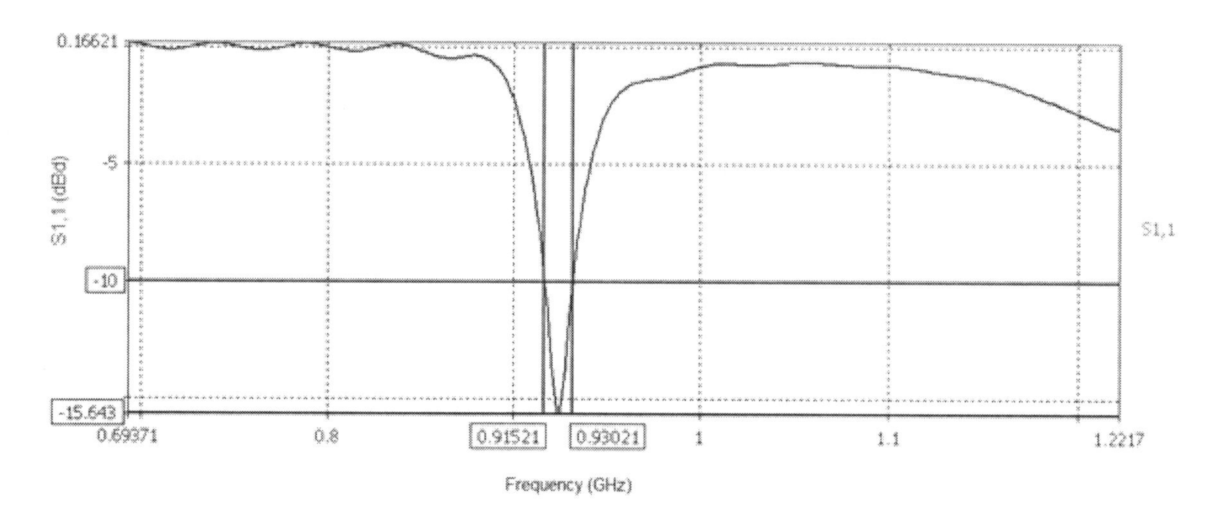

Figure 18. Radiation pattern

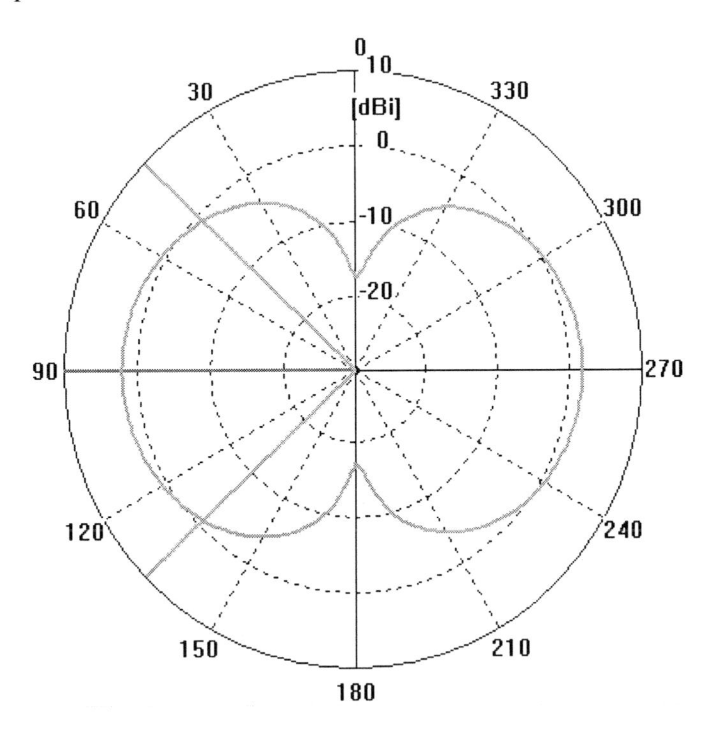

- **Operation Frequency:** 915 MHz
- **Max Dimensions:** Length 9 cm, Wide 2 cm
- **Material:** Cupper
- **Coupling:** Inductive

Figures 16 and 17 show the S_{11} performance parameter relative to the wide variations and the wideband of the designed antenna, and the Figure 18 shows the radiation pattern.

4. THE PROPAGATION MODEL IN RFID SYSTEMS APPLIED TO VEHICULAR IDENTIFICATION

4.1. Statement of the Problem

The RFID System is a wireless communication system recently it is being applied to vehicular identification as shown in Figure 19. The objective of the Radio Frequency Technology applied to vehicular identification is to have all the motor vehicle information inside the memory tag, and actually, it is limited to owner name, driving license and fines. It consists of the reader acting as the transmitter and the receiver device simultaneously, the tag strapped on the car windshield acting as the receiver and as a storage device, and between them, the Radio Channel, influenced by the radio electric and geometry environment characteristics. The Radio Channel has been analyzed extensively in different papers, but no one of them has analyzed the RFID technology applied to vehicular technology. So, a radio channel model is necessary for the effective design of transmitters and receivers, and for determining the positions of the tag and the reader. This motivates the analysis of the propagation environment for outdoor RFID systems, in order to know a feasible RFID application to the vehicular identification. A set of tests were implemented in laboratory and in the field with the objective to achieve RFID Channel Model capable to predict the electric field

Figure 19. The RFID technology applied to the vehicular identification

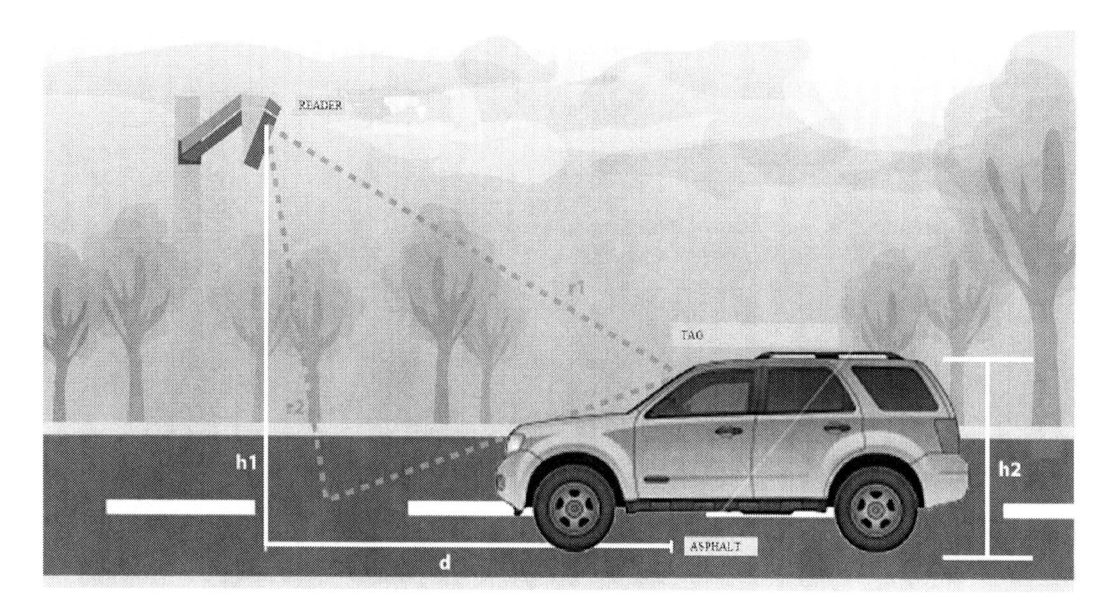

strength in an area around a transmitter, should take into account the structure, materials, objects and human beings included at every location in the area, this is, a deterministic channel model. On the other hand, a stochastic channel model predicts average field strength and the variations around this average. A purely stochastic model does not take into account the details of the radio propagation environment.

For the vehicular identify, the communication is line of sight, however there are other problems, such as: the multipath generated by asphalt and objects nearby to vehicles, as well as climate change. If we consider the RFID System applied to the vehicular identification, this system is surrounded by the ground (asphalt) and walls (buses, vans, sport cars, etc., located at adjacent lanes), from that, it can be modeled as a model, which, combines the deterministic and stochastic approach. As a result from the tests, the feasibility RFID application to vehicular identification, depends of several variables, for example, the mismatch polarization between the tag antenna and the reader antenna, because the scenario's geometry changes frequently and the polarization coupling is very difficult to obtain, and it should, moreover, be remembered, that the vehicle's windshield angle is not always the same; the speed of the mobile, because, the vehicles don´t move with uniform speed at the reader's antenna coverage area; and the incident reflected waves to the tag antenna, coming from the reflecting surface and other objects, this way, in an effort to meet a radio channel model, these parameters must be considered. A radio channel model proposal is the objective of this point, based in multipath environment and variables found in the experimental works, a set of tests were implemented in laboratory and in the field, some results and performance curves will be presented in this part of the chapter, so the reader can evaluate the convenience of the model for use in a particular analysis, in order to know the RFID signal performance and if the RFID application is feasible to the vehicular identification.

4.2. The Environment

The Passive RFID tags answer a reader query with a modulated signal according to the information already stored in the memory of a tag. RFID technology has been in place for many years supporting manufacturing and logistics practices (Landt, 2005). More recently, traffic control, vehicle monitoring and toll road applications have been developed. While indoor RFID applications can be conducted in fairly controlled conditions, vehicular environments are subject to multiple random phenomena. Therefore, in order to ensure a good readability redundant readers are often used and some channel models validated through field observations in different automotive scenarios.

For UHF passive RFID technology, the communication medium is very complex, because there may be objects of dimensions equal to the wavelength of the carrier frequency. This problem and other causing reflections, diffractions and scattering of the electromagnetic waves, degraded the system performance.

The vehicle identification's performance is highly dependent on the scenario's geometry. The two-ray model has been normally used for this analysis. Several factors must be considered, including the reader's antenna position, the location and orientation of the tag on the vehicle's windshield, the inclination of the vehicle's windshield, the radiation pattern of the reader's antenna and the tag's antenna. Other factors are the electrical and physical characteristics of the asphalt, the speed of the moving vehicles and the nearby traffic conditions. All scenarios are analyzed considering the RF signal emitted by the reader, the effect of antenna coupling polarization (reader-tag), the RF signal multi-rays and the dispersion of the generated signals for different vehicle heights. The radio frequency environment and the geometric RF signal channel propagation scenarios have been analyzed in general in different forms, but not in relation to the propagation analysis applied for the vehicular environments.

The search for a practical propagation model in outdoor environments for RFID systems depends heavily on the application at hand. On the road, where vehicles are passively tagged on the windshield and read at high speeds while passing through a portal, a simplified model is proposed which fits the experimental data with variances obtained in the present communication.

RFID performance is highly dependent on the environment characteristics. While, in well controlled conditions, a deterministic analysis provides a good phenomena understanding, a probabilistic approach allows an adequate modeling in scenarios where tight control of the parameters is not feasible.

From a geometric perspective and equipment characteristics must comply with local regulations and technological constraints. For instance, in relation to the downlink, the interrogator transmitted power is regulated. In region 2 the Effective Isotropic Radiated Power (EIRP) (ERM, EMC *Standard for Radio Equipment and Services Part 1: Common Technical Requirement*, ETSI EN 301 489-1) at the reader's antenna is limited to 4 Watts or 36dBm. Additionally, the sensitivity of the tag's integrated circuit imposes a physical limitation with nominal sensitivity values typically in the range from -10dBm to -18dBm depending on the manufacturer. On the other hand, in relation to the uplink direction, limitation factors relates the tag's efficiency of the backscattering signal which is associated with the antenna-chip coupling impedances and with the reader's sensitivity, whose typical levels are in the -80 dBm to -95 dBm range.

Figure 20 illustrates a basic ray scenario where the reader is above ground and mobile experiences a direct ray and reflected ray that depending on the current location and local conditions may be coming from the asphalt, the vehicle's hood or from another vehicle's roof, as well as the tag is exposed to multipath fading and variable delays, it is moving as the mobile moves, that is, the tag is involved in a violent electromagnetic environ-

Figure 20. Radio frequency environment

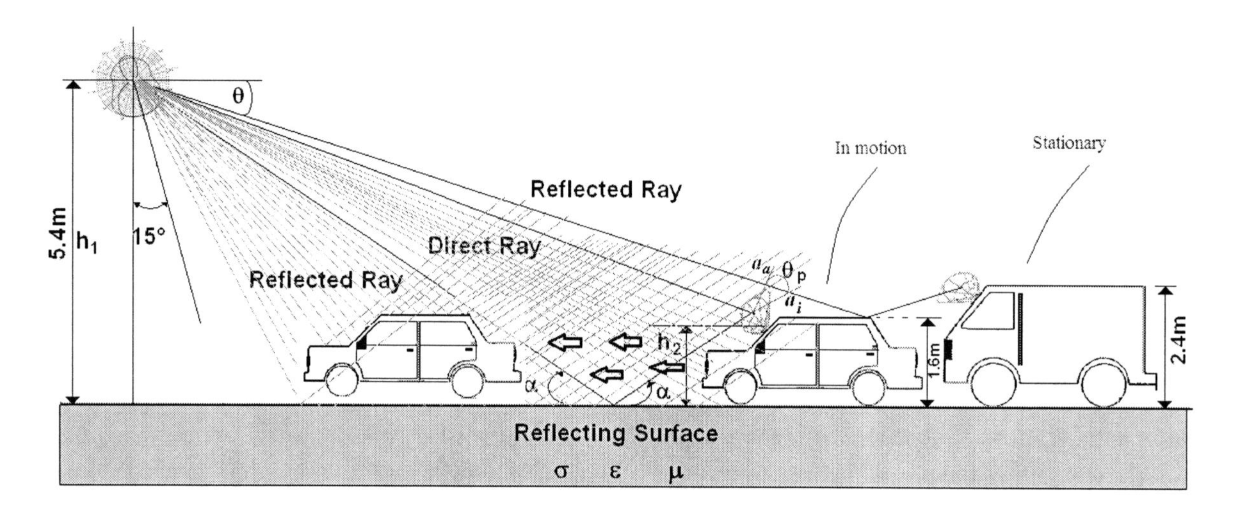

ment, where a ray is a direct ray for a while, and an instant later, now, it is a reflected ray.

4.3. The Deterministic Multi-Ray Model for RFID Technology

The multi-ray model is widely known and can be applied and adjusted to the RFID vehicle identification process. Both uplink and downlink signals can be modeled according to Equation 8 in Box 1.

According to Equation 10, where

$$r_1 = \sqrt{d^2 + \left(h_1 - h_2\right)^2}$$

corresponds to the direct ray length and

$$r_2 = \sqrt{d^2 + \left(h_1 + h_2\right)^2}$$

Box 1.

$$P_r = P_t G_t G_r \left(PLF\right)\left(\frac{\lambda}{4\pi}\right)^2 \left|\frac{1}{r_1}\exp\left(-jkr_1\right) + \sum_{i=1}^{N} \Gamma_T\left(\alpha_i\right)\frac{1}{r_1}\exp\left(-jkr_1\right)\right|^2 \quad (8)$$

where:

P_r, received power

G_t, the transmitting antenna gain

$k = 2\pi/\lambda$, wave number

P_t, transmitted power

λ, wavelength of radio frequency signal

$\Gamma_T\left(\alpha_i\right)$, total reflection coefficient

PLF, polarization loss factor

Figure 21. Multi-ray model geometry

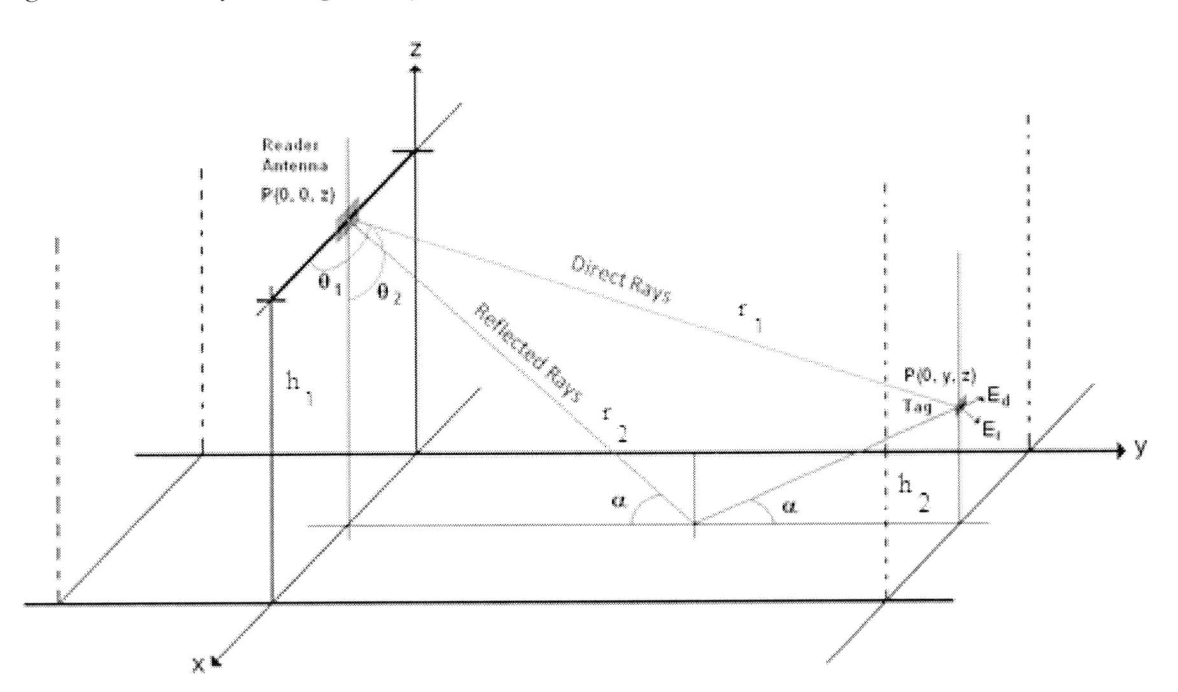

to the reflected ray length. The parameter $\Gamma_T(\alpha_i)$ is the total reflection coefficient, which may be for vertical polarization or horizontal polarization. The incidence angle on the reflecting surface is given as $\alpha = \sin^{-1}\left(h_1 + h_2/r_2\right)$. In the case of multiple rays (see Figure 21), they are added as is specified in Equation 8 for r_i and α_i.

On the other hand, the radio frequency energy reflected ray is affected in amplitude and phase by the total reflection coefficient $\left(\Gamma_T\left(\alpha_i\right)\right)$, which in turn depends on the Fresnel reflection coefficient $\left(\Gamma\left(\alpha\right)\right)$ and the surface roughness attenuation factor $\left(\rho\left(\alpha\right)\right)$. The reflection's general expression is given by Equation 9.

$$\Gamma_T\left(\alpha_i\right) = \Gamma\left(\alpha_i\right)\rho\left(\alpha_i\right) \tag{9}$$

Note that the Fresnel refection coefficient depends on the incident angle $\left(\alpha_i\right)$, and the electrical characteristic (conductivity σ, relative permittivity ε_r, relative permeability μ_r) of the reflecting surface, as well as on the polarization of the incident wave on the surface.

Since the reflected surface has no magnetic parameters $\mu_r = 1$. Then, for a polarized TEM plane wave with the electric field on the incidence surface (vertical polarization) the Fresnel reflection coefficient is given by Equation 10 (Collin, 1985; Jordan, et al., 1968).

$$\Gamma_{\parallel}\left(\alpha\right) = \frac{k_\varepsilon \cos\alpha - \sqrt{k_\varepsilon - \sin^2\alpha}}{k_\varepsilon \cos\alpha + \sqrt{k_\varepsilon - \sin^2\alpha}} \tag{10}$$

For a polarized TEM plane wave with the electric field perpendicular to the incidence surface (horizontal polarization), the Fresnel refection coefficient is given Equation 11.

$$\Gamma_{\perp}(\alpha) = \frac{\cos\alpha - \sqrt{k_{\varepsilon} - \sin^2\alpha}}{\cos\alpha + \sqrt{k_{\varepsilon} - \sin^2\alpha}} \quad (11)$$

The parameter $k_{\varepsilon} = \varepsilon_r - j60\sigma\lambda$ is the complex permittivity given for $\varepsilon_0 = 8.85x10^{-12}$ *(F/m)*, ε_r is the relative dielectric constant of the reflecting surface, $\sigma(Siemens\,/\,m)$ is the conductivity of the reflecting surface and λ is the wavelength of the incident radio frequency signal.

An important impairment in RFID technology is the mismatch between tag and reader antennas, which depends on the antennas' polarization. For vehicular identification this problem is critical as the geometry changes frequently and the polarization coupling cannot be guaranteed. For example, the vehicle's windshield angle is not the same for all makes. Besides, vehicles cross at different speeds by the site where the reader's antenna is placed. The polarization mismatch renders an extra loss. The mutual polarization efficiency is expressed in Equation 12 (Milligan, 2005):

$$P = \frac{1 + |\hat{\rho}_1|^2 |\hat{\rho}_2|^2 + 2|\hat{\rho}_1||\hat{\rho}_2|\cos(\theta_1 - \theta_2)}{\left(1 + |\hat{\rho}_1|^2\right)\left(1 + |\hat{\rho}_2|^2\right)} \quad (12)$$

where:

θ_1, θ_2, polarization phase ratios of the reader's antenna and the tag's antenna respectively

ρ_i, complex polarization ratios

The polarization of the antennas in a given direction is defined as the polarization of the wave radiated in the antenna's direction. The incoming wave electric field can be expressed as $\vec{E}_i = \hat{a}_w E_i$ and the polarization of the electric field in the receiving antenna can be written as $\vec{E}_a = \hat{a}E_a$, where $\hat{a}$ is the unitary vector. The polarization loss effects are quantified by the polarization loss factor (PLF), which is defined according to antenna's polarization and the transmission mode, as in Equation 13 (Balanis, 2005):

$$PLF = |\hat{a}_w \bullet \hat{a}_a|^2 = |\cos\theta_p| \quad (13)$$

From Equation 13, θ_p is the angle between the two unit vectors, which can be obtained according to Equation 10 by

$$\theta_p = \tan^{-1}\left[\frac{h_1 - h_2}{d}\right]$$

where h_1 and h_2 are the height of the reader's antenna and that of the tag's antenna respectively. Symbol d denotes the horizontal separation between antennas.

In this way, the RFID vehicular propagation channel behavior can be characterized by the Equation 8, and according to region 2 international regulations, in the down link, the power transmitted by the reader's antenna to the tag's antenna will be 36 dBm EIRP i.e. P_tG_t*=4Watts* and on the other hand in relation to the uplink, the power transmitted from the tag's antenna to the reader's antenna is determined by the backscattering efficiency η_{tag}, which in turn, is related to the maximum energy transfer or antenna-tag impedances coupling according to Equation 14 (Nikitin, *et al*, 2008).

$$\eta_{tag} = \frac{P_{re-radiated}G_{tag}}{P_a} = \frac{4R_a}{|Z_a + Z_c|^2} \quad (14)$$

Factor $P_{re-radiated}$ represents the re-radiated power, G_{tag}, the tag's antenna gain, P_a, the incident power in the tag's antenna, $Z_a = R_a + jX_a$, the tag's antenna impedance and $Z_c = R_c + jX_c$, the tag's integrated circuit impedance. For the tag, the EIRP is determined by the product $P_{re-radiated}G_{tag}$. The maximum efficiency is $(\eta_{tag} = 1)$, when the impedance coupling is accomplished, and it is an ideal situation. Generally, it is used $\eta_a = 0.7$, because, in mass production some mismatching can occur.

The reflecting surface parameters, conductivity σ and relative permittivity ε_r have influence in the Fresnel reflection coefficient. For the case of asphalt the typical values reported in the literature (Shang, 1999) are: conductivity in the 0.001 to 0.005 range when the asphalt is dry and upwards toward 0.1 when the asphalt is wet. The relative dielectric permittivity is in the 5 - 6 range with dry asphalt and in the 12 - 18 range with wet asphalt.

The downlink analysis is conducted for the central operation frequency of 915 MHz, considering for reflecting surface a typical relative permittivity of $\varepsilon_r = 5$ for dry asphalt, $\varepsilon_r = 15$ for wet asphalt and typical conductivity value of $\sigma = 0.005$ (Siemens/m) according to the place humidity. The results for both a relative permittivity of 5 and a relative permittivity of 15 are shown in Figure 22 for vertical polarization and in Figure 23 for horizontal polarization of Fresnel Reflection Coefficient.

From the figures, it is possible to observe that fading phenomena tend to be more sensitive to reflection coefficient for horizontal polarization than for vertical, and how multipath propagations improve reception levels. For the analyzed propagation channel, best conditions occur for vertical polarization and a wet environment (i.e. $\varepsilon_r = 15$).

Results from the graphics assumed a 100% reflecting surfaces but in practice, this coefficient will vary for the type of pavement which is expected to remain more or less constant (changes can occur due to a wet floor). Regarding side vehicle reflections, they are expected to exhibit faster changes as reflection coefficients will differ for different vehicles, and further random behavior is expected, as sensitivity level is not constant due to the RFID tag (antenna-chip) fabrication processes, impedance coupling and welding practices.

4.4. The Radio Channel in Vehicular Identification as a Dielectric Slab Waveguide Model

The ray tracing technique has been used to model specific environment, in this case, applied to vehicular technology, where the transmitter is modeled as a source of many rays in all directions around it. Each ray is traced as it bounces and penetrates different objects in the environment, including, the ground (asphalt) and walls (buses, vans, sport cars, etc., located at adjacent lanes). Nevertheless, the tracing calculation requires as input a detailed description of the environment, because every point of reflection of each ray from a surface has to be characterized in terms of the surface and geometry material. For the case of the reflections from the ground, the two ray model (Anderson, 1993; Xia, 1993) considers the direct wave and the reflected wave, and the reflection coefficient is based in the electric properties of the reflecting surface and the angle of incidence. In order to alternative way to obtain a RFID Channel Model capable to predict the electric field strength in an area around a transmitter, let consider the RFID System applied to the vehicular identification surrounded by the ground (asphalt) and walls, vehicles located (buses, vans, sport cars, etc.) at adjacent lines of the vehicle under test located, from that, it can be modeled as rough waveguide, a model, which one, combines the deterministic and stochastic approach. This kind of analysis based on dielectric waveguides on the guiding of radio waves in hallway and street environments, tunneling effect or canyon effect, have already been

Figure 22. Propagation channel uplink for $\varepsilon_r = 5$ *and* $\varepsilon_r = 15$ *vertical polarization respectively*

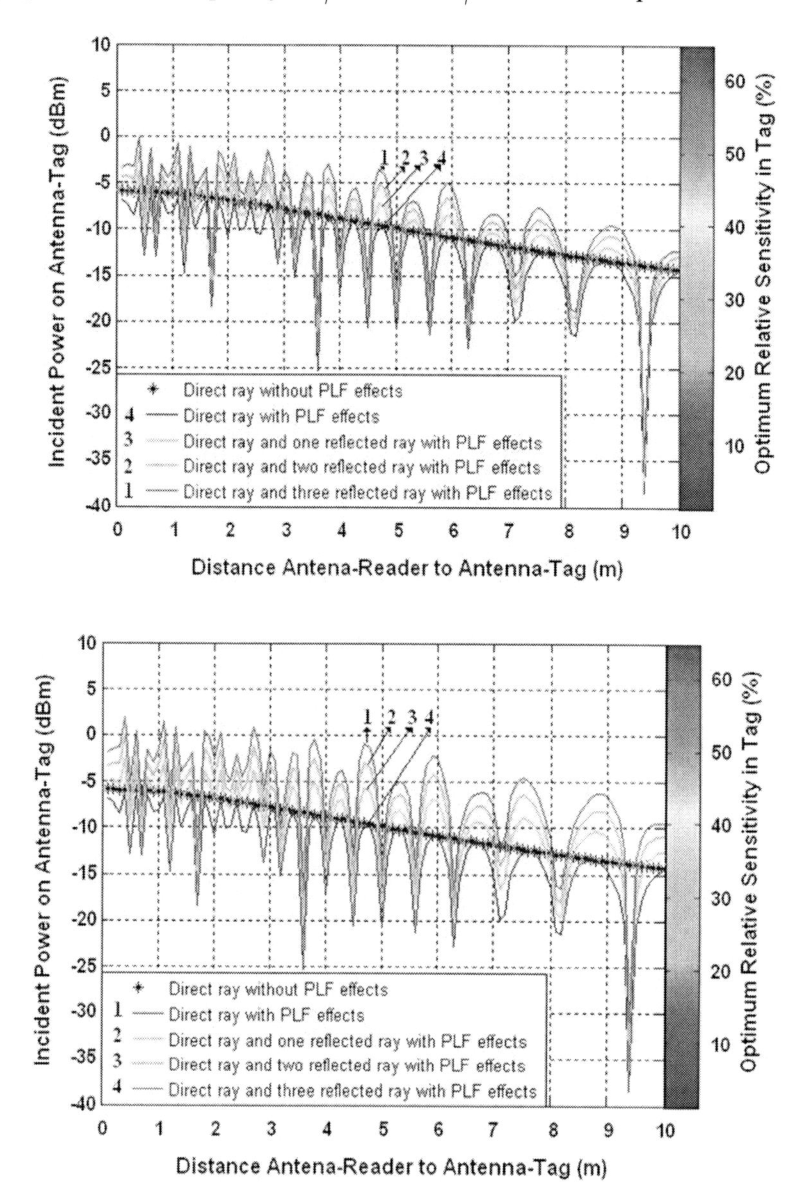

the subject used and explored (Marcuse, 1969; Yamaguchi, 1989), but not applied to vehicular identification.

The model suggested in this section concentrates and it is focused on the lateral walls, where the reflections from the different vehicles are a result of loss dielectric material. The required parameters are the width of the portal, the average height of the vehicles and the average electrical parameters (dielectric constant and conductivity) of the walls (cars). This way, considering the radio channel as waveguide, the waveguide floor is made of uniform material, the waveguide ceiling is the free space and the walls are roughed, formed by the cars at each side and by the spaces between them, additionally, the variations of the electrical

Figure 23. Propagation channel uplink for $\varepsilon_r = 5$ *and* $\varepsilon_r = 15$ *horizontal polarization, respectively*

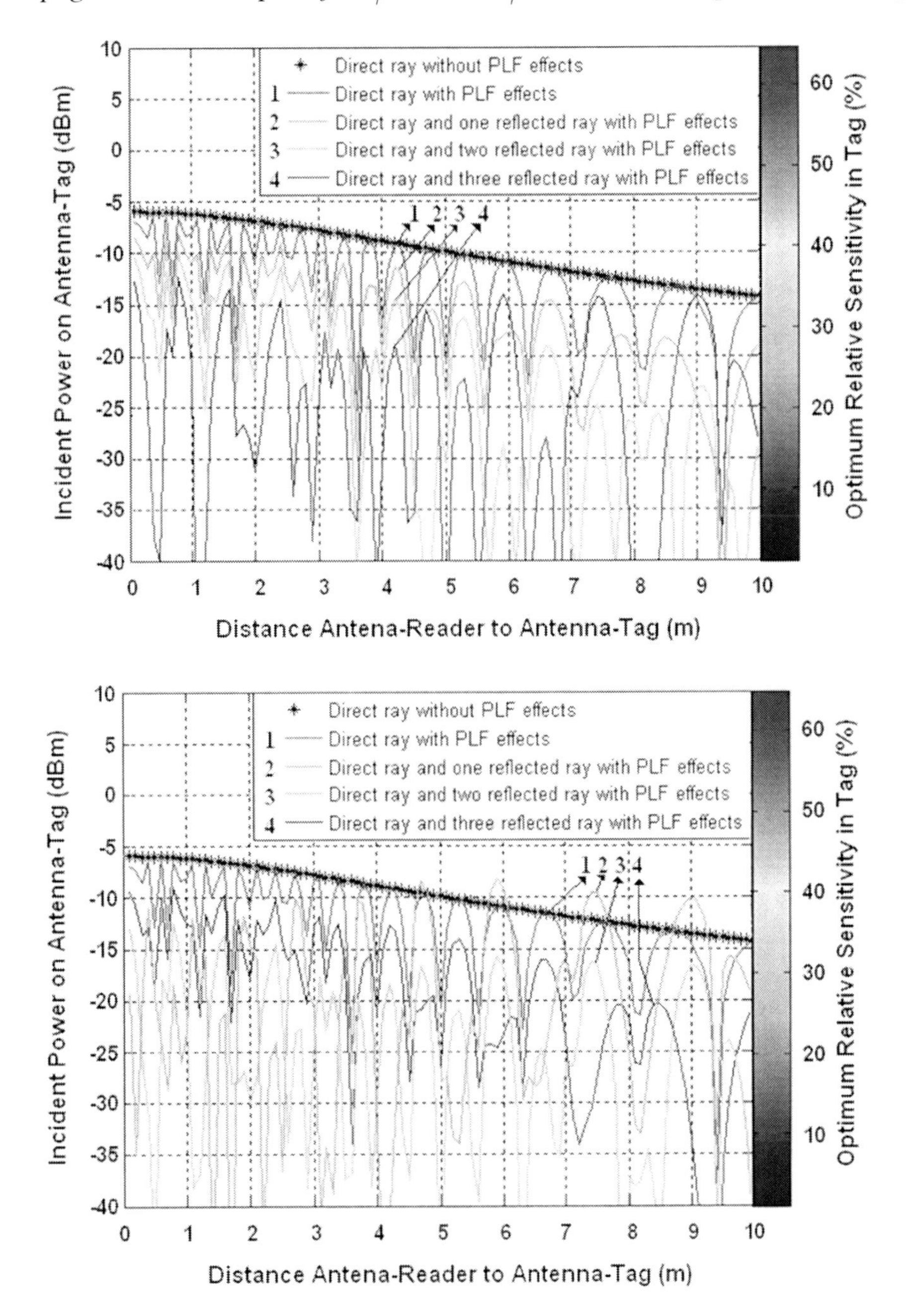

properties among the materials that constitute the walls, increases the effective roughness, as car windows, car wheels, etc.

The walls geometry is made of homogeneous isotropic material with linear properties, the electric permittivity ε_r, and the magnetic permeability μ_r. The wall roughness causes coupling among the propagating modes, causing the production of imperfections and the effects of mode coupling on the dispersion of the transmitted signals (Marcuse, 1969), in this way the objective in a dielectric slab waveguide is to contain energy within

Figure 24. The vehicle under test crossing the portal

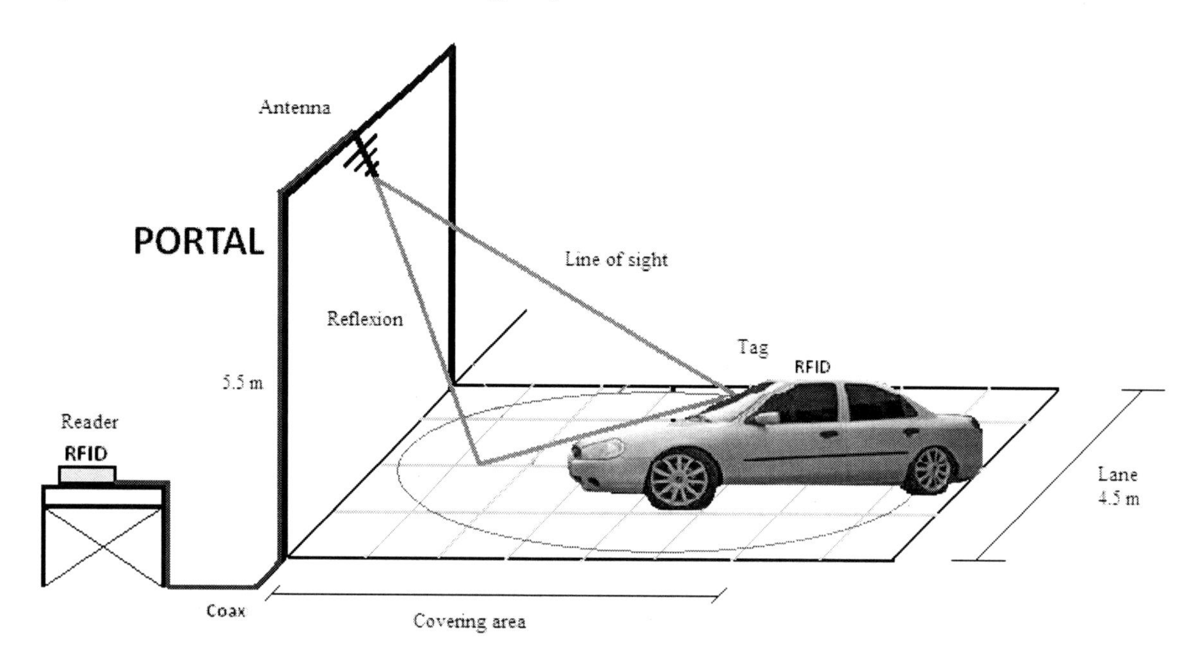

the structure and direct it toward a given direction. Inside of the roughness waveguide, the field is formed with the incident and reflected waves of the dominant mode, caused by the wave bounce back and forth between the walls.

One of the tests on the road consisted of the tag strapped on the windshield of the motor vehicle under test, crossing the portal, a Yagi antenna as receiver-transmitter antenna, the reader, and two buses at side forming the lateral walls. Figure 24 shows the scene of the test, where the width of the portal is approximately 6 meters, with regard to the adjacent vehicles close to 3 meters. RFID operation frequency is 915 MHz, the parameter $\varepsilon = \varepsilon_r - j60\sigma\lambda$ is the complex permittivity given for $\varepsilon_0 \approx 8.85x10^{-12}$ *(F/m)*, and ε_r is the relative dielectric constant of the vehicles reflecting surface, in this case, dry asphalt. Figure 25 shows the result of this test. The performance of the electric field distribution when the mobile vehicle crosses the reader coverage area with the buses forming lateral walls (see Figure 26).

From Figure 27, it is possible to observe that a decrease in power level at increasing the distance from the reader is notorious and it is possible to observe too, that the propagation phenomena inside the walls concentrates the energy in the middle of the covering area.

5. ELECTROMAGNETIC COMPATIBILITY ASPECTS

As shown in Figure 20, the RFID system applied to the vehicular identification operates in a complex electromagnetic environment, where the different tags strapped in the vehicle windshields operate in the same reading covering area causing electromagnetic interferences. Three kinds of interferences can be mentioned, tag-tag interference, numerous readers-tag interference and the reader-reader interference. The first one happens when many tags are energized simultaneously provoking the response of the backscatter signals, creating a bad combination of them, making difficult the tag identification. The mechanisms to avoid this

Figure 25. The RFID system geometry as a dielectric rough slab waveguide

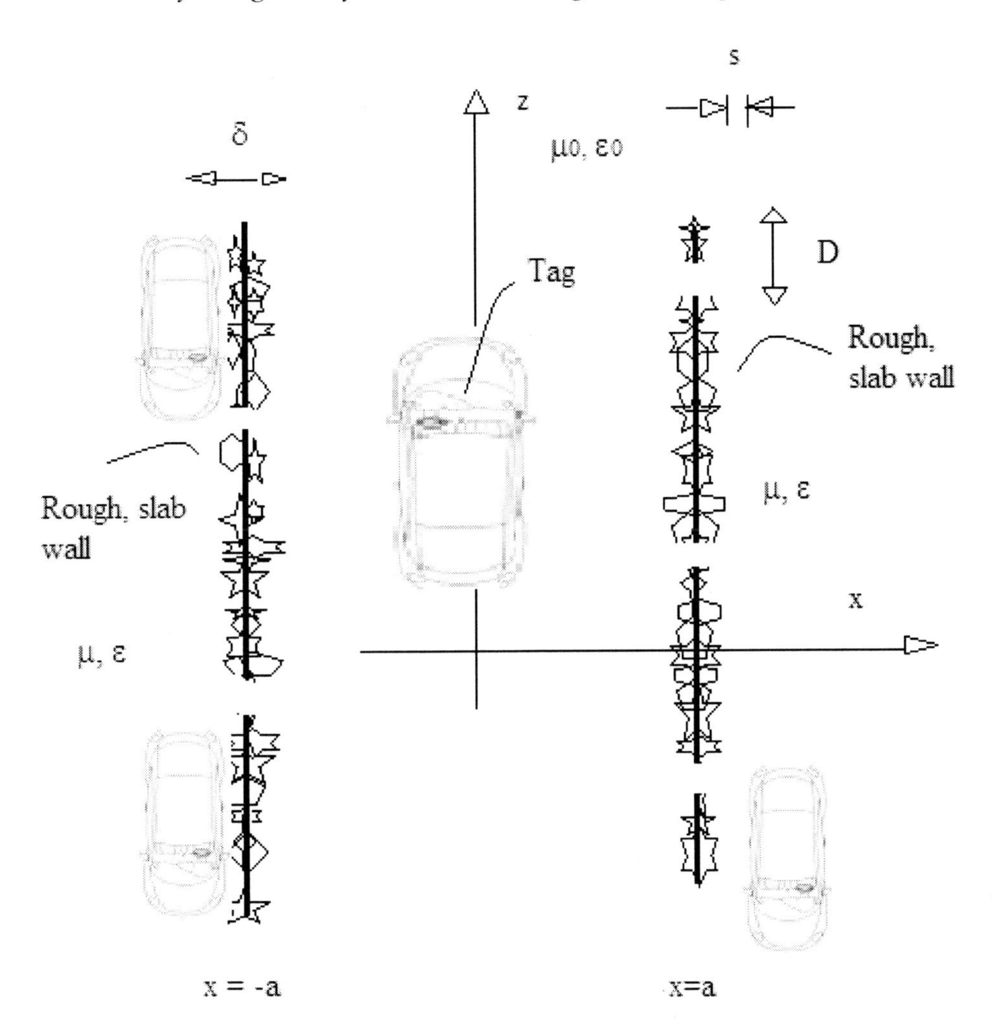

kind of interference problem are the Aloha and Binary Tree Techniques.

The second problem interference happens when a tag is situated inside the covering area of numerous readers and all of them try to establish communications with it.

And the last one happens when two readers are very close to each other creating an overlapping of the covering areas. This situation produces a serious decrease in power level and the no tag detection. This kind of problem interference happens very often and the techniques to avoid it are based on FDMA, TDMA and CSMA.

It is possible to presume the presence of a direct propagation path. However, line of sight shadowing may occur when a tag to be read is in the nominal reader foot print and the tag is behind a tall vehicle that obstructs the direct path. In a practical scenario the tag to front vehicle distance is determined by the car hoot and the safety guard distance that tends to increase with vehicles' speed reducing the likelihood of a mobile being in a shadowing region. In traffic jam conditions the vehicles separation will be less and the speed will also be reduced increasing the number of readings as the mobile stays longer in the interrogator footprint.

Figure 26. The buses form a dielectric rough wall

The tunneling effect, this kind of electromagnetic compatibility problem happens when the reader transmits a RF signal interfering with the adjacent communication tag reader, producing a high decrease in Signal to Interference Ratio in this link. This effect affects the wireless communication systems when the RF signal wavelength is smaller than the dimensions of the objects around the link. For the RFID passive technology UHF Band, 915 MHz frequency operation, $\lambda = 0.327m$ which is smaller than the motor vehicles dimensions. The following Figures 28 and 29 shows the equipment configuration to evaluate the interference effect from one reader to other, and the result of this test, respectively, when the RFID System is under jamming attack.

6. CONCLUDING REMARKS

Precise deterministic models are very useful in point to point link analysis. In vehicular identification the propagation parameters are not always available, because the scenario is continuously changing. Thus, the models for line-of-sight predictions do not consider random parameters of the environment. In this way, the power received in tag is the expected signal strength at different heights for tag traveling along the center of a lane. For simplicity, an isotropic radiator at the center of the lane is placed at a height of 5.5m, with a 4W EIRP. It is assumed, and it can be considered best case results as actual antennas beam patterns will lower the gain in directions other than the main beam, and this isotropic assumption allows spotting fade locations to be avoided and to favor radiation in regions that ensure a good reading. This way reducing to some extent the interference from other exogenous in band radio sources and reduces geometric shadowing region, it should, moreover, remembered that the propagation environments become even less regular in the presence of multiple reflecting paths, and the canyon effect allows higher reception levels when present.

Actually basic information (owner name, license, fines, car robbery report, etc.), is resident in

Figure 27. The average electric strength field distribution

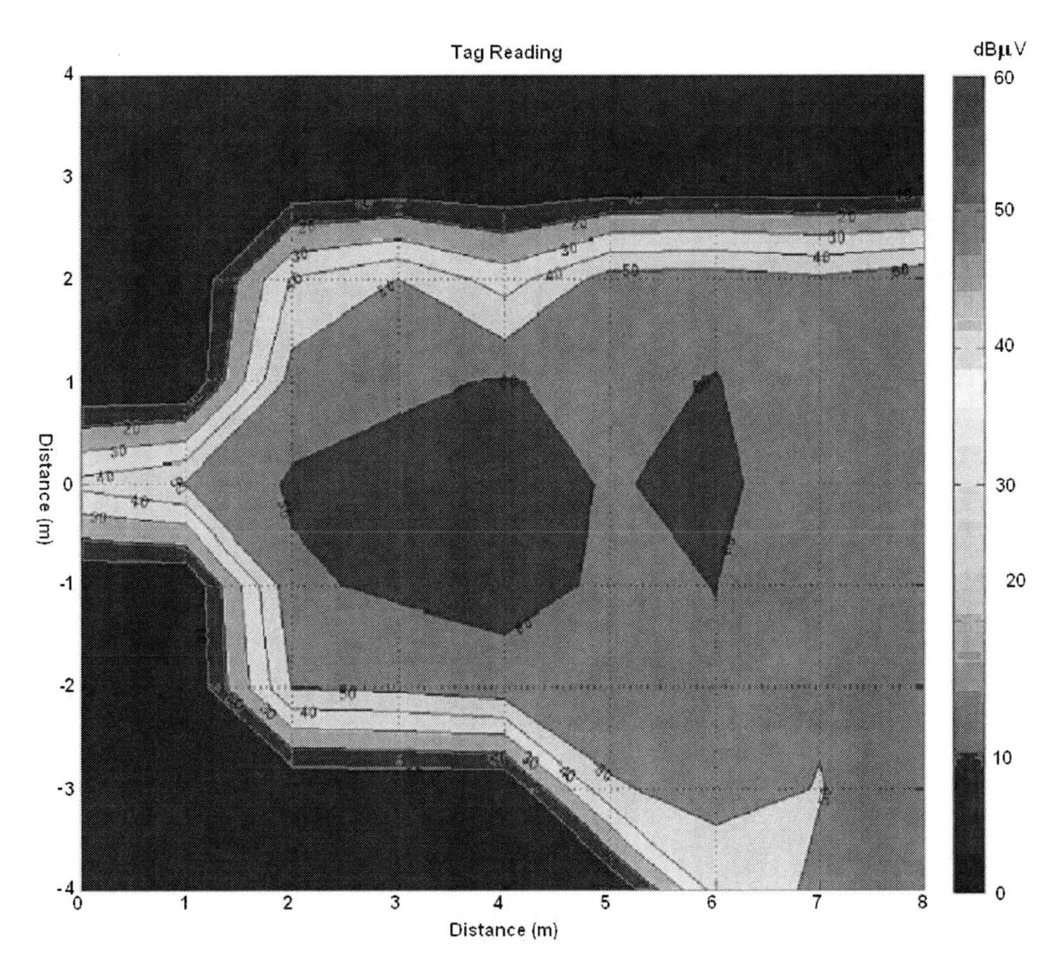

Figure 28. Equipment configuration to evaluate the interference test

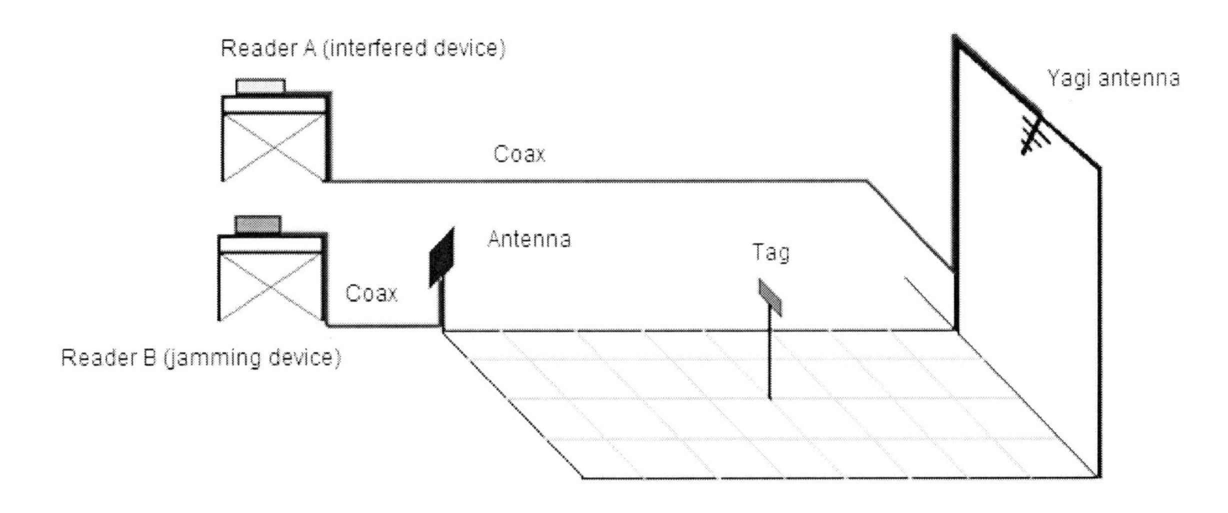

Figure 29. RFID system under jamming attack

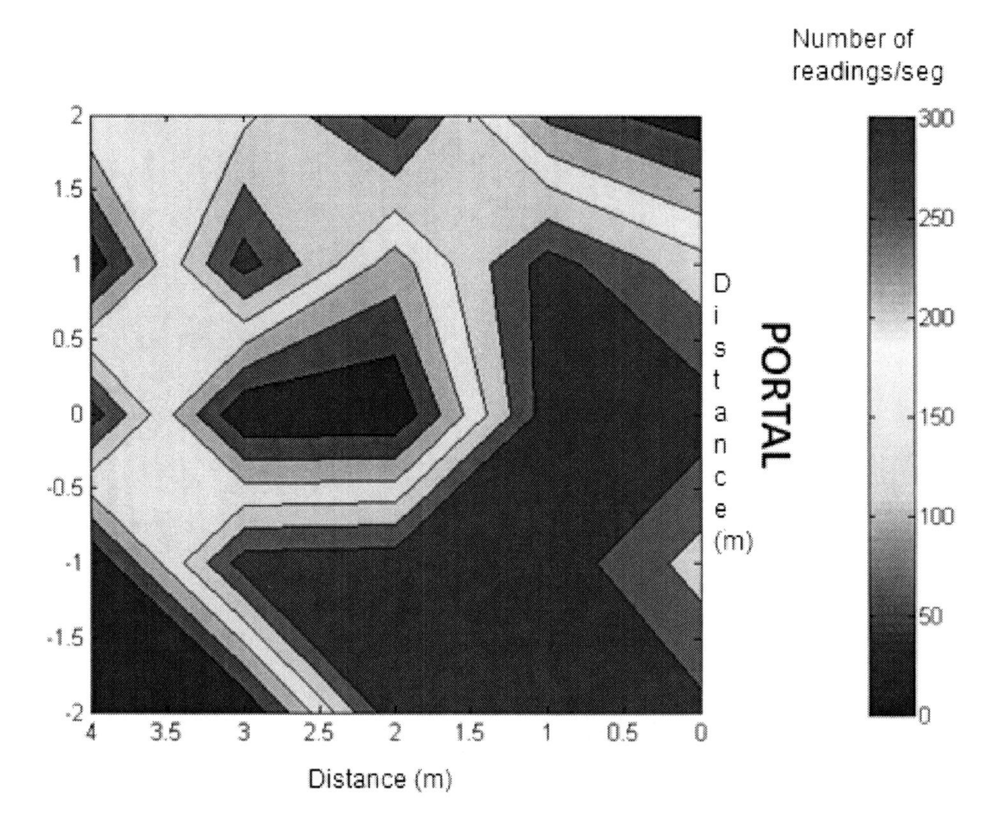

the memory tag, which it could be too much, if what is needed a key word linked with a database only where this information is contained, or it could be so little, if the text information is not enough; for example the owner photograph, image files, video files and other kind of multimedia information. But this need involves a higher data transmission rate and modulation and codification methods capable to transmit and recover the data efficiently under mobility and environment conditions.

The higher data transmission rate means a challenge, due to involve the study of some parameters; for example the sufficient power to energize the passive tag, selected for this application, the tag sensitivity, which in this application, at least, should be detectable 4 meters away from the reader, the necessary backscattered power to recover the data by the reader, also the appropriateness protocol communication, the modulation and codification methods as much the uplink as the downlink.

In the same way, meanwhile the time is essential to acquire the carrier signal and being the RFID system a time sensitive system, then, under the conditions above mentioned the reader must be able to receive the backscattered signal from the tag and recover the information, even when the vehicle is running at high speed making possible an error in the data, as consequence of frequency shifting, multi-paths which alter the carrier amplitude and for that reason some wrong bits in the frame synch or subsequent commands. Thus, under these conditions and for this application the RFID technology must be functional when a higher data transmission rate is necessary if it wants to be a new distribution method of digital information when the vehicle is traveling by road.

The considerations relative to the RFID system applied to vehicular identification, in the same way as the other communication system, must carry out with a requirement set as: Bit rate, maximum acceptable bit error rate, the channel propagation and the current regulations are important factors to state the maximum band width, the maximum signal transmission power, the maximum used power by the detector and the maximum time acquisition by the detector.

Additionally, the sensitivity of the tag's integrated circuit imposes a physical limitation with nominal sensitivity values, typically in the range from -10dBm to -18dBm depending on the manufacturer. In the uplink direction, limitation relates the tag's efficiency of the backscattering signal which is associated with the antenna-chip coupling impedances and with the reader's sensitivity, whose typical levels are in the -80dBm to -95dBm range. In the same way, a portal holding a mid-lane yagi antenna at 5.4 m height with main beam tilted down 15°. This elevation is a common in some scenarios in order to ensure successful readings in the close proximity of the reader. A minimum tag-reader distance clearance of 4 meters was considered. Then, under these conditions, in road applications readers will conduct several enquires while the mobile traverses the footprint and it is said that successful reading occurred when at least one out of the k enquires was read. Thus, the statement of the problem is the relation between the speed of the vehicle, the successful readings and the data frame length.

"Tari" is the time interval reference for the reader tag signaling; during this time the tag must be detected, there are i-th reader interrogations, and k is the maximum number of possible reads while the mobile is on the nominal reader footprint. Also during the interrogation, the reader manages three basic operations, the Select Command, used to determine which groups of tags will respond; the Inventory Command used to identify individual tags from a group, and Access Command used once the tag has singulated and ready to be addressed with individual commands, whereas the tag manages the next states: ready, arbitrate, reply, acknowledge, open, secured and kill, and on the other hand, considering that preamble lasts 100 μs, the frame synch 62.5 μs, the data 0, 25 μs, the data 1, 50 μs and the time interval of the remaining commands, the communication establishment must be able to adjust to the interval time during the vehicle crosses the portal where the reader is installed, considering, that, the speed of the vehicle is going from 40 km/H to 120 km/H, and if the speed keeps constantly, uniform, thus, the vehicle covers the 4 meters, resulting clear that these time intervals are long enough to host several reader tag communication trials. From the speed tests in the road, we have found that at low speed the number of successful readings is higher than high speed.

Finally, multipath propagation, polarization mismatch antennas, speed of the mobile, speed of surrounding objects, ground reflections and the transmission bandwidth of the signal are the most important physical factors in the radio propagation channel influence. The best conditions for the analyzed propagation channel are those with vertical polarization of the reflection coefficient and wet environment. So, a propagation model allows us to predict the power, the electric field spatial distribution, and the polarization state of the wave, in order to optimize the receiver operation.

REFERENCES

Alien. (2011). *RFID ICs.* Retrieved from http://www.alientechnology.com/tags/rfid_ic.php

Altin, N., & Yazgan, E. (2008). A new roof model on randomly placed buildings in mobile communication. *Progress in Electromagnetic Research, 3*(3), 95–101.

Anderson, H. A. (1993). A ray tracing propagation model for digital broadcast systems in urban areas. *IEEE Transaction on Broadcasting, 39*(3).

ATA5590 Datasheet. (2005). *1-kbit UHF R/W IDIC with anti-collision function.* ATMEL Corporation.

Avery Dennison. (2011). *About RFID.* Retrieved from http://www.rfid.averydennison.com/us/index.php

Balanis, C. A. (2005). *Antenna theory analysis and design.* John Wiley & Son.

Bancroft, R. (2009). *Microstrip and printed antenna design* (2nd ed.). SciTech Publishing, Inc.

Blythe, P. (1999). RFID for road tolling, road-use pricing and vehicle access control. *IEEE Colloquium RFID Technology,* London, England, (pp. 8/1-8/16).

Braaten, B., Owen, G., Vaselaar, D., Nelson, R., Bauer-Reich, C., & Glower, J. … Reich, M. (2008). A printed rampart line antenna with a dielectric superstrate for UHF RFID applications. *IEEE RFID 2008 Conference Proceedings,* Las Vegas, NV, April 16-17.

Braaten, B., Scheeler, R., Reich, M., Nelson, R., Bauer-Reich, C., Glower, J., & Owen, G. (2010). Compact metamaterial-based UHF RFID antennas: Deformed omega and split-ring resonator structures. *ACES Journal, 25*(6).

Calabrese, C., & Marrocco, G. (2008). Meander-slot antennas for sensor-RFID tags. *IEEE Antennas and Wireless Propagation Letters, 7,* 5–8. doi:10.1109/LAWP.2007.914123

Collin, R. E. (1985). *Antennas and radiowave propagation.* New York, NY: McGraw-Hill.

Dolukhanov, M. (1971). *Propagation of radio waves.* Moscow, USSR: MIR Publishers.

Finkenzeller, K. (2003). *RFID handbook: Fundamentals and applications in contactless smart cards and identification.* West Sussex, UK: John Wiley and Sons.

Finkenzeller, K. (2010). *RFID handbook: Fundamentals and applications in contactless smart cards, radio frequency identification and near-field communication* (3rd ed.). West Sussex, UK: John Wiley and Sons. doi:10.1002/9780470665121

Iizuka, H., & Hall, P. (2007). Left-handed dipole antennas and their implementations. *IEEE Transactions on Antennas and Propagation, 55*(5), 1246–1253. doi:10.1109/TAP.2007.895568

ISO/IEC 18000-6. (2010). *Radio frequency identification for item management -- Part 6: Parameters for air interface communications at 860 MHz to 960 MHz.*

Jari, P., Pascal, C., Declereq, M., Dehollain, C., & Joehi, N. (2007). *Design and optimization of passive UHF RFID system.* Springer.

Jordan, E. C., & Balmain, K. G. (1968). *Electromagnetic waves and radiating systems.* Englewood Cliffs, NJ: Prentice Hall.

Kumar, A., & Parkash, D. (2010). Planar antennas for passive UHF RFID tag. *Progress in Electromagnetics Research, 19,* 305–327. doi:10.2528/PIERB09121609

Kurokkawa, K. (1965). Power waves and scattering matrix. *IEEE Transactions on Microwave Theory and Techniques*, *13*(2), 194–202. doi:10.1109/TMTT.1965.1125964

Landt, J. (2005). The history of RFID. *IEEE Potentials*, *24*(4), 8–11. doi:10.1109/MP.2005.1549751

Li, H., Lin, J., & Wu, H. (2008). *Effect of antenna mutual coupling on the UHF passive RFID tag detection*. IEEE Antennas and Propagation Society International Symposium, San Diego, CA.

Loo, H., Elmahgoub, K., Elsherbeni, A., & Kajfez, D. (2008). Chip impedance matching for UHF RFID tag antenna design. *Progress in Electromagnetics Research, PIER*, *81*, 359–370. doi:10.2528/PIER08011804

Marcuse, D. (1969). Mode conversion caused by surface imperfections of a dielectric slab waveguide. *The Bell System Technical Journal*, 3187–3215.

Marrocco, G. (2007). Gain-optimized self-resonant meander line antennas for RFID applications. *IEEE Antennas and Wireless Propagation Letters*, *2*, 302–305. doi:10.1109/LAWP.2003.822198

Marrocco, G. (2008). The art of UHF RFID antenna design: Impedance-matching and size-reduction techniques. *IEEE Antennas and Propagation Magazine*, *50*(1), 66–79. doi:10.1109/MAP.2008.4494504

Milligan, T. A. (2005). *Modern antenna design*. John Wiley-IEEE Press. doi:10.1002/0471720615

Monti, G., Catarinucci, L., & Tarricone, L. (2010). Broad band dipole for RFID applications. *Progress in Electromagnetic Research C*, *12*, 163–172. doi:10.2528/PIERC10012606

Nikitin, P. (2005). Power reflection coefficient analysis for complex impedances in RFID tag design. *IEEE Transactions on Microwave Theory and Techniques*, *53*(9), 2721–2725. doi:10.1109/TMTT.2005.854191

Nikitin, P., & Rao, K. V. S. (2008). Antennas and propagation in UHF RFID systems. *IEEE RFID Conference Proceedings,* Las Vegas, NV, April 16-17.

Pais, S., & Symonds, J. (2011). Data storage on a RFID tag for a distributed system. *International Journal of UbiComp (IJU)*, *2*(2), 26–39. doi:10.5121/iju.2011.2203

Phillips Semiconductor. (2006). *Application note*

Popovic, N. (2011). UHF RFID antenna: Printed dipole antenna with a CPS matching circuit and inductively coupled feed. *International Journal of Radio Frequency Identification and Wireless Sensor Networks*, *1*(1), 28–33.

Shang, J. Q., Umana, J. A., Bartlett, F. M., & Rossiter, J. R. (1999). Measurement of complex permittivity of asphalt pavement materials. *Journal of Transportation Engineering*, *125*(4), 347–356. doi:10.1061/(ASCE)0733-947X(1999)125:4(347)

Son, H. W., & Pyo, C. (2005). Design of RFID tag antenna using an inductively coupled feed. *Electronics Letters*, *41*(18), 994–996. doi:10.1049/el:20051536

Xia, H. H., Bertoni, H. L., Maciel, L. R., Lindsay-Stewart, A., & Rowe, R. (1993). Radio propagation characteristics for line-of-sight microcellular and personal communications. *IEEE Transactions on Antennas and Propagation*, *41*, 1439–1446. doi:10.1109/8.247785

XRA00-SBN18I datasheet. (2005). *ST microelectronics.*

Yamaguchi, Y., Abe, T., & Sekiguchi, T. (1989). Radio wave propagation loss in the VHF to microwave region due to vehicle in tunnels. *IEEE Transactions on Electromagnetic Compatibility*, *31*(1). doi:10.1109/15.19912

Ziolkowski, R. W., & Lin, C. (2008). Metamaterial-inspired magnetic based UHF and VHF antennas. *Proceedings of IEEE Antennas and Propagation Society International Symposium Digest,* San Diego, CA.

KEY TERMS AND DEFINITIONS

Aloha Technique: A pioneering computer networking system developed at the University of Hawaii. ALOHAnet provides the first public demonstration of a wireless packet data network. It uses a new method of medium access (ALOHA random access) and experimental UHF frequencies for its operation, since frequency assignments for communications to and from a computer were not available for commercial applications in the 1970s.

Binary Tree Techniques: A tree data structure that keeps data sorted and allows searches, sequential access, insertions, and deletions in logarithmic time.

Communication Protocol: A system of digital message formats and rules for exchanging those messages in or between computing systems and in telecommunications. A protocol may have a formal description.

Effective Isotropically Radiated Power (EIRP): The amount of power that a theoretical isotropic antenna (which evenly distributes power in all directions) would emit to produce the peak power density observed in the direction of maximum antenna gain.

RIFD Vehicular Identification: The RFID technology used to identify a particular vehicle when it passes a particular point.

Chapter 10
Exploring Value-Added Applications of Chipless RFID Systems to Enhance Wider Adoption

Ming K. Lim
Aston University, UK

ABSTRACT

Radio-frequency identification technology (RFID) is a popular modern technology proven to deliver a range of value-added benefits to achieve system and operational efficiency, as well as cost-effectiveness. The operational characteristics of RFID outperform barcodes in many aspects. Despite its well-perceived benefits, a definite rationale for larger scale adoption is still not so promising. One of the key reasons is high implementation cost, especially the cost of tags for applications involving item-level tagging. This has resulted in the development of chipless RFID tags which cost much less than conventional chip-based tags. Despite the much lower tag cost, the uptake of chipless RFID system in the market is still not as widespread as predicted by RFID experts. This chapter explores the value-added applications of chipless RFID system to promote wider adoption. The chipless technology's technical and operational characteristics, benefits, limitations and current uses will also be examined. The merit of this chapter is to contribute fresh propositions to the promising applications of chipless RFID to increase its adoption in the industries that are currently not (or less popular in) utilising it, such as retail, logistics, manufacturing, healthcare, and service sectors.

INTRODUCTION

In the last decade, radio-frequency identification technology (RFID), which is one of the most promising automatic identification and data capture (AIDC) technologies, has been increasingly popular due to its ability to achieve system and operational efficiency, as well as cost-effectiveness. RFID technology uses radio frequency waves to transmit data between data carrying devices (known as RFID tags or transponders) and data receiving devices (RFID readers/interrogators).

DOI: 10.4018/978-1-4666-2080-3.ch010

When the RFID reader receives data from the tag, the data will be passed on to the RFID middleware for use in various applications, in most cases, integrated with other systems (e.g. Enterprise Resource Planning, Warehouse Management System, and Customer Relationship Management). Each RFID tag consists of unique data about the item to which it is attached to, e.g. date of production, shipping detail, expiry date, depending on the intended applications. For instance, when a carton within a shipment is tagged, it can be easily tracked in the supply chain and its information, such as destination, customer detail, value, contents, etc., can also be seamlessly traced to avoid wrong shipments or shrinkages.

In general, there are three types of RFID tags (Hunt *et al.,* 2007; Brown *et al.*, 2007): passive, semi-active and active. Passive tags are most commonly used and they only transmit information to the reader when they receive a radio wave signal from the reader (through the antennas connected to the reader). Semi-active (or sometimes called semi-passive) tags consist of an internal battery that is only used to perform tasks, such as recording temperature readings or moisture levels. The battery does not initiate communication with the reader. In contrast, active tags initiate communication with the reader as specified by the users. They are generally more powerful than passive tags in terms of read range, information content and data transmission rate. RFID tag selection is dependent on its application and the level of technological capability required. In recent years, as more vendors entering the RFID market, more advanced RFID features have been introduced, such as WiFi-, GPS-, sensor-enabled and battery-assisted passive tags. For example, if a GPS- and sensor-enabled RFID tag is attached to a consignment, the haulier is able to accurately monitor its real-time location and status throughout its delivery, and the customer can be certain, from the RFID data, of more accurate arrival time and if the consignment has been tampered or exposed outside optimum temperature range.

RFID technology promotes a new way of identifying objects and its operational characteristics outperform barcodes in many aspects, such as no direct line of sight is required, multiple tags can be read simultaneously, tags are durable and capable of withstanding harsh conditions, and data is re-writable on the tags (Daniel Hunt *et al.*, 2007; Holmes, 2005; McFarlane & Sheffi, 2003). The proven benefits include unique identification of tagged items and status monitoring, improved stock visibility and traceability at any stage in the supply chain, automated inventory counts, automated operations, increased product availability, reduced shrinkages, and so on (Ferrer *et al.*, 2010; Ngai, 2008; Daniel Hunt *et al.*, 2007; Holmes, 2005; Luckett, 2004; Finkenzeller, 2003). Each industry has a unique interest in the technological benefits for their business. It can be seen that the benefits of RFID were well-perceived in the early years when it was introduced, especially by the retail and logistics industries (Holmes, 2005; Finkenzeller, 2003; McFarlane & Sheffi, 2003). Due to its robustness and flexibility in application, the realisation of benefits have spread across manufacturing (robotics and M2M communications), healthcare (real-time location system and patient treatment/safety), pharmaceuticals (anti-counterfeit and status monitoring), and more recently service sector, such as through smart phones, social networks, resorts, amusement parks and restaurants (Wasserman, 2011; Edwards, 2011; Wessel, 2010a; Butner, 2010; O'Connor, 2010; Thuemmier *et al.,* 2009; Wei *et al.*, 2009).

Despite it is proven that RFID technology does bring benefits to businesses in terms of operational efficiency and costs saving, a definite rationale for larger scale adoption is still not so promising. For any technology to be adopted, there are issues that need to be carefully investigated. The issues related to RFID technology are associated with cost, technicality, implementation, integration, standards, privacy, and security (Daniel Hunt *et al.*, 2007; Wu *et al.*, 2005; Weinstein, 2005; Berthiaume, 2004). At present, although RFID technology has been perceived to be reaching the maturity stage of its lifecycle and a substantial number of RFID vendors have entered the market, the system cost is still not as low as predicted. The cost is still one of the main barriers for many potential RFID adoptions.

As far as the cost is concerned, two potential efforts could encourage wider adoption, one of which is new design leading to lower RFID device cost and the other is seeking for promising, high-impacting, and cost-justifying applications (business cases) to give the adopters confidence over satisfactory return on investment (ROI). This coincides with the academic literature that two groups of RFID research contributing to its adoption can be identified. One is related to the technical aspects, such as enhanced security tags (Lee, 2008), improved tracking range (Lee, 2008), advanced tags with lower power consumption (Liu *et al.*, 2006), and standardised authentication protocols (Piramuthu, 2007). The other is associated with the applications, which provide greater contribution to potential wider adoption. This area of research facilitates RFID applications in manufacturing (Waggoner, 2008), inbound operations (Rach, 2008), logistics/supply chain management (Niederman, 2007; Sigala, 2007), warehousing (Garcia *et al.*, 2007) and so on. This chapter will capture these two aspects.

A survey conducted by Aberdeen Group stated that 42% of respondents stated "A Clear Business Case" as a crucial decision criterion of whether to pursue RFID adoption (Aberdeen Group, 2008). The survey also indicated that the main challenge for RFID adoption is proving its ability to maximise competitive agility and responsiveness to changing business requirements with justifiable investment. The value of RFID is its ability to enhance the quality of information collected in terms of its accuracy, timeliness and usage. The benefits are only realised when the information is turned into actions. According to the survey, two actions can be taken to maximise its ROI, which are to incorporate RFID solutions with key operational applications/infrastructure and to integrate RFID data with business analysis and intelligence applications. These two actions will also be explored in this chapter.

The cost of RFID systems can now be seen to be lower in comparison to early years when RFID was first proven to be successful in delivering its benefits. As for an application that requires a large volume of tags (e.g. for item-level tagging), the cost of tags remains a major concern. The cost of tag is largely dependent on its silicon integrated circuit (IC), a.k.a. chip or microchip, and its capability. This has resulted in the development of chipless RFID tags, which is the main subject of this chapter. Chipless RFID tags have been introduced in the market as an alternative option in order to benefit from much lower tag cost (Bacheldor, 2007; RFID Journal, 2004; Moore, 2002). However, the uptake of these chipless tags is still not as widespread as predicted by RFID experts (Das, 2007).

This chapter will explore in-depth the chipless RFID system in terms of its technical and operational characteristics, benefits, and current uses. The merit of this chapter is to contribute fresh propositions to the promising applications of chipless RFID to increase its adoption in the industries that are currently not (or less popular in) utilising it, such as retail, logistics, manufacturing, healthcare and service sectors. This aims at promoting wider adoption of chipless RFID to achieve value-added benefits, operational efficiency and justified ROI.

CHIPLESS RFID

A conventional RFID tag usually consists of an antenna and an integrated circuit (IC) whereby the majority is silicon-based. For passive tags, the antenna is used to receive radio waves from the RFID reader and generate power energy in order for the IC to perform its data processing functions. In addition, the IC is also capable of storing information related to the tag, such as a unique serial number provided by the tag manufacturer and other information defined by the users for intended applications. These capabilities have made RFID technology a robust system to achieve a wide range of benefits across different industry sectors.

In contrast, a chipless RFID tag does not consist of an IC and therefore, the cost of tag can be reduced significantly. This has made chipless tags a cost-effective alternative for item-level tagging. Chipless tags tend to be less sensitive to radio frequency interference and are able to sustain wider temperature range. However, they usually contain very limited data storage and only one tag (or very small number of tags) can be read at one time. Up to present, very limited chipless tags are commercially viable and available; most are still reported as prototypes. Without the presence of IC, no serial number is available in chipless tags. This also means that chipless RFID systems are utilising a different approach to identify individual items in comparison to chip-based RFID systems. Most chipless tags use electromagnetic (EM) properties of materials for identification purposes, i.e. based on these properties they reflect a portion of the radio waves received from the reader to generate a unique return signal pattern and the reader will then interpret this signal pattern to form some kind of a serial number. As the EM property of material for each tag is different (due to the deviation in tag production), therefore the reader is able to differentiate the tags based on these unique return signal patterns. These tag materials could include conductive polymers, fibres, and electronic inks, which can be directly printed on packaging or products. Since these tags are not silicon-based and they have been given a few names, such as "organic" or "polymer-based" tags.

Without IC as a means to uniquely identify each tag, one of the challenges of producing chipless RFID systems is the design of data encoding mechanism. Based on the literature and industrial reports, three categories of chipless tags can be summarised based on data encoding, i.e. time domain reflectometry (TDR), spectral signature, and amplitude/phase backscatter modulation (Preradovic and Karmakar, 2010; Das, 2007). These categories can be further analysed in Table 1 (deriving from Preradovic and Karmakar, 2010; Daniel Hunt *et al.*, 2007) based on their type, advantages and limitations.

THE EXISTING APPLICATIONS OF CHIPLESS RFID

Since the introduction of chipless RFID tags almost a decade ago, a number of applications (predominantly prototypes) have been reported. Amongst these include CrossID's chipless tags, which can be used to protect sensitive documents, such as confidential reports, financial securities and banknotes (RFID Journal, 2004). This chipless RFID system uses nanometric materials, which consist of tiny chemical particles with various degrees of magnetism that resonate when exposed to radio waves. The system uses up to 70 different chemicals, and therefore is able to transmit 70 digits of binary numbers. This system can also be used to prevent unauthorised photocopying and allow only specified type of papers to be used. CrossID tested this system with readers of radio frequency between 3-10 GHz. The advantages of this tag are the ability to print on any object without complicated tag production processes and to work with existing barcode systems without the need to majorly upgrade their application software and databases. Tapemark is another company using nano-resonant materials for their chipless tags for authentication, anti-counterfeiting and security applications (Collins, 2004). This system uses frequencies from 24 GHz to over 60 GHz, with a read range of up to 5 feet. Furthermore, funded by the Australian Research Council, Monash University developed a chipless RFID system for book tagging to automate library check-in system (Foreshew, 2011). The chipless tags developed

Table 1. Types of chipless RFID tags

		Type of Tags	Advantages	Limitations
TDR	Non-Printable	Surface Acoustic Wave (SAW) tags	Low cost/ greater read range/ applicability in positioning applications	Small data storage/ high speed reader required
	Printable	Thin Film Transistor Circuit (TFTC) tags	Smaller tag size/ low power consumption/ more functionality	Higher cost/ low electron mobility (operating up to several Mb)
		Delay line based tags	Benefits akin to SAW tags but printable	Limited no of bits
Spectral Signature	Chemical	Nanometric materials	Low cost/ printable/ robust/ greater data storage/ good tolerances to metal and water	Bigger tag size/ large spectrum requirement/ orientation require-ment/ dedicated RF component/ operate up to a few kHz
		Ink tattoo	Greater read range/ printable	Bigger tag size/ only operate at microwave signal
	Planar Circuits	Capacitively tuned dipoles	Printable on any polymer substrate or laminate	Bigger tag size/ mutual coupling effects between dipoles
		Space filling curve	Compact size due to its resonating property	Small data storage/ significant layout modification for data encoding
		LC resonant	Low cost/ simple design	Restricted in operating range and bandwidth/ 1-bit data storage/ multiple tag collision
		Multiresonator	Support simple spiral data encoding	Bigger tag size/ operate only at UWB range
		Multiresonant diploes	Smaller tag size/ spatial efficiency enhanced	Operate only at UWB range
Amplitude/Phase Backscatter Modulation		Left-hand delay lines	Require less bandwidth/ simple design/ smaller tag size	Smaller data storage/ higher cost/ require higher signal-to-noise ratio
		Remote complex impedance	Printable antenna with a lossless reactance	Less robust in applications
		Stub-loaded patch antenna	More robust (cross-polarization diversity)	Not suitable for circularly polarised reader
		Carbon nanotube loading	Benefit with using nano-technology for sensing	Higher cost

use backscatter technology and come with 24-bit capacity that can support up to 16 million entries.

Inkode introduced chipless RFID tags for protecting higher value items (e.g. meats) in supermarkets (RFID Journal, 2003). It is claimed that the cost of tag can be mass-produced for less than a penny. The tag consists of very tiny metal fibres that reflect radio waves back to the reader with a resonant signature, which can then be converted into a unique serial number. They claimed to have sold 80 million tags to a pharmaceutical company for brand protection. PolyIC introduced printable 13.56MHz HF (high frequency) tags which made with conductive inks using a roll-to-roll printing process. They use common printing processes, such as flexographic, rotogravure, offset or rotary screen. This tag consists of only eight bits of memory and has not been produced to any international standards. The targeted markets are authentication, security and some specific transportation and logistics applications that do not require high tag memory. Somark Innovations is another company that developed chipless RFID for animals tagging by using permanent ink tattoo, which is biocompatible and translucent (Bacheldor, 2007). This system uses very high radio frequency, and multiple tags can be read simultaneously from several feet away.

However, up to present, very limited chipless tags are commercially used. One of the more widely used is known as surface acoustic wave (SAW) tags based on the time domain reflectometry (TDR) approach (Preradovic and Karmakar, 2010). Carinthian Tech Research developed SAW tags, which use piezoelectric crystal to identify the tags (Collins, 2005). They are able to tolerate and monitor temperature above 400°C (where as conventional silicon-based tags will be damaged at this high temperature), as well as very low temperature at around -55°C. These tags are aimed for engineering and industrial applications, e.g. inside furnaces or engines in motion. Each crystal is read-only and transmits a unique ID encoded onto the tag through etching process during tag production. The crystal expands and contracts corresponding to temperature changes, and as a result, these tags are good for monitoring temperatures. When the crystal exposes to high temperature, it will expand and result in delay responding to radio wave signals from the reader. Likewise, the response will be quicker in cold temperature. With this system, an algorithm is used to calculate temperature changes according to the signal delays. These tags operate at 2.4GHz with up to 10 meters read range.

Airgate Technologies (www.airgatetech.com) is another company that produces SAW tags aiming to be used for temperature, torsion and pressure applications. The signal strengths received by the reader can also determine the location of the tag. These tags are operating at 2.4GHz with up to 300 feet. Other companies include RF Saw (www.rfsaw.com) producing SAW tags for NASA to run a pilot to track crewmember supplies, and Advanced Research (www.advresearch.com) using SAW tags for its OpenCrib application to track tools and parts in large storage bins/cribs. Thoronics of Germany has established a SAW design house for custom design of SAW RFID tags. It was also reported that SAW RFID was first appeared in non-stop road tolling in California (USA) and Norway (Harrop, 2006).

There have been some operational improvements and characteristics enhancement since the first generation of chipless RIFD was introduced. InkSure (www.inksure.com) claimed, unlike microscopic fibre tags, their chipless tags are able to support multiple tags reading and are aimed for anti-counterfeiting documents applications and track and trace consumer packaged goods. RF SAW introduced their Global SAW tags to overcome the limitations of IC tags including power requirements and performance limitations. The tags can provide excellent read range while maintaining high speed and accurate readings, and operate legally worldwide. In addition, initial work has also been undertaken by EPCglobal to incorporate their developing standards for SAW capability with ISO (Das, 2007).

For the advancement of chipless RFID, the academic research community has contributed to various aspects of technical development of this system. These include designing better antenna to achieve greater performance (Hu *et al.* 2010; Vyas *et al.*, 2009; Yang *et al.*, 2009; Balbin and Karmakar, 2009a; Mukherjee, 2008), improving tag design to enhance applications (Shrestha *et al.*, 2009; Preradovic *et al.*, 2009; Balbin and Karmakar, 2009b; Zheng *et al.*, 2008), enhancing authentication and anticounterfeiting capabilities (Lakafosis *et al.* 2011), increasing data storage capacity on tags (Preradovic and Karmakar, 2009), enhancing composite properties to improve tag reading (Yang *et al.*, 2009), and developing more robust data coding schemes (Schussler *et al.*, 2009).

The initial driver for the development of chipless RFID is the high tag cost. However, the applications above suggest that chipless tags are also capable of providing other sought after features, such as sustaining very high (and very low) temperatures due to the material used and operating in metal/liquid environment due to the data transmission mechanism. They can also be printed directly on products and packaging and have a long read range that can potentially replace high-cost battery-powered active/passive tags. In addition, chipless RFID also has the ability to measure tag location, position and direction of travel, which are costly or difficult to implement with competing technologies (Harrop, 2006). This has made chipless RFID a good candidate as a passive RFID system (with sensing capability) for various types of applications. With these benefits, it was predicted that the chipless RFID system will dominate the market rapidly especially for item-level tagging (Das, 2007). However, it does not seem to be happening as yet; it has not been well-perceived especially in supply chain applications. The next section will explore how chipless RFID tags could be enhanced in usage in order to promote wider adoption in the market.

VALUE-ADDED APPLICATIONS OF CHIPLESS RFID

The capability and benefits of chipless RFID technology have been well demonstrated through a range of applications, in particular, for authentication, anti-counterfeiting and security applications, animals tagging, as well as temperature and pressure controls. However, its adoption in the RFID market is still not as widely forecasted. It has very limited publications reporting the actual use of chipless RFID systems. Since the chipless RFID is able to drive down the tag cost dramatically and therefore, the biggest opportunity with this system is the ability to enable item-level tagging at a very affordable price in comparison to silicon-based tags. Up to present, its uptake is not so promising.

This section will examine the potential value-added applications of chipless RFID in the industries that are currently not (or less popular in) adopting this technology. Specific emphasis will be placed on those applications that favour chipless RFID capabilities, such as reducing overall RFID costs with much lower tag cost, ensuring authentication and security, sustaining extreme temperature and harsh environment (with the presence of liquid/metal surrounding), printing directly on packaging/products, and so on. In order to make a significant contribution to the chipless RFID adoption in the market, the major industry sectors that are currently the main players in RFID market will be examined. Examples of the existing applications that can be replaced by chipless RFID to achieve the same objectives will be chosen and analysed. Furthermore, potential additional benefits that can be provided by chipless RFID will also be explored. Where appropriate, the applications whereby chipless RFID can most benefit and contribute to higher ROI will also be identified.

Retail

Apparel Industry

This is one of the earlier major industries which tested RFID systems for item-level tagging. American Apparel carried out a number of pilots applying EPC Gen 2 passive tags on garments before delivering to the stores. These have resulted in successful outcomes improving inventory management, increasing sales and tackling theft issues at the stores. They have consumed over 1 million tags on these pilots and will expect to use a few more million when they expand RFID systems to more of their stores (O'Connor, 2008b). Up to present, they have 10 RFID-enabled stores (Swedberg, 2010a). German clothing company Oliver Bernd Freier carried out an RFID pilot to tag garments at their distribution centre and deliver them to three stores. The system has proved its ability to prevent out-of-stock and ease inventory counts (Wessel, 2010b). Korean Basic House employed EPC Gen 2 tags to garments at their manufacturing plants and installed RFID interrogators at their two distribution centres and 159 stores to prevent stock shortages (Swedberg 2010b). They have tagged 2 million pieces of garments and forecasted this number will go up to 10 millions in the following year. Turkish Eren Holding is another clothing company that used EPC RFID to track shipments and take inventory at their factory, distribution centres and stores (O'Connor, 2009a).

The main objectives of using RFID system in this industry revolve around product on-shelf availability, supply chain visibility, inventory management, and shrinkages. Item-level tagging in apparel business is a large market for RFID, and ABI Research forecasted this market will triple by 2014 (ABI Research, 2009). It is also pointed out by the German retailer Kaufhof Warenhaus that RFID will be widely adopted in the apparel industry as soon as tag prices fall further, and the main reason apparel companies are not rolling out RFID at the present is due to the price of tags (Roberti, 2006). This, indeed, is very promising for chipless RFID to enter the apparel market with thorough considerations of how it can operate and what it can offer. Much lower tag price is an obvious incentive for shifting to chipless RFID in this industry due to the requirement of a huge number of tags needed for tagging each piece of garment. Chipless tags, which are almost impossible to be cloned, can also be used for anti-counterfeiting purposes to ensure no garments from the pirated market enter their supply chain. Furthermore, if appropriate, chipless tags can also be printed directly onto the garment brand labels during label printing. This will eliminate the need for an additional process after production to attach the tags to the garments. As the existing issue of lack of standardisation with chipless RFID is concerned, it will be less of a challenge for this industry if the retailer owns the logistics operations, managing the flow from the manufacturing through transportation and storage to the retail outlets.

Food Supply Chain

The U.S. Center for Disease Control and Prevention estimated foodborne diseases to cause 76 million illnesses, 325,000 hospitalisations and 5,000 deaths in the US every year (Mead *et al.*, 2000). Safeguarding the food supply from its point of origin to the point of sale is therefore critical. This has led to tighter regulations and the increasing need for track-and-trace capability in the food supply chain (Roberti, 2009). Some food organisations have tested and implemented RFID systems to improve food safety and shelf life.

The largest Norwegian food supplier, Nortura, employed RFID to track meat from its butchering plant to the store in order to get full traceability throughout the chain (Swedberg, 2008a). Adhesive passive EPC Gen 2 UHF (ultra-high frequency) tags are attached to totes at its butchering plants. The cut meat will be placed on these totes, which will undergo several washing cycles and be kept in cold environment to maintain the meat freshness. The system is able to record the time and location along the process. These totes will eventually be delivered to the stores. In addition to having increased visibility where the meat comes from and where it is produced, Nortura is also interested in the information such as animal age and what it has eaten. A vegetable supplier, Tanimura & Antle, applied RFID tags on plastic containers whereby lettuces (fresh from the farm) will be placed on, and these containers will be packed to form a pallet and shipped to their warehouse. The pallet will then be cooled down to ensure longer shelf life (Wasserman, 2006). This application will also help to improve inventory control. Other similar pilots include Del Monte tagging shipments of fresh pineapple and Duda & Sons tagging celery (Wasserman, 2006).

Wasserman (2006) pointed out some challenges of using RFID for fresh food supply chain, whereby chipless RFID could potentially be able to address. Firstly, the costs are still high for RFID tags and middleware. Secondly, although RFID is capable, but there are still some technical difficulties such as dealing with cases of beverages or food with high water content. Thirdly, roughly 10 percent of all perishable food becomes unusable before it reaches consumers, therefore temperature sensor is essential but this will significantly increase the system cost as a large number of sensing-enabled tags are required. In addition to addressing the challenges above, chipless RFID could also add value by using low-cost ink tattoo on animals. This will make a huge saving when tagging with a massive number of livestock. The ID of tagged animals can then be linked up to the containers transporting their meats down the supply chain and in this way specific information (e.g. animal age, type of food eaten) about the meat's origin can be traced. On the other hand, there are other RFID systems used for food supply chain that would need a substantial amount of data storage that the existing chipless RFID may not able to provide. However, if the tag costs saved could be justified, there will be alternative options how these data can still be captured with chipless RFID, e.g. storing of data captured at each supply chain stage in the shared back-end system.

Beverage Industry

In the beverage business, there are a number of RFID implementations to improve operational/ supply chain efficiency, such as tracking kegs in the supply chain to ensure no counterfeits enter its chain and better stock visibility (Swedberg, 2009a), tracking bottles of liquor for inventory control, pouring management and preventing shrinkages due to dishonest bartenders (Swedberg, 2011a; Wasserman, 2011), and holographic RFID labels for anti-counterfeited liquor in the supply chain (Swedberg, 2008b). These applications would require a tag that is hard to be duplicated in order for it to be used for anti-counterfeit purposes. When it comes to a large number of bottles/kegs to be tagged, the lower cost of chipless RFID could take advantage, not to mention that chipless RFID's EM properties are almost impossible to be duplicated. Chipless tags can also be printed directly on the bottle labels during label printing. This will save cost without having a separate process to print the tags and attach them to the bottles. Furthermore, to maintain the quality of the liquor chipless RFID can also offer sensing capability during transit in the supply chain to ensure the liquor has not been exposed outside the optimum temperature range. The ability to perform in liquid environment will be a mandatory for bottle tagging and chipless RFID is capable of providing this.

Logistics

Logistics is another major industry adopting RFID to achieve a wide range of benefits in operational efficiency and costs saving. This section focuses on RFID applications in the logistics functions of distribution and warehousing.

Distribution

To ensure a smooth goods flow (forward and reverse) in the supply chain, many logistics service providers are using returnable transit items (RTIs) to transport goods. RFID-enabled RTIs can achieve greater benefits, such as providing real-time location information, increasing RTI utilisation, and preventing wrong shipments. Norsk Lastbærer Pool (NLP), established by the Norwegian consumer goods manufacturers and retailers, developed EPC Gen 2 tagged RTIs (plastic pallets and totes) to manage a nationwide pool of goods transfer (Swedberg, 2011b). At present, 150,000 RFID RTIs are in use and NLP is expecting this figure to reach 6 millions in the next few years. In house, NLP uses the RFID system to better manage their RTI wash and repair programme. For business, they will charge their customers rental and handling fees based on the RFID data collected through RTI usage. In Germany, Mars,

Rewe, Deustshe Post and Lufthansa Cargo are using RFID-enabled RTIs with EPC Gen 2 tags, aiming to achieve self-guided logistics systems to enable better automation of goods flow, real-time tracking and less error-prone (Wessel, 2010c). In Canada, JD Smith & Sons tested an RFID system at its distribution centre in Toronto to track loaded pallets leaving its facility and the empty pallets returning for reuse (Swedberg, 2010c). They aim to automate distribution processes, increase pallet utilisation and remove manual and error-prone tasks. The RFID system consists of EPC Gen 2 RFID-enabled plastic pallets and is coupled with GPS and cellular communication technologies on their trucks. This will allow their customers to track their goods in transit. There are also other applications using RFID-enabled reusable plastic pallets (Swedberg, 2009b; Swedberg, 2007a; Swedberg, 2007b).

Chipless RFID could well benefit RTI business in terms of its tag requirement in millions. Chipless tags are also good in sustaining harsh environment, especially as these RTIs may have to undergo a range of processes throughout the supply chain. Chipless RFID's performance will also not be compromised when RTIs pass through environment with the presence of heavy metal or liquid substances (or goods themselves). Furthermore, temperature-sensing applications could be useful for regular monitoring of the temperatures the goods are exposed to in transit. Same as in the retail sector, an open system involving various parties will need some form of standardisation that currently is a drawback for chipless RFID systems.

Warehousing

Artilux NMF is a lighting fixtures manufacturer based in Lithuania using EPC RFID to track the movements of pallets and of raw materials in its warehousing facility (Swedberg, 2010d). They ship 25,000 pallets a year, predominantly in the European market. They use RFID to ensure the right level of quality control is imposed before leaving its warehouse and the right pallets are despatched to the right customers. Similarly, they tag their incoming materials as they arrived to their facility and this will give an accurate account of inventory available in stock to prevent manufacturing disruption. Mitsubishi Electric Asia implemented an RFID system to replace their laborious and inefficient manual operations to track and manage inventory (Bacheldor, 2006). Their electronic products are tagged with EPC Class 1 tags when they arrived at the warehouse to be stored; the inventory data in the back-end system will then be updated accordingly. When the products are shipped out from the warehouse, the tags will be removed and reused. SC Freda, another Lithuanian manufacturer, began employing RFID to track furniture manufactured for IKEA, with the goal of reducing shipping errors and labour costs. Once a piece of furniture is manufactured, it is placed on a pallet and an adhesive EPC Gen 2 tag will be applied on it. The pallet will then be moved to the warehouse for storage. Upon receiving an order, the RFID-enabled forklift will be used to identify the location of pallets in the warehouse and performs order picking accordingly. It is hoped the RFID data will also help to better manage their production schedule to avoid excessive stock-up (Swedberg, 2010e). Indian steel manufacturer Viraj Profiles uses RFID to determine the precise location of their variety of shaped steel profiles in the warehouse and hence, to improve the speed and accuracy of steel profile retrieval and shipping (Friedlos, 2010). A reduction in labour costs has also been achieved. Currently, 20,000 tags have been used for the first run and Viraj will continue to use RFID in their business.

RFID applications in the warehouse can be closed-loop or open systems. Some applications would require tagging on pallets or products that associated to their shipping operations. Chipless RFID will be favourable when the quantity of tagging required is large and only data of unique ID is required. As appropriate, the chipless temperature-sensing capability will be versatile for warehousing applications, as well as its ability to determine the location of tags and to sustain harsh environment in the warehouse.

Manufacturing

Chipless RFID could be more popular for manufacturing applications, assuming these are closed-loop systems that do not involve supply chain partners. BMW employed an RFID-based real-time location system at its assembly plant in Germany to match the cars being assembled with the correct parts and tools needed to perform the jobs (Swedberg, 2009c). As their orders are much bespoke, each car is assembled according to individual client's requirement, with specific seats, interiors, engine parts and accessories. As the cars enter the assembly line, each of them will be tagged (using an ultra-wideband UWB tag) and linked to the back-end system to match the bespoke requirement. Along the assembly line, at each station the tags will be read and the correct tools and parts will be identified to assemble the cars. Marigold Industrial is a protective gloves manufacturer based in Portugal. In order to achieve enhanced visibility of the materials used to manufacture gloves, to decrease time in inventory counts, and to improve its ability to track the materials during manufacturing and the finished products despatching to customers, the company deployed an RFID system based on passive EPC Gen 2 RFID labels (O'Connor, 2009b). As the boxes of materials arrived at their factory they will be tagged and stored. Until these materials are used up in manufacturing, the finished products will be put in new tagged boxes to be stored until the customer order is raised. In this way, the company can better manage their materials flow in the entire factory. However, one issue discovered was that read reliability will sometimes suffer due to cotton liners densely packed into the cases. A Spanish agricultural cooperative, COVAP, uses passive UHF RFID tags (which contain only unique ID numbers) to track legs of premium ham at different points of the production. COVAP saw the potential with RFID to refine its processes, which involve a large number of ham legs (estimated 300,000 each year). By gaining item-level visibility in each production phase, they are able to develop more efficient processes and have a better understanding of the changes in production conditions that will affect the ham's taste and quality. To produce its hams, they will be dried, salted and aged in temperature-controlled cellars. Even a slight temperature change can impact the meat's taste. The RFID system has helped them in maintaining the quality of hams produced.

Chipless RFID could add significant values to manufacturing applications especially in tagging very large amount of items without the needs for user-defined data. With the capability of chipless tags sustaining harsh conditions throughout different manufacturing operations, they could be a good option for tagging the work-in-process directly. The heavy appearance of metal-based or RF distractive environment could be in favour of using chipless RFID. Furthermore, chipless RFID can also be useful if temperature monitoring is required during the manufacturing processes.

Healthcare/Pharmaceutical

Counterfeit drugs are dangerous and may damage the health of consumers as they may contain unapproved or inappropriate quantities of the ingredients. No doubt this will also impact the sales of drug manufacturers. Many governments have been putting in great efforts combating this problem. The US Food and Drug Administration (FDA) is seeking to impose a drug pedigree mandate of a track of trace system for all pharmaceuticals to be electronically tagged in the supply chain. Experts pointed out that this mandate will be expensive for the industry, estimating a cost as high as USD $100,000 per pharmacy store (Bacheldor, 2008a). Large drug manufacturers have tested RFID systems for brand protection, such as GlaxoSmith-

Kline and Pfizer (O'Connor, 2006). In Nigeria, the National Agency for Food and Drug Administration and Control has approved the use of RFID systems to authenticate pharmaceutical products sold in the country (Swedberg, 2010f). In addition to drug authentication, Axxa Pharma (Argentina) has taken the opportunity of employing an RFID system to achieve a range of operational benefits in their drug distribution supply chain (Swedberg, 2010g). Upon receiving drugs in their warehouse, employees will unpack the cases and apply a tag on each high-value item. From there, they will receive the benefits of running an RFID-enabled warehouse as described earlier. They are utilising between 2000 and 3000 tags a month. Parexel International, a biopharmaceutical services organisation, has completed an RFID trial (using HF tags) to determine the technological ability to ensure a drug's temperature has not exceeded safe thresholds while being shipped to participants in clinical trials (Swedberg, 2010h).

The RFID system used in other areas of healthcare could also potentially be replaced with chipless RFID. The blood bank in Mallorca is deploying an RFID system with EPC Gen 2 tags to track bags of blood and its derivatives from donors to hospitals (Swedberg, 2010i). This system aims to enhance the handling process efficiency, so as to ensure the quality of bloods (through constant monitoring for optimum temperature storing the bloods) and safety to recipients (no mistakes in providing the right type of blood). A UHF tag is applied to each blood bag at the donor clinics and will be tracked through the blood treatment process and cold storage, and despatching to the hospitals as requested. They handle a total of approximately 60,000 bags a year. Another fascinating RFID application in healthcare is Geisinger Medical Center's RFID-enabled robots to securely transport controlled substances and high-value pharmaceuticals throughout its facility, and to provide a real-time data showing who sent and received the drugs (Swedberg, 2010j). In this application, the robots are equipped with RFID readers that can capture ID from any passive EPC Gen 2 UHF tags on the drug containers. This has greatly reduced their staff time transporting the medications and improved operational efficiency.

In healthcare/pharmaceutical applications, chipless RFID will benefit the needs for anti-counterfeiting purposes. The capability of temperature and temper sensing as well as performing under liquid surrounding could also be well-perceived in this industry. For drugs and bloods tagging, the tags required can reach millions and for a fraction of the price of chip-based tags, chipless tags will be financially justified with promising ROI.

Service Sector

In recent years, the RFID technology has gained remarkable attention in the service sector, such as resorts, amusement parks, restaurants, near-field communication (NFC) smart phones, and social networks. Most of these applications would require high volume of tags, which make chipless RFID a cost-justifiable option.

For leisure business, more and more RFID applications can be seen to increase their customer services and improve their ways of running their businesses. A recent article (Wasserman, 2011) has described a series of companies in leisure business utilising RFID systems in recent years. Jay Peak Resort uses RFID-enabled plastic cards to serve as a pass to give access to the lifts, their room, the water park and the ice rink. This will potentially be used to pay for meals in the restaurant, drinks at the snack bar, and video games in the arcade. Great Wolf Resorts are also using RFID-enabled wristbands to achieve the functions above in their water park business. In addition, KeyLime Cove Water Resort is using RFID wristbands to allow their guests to load credits associated with the wristband accounts so that they can make purchases. Other resorts using RFID-enabled cards/wristbands include Northstart Ski Resort, Vail Resorts, Revelstoke Ski Resort, Italian Alpine Report, and Malaysian Sunway Lagoon. To achieve operational benefits, Universal Studios Singapore installed RFID systems to triggers scary sound effects in each of the coaster vehicles.

Counterfeiting is a serious revenue losing problem for many concert, event and sports promoters. Front Gate Tickets, a provider of promoting, ticketing and managing services for live events, introduces an RFID system to allow concert producers to track the flow of ticket holders entering their events. Furthermore, the system also offers a variety of additional services, such as VIP access, social networking for ticket holders, and demographic information regarding ticket buyers (Swedberg, 2011c). The system utilises RFID-enabled wristbands (with embedded HF inlays) for the audience and the wristbands will be read as they enter the venue. This application is expected to acquire tens of thousands of tags for each event; therefore the demand for tags could be in excess of a million a year. Goldenvoice, which promotes the annual three-day Coachella Valley Music and Arts Festival, held in California with 75,000 attendees is also using RFID-enabled wristbands to battle counterfeiting problems. With the newly opened Red Bull Arena in New Jersey, 6,000 full-season ticketholders were issued RFID cards that not only activate turnstiles to allow admission but also can be loaded with stored value on the team's website to make purchases at concession stands.

As these applications only use very basic tags with unique ID numbers without additional data storage capacity required, chipless tags can do just the same with much lower tag cost. This will be a good initiative to explore for chipless technology. As chipless tags are very difficult to be cloned, the anti-counterfeit feature of chipless tags could also be an appealing selling point for the applications above.

Apart from leisure industry, chipless RFID could also be a good option for Dutch Forensic Institute, a government agency that collects and evaluates crime-scene evidence from around the Netherlands. They use an RFID track-and-trace system for the 100,000 pieces of evidence they collect every year, such as guns, knives, cigarette butts, and hair samples. Each of these pieces of evidence will be placed in a plastic bag at crime scenes and labelled with an EPC Gen 2 RFID tag (Wessel, 2008a). Chipless RFID will also benefit the new California Crime Lab which uses passive RFID tags keep tabs on evidence as technicians perform DNA/drug tests, fingerprint analysis, toxicology and other procedures. Each year, they manage approximately 40,000 items of evidence.

There is an increasing interest in using smart cards and mobile phones equipped with NFC inlays for a range of applications, such as payments, unlocking doors, retrieving information, etc. In Nice (France), residents have begun to use NFC-enabled cards or mobile phones to function as tickets on the city buses and trains, as well as to earn loyalty points for shopping at specific merchants (Swedberg, 2010k). The authority in Frankfurt (Germany) is expanding its use of NFC technology to allow passengers to use mobile phones to purchase and store tickets and to check schedules. They plan to install NFC systems at 700 bus, tram and train stops, as well as at the city's airport, providing Frankfurt the widest NFC-enabled transit network in the country (Wessel, 2007). Other NFC applications include Valentines Resort and Marina in Bahamas that installed an RFID system to allow guests to use NFC-enabled mobile phones to unlock their doors and to secure in-room safes (Wasserman, 2011), and the Store Logistics and Payment with NFC (StoLPaN) consortium in Italy that introduced a new service in the Italian Alps to enable tourists visiting the ski resorts and hotels to obtain weather reports and other information, receive coupons and promotions, and make reservations using their NFC-enabled mobile phones or smart cards (Bacheldor, 2008b). Mobile phone business is a very large market, if NFC applications have become a norm in this market and well-perceived by the service sector, billions of RFID inlays will be required. As NFC mobile phone applications so far are for identification purposes, chipless tags could well benefit from it. They can be printed directly on the appropriate component in the phone.

This section describes how chipless RFID could potentially replace the existing RFID applications. However, as with any other technological replacement, replacing chip-based RFID with chipless will require thorough analysis and pilots to be carried out to ensure the technical and operational implications are thoroughly assessed. Trade-off might need to be considered, i.e. price reduced against technical limitations.

CONCLUDING REMARKS AND FUTURE DEVELOPMENT

One of the main barriers for many potential large-scale RFID implementations is the cost of tags. This drives the development of an alternative transponder called chipless tag which does not contain an integrated circuit (IC). Without ICs, it is possible to significantly reduce the cost of tags and to potentially dominate the RFID market especially for item-level tagging. It was predicted by RFID experts that chipless RFID will rapidly take over chip-based RFID in some major applications (Das, 2007). It is also pointed out that for item-level tagging, what is needed is a simple RFID tag that is inexpensive, small and easily to be attached to an item (RFID Journal, 2006). In addition to much lower tag cost, chipless RFID could well be a solution to wider RFID adoption with additional benefits, such as the ability to sustain very high (and very low) temperature, temperature-sensing capability, hard to be cloned, can be printed directly onto products or packaging, work well in metal/liquid environment, and so on. Despite the capability of chipless RFID delivering benefits beyond what is commonly required for item-level tagging, it is still not being widely adopted in the market as forecasted.

The chipless RFID system does not transfer unique serial numbers; it uses the EM properties of the tag materials to reflect the radio waves back to the reader in a unique pattern. In this way, the reader is able to differentiate the tags. As the EM property of each tag is different (due to the deviation in tag production), therefore chipless tags are almost impossible to be cloned. For this very reason of uniqueness, chipless tags can replace IC tags for a fraction of the price to have this security feature. However, without a unique serial number and the capability to store user-definable data, they become less popular in supply chain applications (i.e. in logistics, warehousing and manufacturing) with an open system environment involving various parties in the chain. Many of the existing supply chain RFID applications would require these features, which are not provided by the current chipless RFID market.

Furthermore, up to present, the chipless technology does not meet open standards for use by many service providers in the supply chain and this is a hitch for wider adoption in open system applications. This also leads to different (proprietary) data transfer/encoding protocols being developed. The current data transfer protocols tend to compromise simultaneous reading of multiple tags accurately, which is ultimately important for supply chain applications, e.g. reading items at a rate of 100 units per second. As the data transfer is not standardised, unique reader design is mandatory and the same bespoke reader must be used at each supply chain party. All these have formed a great challenge for chipless RFID system to be widely adopted in the supply chain open system applications.

This chapter explored in detail the chipless RFID's technical and operational characteristics, as well as the benefits, limitations and current uses. The main emphasis of this chapter is to examine the current applications of the main industries using chip-based RFID systems and how chipless RFID can also offer the same benefits with much lower tag cost. In addition, further investigation was also carried out to explore potential value-added applications of chipless RFID to justify financial values for adoption. The industries explored are retail, logistics, manufacturing, healthcare and service sectors. Based on the analysis, it can be gathered that chipless RFID is well suited for closed-loop applications that require a large number of item-level tagging and do not require much of data storage capacity. Although this chapter has discussed the examples of RFID applications could well benefit from chipless RFID, trials need to be carried out to ensure the technical and operational aspects of chipless RFID are tested and thoroughly assessed.

The new generation of chipless RFID tags must possess the critical characteristics that are most sought after in the market and overcome the limitations that may have hindered the widespread of potential uptake. First of all, much effort needs to be put in to develop standards for the use of chipless systems for major RFID applications (e.g. in logistics and retail), especially involving supply chain parties (i.e. open systems). For instance, there was a noticeable widespread of adoption of UHF RFID systems once the ISO 18000 standards were developed and well-perceived. Chipless RFID could receive the same attention if there is a standard developed to support its adoption for open systems in supply chain applications. This standard will also result in more standardised design of readers. Secondly, more research should be stirred for developing larger data storage on tags to enable incorporation of user-definable data to provide greater uses. Thirdly, as RFID is emerging more sought after features should be developed, such as operating at popular frequencies as conventional RFID and the ability to integrate with Wi-Fi, Infra-red or GPS systems. Finally, the ability to print directly on packaging is one of the selling points, more research should be focussing on specific type of applications for targeted market groups (e.g. for pharmaceutical packaging to meet government mandates on drug e-pedigree system and distribution).

REFERENCES

Aberdeen Group. (2008). *Where RFID meets ROI: Beyond supply chains.*

Bacheldor, B. (2006). *Mitsubishi Electric Asia switches on RFID.* Retrieved March 9, 2011, from http://www.rfidjournal.com/article/view/2644

Bacheldor, B. (2007). *RFID tattoos for livestock.* Retrieved February 6, 2011, from http://www.rfidjournal.com/article/view/3079

Bacheldor, B. (2008a). *Drug pedigree mandate could be expensive.* Retrieved March 12, 2011, from http://www.rfidjournal.com/article/print/4179

Bacheldor, B. (2008b). *Italian alpine resort takes near field communications for ride.* Retrieved March 14, 2011, from http://www.rfidjournal.com/article/view/4463

Balbin, I., & Karmakar, N. (2009a). Novel chipless RFID tag for conveyor belt tracking using multi-resonant dipole antenna. *European Microwave Conference, 1-3,* 1109-1112.

Balbin, I., & Karmakar, N. (2009b). Phase-encoded chipless RFID transponder for large-scale low-cost applications. *IEEE Microwave and Wireless Components Letters, 19*(8), 509–511. doi:10.1109/LMWC.2009.2024840

Berthiaume, D. (2004). Standards ease global RFID adoption. *Retail Technology Quarterly,* May 2004.

Butner, K. (2010). The smarter supply chain of the future. *Strategy and Leadership, 38*(1), 22–31. doi:10.1108/10878571011009859

Collins, J. (2004). *RFID fibers for secure applications.* Retrieved February 27, 2011, from http://www.rfidjournal.com/article/view/845

Collins, J. (2005). *New tags use crystal, not silicon.* Retrieved March 1, 2011, from http://www.rfidjournal.com/article/view/1967

Daniel Hunt, V., Puglia, A., & Puglia, M. (2007). *RFID - A guide to radio frequency identification*. John Wiley & Sons. doi:10.1002/0470112255

Das, R. (2007). Chipless RFID. *Adhesives & Sealants Industry*, May 2007, 47-48.

Edwards, J. (2011). RFID: The next stage. *RFID Journal*, Nov/Dec 2011, 12-19.

Ferrer, G., Dew, N., & Apte, U. (2010). When is RFID right for your service? *International Journal of Production Economics*, *124*(2), 414–425. doi:10.1016/j.ijpe.2009.12.004

Finkenzeller, K. (2003). *RFID handbook: Fundamentals and applications in contactless smart cards and identification*. London, UK: John Wiley & Sons.

Foreshew, J. (2011). *Chipless tracker to transform the library industry*. Retrieved August 27, 2011, from http://www.theaustralian.com.au/australian-it/chipless-tracker-to-transform-libraries/story-e6frgakx-1226106169778

Friedlos, D. (2010). *Stell products maker sees ROI in six months*. Retrieved March 9, 2011, from http://www.rfidjournal.com/article/view/7448/1

Friedlos, D. (2011). *South Korean consortium launches EPC Gen 2 reader for mobile phones*. Retrieved February 5, 2011, from http://www.rfidjournal.com/article/print/8155

García, A., & Chang, Y. (2007). RFID enhanced MAS for warehouse management. *International Journal of Logistics: Research & Applications*, *10*(2), 97–107. doi:10.1080/13675560701427379

Harrop, P. (2006). *New advances in RFID help foot traceability*. Retrieved March 7, 2011, from www.idtechex.com/documents/downloadpdf.asp?documentid=1279

Holmes, P. (2005). *Will tags get out into the supply chain?* Works Management.

Hu, S. M., Zhou, Y., Law, C. L., & Dou, W. B. (2010). Study of a uniplanar monopole antenna for passive chipless UWB-RFID locatization system. *IEEE Transactions on Antennas and Propagation*, *58*(2), 271–278. doi:10.1109/TAP.2009.2037760

Journal, R. F. I. D. (2003). *1-cent RFID tags for supermarkets*. Retrieved February 27, 2011, from http://www.rfidjournal.com/article/view/363

Journal, R. F. I. D. (2004). *Firewall protection for paper documents*. Retrieved February 27, 2011, from http://www.rfidjournal.com/article/view/790

Journal, R. F. I. D. (2006). *PolyIC announces printable RFID prototypes*. Retrieved February 27, 2011, from http://www.rfidjournal.com/article/print/6589

Lakafosis, V., Traille, A., Hoseon, L., Gebara, E., Tentzeris, M. M., DeJean, G. R., & Kirovski, D. (2011). RF fingerprinting physical objects for anticounterfeiting applications. *IEEE Transactions on Microwave Theory and Techniques*, *59*(2), 504–514. doi:10.1109/TMTT.2010.2095030

Lee, D. (2008). RFID–based traceability in the supply chain. *Industrial Management & Data Systems*, *108*(6), 713–725. doi:10.1108/02635570810883978

Lee, Y. K. (2008). Elliptic-curve-based security processor for RFID. *IEEE Transactions on Computers*, *57*(11), 1514–1527. doi:10.1109/TC.2008.148

Liu, D. S., & Zou, X. C. (2006). Embeded EEPROM memory achieving lower power - New design of EEPROM memory for RFID tag IC. *IEEE Circuits and Devices Magazine*, *22*(6), 53–59. doi:10.1109/MCD.2006.307277

Luckett, D. (2004). The supply chain. *BT Technology Journal*, *22*(3), 50–55. doi:10.1023/B:BTTJ.0000047119.22852.38

McFarlane, D., & Sheffi, Y. (2003). The impact of automatic identification on supply chain operations. *International Journal of Logistics Management, 14*, 1–17. doi:10.1108/09574090310806503

Mead, P. S., Slutsker, L., Dietz, V., et al. (2000). *Food-related illness and death in the United States*. Retrieved March 9, 2011, from http://www.cdc.gov/ncidod/eid/vol5no5/mead.htm

Moore, B. (2002). RFID: The chips are down. *Material Handling Management, 57*(4), 17.

Mukherjee, S. (2008). Antennas for chipless tags based on remote measurement of complex impedance. *European Microwave Conference, 1-3*, 1728-1731.

Ngai, E. (2008). RFID: Technology, applications, and impact on business operations. *International Journal of Production Economics, 112*(2), 507–509. doi:10.1016/j.ijpe.2007.05.003

Niederman, F. (2007). Examining RFID applications in supply chain. *Communications of the ACM, 50*(7), 93–101. doi:10.1145/1272516.1272520

O'Connor, M. C. (2006). *GlaxoSmithKline tests RFID on HIV drugs*. Retrieved March 12, 2011, from http://www.rfidjournal.com/article/view/2219

O'Connor, M. C. (2008a). *Wal-Mart, DOD point to sustained progress*. Retrieved February 5, 2011, from http://www.rfidjournal.com/article/articleview/4407/

O'Connor, M. C. (2008b). *American Apparel expands RFID to additional stores*. Retrieved March 8, 2011, from http://www.rfidjournal.com/article/view/4510/1

O'Connor, M. C. (2009a). *RFID trims costs for retailer of Lacoste, CK, Burberry*. Retrieved March 7, 2011, from http://www.rfidjournal.com/article/view/4626

O'Connor, M. C. (2009b). *Marigold industrial gets a better grip on glove production, inventory*. Retrieved March 9, 2011, from http://www.rfidjournal.com/article/print/5004

O'Connor, M. C. (2010). *RFID finds flavor at Izzy's ice cream shop*. Retrieved February 5, 2011, from http://www.rfidjournal.com/article/view/7651

Piramuthu, S. (2007). Protocols for RFID tag/reader authentication. *Decision Support Systems, 43*(3), 897–914. doi:10.1016/j.dss.2007.01.003

Preradovic, S., Balbin, I., Karmakar, N. C., & Swiegers, G. F. (2009). Multiresonator-based chipless RFID system for low cost item tracking. *IEEE Transactions on Microwave Theory and Techniques, 57*(5), 1411–1419. doi:10.1109/TMTT.2009.2017323

Preradovic, S., & Karmakar, N. (2009). Design of fully printable planar chipless RFID transponder with 35-bit data capacity. *European Microwave Conference, 1-3*, 13-16.

Preradovic, S., & Karmakar, N. C. (2010). Chipless RFID: Bar code of the future. *IEEE Microwave Magazine*, (December): 87–97. doi:10.1109/MMM.2010.938571

Rach, N. M. (2008). RFID applications spread in upstream operations. *Oil & Gas Journal, 106*(27), 37–44.

Research, A. B. I. (2009). *Market for item-level RFID in fashion apparel and footwear will nearly triple by 2014*. Retrieved March 7, 2011, from http://www.abiresearch.com/press/1489-Market+for+Item-Level+RFID+in+Fashion+Apparel+and+Footwear+Will+Nearly+Triple+by+2014

Roberti, M. (2004a). *Wal-Mart begins RFID rollout*. Retrieved February 6, 2011, from http://www.rfidjournal.com/article/view/926

Roberti, M. (2004b). *DOD releases final RFID policy*. Retrieved February 6, 2011, from http://www.rfidjournal.com/article/view/1080

Roberti, M. (2006). *RFID is fit to track clothes*. Retrieved March 7, 2011, from http://www.rfidjournal.com/article/view/2195

Roberti, M. (2009). *Restoring confidence in the food chain*. Retrieved March 8, 2011, from http://www.rfidjournal.com/article/view/4621/1

Roberti, M. (2010). *Wal-Mart relaunches EPC RFID effort, starting with men's jeans and basics*. Retrieved February 6, 2011, from http://www.rfidjournal.com/article/view/7753

Schussler, M., Mandel, C., Maasch, M., Giere, A., & Jalcoby, R. (2009). Phase modulation scheme for chipless RFID and wireless sensor tags. *Asia Pacific Microwave Conference,* Vol. 1, (pp. 5229-5232).

Shrestha, S., Balachandran, M., Agarwal, M., Phoha, V. V., & Varahramyan, K. (2009). A chipless RFID sensor system for cyber centric monitoring applications. *IEEE Transactions on Microwave Theory and Techniques*, *57*(5), 1303–1309. doi:10.1109/TMTT.2009.2017298

Sigala, M. (2007). RFID applications for integrating and informationalizing the supply chain of foodservice operators: Perspectives from Greek operators. *Journal of Foodservice Business Research*, *10*(1), 7–29. doi:10.1300/J369v10n01_02

Swedberg, C. (2006). *Marks & Spencer to tag items at 120 stores*. Retrieved February 5, 2011, from ttp://www.rfidjournal.com/article/articleview/2829/1/1/

Swedberg, C. (2007a). *Growers and grocers get into plastic pallet pool*. Retrieved March 9, 2011, from http://www.rfidjournal.com/article/view/3821

Swedberg, C. (2007b). *RFID sweetens imperial's shipping process*. Retrieved March 9, 2011, from http://www.rfidjournal.com/article/view/3720/1/1

Swedberg, C. (2008a). *Norwegian food group Nortura to track meat*. Retrieved March 8, 2011, from http://www.rfidjournal.com/article/view/4208/1.

Swedberg, C. (2008b). *Cosmetics and liquor companies assess Toppan Printing's holographic RFID labels*. Retrieved March 7, 2011, from http://www.rfidjournal.com/article/view/4356

Swedberg, C. (2009a). *New Belgium brewing rolls out RFID to track kegs*. Retrieved March 7, 2011, from http://www.rfidjournal.com/article/view/4925

Swedberg, C. (2009b). *PLS uses RFID to track pallets, containers*. Retrieved March 9, 2011, from http://www.rfidjournal.com/article/view/5043

Swedberg, C. (2009c). *BMW finds the right tool*. Retrieved March 12, 2011, from http://www.rfidjournal.com/article/view/5104/1

Swedberg, C. (2010a). *American Apparel adds RFID to two more stores, switches RFID software*. Retrieved March 7, 2011, from http://www.rfidjournal.com/article/view/7313

Swedberg, C. (2010b). *Korean clothing company adds RFID to its supply chain*. Retrieved March 7, 2011, from http://www.rfidjournal.com/article/view/7360

Swedberg, C. (2010c). *Axios MA launches tagged pallets and real-time tracking solution*. Retrieved March 9, 2011 from http://www.rfidjournal.com/article/view/8014/1

Swedberg, C. (2010d). *RFID illuminates Lithuanian lamp manufacturer*. Retrieved March 9, 2011, from http://www.rfidjournal.com/article/view/7419/1

Swedberg, C. (2010e). *Lithuanian manufacturer tracks IKEA-bound furniture*. Retrieved March 9, 2011, from http://www.rfidjournal.com/article/view/7978/1

Swedberg, C. (2010f). *Nigerian drug agency opts for RFID anticounterfeiting technology*. Retrieved March 12, 2011, from http://www.rfidjournal.com/article/view/7856

Swedberg, C. (2010g). *RFID boosts profit margin, safety for Axxa Parma*. Retrieved March 12, 2011, from http://www.rfidjournal.com/article/view/7823/1

Swedberg, C. (2010h). *Parexel tests system to track temperature of test drugs*. Retrieved March 12, 2011, from http://www.rfidjournal.com/article/view/7848/1

Swedberg, C. (2010i). *RFID to take the chill out of frozen plasma tracking*. Retrieved March 13, 2011, from http://www.rfidjournal.com/article/view/7632/1

Swedberg, C. (2010j). *Hospital robot tracks controlled substances, high-value meds*. Retrieved March 13, 2011, from http://www.rfidjournal.com/article/view/7825

Swedberg, C. (2010k). *Cityzi seeks to spur adoption of NFC RFID technology*. Retrieved March 14, 2011, from http://www.rfidjournal.com/article/view/7650

Swedberg, C. (2011a). *Beverage metrics serves up drink-management solution*. Retrieved March 7, 2011, from http://www.rfidjournal.com/article/view/8237/1

Swedberg, C. (2011b). *Norsk Lastbaerer Pool inserts RIFD into Norwegian food chain*. Retrieved March 9, 2011, from http://www.rfidjournal.com/article/view/8137/1

Swedberg, C. (2011c). *Payback could be a lollapalooza for concert promoters*. Retrieved March 9, 2011, from http://www.rfidjournal.com/article/view/8199/1

Thuemmier, C., Buchanan, W., Fekri, A., & Lawson, A. (2009). Radio frequency identification (RFID) in pervasive healthcare. *International Journal of Healthcare Technology and Management*, *10*(1/2), 119. doi:10.1504/IJHTM.2009.023731

Vyas, R., Lakafosis, V., Rida, A., Chaisilwattana, N., Travis, S., Pan, J., & Tentzeris, M. M. (2009). Paper-based RFID-enabled wireless platforms for sensing applications. *IEEE Transactions on Microwave Theory and Techniques*, *57*(5), 1370–1382. doi:10.1109/TMTT.2009.2017317

Waggoner, M. (2008). Application of RFID technology in the manufacturing process. *Plant Engineering*, *62*(4), 45–47.

Wasserman, E. (2006). *Keeping fresh foods fresh*. Retrieved March 8, 2011, from http://www.rfidjournal.com/article/print/2137

Wasserman, E. (2011). RFID serves up benefits for guests and hosts. *RFID Journal*, Nov/Dec 2011, 20-23.

Wei, Z., Kapoor, G., & Piramuthu, S. (2009). RFID-enabled item-level product information revelation. *European Journal of Information Systems*, *18*(6), 570–577. doi:10.1057/ejis.2009.45

Weinstein, R. (2005). RFID: A technical overview and its application to the enterprise. *IT Prof Magazine*, May/June 2005.

Wessel, R. (2007). *Frankfurt widens its NFC-enabled transit network*. Retrieved March 14, 2011, from http://www.rfidjournal.com/article/view/3755

Wessel, R. (2008a). *Dutch Forensic Institute uses RFID to control crime evidence*. Retrieved March 13, 2011, from http://www.rfidjournal.com/article/view/4410

Wessel, R. (2010a). *RFID helps Medlog monitor pharmaceutical cold chain*. Retrieved February 5, 2011, from http://www.rfidjournal.com/article/view/7494

Wessel, R. (2010b). *German clothing company s.Oliver puts RFID to the test*. Retrieved March 7, 2011, from http://www.rfidjournal.com/article/view/8013/1

Wessel, R. (2010c). *Mars, Rewe, Deustsche Post and Luthansa cargo work on SmaRTI*. Retrieved March 9, 2011 from http://www.rfidjournal.com/article/view/8095/1

Wu, N. C., Nystrom, M. A., & Lin, T. R. (2005). Challenges to global RFID adoption. *Technovation, 26*, 1317–1323. doi:10.1016/j.technovation.2005.08.012

Yang, L., Staiculescu, D., Zhang, R., Wong, C. P., & Tenzeris, M. M. (2009). A novel "green" fully-integrated ultrasensitive RFID-enabled gas sensor utilizing inkjet-printed antennas and carbon nanotubes. *IEEE Antennas and Propagation Society: International Symposium and USNC/URSI National Radio Science Meeting,* Vol. 1, (pp. 6804-6807).

Yang, L., Zhang, R. W., Staivulescu, D., Wong, C. P., & Tentzeris, M. M. (2009). A novel conformal RFID-enabled module utilizing inkjet-printed antennas and carbon nanotubes for gas-detection applications. *IEEE Antennas and Wireless Propagation Letters*, 8653–8656.

Zheng, L. L., Rodriguez, S., Zhang, L., Shao, B. T., & Zhang, L. R. (2008). Design and implementation of a fully reconfigurable chipless RFID tag using inkjet printing technology. *IEEE International Symposium on Circuits and Systems,* Vol. 1-10, (pp. 1524-1527).

KEY TERMS AND DEFINITIONS

Chipless RFID: Radio Frequency tags without a microchip.

Radio Frequency Identification (RFID) Technology: A system that uses radio waves to transmit information.

RFID Tag: A microchip tag containing data.

Technology Adoption: Applying a new technology to an existing system/infrastructure.

Value-Added Application: An application that adds value to a system to extend the benefits it could offer.

Chapter 11
Potential Impact of RFID-Based Tracing Systems on the Integrity of Pharmaceutical Products

Michele Maffia
University of Salento, Italy

Luca Mainetti
University of Salento, Italy

Luigi Patrono
University of Salento, Italy

Emanuela Urso
University of Salento, Italy

ABSTRACT

Radio Frequency Identification (RFID) is going to play a crucial role as auto-identification technology in a wide range of applications such as healthcare, logistics, supply chain management, ticketing, et cetera. The use of electromagnetic waves to identify, trace, and track people or goods allows solving many problems related to auto-identification devices based on optical reading (i.e. bar code). Currently, high interest is concentrated on the use of Radio Frequency (RF) solutions in healthcare and pharmaceutical supply chain, in order to improve drugs flow transparency and patients' safety. Unfortunately, there is a possibility that drug interaction with electromagnetic fields (EMFs) generated by RF devices, such as RFID readers, deteriorate the potency of bioactive compounds. This chapter proposes an experimental multidisciplinary approach to investigate potential alterations induced by EMFs on drug molecular structure and performance. To show the versatility of this approach, some experimental results obtained on two biological pharmaceuticals (peptide hormone-based) are discussed.

DOI: 10.4018/978-1-4666-2080-3.ch011

INTRODUCTION

The auto-identification solutions have recently seen explosive interest from a wide range of application sectors, such as manufacturing, retail, logistics, ticketing, healthcare, and pharmaceuticals, in order to improve transparency in goods flows in the international markets. These needs are substantially contributing to assert the important concept of the Internet of Things (IoT) (Thiesse et al., 2009). It represents the future vision of the Internet, a worldwide network composed of uniquely addressable interconnected objects, able to collect any valuable information about the same objects, using them in various applications during their life cycle. Furthermore, the growing of goods' counterfeiting imposes to the governments of many countries to define and apply effective strategies to face this serious problem. Among several scenarios, the pharmaceutical supply chain, with millions of medicines moving around the world, represents a very interesting and challenging test case. Indeed, several international institutions (e.g. Food and Drug Administration, European Medicines Agency, European Federation of Pharmaceutical Industries and Associations, GS1) are recently encouraging the use of innovative solutions for the auto-identification in healthcare and pharmaceuticals, to improve the patient safety and enhance the efficiency of the pharmaceutical supply chain, with a better worldwide drug traceability.

The Radio Frequency Identification (RFID) is an emerging auto-identification technology that promises to solve many problems related to almost obsolete optical solutions such as bar code (Finkenzeller, 2003). A communication based on the use of EMFs is able to overcome constraints as Line-of-Sight (LoS), low read rate, no bulk reading, etc. In addition, recent technological enhancements allowed realizing new types of RFID tags whose cost is very low and comparable with bar code one. These particular tags are known with the generic term "chipless" (Preradovic, 2009) because they exploit RF energy to communicate data but do not use a silicon microchip to store a serial number.

Unfortunately, there are still several obstacles that are limiting the deployment of RFID technology in healthcare organizations and in the pharmaceutical supply chain. One of these, very interesting from scientific point of view, is related to the evaluation of potential exposure effects to an EMF on drugs. It is well known that the pharmaceutical and healthcare sectors are characterized by a rich set of national and international rules (e.g. GAMP, GMP, FDA CRF21, etc.) that strictly regulate the use of new information and communication technologies.

The RF exposure effects on materials have been broadly classified into two categories: thermal effects and non-thermal effects (Uysal et al., 2010). Thermal effects are defined as those that stem from an appreciable increase in the temperature of products under RF exposure, comparable to a change in temperature caused by other heating sources. The entity of thermal effects mainly depend on the frequency of the EM source, the dielectric constant, the water content and the overall thickness of the exposed materials. Furthermore, a bulk of scientific literature focuses on the effects of temperature on drug stability.

In contrast, non-thermal effects, arising even in absence of any appreciable increase in the material temperature upon exposure to RF, are obscure. The rise of temperature generated by the high penetration power of high-frequency radiations may mask non-thermal effects, especially in liquid and metal containing samples. So, most studies in this field privileged the use of low frequency or high frequency/low power radiations, having a negligible impact on temperature (De Pomerai et al., 2003; Bismuto et al., 2003). As a turning point in this sense, Cox et al. (2006) described a tightly controlled approach to analyze non-thermal effects on drugs of RF radiation over a broad range of frequencies, at initial sample temperatures. More recently, a methodology to generate powerful RF EMFs (HF, UHF) in a temperature-controlled environment has been described (Uysal et al., 2010). The intent was to determine the impact on purity and potency of selected protein biopharmaceuticals.

The goal of this chapter is delineating an experimental framework to evaluate potential exposure effects of the EMFs generated by RFID devices working in UHF band on biological pharmaceuticals. Heterogeneous skills (engineering, chemistry and physiology) have been recruited to draw up a suitable experimental protocol for this analysis, consisting of three main steps: (i) simulation of drug exposure to EMFs generated by RFID devices in a test environment reproducing the pharmaceutical supply chain; (ii) investigation of potential drug structural changes by High Pressure Liquid Chromatography (HPLC) technique; and (iii) analysis of performance by *in vitro* functional assays. The proposed protocol has been applied on two different commercial drugs, whose active principles are human recombinant insulin and the Follicle Stimulating Hormone (FSH). Experimental results did not unravel any changes in the structure and performance of both drug types following RF exposure in UHF band.

RELATED WORKS

The possibility that the exposure of drugs to EMFs generated by RFID devices can deteriorate the molecular structure and potency of bioactive compounds is a very challenging research topic. Unfortunately, this aspect has been for a long time disregarded by international governments, mainly engaged in founding a supportive environment to RFID users through the establishment of uniform operating conditions (i.e. working frequencies) and regulatory measures. The U.S. FDA has recognized the need for a deep investigation into potential impacts of RFID technology on the molecular structure of drugs by encouraging the initiation of pilot research projects that depict the possible consequences on drug quality of occasional time overexposure to RFID emitting devices (U.S. FDA, 2004; Bassen, 2005; Seevers, 2005; Barchetti et al., 2009; Barchetti et al., 2010; Acierno et al., 2011). Unfortunately, at this moment literature does not offer a great deal of detailed research reports on RF-related risk that reassure the pharmaceutical distributors on RFID-based "track and trace" technologies (Bassen, 2005; Fortune et al., 2010).

Previous research reports on the assessment of RF-related risk can be classified into two categories: (i) development of hardware to simulate the exposure of drugs to RF radiation; (ii) experimental evaluation of the impact of EMFs on drug molecular integrity.

The first group includes some tentative proposals mainly characterized by an engineering approach, aimed to evaluate non-thermal effects on protein structures of biopharmaceuticals by constant exposure to a high power RF energy, at different wavelengths. The study performed by Uysal et al. (2010) summarizes the main results obtained about the RF radiation effects on various biological drugs – such as vaccines, hormones and immunoglobulins – testing five different frequencies (i.e. 13.56 MHz, 433 MHz, 868 MHz, 915 MHz, and 2.4 GHz), typically used for commercial RFID devices and setting the radiation power at 8 W, that is twice the equivalent isotropically radiated power (EIRP) allowed by U.S. Federal Communication Commission (FCC). A further contribution of the same work is related to the setup of an experimental protocol defined to provide a fundamental and universally applicable methodology which combines the hardware to generate and radiate high power RF signals with analysis techniques, such as high-pressure liquid chromatography and immunoelectrophoresis, able to evaluate potential damages on the drug molecular structure. The exposure environment was made realizing a temperature controlled dark anechoic chamber. The experimental activities have involved some pharmaceutical companies (e.g. Abbott and Pfizer) that provided the biopharmaceuticals to be tested. This is very important because it demonstrates the real interest of the pharmaceutical market in the RFID technology like tagging and tracing systems in the whole supply chain. The exposure time was 24 hours. The experimental results demonstrated that none of three major protein groups showed degradation in their structural properties excluding any non-thermal effect due to RF exposure.

Another work by (Cox et al., 2006) focused on building an ad hoc RF exposure environment. It describes a method and an apparatus for accurately maintaining sample temperature during exposure to EMFs, under conditions where in the absence of such control the sample temperature would increase substantially. Two different frequencies (2.45 GHz and 915 MHz) have been used to evaluate RF effects on the sample temperature at two different radiation power, 22 W and 4 W. This study demonstrated that the RF radiation increases sample temperatures but it is possible to control this induced-heating allowing an investigation of RF-induced non-thermal effects in drug aqueous solutions.

Always in the first group, Bassen et al. (2007) describe the development of hardware and software systems able to perform exposure studies on the effects of RF EMFs on solid and liquid pharmaceuticals and biologics. The proposed apparatus is set in order to use HF and UHF bands and it can expose drug samples to uniform electric and magnetic fields at levels that are much higher than those experienced by commercial RFID readers at a 0.2 m distance. The UHF component includes a commercial circularly polarized antenna, a microwave generator with pulse modulation, a high power amplifier, a plastic foam enclosure, and a fiber optic temperature monitoring. Instead, the HF component includes an ad hoc Helmholtz coil pair, an HF signal generator with pulse modulation, an HF amplifier, an RF impedance matching device, and the same enclosure and thermometry system as in the UHF system. The apparatus is able to irradiate drugs both in their retail primary packages and in 54 mm diameter culture dishes. In the first case, exposures in the primary container (e.g. vial, blister, bottle, aluminium sachet, etc.) allow to analyze the interactions of RF fields with the packaging materials and container geometry. In contrast, exposures of drugs in culture dishes assure a uniform induced electric field. The experiments in UHF band were performed setting the radiation power to 20 W EIRP, over 5 times the FCC limits. The proposed system allows controlling, through computer monitoring, power, drug temperature, and air temperature, during the exposure.

The second category of researches reports the attempt to dissect the mechanisms and consequences of interaction between RF radiation in a range of frequencies experienced by mobile phone users (humans) and protein macromolecules. De Pomerai et al. (2003) analysed the propensity of Bovine Serum Albumin (BSA) solutions to aggregate upon 3 to 48 h microwave exposure at 1.0 GHz, 0.5 W. The light scattering intensity was measured at 320 nm as a probe of protein aggregation. Results evidenced a scalar and time-dependent increase in the percentage changes of light scattering, indicating the occurrence of irreversible aggregation events in microwave-exposed BSA solutions. Such finding was confirmed on bovine insulin samples exposed to microwave radiation for 24 h at 60°C. These experimental conditions were adopted to accelerate phenomena that result extremely slow under physiological conditions. Electron microscopy staining allowed detecting amyloid fibrils on exposed protein samples, confirming the microwave potential to promote aggregation phenomena. A contemporary research by Bismuto et al. (2003) did not evidence any influence of microwave radiation (1.95 GHz, 2.5 h) on a *T. tynnus* myoglobin preparation. Absorption spectroscopy measurements, circular dichroism analysis, and frequency domain fluorimetry confirmed that myoglobin maintained its binding ligand function and conformational substates.

Different factors could explain these contrasting findings, ranging from individual molecular susceptibility to EMFs, to inhomogeneity of experimental conditions (exposure circuits, length of treatment) and sensitivity of study techniques. The complexity of such question assumes a daunting proportion if we consider that several pharmaceuticals are protein-based and at the moment the few available literature reports disregard the potential effects of the EMFs generated by commercial RFID devices, privileging the analysis of effects linked to RF emissions in the mobile communication frequency range. As already mentioned, it was only recently that tightly controlled methodologies to generate powerful RF EMFs (HF, UHF) under steady temperature have been reported, with the aim to demonstrate the impact of commercial RFID devices on characteristics of protein biopharmaceuticals (Cox et al. 2006; Uysal et al., 2010). This kind of research approach is still *in embryo* and a unified experimental strategy does not exist, although the adoption of measures by drug manufacturers reassuring over a massive use of RFID technologies is now perceived as imperative for its potential to neutralize the growing scourge of counterfeiting and guarantee consumers' safety.

The hidden objective of this chapter is proposing a strategy to evaluate the effects of RFID technologies on pharmaceutical product integrity, through the conjugation of technological resources placed at our disposal by the engineering and modern biotechnological sectors. This attempt arises from the firm belief that the high potential resources of a multidisciplinary approach, if properly channeled into a phased program, can return high quality results and offer the necessary guarantees.

A multidisciplinary strategy for an in-depth study of the impact of UHF RF EMFs on a commercial insulin-based pharmaceutical preparation has been the object of recent works from our research group that unequivocally demonstrated the innocuity of in-use RFID devices for the bioactive compound (Acierno et al., 2010a; 2010b; 2010c).

EXPERIMENTAL ASSESSMENT OF RFID-RELATED RISK: THE IMPORTANCE OF A MULTIDISCIPLINARY STRATEGY

The potential effects of EMFs emitted by commercial RFID devices upon tagged pharmaceuticals and categorized as "non-thermal" have been addressed by a scarce scientific literature, with only a few relevant results (Uysal et al., 2010). Most studies are not directly comparable, because of the different research designs and methods applied. The consequent absence of a uniform and reassuring point of view about the RF-related risk has nourished the drug manufacturers' distrust toward the RFID-based "trace and track" technologies. Grounded on the need to develop a standardized procedure for the evaluation of RFID device-related risk on drugs, the integrated protocol described and discussed in this chapter brings together a broad set of skills in heterogeneous areas, in order to create a consistent test bed environment for testing on individual drugs.

Before introducing the protocol aimed at verifying possible loss of drug potency under RF emissions, it can be observed that in compliance with the recommendations of most pharmacopoeias, the quality control processes applied on finished biology and biotechnology products in the pharmaceutical industry are substantially based on physicochemical assays. A major weakness is that they are often unable to fully enlighten possible alterations of the molecular steric arrangements. In fact, a preliminary sample denaturation is foreseen by almost all these techniques, leading to the breaking of molecular weak chemical bonds (hydrogen, ionic), that determine the unique three-dimensional shape (tertiary structure) of a specific compound and consequently its biological identity.

Referring to peptide compounds, only alterations in their primary structure can be evidenced. This consideration reinforces the concept that drug-specific bioassays must be included to provide an actual measure of the drug ability to exert a defined biological effect. To meet this exigency, given the high variability of drug elicited physiological responses, the experimental framework must maintain sufficient flexibility.

Based on these premises, the elaborated protocol extends far beyond the adoption of chromatographic and spectroscopic techniques, even if sophisticated, thinking the inspection of functional properties as an integral part of the procedure. The rationale of the proposal is that under highly standardized experimental conditions, an operator can predict both drug structural and functional changes associated with an occasional time-overexposure to RFID-linked UHF radiation. The introduction of different perspectives and methods into a structured multidisciplinary approach helps the operators to cover all eventualities.

So, the adoption of the presented protocol by pharmaceutical manufacturers can be credited with providing exhaustive knowledge for decision-making on the commercial introduction of tagged products.

The most relevant trait of our quality control program is the multidisciplinarity, i.e. a non-integrative mixture of disciplines as it matured from cooperation among different professional profiles, each one supervising a single step of the procedure in tight cooperation with other operators. The proposed overall approach, shown in Figure 1 consists of three main steps carried out exploiting heterogeneous skills.

The first step consists of the simulation of an occasional time-overexposure (up to 24 h) of drug samples to the EMFs generated by RFID devices in a laboratory asset created to simulate a typical supply chain environment. Information and communication engineering competencies are required in this phase. At the second step, drug samples exposed to radiation are subjected to a chromato-

Figure 1. Experimental strategy to evaluate the impact of RFID technology on drugs

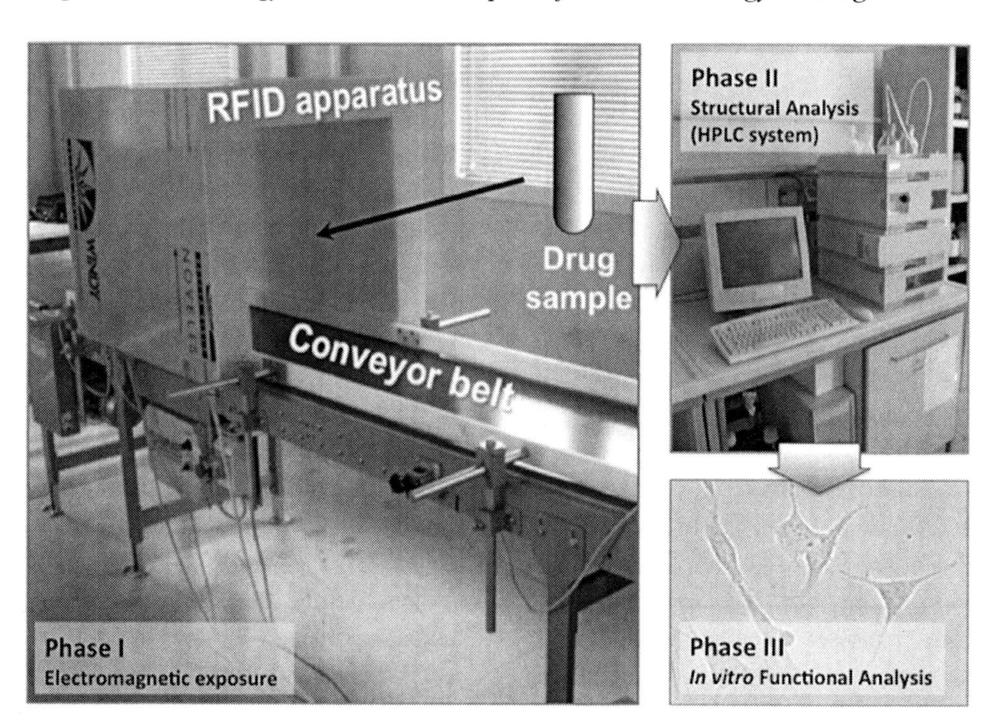

graphic investigation of possible RF-induced structural alterations. Third and last, compound specific functional properties are examined and quantified by *in vitro* bioassays.

We notice that in compliance to the 3Rs (Replacement, Refinement and Reduction) ethical principle, functional assays discussed in this chapter are intended under a revised form – cell culture based – with respect to the old practice ones – animal based (Russell and Burch, 1959). The design of sensitive, reproducible and drug-specific functional tests requires a deep knowledge of cell physiology, as well as wide methodological competencies (gene/protein expression, metabolic and functional profiling) and a skillful use of statistical tools. The use of cell culture-based tests shows significant benefits over traditional animal testing, in terms of reliability and considerably reduced test variability. They allow a strict control over all the experimental conditions and so can be optimized and validated, to achieve high specificity, sensitivity, and reproducibility.

Interestingly, embracing this concept, during the last years the USP (U.S. Pharmacopoeia) standards setting bodies have acknowledged the need to update chapters dedicated to drug-specific bioassays, by including in vitro tests based on the use of engineered cell lines, that progressively will replace animal based assays (Morris et al., 2009). As for insulin, a metabolic test based on the ability to stimulate glucose uptake by adipocytes differentiated from 3T3L1 fibroblasts has been submitted for experts' evaluation as a candidate substitute of rabbit blood sugar assay (Morris et al., 2009).

The use of cell culture based functional assays is inspired by some basic physiological considerations. Drugs generally interact with specific protein targets (receptors) located on cell membranes or intracellularly, to elicit a cell response (i.e. gene transcription, mRNA transduction, enzymes activation, apoptosis, cell proliferation, metabolic alteration).

When performing quality controls on pharmaceuticals subject to a deterioration risk, the efficiency of this interaction should be checked. Where an anyway induced molecular disruption should reduce the drug-receptor affinity, we would observe a parallel loss of pharmaceutical efficacy, quantifiable as a reduction in the entity of these responses.

In detail, two basic requirements should be fulfilled, that is the preservation of potency and safety. The potency of a drug is determined by both the binding affinity toward its receptor, and the efficiency by which a molecule, once bound to its target site, initiates a series of events leading to a measurable response. The drug-receptor affinity is strongly dependent on the structure of molecule and even slight modifications may affect the strength of binding, leading to loss of expected biological effects.

In vitro functional assays can helpfully complement the structural ones, based on cellular drug targets and metabolism. For the latter, cell cultures represent easy-access models to monitor targeted processes and quantify responses, keeping in mind that the drug-receptor interaction activates a variety of intracellular responses, investigable through highly diversified techniques. They can be summarily grouped as in the list below:

- **Proteomics:** Useful to assess the phosphorylation status (activation) of receptors or effectors downstream of them (Western blot), to quantify protein amounts following the activation/repression of specific gene expression (Western blot, ELISA), etc.
- **Molecular Biology Tools:** Useful for a semi-quantitative determination of gene transcripts (Real Time PCR).
- **Metabolic Assays:** Allowing a rapid measurement of proteins in small amounts, a quantitative evaluation of cell proliferation/cytotoxicity (i.e. MTT colorimetric test), etc.

- **Biochemical Assays:** For a quantification of drug metabolites and detection of alterations of protein/lipid/carbohydrate metabolism.

Some of these techniques will be recalled and thoroughly explained in the following section dedicated to the application of the proposed study model to two commercial pharmaceuticals.

A Brief Introduction to Case Studies

In order to concretely illustrate the architecture and flexibility of protocol, two liquid pharmaceutical preparations, ActrapidTM (100 I.U. ml^{-1}; Novo Nordisk, Bagsvaerd, Denmark) and Gonal-F (450 I.U./0.75 ml; Merck Serono S.p.A.), have been considered as objects of comprehensive studies. Their active principles are, respectively, recombinant human insulin and Follicle Stimulating Hormone (FSH). Insulin is a 51 amino acid peptide hormone, made up of two chains bound together by two disulfide bonds. FSH is a heterodimeric glycopeptide hormone consisting of two subunits (α, β), linked by two disulfide bonds.

The peptide nature of bioactive compounds, in the presence of structural alterations, can make possible a loss of the tertiary/quaternary structure, or even self-aggregation phenomena, leading to a loss of drug performance and potential risks to health. To manage this issue, an analysis of drug performance was conducted, following the scheme already depicted. To investigate the intra- and inter-lot consistency of analysis results, drug samples from different packaging and production lots were included in each study. For every batch, as many cartridges as UHF time period exposures to be tested were used (Table 1).

Half a content of each cartridge was aseptically collected under a laminar flow hood into sterile polypropylene tubes to be subjected to RFID device EMFs, for the indicated time periods.

Table 1. Design of product exposure to UHF EMFs

Product (Active Principle)	Lot Numbers	Nr Vials/ Batch	UHF Time Exposure
ActrapidTM (insulin)	XS63630 XS63631 XS64223	4	5 min, 1, 6, 24 h
Gonal-F (FSH)	Y16B1866 Y16B5341	2	1, 6 h

Remaining quantities were regularly stored to be used as controls. A small fraction of control and treated samples was destined to the structural analysis by Reverse Phase-High Pressure Liquid Chromatography (RP-HPLC), while the remaining was taken for biological assays.

MATERIALS

Chemicals and Reagents

ActrapidTM (100 I.U. ml-1; Lot numbers XS63630, XS63631, XS64223) was purchased from Novo Nordisk (Bagsvaerd, Denmark). Gonal-F (450 I.U./0.75 ml; Lot numbers Y16B1866, Y16B5341) was from Merck Serono S.p.A. Bagsvaerd (Denmark) plant. CH3CN (HPLC grade) and extra-pure K2HPO4 were from Merck (Darmstadt, Germany). Cell culture media, fetal bovine serum (FBS), L-glutamine, antibiotics and 3-(4,5-Dimethylthiazol-2-yl)-2,5-diphenyltetrazolium bromide (MTT) were purchased from Euroclone Life Science (Pero, MI, Italy). Hormones were obtained from Sigma Aldrich (Milan, Italy). Protease inhibitors were from Sigma-Aldrich Milan (Italy) plant. Antibodies were obtained from Santa Cruz Biotechnology Milan (Italy) plant. The Enhanced Chemiluminescence (ECL) detection system was purchased for Amersham Biosciences Milan (Italy) plant.

Cell Culture Models: Strategies of Choice

ActrapidTM. PCCl3 cells, a rat fully differentiated thyroid cell line, was selected to test the biological activity of ActrapidTM insulin through the quantitative evaluation of the *in vitro* mitogenic activity. In fact, it is well known that PCCl3 cell proliferation is regulated by insulin in a stimulatory way, following the interaction with a specific membrane receptor (IR) (Avruch, 1998). It was also considered that insulin binds with low affinity to the insulin-like growth factor receptor (IGF-R), promoting the same effect, although with a different potency (Sasaoka et al., 1996). It has been shown that structural alterations may enhance the affinity of insulin toward IGF-R, resulting in an aberrant proliferative response, with high tumorigenic potential (Shukla et al., 2009; Baserga et al., 1997). To take into account this aspect, the experimental study was extended to a human breast carcinoma cell line MCF7, exhibiting a high IGF-R/IR expression ratio (Shukla et al., 2009).

Gonal-F. FSH peptide hormone, upon binding to its receptor (FSHR) on the outer plasma membrane, promotes the activation of signal transduction cascades, i.e. PI3K/Akt, stimulating the proliferation of ovarian cancer cells (Huang et al., 2008). The serine/threonine protein kinase Akt is a downstream effector of FSH and its phosphorylation status can be monitored as an index of activation. To evaluate the ability of irradiated drug to initiate the PI3K/Akt pathway, a human ovarian carcinoma cell line (Skov-3) has been chosen as a model, being responsive to FSH. An experimental framework has been designed to semi-quantitatively detect levels of phospho-Akt in response to the drug treatment, assessing the ability of irradiated Gonal-F to stimulate cell growth.

Cell Culture Conditions. Rat PCCl3 thyroid cells were cultured in Coon's modified Ham's F12 medium, supplemented with 5% FBS, a mix of six hormones [1 μg/ml insulin, 3.62 μg/ml hydrocortisone, 5 μg/ml apo-transferrin, 20 ng/ml GHL, 10 ng/ml somatostatin, 1 mU/ml TSH (thyroid stimulating hormone)], 2 mM L-glutamine, 0.05 mg/ml gentamicin, 5000 U/ml/5 mg/ml penicillin/streptomycin. Human MCF7 breast carcinoma cells and human Skov-3 ovarian carcinoma cells were grown in DMEM (Dulbecco's modified Eagle's medium), supplemented with 10% FBS, 2 mM L-glutamine and 5000 U/ml/5 mg/ml penicillin/streptomycin.

METHODS

RF Exposure Phase: Setup of a Laboratory Test Environment

The test environment set up to carry out the experimental analysis of the potential effects of UHF RFID technologies on biological drugs is reported in Figure 1. It consists of a conveyor belt, equipped with one UHF RFID reader, the Impinj's Speedway. Its operating frequency range is 865-956 MHz. Its maximum transmission RF power was set to +30 dBm (1 Watt) for the exposure campaigns. Furthermore, the line is characterized by a metallic tunnel equipped with four near field UHF antennas of the same type: the Impinj's CS-777 Brickyard. Each reader antenna is in the centre of each tunnel side. The width of the tunnel is 0.6 m. In order to obtain some indications about the intensity of the electric field inside the tunnel – where the drug sample is located to be irradiated – the PMM 8053A analyzer was used. It was equipped with an electric field probe PMM EP-330 able to work in the frequency range from 100 KHz to 3 GHz. Time period exposures for each preparations were as follows: 5 min, 1, 6 and 24 h for ActrapidTM; 1 and 6 h for Gonal-F. For further details on the design of the exposure campaign the reader is referred to Table 1.

Structural Investigation of RF-Induced Molecular Damage

Loss of drug structural integrity following the exposure to RF radiation was investigated by RP-HPLC (Reverse Phase-High Performance Liquid Chromatography), in both cases.

Basic Theory. Chromatographic techniques enable the separation, identification and quantification of substances in a mixture, as well as the detection of degradation products, combining the rapidity of execution with a high sensitivity and reproducibility.

In a typical HPLC system, a high-pressure pump pushes a solvent (mobile phase) at a controlled flow rate through the analytical column, filled with the so-called stationary phase, that interacts at various degrees with analytes dissolved in a sample. An injector introduces the sample into the continuously flowing mobile phase that delivers it to the column. Compounds with the lowest affinity toward the stationary phase move faster through the column and exit the first. A detector continuously registers specific chemical/physical properties of the column eluate. One of the most diffused detection principles in HPLC system is the UV (ultraviolet) absorbance of single compounds in the effluent. A computer data station records the signal needed to generate a chromatogram, a plot of detection signal intensity variation in detection versus the retention time. It appears as a series of peaks, each one corresponding to a specific molecule. The time spent by a compound to elute from the column (retention time), falling within a pre-defined window, is a well-established criterion for molecular identification. Where a molecule should loss its integrity, a shift in the retention time or the appearance of additional peaks corresponding to degradation products would be observed. A reduction of the main peak area value, confirming the occurrence of degradation phenomena, would also be expected.

Reverse-phase chromatography is the variant of choice for the quality control of peptide molecules, since effective in separating protein molecules by virtue of its high resolving power and the quick elution of polyelectrolytes. For this variant, the stationary phase is relatively non polar and the solvent is polar with respect to the sample. Polar molecules will obviously travel through the column more quickly.

Analysis Conditions. Drug samples were run through an Agilent 1100 HPLC system, equipped with a Vydac 218TP54 column and coupled with an UV detector, at 25°C.

ActrapidTM samples were analysed under isocratic conditions (mobile phase: 0.05M KH_2PO_4-CH_3CN 65:35, v/v; pH 2.4, flow rate 1.0 ml min^{-1}) (Hoyer et al., 1995), with UV detection at 230 nm.

For Gonal-F drug samples, a gradient elution method (mobile phase A: 0.04M phosphate buffer, B: CH_3CN – 15 to 50% v/v of B in 40 min, pH 7.0, flow rate 0.5 ml min^{-1}) was applied, with UV detection at 220 nm (Loureiro et al., 2006).

For both pharmaceutical preparations, the following parameters related to the bioactive principle were regularly recorded for each chromatogram: (i) the retention time (RT, min), critical for identification purposes; (ii) the main peak area (PA, mAU*s), proportional to the sample content of the main compound.

Functional Bioassays

For both ActrapidTM insulin and Gonal-F FSH, the *in vitro* mitogenic potency of bioactive compounds was measured as a functional parameter, to confirm the maintenance of drug potency after RF exposure. For Gonal-F, the drug ability to stimulate the phosphorylation of Akt kinase enzyme was additionally tested by a Western Blotting analysis.

Proliferation Assay: Basic Theory and Practice

Basic Theory. A simple colorimetric test, namely the MTT proliferation assay, was used to quantify the cell growth in response to drug treatments, with some modifications respect to the original version (Mosmann, 1983). MTT (3-(4,5-dimethylthiazol-2-yl)-2,5-diphenyl tetrazolium bromide) is a yellow and membrane-permeable tetrazolium salt that upon reduction by mitochondrial dehydrogenases is converted to purple formazan crystals. They are then dissolved by organic solvents, such as dimethylsulfoxide (DMSO) or isopropanol, and the light absorbance quantification of the purple product allows the estimation of viable cell number.

MTT assay is extensively used for high throughput screenings and it is adaptable to both adherent and suspension cells. It is safe, as it does not involve radioactivity, and economic.

Protocol. Cells were seeded into 96-well trays at a density of 4-6*10^3 cells/well and, after 24 h incubation with 0.2% FBS hormone-free media ("serum starvation"), drugs were administered at the lowest concentration able to induce an appreciable proliferation on each cell line (5 nM ActrapidTM insulin, 10 mIU/ml Gonal-F FSH).

After 24 h treatment, 0.5 mg/ml MTT was added to cultures for 3 h at 37°C. The formazan product, solubilized by dimethyl sulfoxide, was measured spectrophotometrically at 570 nm (background correction at 690 nm). The proliferation of treated cultures was expressed as a percentage with respect to untreated cells, assuming the related mean absorbance value as a reference (basal growth).

Results of MTT tests were expressed as mean ± standard error of three separate experiments, with four/six replicates at least for each treatment.

Western Blotting Analysis: Basic Theory and Practice

Basic Theory. Western blotting is a powerful analytical tool useful to detect and quantify specific proteins in a mixture. In principle, proteins dissolved in a complex sample (i.e. cell or tissue extract) are separated by polyacrylamide gel electrophoresis based on their molecular weight and then transferred from the gel to a nitrocellulose sheet. A specific primary antibody is used to recognize and locate the target protein on the membrane. The sheet is then incubated in the presence of an enzyme-conjugated (generally horseradish peroxidase or alkaline phosphatase) secondary antibody that interacts with the stem of primary one. When luminol is added for detection, the peroxidase catalyzes the chemiluminescent oxidation of this substrate that being in an excited configuration, decays to the ground state emitting light at 428 nm wavelength. When the membrane is placed against an autoradiographic film, discrete bands appear, corresponding to the proteins specifically identified by the primary antibody. The size and the intensity of such bands tightly relate to the sample protein amount and can be quantified by densitometric analysis on scanned film images.

Protocol. The FSH-induced phosphorylation of Akt enzyme was analyzed as follows. Skov-3 cell cultures were treated for 30 min with 40 mIU/ml Gonal-F FSH. Experimental conditions were determined based on limits of detection of increased Akt phosphorylation.

Total cell lysates were prepared from treated cultures in ice-cold modified RIPA buffer (50 mM Tris-HCl, pH 7.4, 1% NP-40, 0.25% Na-deoxycholate, 150 mM NaCl, 1 mM EDTA, 1 mM Na_3VO_4, 1 mM NaF) containing a protease inhibitor mixture. Samples were centrifuged at 14,000 g for 15 minutes and supernatants collected. 30 μg protein aliquots were loaded and run onto 10% SDS (Sodium Dodecyl Sulfate) polyacrylamide gels. The separated protein bands were transferred onto a nitrocellulose membrane by Trans-Blot® SD semi-dry cell (Bio-Rad), at 15 W for 30 min. After blocking with PBS (Phosphate Buffered Saline)-0.1% Tween containing 5% non-fat milk, the membranes were incubated with anti-Akt (total Akt; 1:1,000), anti-pAkt (phosphorylated Akt; 1:500) and 1:10,000 anti-β tubulin (housekeeping gene) primary antibodies overnight at 4°C. They were then incubated with peroxidase-labeled secondary antibodies at 37°C for 1 hour.

After washes in PBS-0.1% Tween, protein bands were detected by a horseradish peroxidase-based chemiluminescent detection system (ECL) and their intensity on autoradiographs quantified by *ImageJ* software (1.44I version). For each experimental condition, Akt phosphorylation level was determined by calculating the pAkt/Akt ratio of intensities, after normalization against β tubulin.

Results were presented as average value ± standard error of five separate experiments, with two replicates for each treatment.

Statistics

Graphic and statistical elaborations were carried out by GraphPad Prism software, version 5.00 (San Diego, California, USA). Comparisons were performed by one-way ANOVA (Bonferroni post-test). Statistical significance was set at $p<0.05$ (*).

EXPERIMENTAL RESULTS

Case Study I: ActrapidTM Summary

RP-HPLC chromatograms were preliminarily recorded from untreated drug aliquots from all available cartridges (a total of twelve, four for each of three production lots; see Table 1). For each sample, both the average retention time and the main peak area (n=3) related to insulin were registered. In fact, as mentioned, both values allow an estimation of possible compound structural changes (shift in retention time), and/or reductions in the amount of the bioactive compound (decrease in the main peak area).

The chromatogram in the upper panel of Figure 2 is representative of all control samples (RT 5.834 min; PA 2,59886 e^4 mAU*s). A chromatogram obtained from a 24 h RF exposed aliquot, representative of all treated samples (even if for shorter periods), is shown in the lower panel of Figure 2 (RT 5.834 min; PA 2,78351 e^4 mAU*s).

For all time points tested, chromatograms resulted perfectly superimposable on those registered from untreated drug aliquots taken from the same cartridges. We concluded that insulin structural integrity (primary sequence, disulfide bonds) was preserved under prolonged RF exposure. Lot-to-lot variation was evidently negligible.

The drug proliferative potency as a reference functional parameter was then evaluated by MTT assay for each time point. Shortly, MCF7 and PCCl3 cultures were grown for 24 h in the presence of 5 nM control and 5 min, 1, 6, 24 h UHF RF exposed ActrapidTM, assuming as a reference the growth rate of cells cultured in 0.2% FBS media. Results of proliferation assays for all the three ActrapidTM lots employed are summarized in Figure 3. To minimize misinterpretations due to an intra-lot variability, for each time exposure, regular and irradiated drug samples were taken from the same vial.

It is evident for both cell lines that the entity of response to regular drug is not significantly different among vials from the same lot and, the most important, the internal comparison between the effects of control and UHF exposed drug does not reveal statistically significant differences, whatever is the length of exposure. It can be observed that the ActrapidTM-induced fold-increase in cell number slightly varies from lot to lot, and the entity of variation seems to be dependent on specific cell type response. MCF7 cells resulted less sensitive to lot-to-lot variation (1.8 to 2.3 fold increase range), while PCCl3 cells exhibited a more accentuated variability (1.9 to 3.2 fold increase range). Whatever the sources of lot to lot variation in the entity of ActrapidTM-induced cell growth (i.e actual cell growth conditions, slight differences in the number of seeded cells), they did not influence the impact of radiation on the performance of the active principle.

Figure 2. Chromatographic analysis of control (upper panel) and 24 h UHF RF-exposed (lower panel) ActrapidTM samples. RP-HPLC profiles are representative of, respectively, untreated drug samples taken from available vials and all radiofrequency exposed ones, independently on time points considered

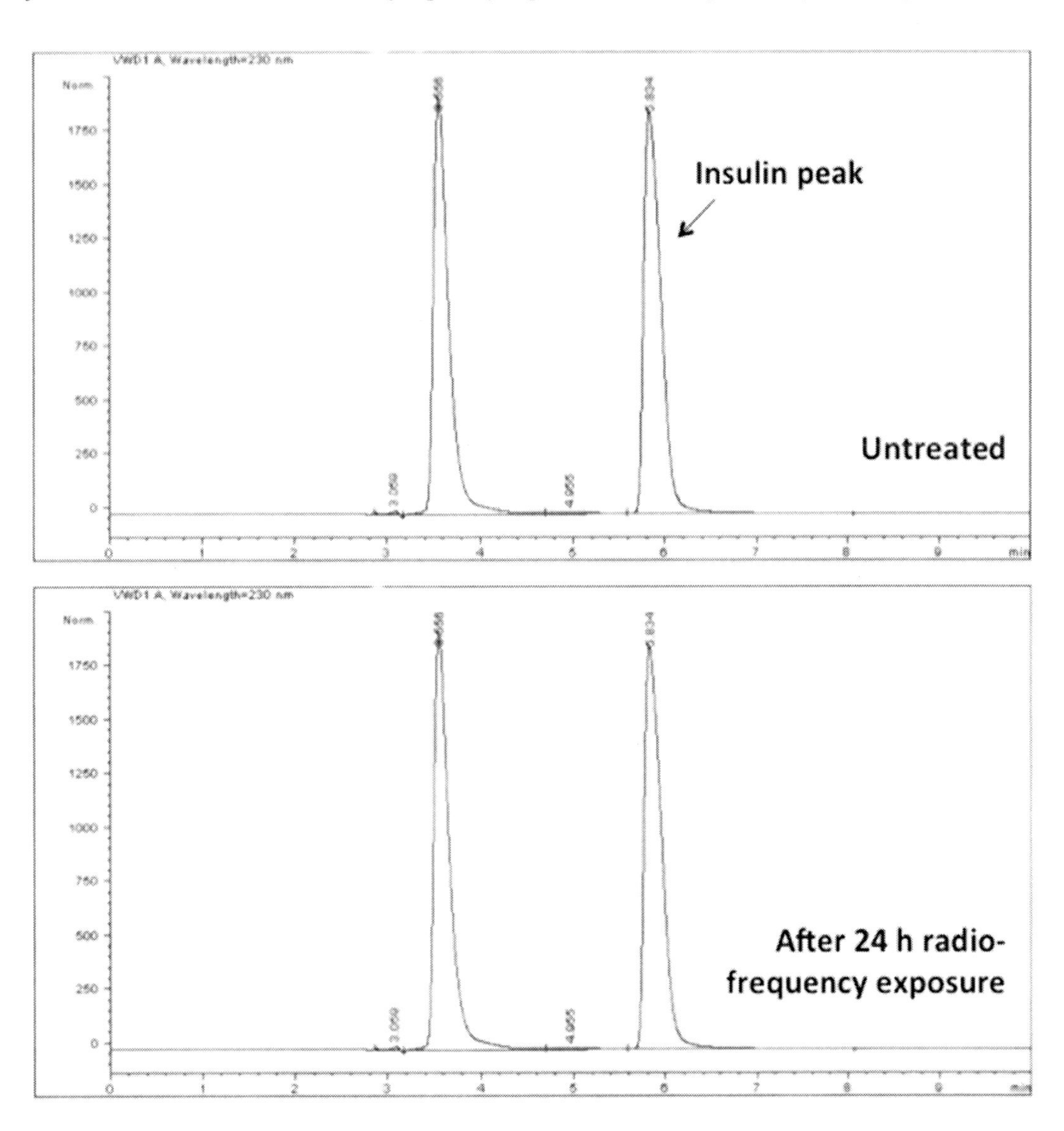

Case Study II: Gonal-F Summary

The chromatographic analysis on 1-6 h RF-exposed Gonal-F samples was conducted following the scheme commented for ActrapidTM. First, in order to define both the retention time and the main peak area values associated with Gonal-F FSH, RP-HPLC phase gradient chromatograms obtained from control aliquots taken from all cartridges (a total of four, two for each production lot) were processed (Figure 4, upper panel).

Comparing the chromatographic profiles registered from drug aliquots subjected to RF emissions with those obtained from control samples,

*Figure 3. Mitogenic effect of ActrapidTM on cultured cells before and after drug exposure to UHF radiation. MCF7 (A-C) and PCCl3 (D-F) cells were treated with 5 nM regular (white bar) and 5 min, 1, 6, 24 hr UHF irradiated (black bar) ActrapidTM for a day. For each exposure time, control and treated drug samples were from the same vial and separate vials were used for the indicated time period exposures. The induced growth rate was quantified respect to spontaneous proliferation (grey bar). Product samples were from three production lots (nr): XS63630 (A, D); XS63631 (B, E); XS64223 (C, F). The proliferative potency of ActrapidTM was preserved even at the longest time point studied. Data points represent the mean ± st.err. of three separate experiments (**p<0.01, Student's t test).*

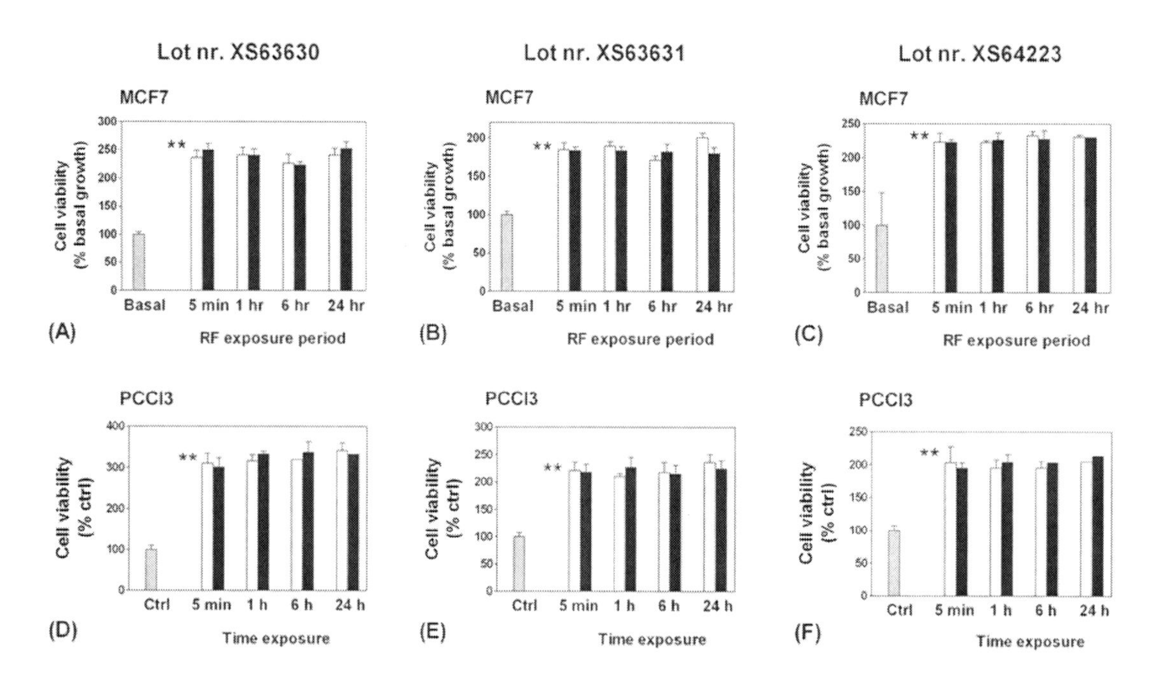

did not reveal any alterations of structural properties, independently on the length of exposure and the lot they were from (Figure 4, lower panel).

In conclusion, UHF RF exposure up to 6 h neither affected the primary structure of Gonal-F FSH nor disrupted the disulfide bonds.

In order to investigate the functional integrity of treated drug samples, the proliferative potency of FSH bioactive compound has been assumed as a measure of a proper interaction with transmembrane receptors (FSH-R). This hormone has been previously shown to exert a proliferative effect on Skov-3 ovary cancer cells by MTT colorimetric assay (Huang et al., 2007).

Shortly, cell cultures were treated for 24 h in the presence of 40 mU/ml regular and 1 to 6 h UHF exposed Gonal-F. The basal cell growth rate was assumed as a reference to quantify FSH stimulating action by MTT tests.

As illustrated in Figure 5, representative of both production lots, the impact of UHF radiation on drug performance was negligible, even for the longest time point tested. Both control and 1-6 h treated drug samples induced an average 41.10 ± 6.78% (mean ± st.err.; n=3) increase in cell proliferation, allowing to exclude an impairment in Gonal-F FSH mitogenic potency. To sustain the reproducibility of MTT assay, it is to be specified

*Figure 4. RP-HPLC analysis of control (upper panel; RT 25.154 min/PA 3754,974 mAU*s) and 6 h UHF RF-exposed (lower panel; RT 25.109 min/PA 3768,533 mAU*s) Gonal-F samples. In the cited order, chromatograms are representative of (respectively) untreated drug aliquots collected from available vials and all RF exposed samples, independently on the length of treatment and production lot tested.*

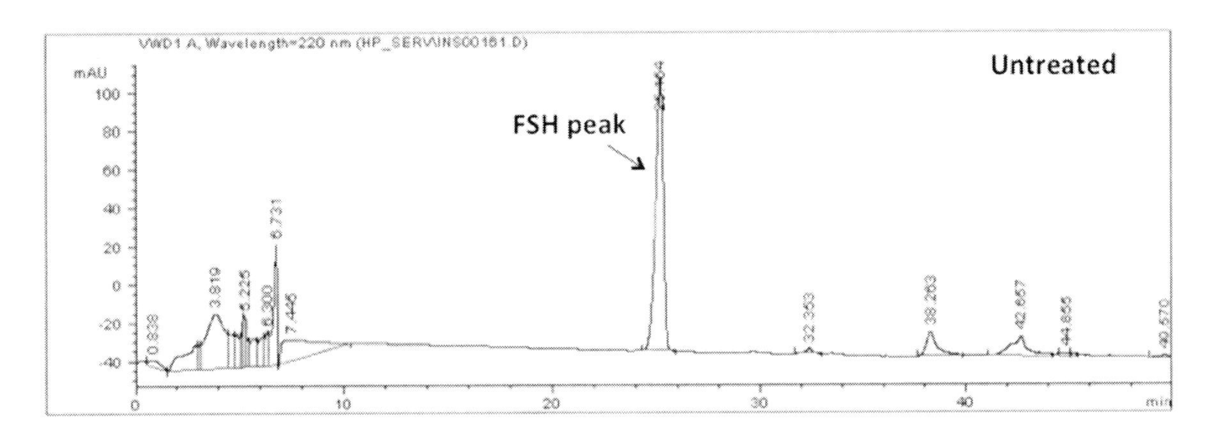

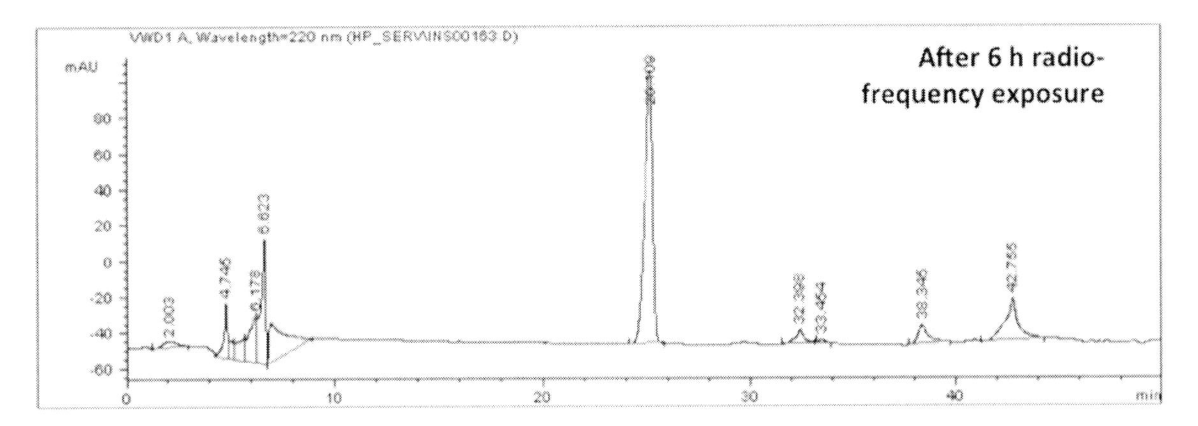

that intra-assay CVs (Coefficient of Variation) were all in an acceptable range, while inter-assay CVs ranged from 4.60 to 10.20%.

The interaction of FSH with its own receptor (FSH-R) expressed by granulosa and Sertoli cells, induces a morphological shift in such protein, coupled to the activation of PI3K, a kinase enzyme which in turn leads to the phosphorylation of Akt serine-threonine kinase (PI3K/Akt transduction pathway), involved in the hormone physiology.

The ability of UHF RF exposed Gonal-F FSH to induce Akt phosphorylation at Serine 473 amino acid residue was evaluated in Skov-3 ovary cancer cells.

The time-dependence of Akt phosphorylation was preliminarily monitored over 60 min to determine the experimental conditions yielding an appreciable increase of pAkt amount. Based on these preliminary tests, the bioactivity of Gonal-F FSH was evaluated after 30 min treatment with 40 mU/ml. Protein extracts obtained from treated cells were examined by Western Blotting to quantify phosphorylated Akt with respect to the total amount of kinase. Analysis performed on untreated Gonal-F from all available vials demonstrated a stimulatory effect on Akt phosphorylation, that resulted increased by 111 ± 13% (mean ± st.err.; n=5) with respect to the basal levels (Figure 6).

*Figure 5. Proliferative potency of Gonal-F FSH before and after drug exposure to UHF radiation. Skov-3 cells were treated for 24 h with 10 mIU/ml regular (white bar) and 1-6 h UHF irradiated (black bar) Gonal-F. For each time point, control and treated drug samples were from the same vial and separate vials were used for the indicated time period exposures. The growth rate was quantified respect to spontaneous growth (grey bar). Graph, although referred to lot Y16B1866 is representative of both production lots tested. Intra-assay CVs (%) are indicated on each bar. Data points represent the mean ± st.err. of three separate experiments (**p<0.01, ANOVA).*

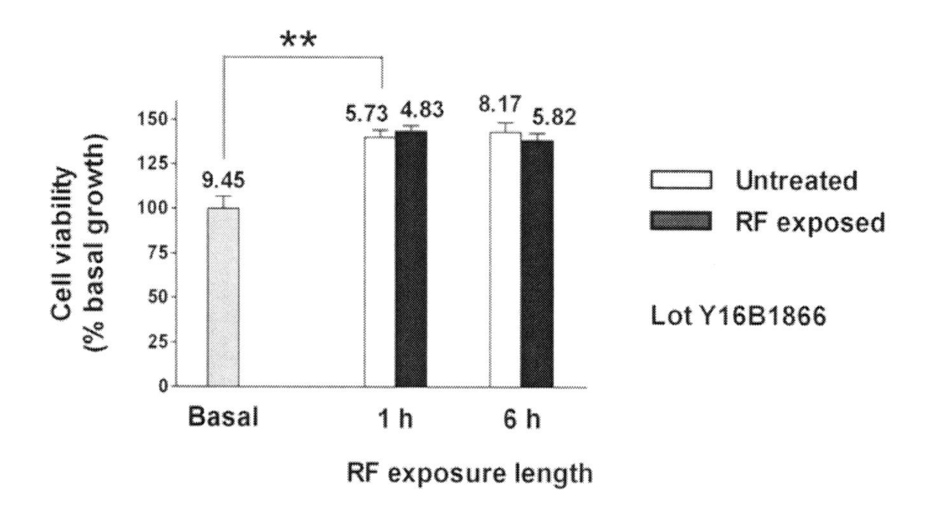

When cell cultures were treated with 1-6 h RF exposed drug aliquots recovered from the same vials of untreated ones, the amount of pAkt was increased by the same percentage (108 ± 12% increase; basal vs treated, $p>0.05$; Figure 6). This indicated that both 1 and 6 hour RF exposed Gonal-F retained the ability to activate the PI3K/Akt pathway.

On the whole, structural and functional results demonstrate the innocuity of UHF RFID devices towards the pharmaceutical considered, even if the proliferation tests are preferable for the routine practice because of considerably lower inter-assay CVs (indicated in Figure 5).

It must be conclusively observed that the use of discussed *in vitro* functional methods allows to circumvent the severe limitations (low sensitivity, variability up to 20%) associated to the Steelman-Pohley *in vivo* assay, for a long time considered the mainstay for determining the FSH activity in pharmacopoeial monographs (Steelman and Pohley, 1953). It is also true that with the advent of recombinant technologies, that guarantee a high consistency of FSH physico-chemical profile, convenient chromatography techniques (size exclusion high-performance liquid chromatography (SE-HPLC), RP-HPLC) have been introduced to test drug quality at industrial scale (Driebergen et al., 2003; Almeida et al., 2011), but in no cases they provide information on FSH biological activity.

CONCLUSION AND FUTURE WORKS

This chapter discussed limits, benefits, and experimental results related to the use of RFID technology in tracing and tracking systems in the pharmaceutical supply chain; it stressed that the paucity of reports on the impact of RFID on drug quality, as well as the inhomogeneous criteria for

*Figure 6. Modulation of Akt phosphorylation in Skov-3 cells by RF exposed Gonal-F FSH. Upper panel, Western blot analysis of total and phosphorylated Akt levels in cell lysates prepared from cultures treated for 30 min with 40 mIU/ml regular (ctrl) and RF exposed (exp) Gonal-F FSH. A β-tubulin blot has been included to show equal protein loading. Lower panel, densitometric quantification of pAkt levels after normalization over total Akt (pAkt/Akt ratios). Band intensities have been determined by ImageJ software (1.0 version). Inter-assay CVs are reported on each bar. Results are presented as mean ± st.err. of five distinct experiments (*p<0.05; ANOVA).*

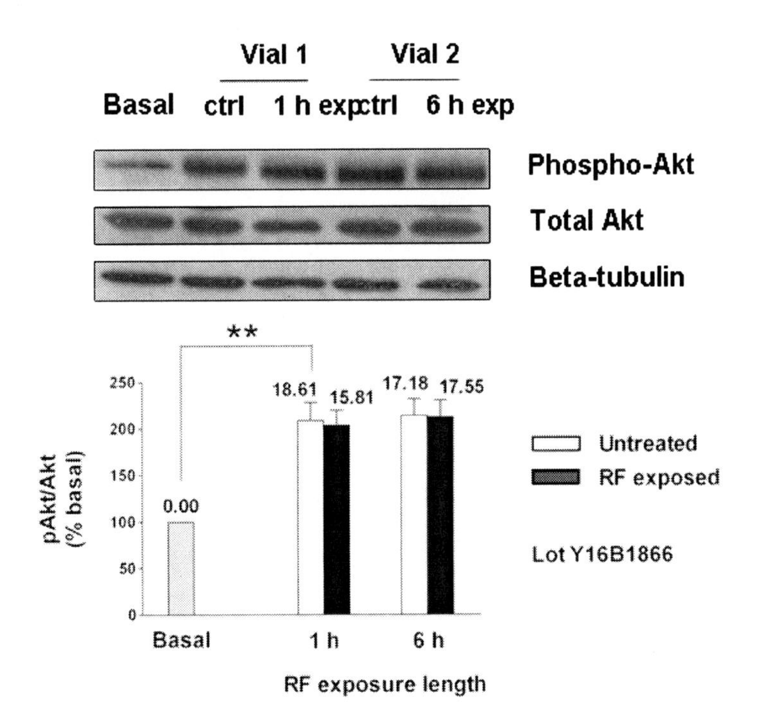

the evaluation and the research designs emerging from literature data, raise the need for a standardized procedure that can be regularly executed prior to introduce tagged products into the commercial distribution. Given the varying susceptibility of bioactive compounds to UHF-induced damage and the extreme diversity of drug therapeutic targets, setting up of an efficient analysis protocol is a very complex issue.

In this chapter, a three-phased experimental model has been presented, built on the convergence of multiple competencies, to help pharmaceutical manufacturers to explore the RFID-related risk for individual commercialized drugs. According to the protocol, drug samples are preliminarily time overexposed to the RF radiation, to simulate an accidental prolonged stay of a product in the area covered by common RFID devices. Then, such samples are subjected to structural (RP-HPLC) and *in vitro* functional analysis, to verify the efficiency of docking and interaction of drug bioactive compounds with specific cell targets (receptors). Where an HPLC analysis may fail to reveal breaking/alterations of the intramolecular weak interactions, a functional assay can evidence the consequent incorrect steric arrangement. In fact, the magnitude of cell response strictly depends upon goodness of the drug fitting to its own receptor.

The high value of the experimental framework clearly emerges through the illustrated case studies, both referred to the effects of a prolonged exposure of liquid and peptide-based pharmaceutical preparations to an UHF RF EF. The results obtained allowed excluding deteriorating effects from RFID devices on the integrity of both products and strongly encourage a larger adoption of this technology in the pharmaceutical distribution sector, supported by case-specific investigations of potential RF-induced effects.

It must be specified that the approach described in this chapter does not aim to be faultless, but the authors just wanted to outline the potential of a multidisciplinary approach. Efforts are now directed towards a refinement of test bed machinery, to experimentally recreate a setting where the tagged drug can occasionally be exposed to higher values of irradiative power. In order to evaluate the gap between normal operating conditions (i.e. no damages) and critical operating conditions (i.e. changing of molecular structure and so degradation in potency of the same drug), the authors are now working to develop a system to vary the power of RF emissions by a scalar way during the drug exposure phase. This new apparatus will be able to generate EMFs with an irradiative power higher than that of commercial RFID readers. In parallel, the design strategy of *in vitro* bioassays will be revised to allow the verification of a compound bioactivity in close relation to the expected therapeutic effects. In fact, it can be observed that a substance triggers a variety of cellular events, but a limited number of them correlate with systemic effects. As an example, the main metabolic action of insulin consists of lowering the blood glucose levels by stimulating the sugar uptake by muscle and adipose cells. Upon this premise, the ideal *in vitro* functional assay should assess the hormone ability to stimulate glucose uptake by insulin-sensitive cultured cells, for example adipocytes. The conclusive step in such a phased work is the establishment of an industry-scale platform for the assessment of the RFID-related risk, an objective that can be pursued only by the coordination of multidisciplinary skills.

The case studies summarized in this book chapter clearly demonstrate how this kind of approach, whose main feature is versatility, is an indispensable condition to achieve high-quality analysis results. Furthermore, the proposed protocol represents a comprehensive response to the pharmaceutical market request for a wider and safe adoption of RFID "track and trace" systems through the distribution networks.

REFERENCES

Acierno, R., Carata, E., De Pascali, S. A., Fanizzi, F. P., Maffia, M., Mainetti, L., & Patrono, L. (2010a). Potential effects of RFID systems on biotechnology insulin preparation: A study using HPLC and NMR spectroscopy. In (IEEE) *Proceedings of the 5th International Conference on Complex Medical Engineering* (pp. 198-203).

Acierno, R., De Pascali, S. A., Fanizzi, F. P., Maffia, M., Mainetti, L., Patrono, L., & Urso, E. (2010b). Investigating potential effects of RFID systems on the molecular structure of the human insulin. In *Proceedings of the 5th Cairo International Biomedical Engineering Conference* (pp. 192-196).

Acierno, R., De Riccardis, L., Maffia, M., Mainetti, L., Patrono, L., & Urso, E. (2010c). Exposure to electromagnetic fields in UHF band of an insulin preparation: Biological effects. In (IEEE) *Proceedings of Biomedical Circuits and Systems Conference* (pp. 78-81).

Acierno, R., Maffia, M., Mainetti, L., Patrono, L., & Urso, E. (2011). RFID-based tracing systems for drugs: Technological aspects and potential exposure risks. In (IEEE) *Proceedings of Topical Conference on Biomedical Wireless Technologies, Networks, and Sensing Systems* (pp. 87-90).

Almeida, B. E., Oliveira, J. E., Damiani, R., Dalmora, S. L., Bartolini, P., & Ribela, M. T. (2011). A pilot study on potency determination of human follicle-stimulating hormone: A comparison between reversed-phase high-performance liquid chromatography method and the in vivo bioassay. *Journal of Pharmaceutical and Biomedical Analysis*, *54*(4), 681–686. doi:10.1016/j.jpba.2010.10.018

Avruch, J. (1998). Insulin signal transduction through protein kinase cascades. *Molecular and Cellular Biochemistry*, *182*(1-2), 31–48. doi:10.1023/A:1006823109415

Barchetti, U., Bucciero, A., De Blasi, M., Mainetti, L., & Patrono, L. (2009). Implementation and testing of an EPCglobal-aware discovery service for item-level traceability. In (IEEE) *Proceedings of International Conference on Modern Ultra Telecommunications* (pp. 1-8).

Barchetti, U., Bucciero, A., De Blasi, M., Mainetti, L., & Patrono, L. (2010). RFID, EPC and B2B convergence towards an item-level traceability in the pharmaceutical supply chain. In (IEEE) *Proceedings of International Conference on RFID-Technology and Applications* (pp. 194-199).

Baserga, R., Hongo, A., Rubini, M., Prisco, M., & Valentinis, B. (1997). The IGF-I receptor in cell growth, transformation and apoptosis. *Biochimica et Biophysica Acta*, *1332*(3), 105–126.

Bassen, H. (2005). Liquid pharmaceuticals and 915 MHz radiofrequency identification systems, worst-case heating and induced electric fields. *RFID Journal.* White Paper Library.

Bassen, H., Seidman, S., Rogul, J., Desta, A., & Wolfgang, S. (2007). An exposure system for evaluating possible effects of RFID on various formulations of drug products. In (IEEE) *Proceedings of International Conference on RFID* (pp. 191-198).

Bismuto, E., Mancinelli, F., d'Ambrosio, G., & Massa, R. (2003). Are the conformational dynamics and the ligand binding properties of myoglobin affected by exposure to microwave radiation? *European Biophysics Journal*, *32*(7), 628–634. doi:10.1007/s00249-003-0310-2

Cox, F. C., Sharma, V. K., Klibanov, A. M., Wu, B. I., Kong, J. A., & Engels, D. W. (2006). A method to investigate non-thermal effects of radio frequency radiation on pharmaceuticals with relevance to RFID technology. In (IEEE) *Proceedings of International Conference Engineering in Medicine and Biology Society: Vol. 1* (pp. 4340-4343).

De Pomerai, D. I., Smith, B., Dawe, A., North, K., Smith, T., & Archer, D. B. (2003). Microwave radiation can alter protein conformation without bulk heating. *FEBS Letters*, *543*(1-3), 93–97. doi:10.1016/S0014-5793(03)00413-7

Driebergen, R., & Baer, G. (2003). Quantification of follicle stimulating hormone (follitropin alfa): Is in vivo bioassay still relevant in the recombinant age? *Current Medical Research and Opinion*, *19*(1), 41–46. doi:10.1185/030079902125001344

Finkenzeller, K. (2003). *RFID handbook, fundamentals and applications in contact-less smart cards and identification.* New York, NY: J. Wiley & Sons.

Fortune, J. A., Wu, B. I., & Klibanov, A. M. (2010). Radio frequency radiation causes no non thermal damage in enzymes and living cells. *Biotechnology Progress*, *26*(6), 1772–1776. doi:10.1002/btpr.462

Hoyer, L., Nolan, P. E. Jr, LeDoux, J. H., & Moore, L. A. (1995). Selective stability-indicating high-performance liquid chromatographic assay for recombinant human regular insulin. *Journal of Chromatography. A*, *699*(1-2), 383–388. doi:10.1016/0021-9673(95)00086-3

Huang, Y., Hua, K., Zhou, X., Jin, H., Chen, X., & Lu, X. (2008). Activation of the PI3K/AKT pathway mediates FSH-stimulated VEGF expression in ovarian serous cystadenocarcinoma. *Cell Research, 18*(7), 780–791. doi:10.1038/cr.2008.70

Huang, Y., Zhao, Y. Q., Su, M., Gao, S. J., Jin, H. Y., & Feng, Y. J. (2007). Effects of follicle stimulating hormone on proliferation, apoptosis, migration and invasion of ovarian carcinoma cells: An in vitro experiment. *Zhonghua Yi Xue Za Zhi, 87*(35), 2512–2514.

Loureiro, R. F., de Oliveira, J. E., Torjesen, P. A., Bartolini, P., & Ribela, M. T. (2006). Analysis of intact human follicle-stimulating hormone preparations by reversed-phase high-performance liquid chromatography. *Journal of Chromatography. A, 1136*(1), 10–18. doi:10.1016/j.chroma.2006.09.037

Morris, T. S., Singer, R., Ambrose, M. R., & Hauck, W. W. (2009). Biological potency assays are key to assessing product consistency. *BioPharm International, 22*(6).

Mosmann, T. (1983). Rapid colorimetric assay for cellular growth and survival: Application to proliferation and cytotoxicity assays. *Journal of Immunological Methods, 65*(1-2), 55–63. doi:10.1016/0022-1759(83)90303-4

Preradovic, S., Balbin, I., Karmakar, N. C., & Swiegers, G. F. (2009). Multiresonator-based chipless RFID system for low-cost item tracking. *IEEE Transactions on Microwave Theory and Techniques, 57*(5:2), 1411–1419.

Russell, W. M. S., & Burch, R. L. (1959). *The principles of humane experimental technique*. London, UK: Methuen.

Sasaoka, T., Ishiki, M., Sawa, T., Ishihara, H., Takata, Y., & Imamura, T. (1996). Comparison of the insulin and insulin-like growth factor 1 mitogenic intracellular signaling pathways. *Endocrinology, 137*(10), 4427–4434. doi:10.1210/en.137.10.4427

Seevers, R. H. (2005). *Report to FDA of PQRI RFID working group. Office of regulatory affairs, guidance for industry, prescription drug marketing act* (pp. 2-4/14).

Shukla, A., Enzmann, H., & Mayer, D. (2009). Proliferative effect of Apidra (insulin glulisine), a rapid-acting insulin analogue on mammary epithelial cells. *Archives of Physiology and Biochemistry, 115*(3), 119–126. doi:10.1080/13813450903008628

Steelman, S. L., & Pohley, F. M. (1953). Assay of the follicle stimulating hormone based on the augmentation with human chorionic gonadotropin. *Endocrinology, 53*(6), 604–616. doi:10.1210/endo-53-6-604

Thiesse, F., Floerkemeier, C., Harrison, M., Michahelles, F., & Roduner, C. (2009). Technology, standards, and real-world deployments of the EPC network. *IEEE Internet Computing Magazine, 13*(2), 36–43. doi:10.1109/MIC.2009.46

U.S. FDA. CPG. Sec. 400.210. (2004). *Radiofrequency identification feasibility studies and pilot programs for drugs*. November 2004.

Uysal, I., De Hay, P. W., Altunbas, E., Emond, J. P., Rasmussen, R. S., & Ulrich, D. (2010). Non-thermal effects of radio frequency exposure on biologic pharmaceuticals for RFID applications. In *Proceedings of IEEE International Conference on RFID* (pp. 266-273).

ADDITIONAL READING

Angeles, R. (2005). RFID technologies: Supply-chain applications and implementation issues. *Information Systems Management, 22*(1), 51–65. doi:10.1201/1078/44912.22.1.20051201/85739.7

Attaran, M. (2007). RFID: An enabler of supply chain operations. *Supply Chain Management, 12*(4), 249–257. doi:10.1108/13598540710759763

Bertocco, M., Dalla Chiara, A., & Sona, A. (2010). Performance evaluation and optomization of UHF RFID systems. In (IEEE) *Proceedings of International Instrumentation and Measurement Technology Conference* (pp. 1175-1180).

Borriello, G. (2005). RFID: Tagging the world. *Communications of the ACM, 48*(9), 34–37. doi:10.1145/1081992.1082017

Bottani, E. (2008). Reengineering, simulation and data analysis of an RFID system. *Journal of Theoretical and Applied Electronic Commerce Research, 3*(1), 13–29.

Bottani, E. (2009). On the impact of RFID and EPC Network on traceability management: A mathematical model. *International Journal of RF Technologies: Research and Applications, 1*(2), 95–113. doi:10.1080/17545730802209104

Bottani, E., & Rizzi, A. (2008). Economical assessment of the impact of RFID technology and EPC system on the fast-moving consumer goods supply chain. *International Journal of Production Economics, 112*(2), 548–569. doi:10.1016/j.ijpe.2007.05.007

Catarinucci, L., Colella, R., De Blasi, M., Patrono, L., & Tarricone, L. (2010). Improving item-level tracing systems through ad hoc UHF RFID tags. In (IEEE) *Proceedings of Radio and Wireless Symposium* (pp. 160-163).

Curtin, J., Kauffman, R., & Riggins, F. (2007). Making the 'MOST' out of RFID technology: A research agenda for the study of the adoption, usage and impact of RFID. *Information Technology Management, 8*(2), 87–110. doi:10.1007/s10799-007-0010-1

De Blasi, M., Mighali, V., Patrono, L., & Stefanizzi, M. L. (2010). Performance evaluation of UHF RFID tags in the pharmaceutical supply chain. In *Proceedings of 20th Tyrrhenian International Workshop on Digital Communications- The Internet of Things* (pp. 283-292).

FDA. (2004-2). *Radiofrequency identification feasibility studies and pilot programs for drugs: Guidance for FDA staff and industry, compliance policy guides.* Sec. 400.210, Radiofrequency Identification Feasibility Studies and Pilot Programs for Drugs, November 2004. Rockville, MD: U. S. Department of Health and Human Services, Food and Drug Administration.

FDA. (2004-1). *Combating counterfeit drugs - A report of the Food and Drug Administration.* Rockville, MD: U. S. Department of Health and Human Services, Food and Drug Administration.

Gandino, F., Montrucchio, B., Rebaudengo, M., & Sanchez, E. R. (2007). Analysis of an RFID-based information system for tracking and tracing in an agri-food chain. In *Proceedings of 1st Annual RFID Eurasia* (pp. 1-6).

Garfinkel, S. L., Juels, A., & Pappu, R. (2005). RFID privacy: An overview of problems and proposed solutions. *IEEE Security & Privacy Magazine, 3*(3), 34–43. doi:10.1109/MSP.2005.78

Gaukler, G. M., Seifert, R. W., & Hausman, W. H. (2007). Item-level RFID in the retail supply chain. *Production and Operations Management, 16*(1), 65–76. doi:10.1111/j.1937-5956.2007.tb00166.x

Heinrich, C. (2005). *RFID and beyond.* Indianapolis, IN: Wiley.

Kelepouris, T., Pramatari, K., & Doukidis, G. (2007). RFID-enabled traceability in the food supply chain. *Industrial Management & Data Systems*, *107*(2), 183–200. doi:10.1108/02635570710723804

Kirschvink, J. L. (1996). Microwave absorption by magnetite: A possible mechanism for coupling nonthermal levels of radiation to biological systems. *Bioelectromagnetics*, *17*, 187–194. doi:10.1002/(SICI)1521-186X(1996)17:3<187::AID-BEM4>3.0.CO;2-#

Mant, C. T., & Hodges, R. S. (1991). *High-performance liquid chromatography of peptides and proteins: Separations, analysis, and conformation*. Boston, MA: CRC Press.

Marrocco, G. (2008). The art of UHF RFID antenna design: Impedance matching and size-reduction techniques. *IEEE Antennas and Propagation Magazine*, *50*(1), 66–79. doi:10.1109/MAP.2008.4494504

Ngai, E. W. T., Moon, K. L. K., Riggins, F. J., & Yi, C. Y. (2008). RFID research: An academic literature review (1995–2005) and future research directions. *International Journal of Production Economics*, *112*(2), 510–520. doi:10.1016/j.ijpe.2007.05.004

Ohkubo, M., Suzuki, K., & Kinoshita, S. (2005). RFID privacy issues and technical challenges. *Communications of the ACM*, *48*(9), 66–71. doi:10.1145/1081992.1082022

Penafiel, L., Litovitz, T., Krause, D., Desta, A., & Mullins, J. M. (1998). Role of modulation on the effect of microwaves on ornithine decarboxylase activity in L929 cells. *Bioelectromagnetics*, *18*(2), 132–141. doi:10.1002/(SICI)1521-186X(1997)18:2<132::AID-BEM6>3.0.CO;2-3

Phillips, T., Karygiannis, T., & Kuhn, R. (2005). Security standards for the RFID market. *IEEE Security & Privacy Magazine*, *3*(6), 85–89. doi:10.1109/MSP.2005.157

Porcelli, M., Cacciapuoti, G., Fusco, S., Massa, R., d'Ambrosio, G., & Bertoldo, C. (1997). Non-thermal effects of microwaves on proteins: thermophilic enzymes as model system. *FEBS Letters*, *402*(2-3), 102–106. doi:10.1016/S0014-5793(96)01505-0

Prater, E., Frazier, G. V., & Reyes, P. M. (2005). Future impacts of RFID on e-supply chains in grocery retailing. *Supply Chain Management: An International Journal*, *10*(2), 134–142. doi:10.1108/13598540510589205

Rao, K. V. S., Nikitin, P. V., & Lam, S. F. (2005). Antenna design for UHF RFID tags: A review and a practical application. *IEEE Transactions on Antennas and Propagation*, *53*(12), 3870–3876. doi:10.1109/TAP.2005.859919

Rappold, J. (2003). The risks of RFID. *Industrial Engineer*, *35*(11), 37–38.

Sbrenni, S., Piazza, T., & Macellari, V. (2010). Use of RFID technology and CCOW standard for patient traceability. In *(IEEE) Proceedings of the 18th International Conference on Software, Telecommunications and Computer Networks (SoftCOM 2010)* (pp.17-20).

Staake, T., Thiesse, F., & Elgar, F. (2005). Extending the EPC network: the potential of RFID in anti-counterfeiting. *Proceedings of ACM Symposium on Applied Computing*. New York, NY: ACM Press.

Tajima, M. (2007). Strategic value of RFId in supply chain management. *Journal of Purchasing and Supply Management*, *13*(4), 261–273. doi:10.1016/j.pursup.2007.11.001

Tzeng, S.-F., Chen, W. H., & Pai, F.-Y. (2008). Evaluating the business value of RFID: Evidence from five case studies. *International Journal of Production Economics*, *112*(2), 601–613. doi:10.1016/j.ijpe.2007.05.009

Uysal, D. D., Emond, J.-P., & Engels, D. W. (2008). Evaluation of RFID performance for a pharmaceutical distribution chain: HF vs. UHF. In (IEEE) *Proceedings of International Conference on RFID: Vol. 1* (pp. 27-34).

Vue, D., Wu, X., & Bai, J. (2008). RFID application framework for pharmaceutical supply chain. In (IEEE) *Proceedings of International Conference on Service Operations and Logistics, and Informatics* (pp. 1125–1130).

KEY TERMS AND DEFINITIONS

Bioassay: Qualitative/quantitative determination of the biological activity of a compound in a sample.

Cell-Based Functional Assay: *In vitro* assay quantifying the biological responses elicited by a ligand through the interaction with specific receptors expressed by cultured cells.

Follicle Stimulating Hormone (FSH): A hormone stimulating the growth of ovarian follicles and the secretion of estrogens in women, and spermatogenesis in men.

High Performance Liquid Chromatography (HPLC): An analytical technique enabling the separation, identification and quantification of substances in a mixture, as well as the detection of degradation products.

Insulin: Peptide hormone able to decrease the blood levels of glucose, by promoting its cellular uptake and storage and inhibiting the liver pathways of sugar release (glycogenolysis and gluconeogenesis).

Pharmaceutical Supply Chain: Network of subjects (manufacturers, wholesalers, retailers), moving drugs from the point of production to consumers.

Receptor: Molecule (often a protein) embedded in the plasma membrane or cytosolic, that recognizes and binds a particular extracellular signaling compound (i.e. hormone, transmitter, drug); such interaction activates a series of intracellular biochemical events, leading to a cell response.

RF Exposure Risks: Potential risks for the structural integrity of objects exposed to the electromagnetic fields generated by RF devices.

Radio Frequency Identification (RFID): Innovative wireless auto-identification technology adopted in "tracing and tracking" systems.

Test Environment: Assembly of laboratory resources (hardware, software) in a controlled environment, to simulate an applicative scenario.

Chapter 12
5.8 GHz Portable Wireless Monitoring System for Sleep Apnea Diagnosis in Wireless Body Sensor Network (WBSN) Using Active RFID and MIMO Technology

Yang Yang
Monash University, Australia

Abdur Rahim
Monash University, Australia

Nemai Chandra Karmakar
Monash University, Australia

ABSTRACT

Sleep apnea is a severe, potentially life-threatening condition that requires immediate medical attention. In this chapter, a novel wireless sleep apnea monitoring system is proposed to avoid uncomfortable sleep in an unfamiliar sleep laboratory in traditional PSG-based wired monitoring systems. In wireless sleep apnea monitoring system, signal propagation paths may be affected by fading because of reflection, diffraction, energy absorption, shadowing by the body, body movement, and the surrounding environment. To combat the fading effect in WBSN, the MIMO technology is introduced in this chapter. In addition, the presented active RFID based system is composed of two main parts. The first is an on-body sensor system; the second is a reader and base station. In order to minimize the physical size of the on-body sensors and to avoid interference with 2.4 GHz wireless applications, the system is designed to operate in the 5.8 GHz ISM band. Each on-body sensor system consists of a physiological signal detection circuit, an analogue-to-digital convertor (ADC), a microcontroller (MCU), a transceiver, a channel selection bandpass filter (BPF), and a narrow band antenna.

DOI: 10.4018/978-1-4666-2080-3.ch012

1. INTRODUCTION

The origin of radio frequency identification technology (RFID) can be traced back to a hundred years ago. In the late 18th and early 19th centuries, researchers started investigating radio waves and their propagation. Inventions such as the identification, friend or foe (IFF) transponder for aircraft identification (UK, 1915) and the espionage tool (Theremin) for retransmitted incident radio waves with radio information (Soviet Union, 1945), can be seen as the predecessors of RFID technology. The breakthrough in RFID technology was made at the end of the 20th century. Based on the dramatic development, RFID technology in the applications of healthcare and wireless patient-monitoring have been widely investigated in many countries over recent years. A number of achievements have been introduced in the following papers: Li et al. (Li, Liu, Chen, Wu, Huang, & Chen, 2004) proposed a RFID-based mobile healthcare service system, and the basic RFID system structure and operational procedure are set out in the article. Sangwan et al. (Sangwan, Qiu, & Jessen, 2005) proposed a real-time RFID system for hospital management of facilities, patient monitoring and service organization. The benefits of the utilization of RFID technology in hospitals have also been analyzed. Park et al. (Park, Seol, & Oh, 2005) introduced the new concept of the combination of RFID, mobile and web technology in healthcare services systems. The proposed system employs mobile devices to extend the connectivity to all ubiquitous environments so that the RFID recognizer can identify the patients who visit the hospital and automatically allocate the '1 on 1' service.

Besides the great potential in healthcare service management, RFID also has its advantages in wireless sensor design. Alippi et al. (Alippi & Vanini, 2005) provide a methodology for reducing power consumption by implementing passive RFID tags. The idea of reducing energy consumption by diminishing the received data instances and data resolution depth provides a possible strategy for developers of modern RFID sensors. A joint-research group from Intel Research, Seattle, USA, and the University of Washington, USA, has made progress in RFID sensing which has accelerated the pace of development of battery- free RFID sensors (Philipose, Smith, Jiang, Mamishev, Roy, & Sundara-Rajan, 2005). Marrocco (Marrocco, 2007) has presented a design for UHF RFID planar antennas for biomedical monitoring. This design considers the issues of passive power supply and the effects of radiation on the human body. Another recent achievement in passive RFID sensing has been achieved by Sample et al. (Sample, Yeager, Powledge, & Smith, 2007). Their wireless identification and sensing platform (WISP) is integrated with sensors of light, temperature and rectified voltage. However, this platform still has limitations for human applications. Occhiuzzi and Marrocco (Occhiuzzi & Marrocco, 2010) have published a paper on their research on wireless monitoring by passive RFID systems. Their investigation of the feasibility of UHF band RFID for wireless monitoring of human body movement provides very useful information for other researchers in this frontier. Their particular design for monitoring leg movement during sleep offers significant ideas for sleep monitoring engineers.

Moreover, from the existing research on wireless sleep apnea monitoring systems, it is known that physiological signals such as Electrocardiography (ECG), Electromyography (EMG), Electroencephalography (EEG), Electrooculography (EOG), etc. can be sent wirelessly to the remote base station only in the Single Input Single Output (SISO) environment under the WBSN scenario. However, a wireless signal is severely sensitive to fading. Particularly in WBSN communications, propagation paths can experience fading due to energy absorption, reflection, diffraction, shadowing by the body, body postures (Yazdandoost & Sayrafian-Pour, 2010), body movement, polarization mismatch and scattering of electromagnetic signals due to the body and the surrounding environment (Khan, 2009). The Multiple Input

Multiple Output (MIMO) system uses multiple antennas at the transmitter and receiver to combat such fading as well as to produce significant capacity and diversity gains over conventional SISO systems using the same bandwidth and power. MIMO systems have recently proven to be an attractive option for WBSN, in which body shadowing and user motion lead to multiple rapid changes in the channel characteristics (Bellon, Cabedo-Fabres, Antonino-Daviu, Ferrando-Bataller, & Penaranda-Foix, 2011). Moreover, the MIMO system enhances the maximum data rate and can serve a large number of users compared to the SISO system. Hence, the application of MIMO in sleep apnea monitoring system will be of great interest for WBSN. Moreover, sleep apnea patients have various levels of mobility, for example, walking, wheel chairing, eating, sitting, twisting, turning which lead to dynamic environment in WBSN. As for a wireless body area network, movements and the low power constraints of sensor nodes can lead to severe losses. The application of MIMO in such environments is an active area of research. To the best of our knowledge, MIMO in dynamic WBSN for sleep apnea monitoring system has not been used yet. Hence, in this chapter, the MIMO technique using space time coding for transmitting physiological data from various sensors mounted on the body to the base station to achieve diversity in WBSN has been proposed.

The objective of this chapter is to introduce an application of RFID in WBSN applications including system design, antenna design and relative issues during system design and development. Moreover, the application of MIMO technology in WBSN has been introduced. Section 2 generally shows the background of RFID application in WBSN. Section 3 portrays the 5.8 GHz portable wireless monitoring system. Section 4 shows the 5.8 GHz antenna tag design and its propagation characteristics in wireless WBSN environment. Section 5 systematically outlines the MIMO application in WBSN with MIMO antenna design for correlation coefficient measurement.

2. BACKGROUND

2.1 Radio Frequency Identification

RFID can be simply defined as radio frequency waves based identification technology employing the process and physical infrastructure under certain predefined protocol definition, where the information can be quickly and reliably transferred from a device (tag) to a reader system (Banks, Hanny, Pachano, & Thompson, 2007). It is free of contact and capable of transmitting simple signals for identifying items, tracking assets, conducting surveillance, controlling access, collecting toll, security checking and many other significant implementations. According to a simple classification (Glover & Bhatt, 2006), these applications can be categorised into five different groups: access control, tag and ship, pallet and carton tracking, track/trace and smart shelf. For different applications, the RFID tags have to be specially designed in terms of its application environment regarding the profile of the target. The RFID reader system is usually located within a short distance of the target when a detecting behaviour is required.

2.2 Wireless Body Sensor Network

Wireless Body sensor network (WBSN) is also called wireless body area network (WBAN), which forms one or more radio frequency (RF) based local communication networks inside the body or around a body area through a series of sophisticated pervasive wireless computing devices (Latre, Braem, Moerman, Blondia, & Demeester, 2011). The wearable sensors combine novel advances of ultra-low-power, compact size, lightweight, high efficiency, low cost and intelligent organization

(Chen, Gonzalez, Vasilakos, Cao, & Leung, 2010). The organization of these sensors is subjected to the requirements, purpose and environment of the application. According to a recent survey (Cao, Leung, Chow, & Chan, 2009), there is an increasing demand in health monitoring and diagnostic analysis via WBSN related technologies, which can be applied for either obesity or other chronic diseases. Generally, a WBSN system consists of sensors design, communication within body area network, communication between WBSN and central node, communication between central node and internet or base station. As is shown in Figure 1, the WBSN is formed by a number of physiological sensors controlled by a central node. The central node wirelessly collects the bio-information from these sensors and then transmits to the base station in the hospital or the patient's doctor.

In traditional BSN, the physiological sensors are connected via cables to the base station. These sensors are accordingly identified by their connections. Therefore, each connected cable on the sensor is deemed as the sign of identity which clearly indicates the signal source to the central base station. Taking the sleep monitoring system polysomnography (PSG) as an example, a typical PSG system requires a minimum of 22 wire attachments to the patient. This integrated system monitors the patient's airflow, heart rate, blood pressure, blood oxygen, brain wave patterns, eye movements and respiratory muscle movements. The key physiological parameters measured by PSG include ECG, EEG, EMG and EOG. The system also deals with cardio-respiratory variables and limb movements measured on the anterior tibialis and penile plethysmography (Kent & Billiard, 2003).

2.3 Investigation on Existing Radio Telemetry Monitoring Technologies in Wireless Body Sensor Network

The previous introduction in section 1 has introduced some RFID based wireless monitoring sensors/networks development for sleep apnea diagnosis. However, to our best of the knowledge, the research on 5.8 GHz wireless monitoring

Figure 1. Architecture of a typical WBSN communication system

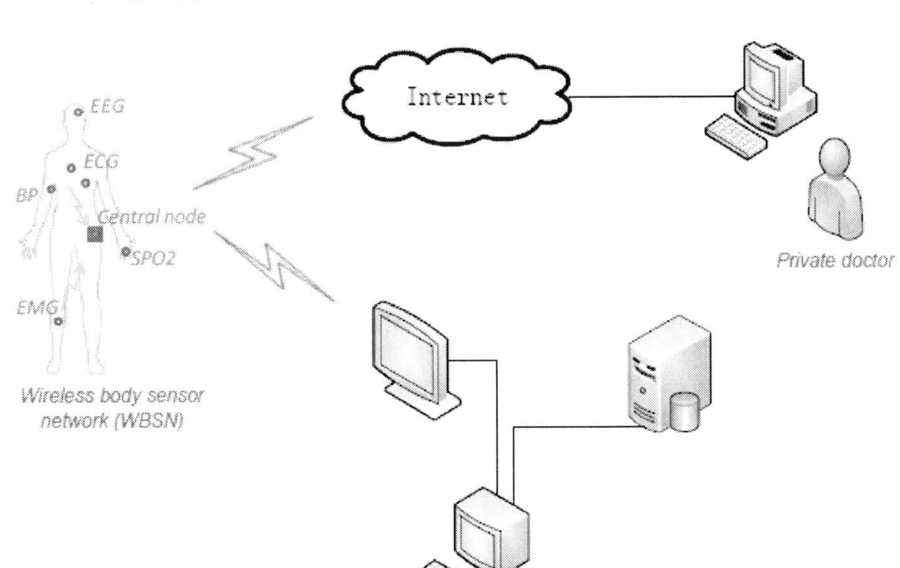

technology for sleep apnea diagnostics is still in the begging stage and there is still a huge gap for the researchers to explore. It is noteworthy that the researches in the similar fields using telemetry monitoring technologies have turned up for a couple of years since 21 century. To have an overview of the existing projects and their progress, a comprehensive study has been given in Table 1.

Table 1 lists the researches around the world investigating the wireless monitoring technologies in biomedical projects. Among these 7 projects, 5 projects occupy the frequency band of 2.4 GHz or lower and only one employs 5.8 GHz band. In terms of different wireless technologies, one project uses Bluetooth module with bulky size and mobile phone assistance (Lam, Wong, Wong, Wong, & Mow, 2009)., one project utilizes ZigBee transceiver relying on mobile phone/PDA assistance (Hu, Wang, Yu, Liu, & Qin, 2008), one applies Wi-Fi technology with PDA (Rodrigues, Estevao, Malaquias, Santos, Gouveia, & Simoes, 2007) and four use RF transceiver modules (Wang, Noel, Fong, Kamoua, & Tang, 2006; Yazicioglu, Merken, Puers, & Van Hoof, 2007; Brown, Grundlehner, van de Molengraft, Penders, & Gyselinckx, 2009). The project using 5.8 GHz transceiver can only be used for body movement and respiration monitoring due to its design limitation of contract-free (Zaffaroni, de Chazal, Heneghan, Boyle, Mppm, & McNicholas, 2009).

All the projects listed in Table 1 have not been fully finished (until October 2010) due to the different design hurdles which have been short-listed in the table. Even though there is one research group exploring the 5.8 GHz wireless monitoring system, their design cannot be fully applied in sleep apnea monitoring. The proposed design in this chapter is able to reduce the size of circuit without assistance from mobile phone or PDA and we are also one of the pioneer groups exploring 5.8 GHz wireless monitoring technology for sleep apnea diagnosis. Currently, we have finished the first prototype for concept approval. The second prototype for implementation is under developing and its design structure and some achieved results have been introduced in section 3, 4 and 5.

2.4 Advantages of Applying 5.8 GHz ISM Band

There are two main advantages of applying 5.8 GHz band during the system design: firstly, 5.8 GHz wireless system can effectively avoid frequency interference from crowded 2.4 GHz ISM band. The 2.4 GHz ISM band has been severely occupied by various wireless devices equipped with Bluetooth, ZigBee, Wi-Fi or 2.4 GHz transceivers. These standard wireless protocols have not been applied in 5.8 GHz band so that the free 5.8 GHz ISM band is an ideal option for most of the wireless system developments for research purpose. Moreover, by using the 5.8 GHz band, the software developer can significantly reduce the complexity of the communication protocol and software program for avoiding interference with other signals. This can reduce power consumption of the microcontrollers and transceivers and therefore extend the operating hours of the portable on-body devices; secondly, the passive circuits (such as antennas and microstrip filters) in 5.8 GHz system are relative smaller compared with the ones in 2.4 GHz system. Since the portable on-body sensor device requires a compact configuration, the 5.8 GHz band is prior to the 2.4 GHz band in the aspect of size reduction. This has been proved in the sections of antenna design (Figure 8a). Even though 5.8 GHz ISM band is relatively free and clean, there still exists the possibility that 5.8 GHz ISM band becomes heavily occupied by various wireless devices in the future. To deal with the interference problem, more complex communication protocols, software programs or digital filters can be applied. This will take more power consumption in signal processing devices (e.g. microcontrollers, transceiver devices) during the computation process. Consequently, the battery life of the portable on-body device will be unavoidable reduced.

Table 1. Researches applying radio telemetry monitoring technologies in WBSN during 2006 – 2010

Author(s)	Wireless Technology	Frequency Band	Physiological Parameters	Organisation	Year	Drawbacks/Progress of the Project
Yazicioglu, et al.	RF transceiver	2.4 GHz	EEG, EOG, EMG	Université Libre de Bruxelles, IMEC (Belgium)	2007	• Need future clinical validation on a population of healthy and non-healthy patients • The system is specially designed for sleep staging; feasibility in sleep apnea monitoring is still in need of investigation • 2.4 GHz only
Wang, et al.	RF trans-ceiver	868 MHz 900 MHz 433 MHz 315 MHz 2.4 GHz	ECG,EMG,EEG	Stony Brook University (USA)	2006	• Need to establish an efficient sensor network for integration of the on-body sensors • Can only work at limited frequency bands (2.4 GHz at most) • The capable sampling rate of MDA300[1] is far below the human body's biopotential frequency • FCC[2] regulation limits the transmission rate
Hu, et al.	ZigBee (CC2430)	2.4 GHz	ECG,EEG, Heart Rate, Breathing Rate, Blood Pres-sure	Guangzhou University (China)	2008	• Project plan only • Still under hardware development • No testing results/data • No system validation • 2.4 GHz only
Lam, et al.	Bluetooth	2.4 GHz	Hear Rate, SPO2, Breath Rate, PPG	HKUST Fok Ying Tung Graduate School (China) and Hong Kong University of Science and Technology (Hong Kong)	2009	• Not power efficient • Limited to mobile cooperated application • No system validation • No clinical testing • 2.4 GHz Bluetooth only • Have not declare feasibility of the application in ECG/EEG detection
Brown, et al.	RF transceiver (Nordic nRF24LOI)	2.4GHz	ECG, Respiration, Skin Conductance and Skin Temperature	Holst Centre, Eindhoven, The Netherlands	2009	• No introduction about user interface • Brief description of the system structure • No system evaluation and validation • No testing results and data analysis • Designed for autonomic nervous system responses • Transmitting at 2.4 GHz only
Rodrigues, et al.	WiFi, UMTS, GPRS	-	Pulse oximetry, video, audio recordings	1.ISA – Remote Management Systems, Coimbra, Portugal; 2.Uni-versity of Coimbra, Portugal; 3.Pediatrics Hospital of Coimbra Avenida Bissaya Barreto	2007	• Project planning only; not too much detail for hardware solution • Need PDA or mobile phone for assistance • Cable is used between bio-sensor and PDA

continued on following page

Table 1. Continued

Author(s)	Wireless Technology	Frequency Band	Physiological Parameters	Organisation	Year	Drawbacks/Progress of the Project
Zaffaroni, et al.	Ultra low-power RF transceiver	5.8 GHz	Movement and respiration of the subject	Biancamed Ltd, University College Dublin and St Vincent's University (Ireland)	2009	• Short sensing range: 0.3 to 1.5 m • Measure actimetry[3] and compare its performance to wrist-actimetry • Can only be used for respiration detection by measuring the actimetry • Cannot be used for biopotential signal acquisition (ECG,EEG,EOG, oxygen saturation)

2.5 Benefits of RFID in WBSN

The advantages of RFID can be summarised as cost reduction, improved efficiency and better visibility of product. These benefits also apply to WBSN based biomedical applications. Besides these advantages, RFID can easily be used to identify different sensor devices in WBSN without collision by setting the unique resonant frequency for each tag and labelling it to its corresponding sensor device. These RFID tags can be easily controlled by reader device for base station to access the on-body sensor unit in WBSN. Therefore, the traditional cable based WBSN system, such as PSG monitoring system for sleep disease, can be replaced by a fully wireless monitoring system which will significantly reduce the uncomfortable feeling from the patient during monitoring process.

2.6 MIMO System in WBSN

The high data rate and reliable transmission between body-worn wireless devices and sensors in patient monitoring systems, sports and entertainment and military applications necessitate the use of multiple antennas for the WBSN channels (Khan & Hall, 2010). MIMO systems have recently proven to be an attractive option for Wireless Body Area Networks, in which body shadowing and user motion lead to multiple rapid changes in the channel characteristics. In this kind of networks, multiple antennas can be used in combination with space-time coding, to save transmit power or to reduce the probability of link failure due to body shadowing (Bellon et al., 2011).To the best of our knowledge, no research exists on MIMO-based sleep apnea monitoring systems. However the application of MIMO in PAN and WBSN has been investigated in other applications which are presented in this section. A PAN is the interconnection of information technology devices within a range of 10 meters (Personal area network, 2011). Figure 2 shows the general configuration of MIMO for single sensor node in sleep apnea monitoring system.

In (Neirynck, Williams, Nix, & Beach, 2006; Johansson, Karedal, Tufvesson, & Molisch, 2005), the MIMO channel was measured for PAN. Neirynck et al. of (Neirynck et al., 2006) concluded that, despite the possible existence of a LOS, it is worth considering MIMO for PAN applications. In addition, the MIMO propagation channel was analyzed at 2.6 GHz in (Johansson et al., 2005) for both LOS and NLOS scenarios. The MIMO system in WBSN has been analysed in (Neirynck, Williams, Nix, & Beach, 2007), where the authors report that, regardless of the Line-of-sight (LOS) operation in PAN, MIMO is shown to offer a considerable increase in the capacity of the system when compared with a conventional SISO system. Furthermore, body shadowing and user motion are shown to lead to multiple rapid changes in the channel characteristics. However, MIMO is able to offer performance enhancement in this case.

Neirynck et al. in (Neirynck et al., 2007) assumed that the high probability of a line-of-sight

Figure 2. General MIMO configurations in WBSN communication system

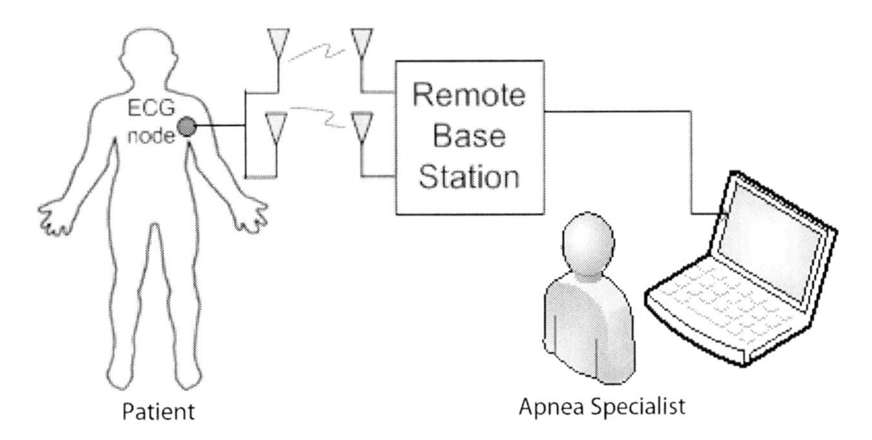

(LOS), and the subsequent correlation between the sub-channels, prevents MIMO from offering throughput benefits in a WBSN environment. The authors also noted that despite the short range, LOS is far less common than expected. Moreover, misalignment of directional antennas, which are necessary to reduce the amount of radiation exposed to the user, means that the LOS component is often not dominant (Neirynck et al., 2007). Dries et al. noted that MIMO can be used to enhance throughput in a WBSN and also concluded that the presence of the multiple antennas in combination with space-time coding will reduce the probability of link failure due to the body shadowing (Neirynck et al., 2007). Moreover, some research studies (Neirynck, Williams, Nix, & Beach, 2005; Sakaguchi, Chua, & Araki, 2005) have pointed out that despite the LOS link, the capacity increase with MIMO is significant in the Rician fading environment.

Khan et al. in (Khan & Hall, 2009) found that the ever-increasing use of wireless devices in personal health care, entertainment, security and personal identification, fashion, and personalized communications drives research to establish more reliable and efficient link between the devices mounted on the body. MIMO is suitable for a very highly fading or dynamic channel in wireless communication. WBSN is also a dynamic environment due to following reasons:

- In most medical conditions, doctors recommend patients to move or walk, as much as they can tolerate, in order to improve their health. Hence Sleep Apnea patient have various levels of mobility, for example, walking, wheel chairing, eating, sitting, twisting etc., thus lead to dynamic environment in WBSN.
- Based on the posture and the positions of the antennas, part of the body may shadow the line-of-sight (LOS) path. Movement of the body changes the orientation of the antennas, and the shadowing conditions (Takada et al, 2008) thus create dynamic scenario. Moreover, several studies (Fort, Desset, De Doncker, Wambacq, & Biesen, 2006; Smith, Hanlen, Zhang, Miniutti, Rodda, & Gilbert, 2009; Smith, Hanlen, Miniutti, Zhang, Rodda, & Gilbert, 2008) indicated that body movement creates a dynamic WBSN channel.
- WBSN nodes experience different channel condition, which can vary dynamically on a time-scale within the same order of magnitude of the data transmission time

(Shi, Medard, & Lucani, 2010). Moreover, in case of WBSN communication, propagation paths can be affected by fading due to energy absorption, reflection, diffraction, shadowing by the body, body postures, and the surrounding environment (Yazdandoost et al, 2010; Khan, 2009). Hence all these factors contribute to make a dynamic WBSN system.

Based on the literature review discussed above, more research is required on wireless sleep apnea monitoring systems using MIMO due to dynamic environment of WBSN. The application of MIMO technology in sleep apnea monitoring system will provide a number of benefits including high spectral efficiency, high throughput, reliability and diversity gain. Hence, it will offer substantial improvement to the existing systems proposed in (Fensli, Gunnarson, & Gundersen, 2005; Tafa & Stojanovic, 2006; Lo, Thiemjarus, King, & Yang, 2011; Shnayder, Lorincz, Fulford-Jones, & Welsh, 2005; Baker, Bones, & Lim, 2006). In particular, for apnea monitoring systems in WBSN, MIMO technology will enable the successful transmission of physiological signals by overcoming several challenges such as reflection, diffraction, body movement etc. Diversity is an effective way to overcome fading effect in WBSN. Spatial diversity is achieved by using MIMO technique (Khan, 2009).This technology provides an enhanced performance by using the spatial diversity gain of the signals in a fading environment. Therefore, it is worth undertaking substantial research on MIMO in WBSN systems for wireless transmission of physiological signals to the remote base station.

For diversity and MIMO applications, the set of correlations, between the signals received by the target antennas on the same side of the wireless link, is an important figure of merit. Usually, the envelope correlation coefficients are presented to evaluate some of the diversity capabilities of a multi-antenna system. Moreover, in order to investigate the performance of MIMO in WBAN, some parameters need to be considered. The envelope correlation between antenna elements is one of the most important parameter because it is related to the spectral efficiency, diversity gain and capacity (Votis, Tatsis, & Kostarakis, 2010) of the MIMO system. For example, the capacity of the MIMO system may be reduced if the received signals by the antenna elements in the system are highly correlated. The performance of the MIMO is superior when correlation coefficient approaches to zero (Votis et al., 2010). Therefore the correlation coefficient represents a performance benchmark of WBSN system.

3. A PORTABLE WIRELESS MONITORING SYSTEM FOR SLEEP APNEA DIAGNOSIS

3.1 System Architecture

In this section, an active RFID based 5.8 GHz wireless monitoring system is introduced. Figure 3 shows the diagram of the proposed portable wireless monitoring system for sleep apnea diagnosis. It is clear that the system is composed of two main parts: (i) the on-body sensors and (ii) the reader system and base station. The on-body sensors include wireless electrocardiogram (ECG), electroencephalogram (EEG), electrooculogram (EOG), electromyography (EMG), air flow and oxygen saturation (SpO_2) transducers operating in the 5.8 GHz ISM band. Each wireless on-body sensor occupies a unique channel controlled by the transceiver and selected by the specially designed high selectivity BPF. During the monitoring process, these physiological parameters are continuously transmitted to the reader system which is located near the patient. The reader system consists of six channels; each of them corresponds to a wireless on-body sensor. The channel selection, filtering, tag identification and signal integrity are fulfilled in this part. After these processes, the base band

Figure 3. Diagram of the portable wireless monitoring system for sleep apnea diagnosis based on active RFID technology

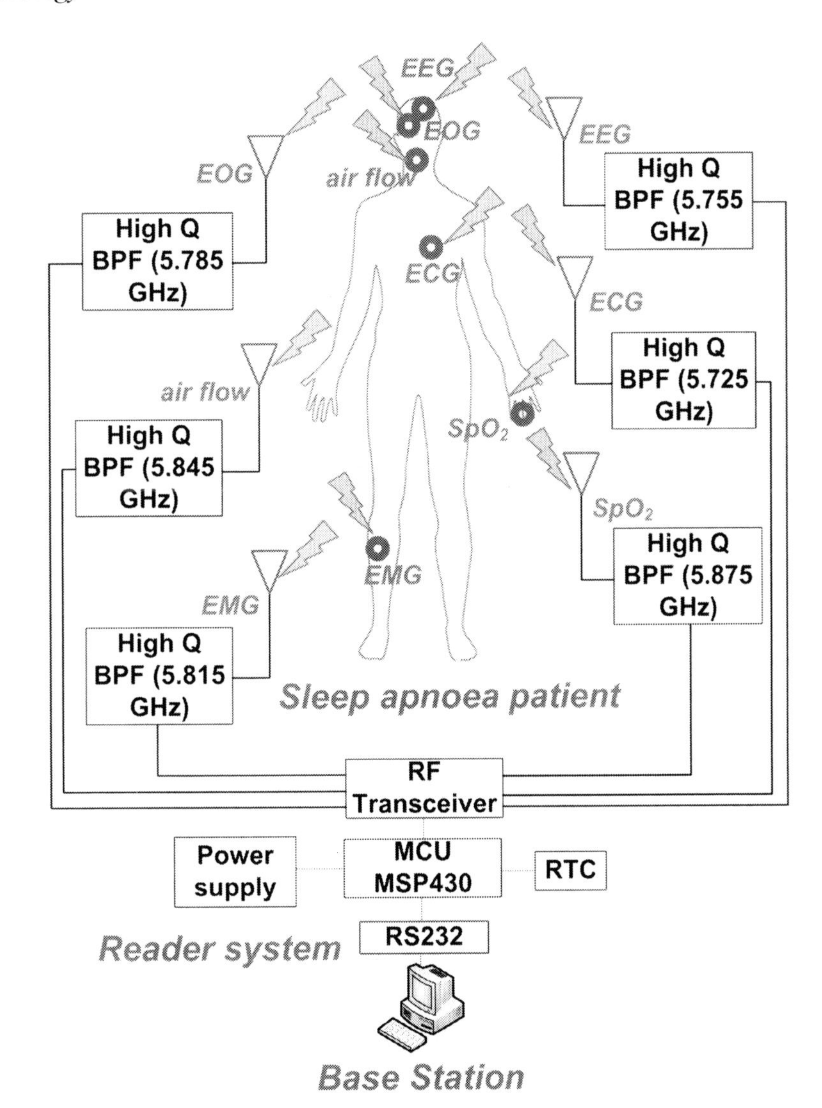

physiological parameters are synchronized and sent to the base station via RS232.

In this system, the physical layer contains two principal parts: wireless on-body sensor system and the reader system. Based on the physical layer, the software focuses on two tasks: (i) RFID response signals transmission and (ii) physiological signals transmission. These two signals are included in each single packet for the communication between the reader and on-body sensors system.

3.2 RFID Tag System

The proposed tag system is integrated with on-body physiological sensors. In this paper, the ECG based design is taken as an example to introduce the proposed system design. It can be seen from Figure 4 that the on-body sensor system consists of an ECG detection circuit, an analogue-to-digital converter (ADC), a microcontroller (MCU), a transceiver, a passive high selectivity BPF and a narrow band antenna. The tag system mainly deals

Figure 4. Block diagram of the proposed portable on-body sensor/tag system for ECG detection

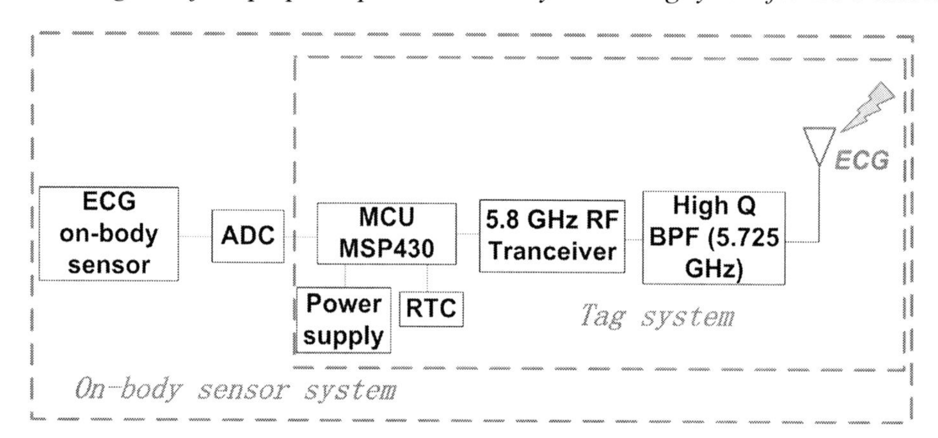

with RF signal identification. Integrated with the ECG sensor and the ADC, this portable wireless on-body ECG sensor can deal with active RFID and ECG signal transmission at the same time. The MCU used in this project is MSP430FR572 produced by Texas Instruments due to the merits of ultra low-power consumption, low profile and sufficient build-in memory (Texas Instruments, 2011). The transceiver can convert the base band signal to the 5.8 GHz ISM band. The passive high selectivity BPF is used for channel selection purpose. The detailed design information is enumerated in (Yang, Roy, Karmakar, & Zhu, 2011).

3.3 On-Body Sensor System

The proposed portable wireless on-body sensor embodies the merits of compact, light weight, low-power and reliable. It contains four interleaved functional blocks as shown in Figure 5: (i) microstrip circular patch antenna, (ii) passive high selectivity BPF, (iii) transceiver, MCU and ECG detection and (iv) power management layer. The on-body unit is in cylinder shape with the bottom circle of 24 mm in diameter and height of 15 mm. Since the development of the on-body sensor system is not fully completed, this section will only present system architecture and the achievements up to date. Following is the description of each layer of this sensor system.

Layer 1: 5.8 GHz Circular Patch Antenna

The 5.8 GHz antenna is etched on the top layer (layer 1) of the proposed prototype. The antenna is shaped into circular patch to save space and its round shape can give an equal radiation pattern at its facing side. The advantages of applying this antenna tag can be summarized as: (i) low profile, (ii) low cost and (iii) symmetrical radiation distribution at its facing side.

Layer 2: Passive High Selectivity BPF

The second layer contains a passive circularly shaped high selectivity BPF operating in the 5.8 GHz ISM band. It is connected to the top layer antenna via 50 ohm coaxial cable. This filter is a double-layer printed microstrip filter. It is composed of a ring resonator and two types of defected ground-plane structure (DGS) slots. The ring resonator is used to generate a narrow passband while the two DGSs are used to increase the selectivity for channel selection and suppress the undesired high order harmonics. In the previous work (Yang et al., 2011), the filter was designed for a 2.4 GHz system. By reducing the size of the ring resonator and accordingly tuning the dimensions of the two DGSs, the required 5.8 GHz passive high selectivity BPF can be easily realized.

Figure 5. Configurations of the proposed wireless on-body sensor

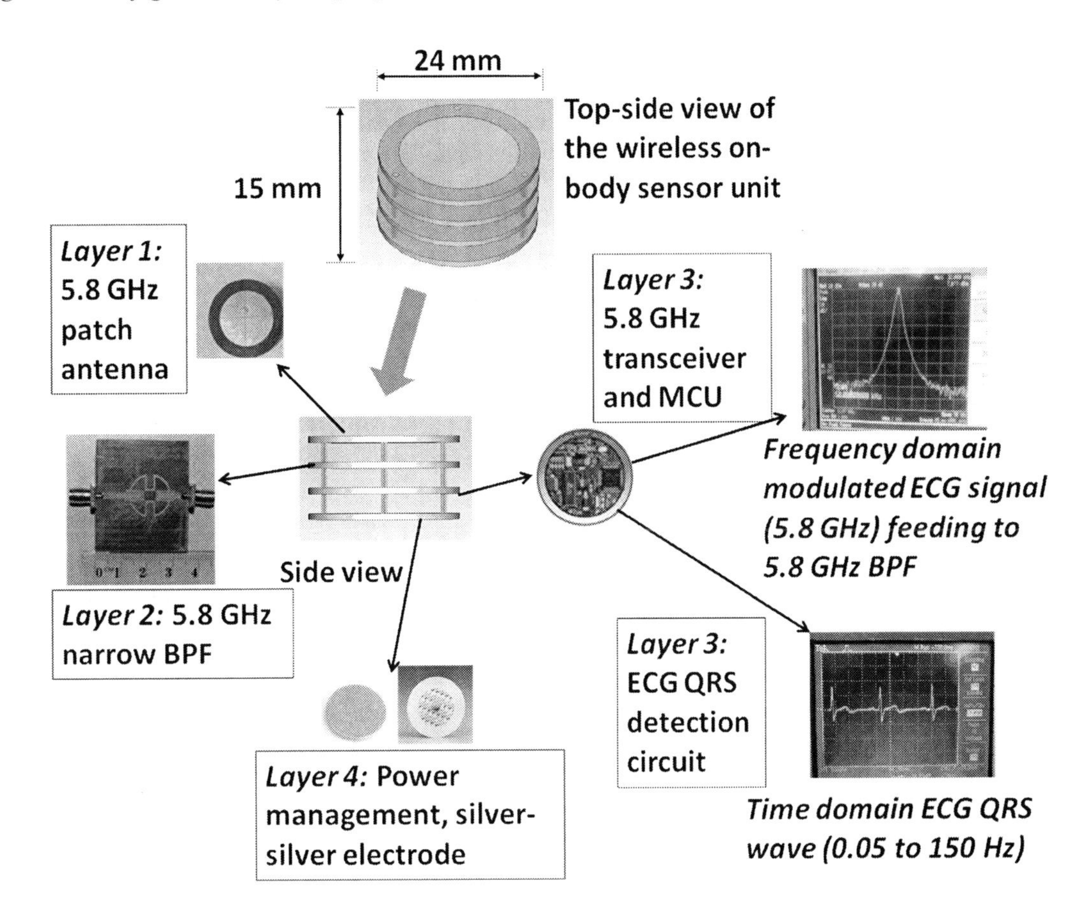

Layer 3: Integrated ECG, MCU and Transceiver Circuit

Layer 3 mainly contains a bio-medical signal acquisition circuit (for example, an ECG circuit), a MCU and a transceiver. This layer is highly integrated with active components. For the purpose of saving battery life, the operational amplifiers, MCU (with build-in ADC) and transceiver (with build-in power amplifier (PA)) are all chosen to be of low power consumption. By using the build-in ADC and PA, the circuit can be successfully integrated within the circular printed circuit board with a diameter of 24 mm.

Layer 4: Electrode and Power Management

The bottom layer contains two cell batteries and voltage regulator which can offer 3 V power supply for the active components in layer 3. The silver to silver electrode is also included in the bottom of this layer for ECG detection. In transmitter mode, the overall current consumption of the on-body sensor is about 80 mA which means that the wireless on-body sensor system can continuously operate for 3.75 hrs by using 2 of 150 mAh cell batteries. This allows enough physiological information from different sleep stages of the monitored patient to be collected.

3.4 Active RFID Reader System

As described previously, the RFID reader system contains six channels. For each channel, the passive high selectivity BPF, transceivers and the MCU are utilized. To differentiate different channels, the uniquely designed BPFs are applied from 5.725 GHz to 5.875 GHz. The frequency span between every adjacent BPF centre frequencies is 30 MHz, resulting in an extremely stringent requirement for the filter performance. The authors' recent design achieves the similar goals, meaning that it can be applied in present research (Yang et al., 2011).

3.5 Flowchart of Designed System

The flowchart of the proposed system is shown in Figure 6. The base station controls the reader system directly via RS232. The wireless on-body sensor system is activated when the reader system starts recording. Once the recording is stopped, the on-body sensor system needs to be turned off or switched into sleep mode to minimize the power consumption. On the other hand, the base station activates the reader system for commencing recording the physiological parameter from the identified channel; both the on-body sensor and reader system need to be initialized. For the reader system, after receiving a waking-up command, the MCU classifies the received packet and split it into command and data sections, and then format a new packet integrated with a unique identity and send it to the on-body sensor system. From the on-body sensor system side, the sleep mode is interrupted once a waking-up command is received from the reader system. Once the received signal is identified, the new packets combining both unique ID and physiological data are formatted. These packets are continuously sent to the reader side until a mismatched ID packet signal is received from the reader system.

Figure 6. Flow chart of the on-body sensor system (Yang, Karmakar, & Zhu, 2011)

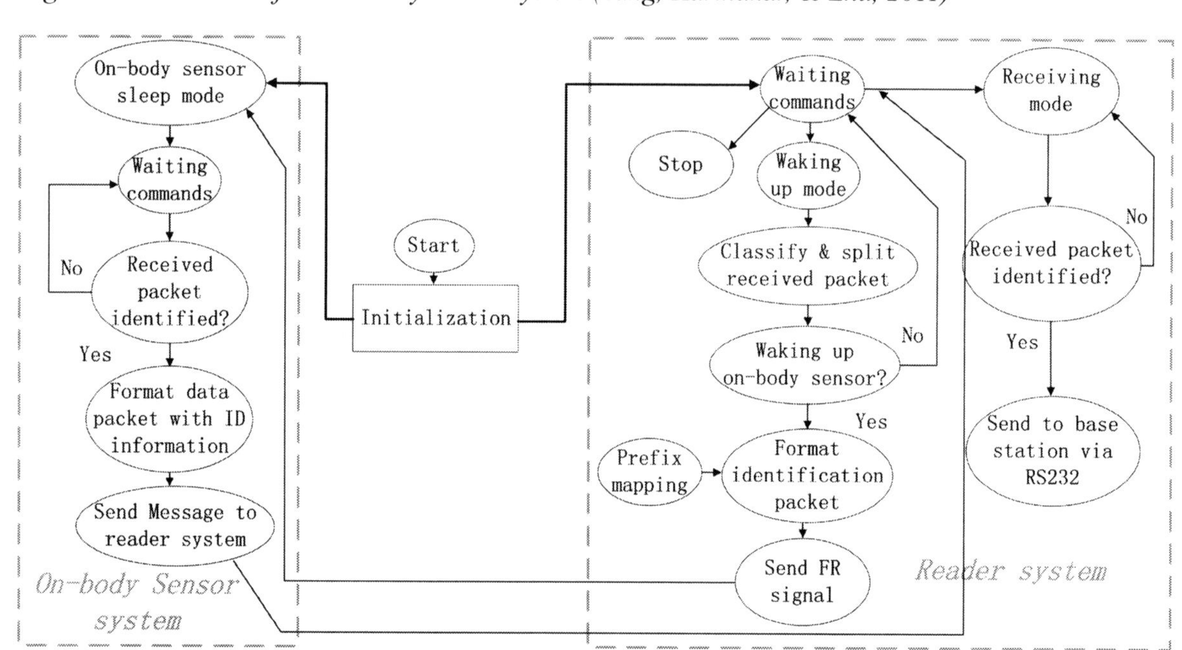

4. 5.8 GHZ CIRCULAR PATCH ANTENNA TAG AND PROPAGATION CHARACTERISTICS OF 5.8 GHZ RADIO WAVE IN WIRELESS BODY SENSOR NETWORK ENVIRONMENT

4.1 Design of 5.8 GHz Circular Patch Antenna Tag

Antenna design in wireless communication system relies on the required system specifications and sometimes is subjective to the implementation environment. In the application of sleep apnea monitoring, the required transmitting antenna should satisfy the following three specifications: a. the transmitting antenna has an omnidirectionally and evenly distributed radiation pattern. This can ensure an effective wireless communication no matter the patient sleeps on his side or back; b. the designed antenna should comply with the shape of the on-body device, be space-saving and ensure the convenience for the monitored patient during the sleep period; c. high front to back ratio of the radiation pattern is required in this project since the antenna is designed sitting on the top layer of the on-body sensor device with its back towards the human body. The antenna characteristics of high front to back ratio will significantly reduce the unnecessary power radiated to the human body. Among various antenna types, circular patch antenna as one of monopole antenna is the best candidate satisfying all the three points mentioned above. The circular shape used in this design is to be conformal to the geometric shape of the on-body sensor (4-layers cylinder shape). Compared with dipole, horn or other conventional antennas, circular patch antenna has an evenly and omnidirectionally distributed radiation pattern. Dipole antennas, no matter in half-wave, folded or other formats, will not be able to fulfill the required specifications mentioned above.

In modern wireless communication systems, there are various circular patch antennas designed for different applications. A broad band circular patch antenna was reported in (Garima, Bhatnagar, Saxena, & Saini, 2009) for space communication purpose. By using the diamond shaped slot, its bandwidth was extended to 1.12 GHz. This design is not applicable for point-to-point wireless communication systems which require narrow bandwidth performance. In (Abd-Elrazzak & Al-Nomay, 2004), the circular patch antenna was designed for 2.4 GHz Bluetooth and 5.2 GHz high performance local area network (HIPERLAN). This requires dual-band performance for the antenna so that this design is not applicable for point-to-point wireless communication systems. In this wireless monitoring system for sleep apnea diagnosis, the conventional circular patch antenna is adopted due to its advantages in shape and performances.

4.1.1 Design Procedure

The design of a circular patch antenna tag is based on the requirements of the application such as resonant frequency, operational range, bandwidth, impedance matching and radiation pattern. For research purpose, we selected 5.8 GHz as the resonant frequency, which is recognized as in the ISM band by the Australian regulatory body. Regarding the WBSN application, the operational range of the transmitting and receiving antenna pairs is about 0.5 to 5 meters. This is also affected by the antenna gain and signal power to be transmitted. The bandwidth of the signal is about 200 Hz, which is sufficient to be covered by the coaxial probe feeding method. Impedance matching is important here to ensure the efficiency of power transmission. The 50 ohm SMA connector is proposed to be mounted at the feeding port on the patch. By using a feeding line with a 50 ohm SMA connector, the impedance input can be perfectly matched. The radiation pattern is also important in this design because the on-body wireless device might not be able to face the receiving antenna directly due to possible movement of the human body during monitoring process. Therefore, a good antenna should have an acceptable radiation pattern. The antenna design procedure is set out in Figure 7.

4.1.2 Prototype of 5.8 GHz Circular Patch Antenna Tag

Given the dielectric constant, substrate thickness and specified resonant frequency, the effective antenna radius can be calculated by (Balanis, 2005):

$$a = \frac{\mathrm{F}}{\left\{1 + \frac{2h}{\pi \mathrm{F} \varepsilon_r}[\ln\left(\frac{\pi \mathrm{F}}{2h}\right) + 1.7726]\right\}^{1/2}} \tag{1}$$

where

$$\mathrm{F} = \frac{8.791 \times 10^9}{f_r \sqrt{\varepsilon_r}} \tag{2}$$

In Equations 1 and 2, h (in cm) is the height of the substrate, ε_r is the dielectric constant of the substrate, fr is the resonant frequency and a (in cm) is the designed radius of the proposed circular patch antenna.

Figure 8 shows the prototype of the proposed microstrip circular patch antenna tag. It can be seen that 5.8 GHz circular patch antenna saves 83.4% area compared with the 2.4 GHz model. As mentioned in previous section, the 5.8 GHz antenna satisfies the system specification of compact size and therefore it is an ideal candidate for this application. In (b) of Figure 8, a is the radius of the top patch, b is the radius of the substrate and ρ is the distance between the feed point to the centre of the top patch. The bottom ground patch has the same area as the substrate. The heights of the top and bottom patches are the same represented by t. Also, h is the height of the substrate. t and h are determined by the materials used for antenna fabrication. In this design, the adopted substrate is Taconic TLX-0 with a dielectric permittivity of 2.45. The proposed dimensional parameters of the 5.8 GHz circular patch antenna tag are shown in Table 2.

4.1.3 *Antenna Gain and Radiation Pattern*

Antenna gain can be evaluated by Friis' transmission equation (Balanis, 2005):

$$\frac{P_r}{P_t} = e_{cdt} e_{cdr} \left(1 - |\Gamma_t|^2\right)\left(1 - |\Gamma_r|^2\right) \frac{\lambda^2 D_t(\theta_t, \phi_t) D_r(\theta_r, \phi_r)}{(4\pi R)^2} \left|\rho_t \cdot \rho_r\right|^2 \tag{3}$$

where e_t is the radiation efficiency and $D_t(\theta_t, \phi_t)$ is the directivity of the transmitting antenna, e_r is the radiation efficiency and $D_r(\theta_r, \phi_r)$ is the directivity of the receiving antenna, P_r is the received power from the receiving tag, P_t is the transmitted power from the transmitting tag, R is the distance between the two antenna tags and λ is the signal wavelength in free space. Assuming the reflection efficiencies are unity for the transmitting and receiving antennas and the polarization loss factor and polarization efficiency are unity at the receiving antenna, Equation 3 can be represented as:

$$\frac{P_r}{P_t} = e_{cdt} e_{cdr} \left(1 - |\Gamma_t|^2\right)\left(1 - |\Gamma_r|^2\right) \frac{\lambda^2 D_t(\theta_t, \phi_t) D_r(\theta_r, \phi_r)}{(4\pi R)^2} \left|\rho_t \cdot \rho_r\right|^2 \tag{4}$$

where e_{cdt} and e_{cdr} radiation efficiency for transmitting antenna and receiving antenna respectively, Γ is voltage reflection coefficient at the input terminals of the antenna (Γ_t is for transmitting antenna and Γ_r is for receiving antenna) and

Figure 7. Design procedure of RFID antenna tag

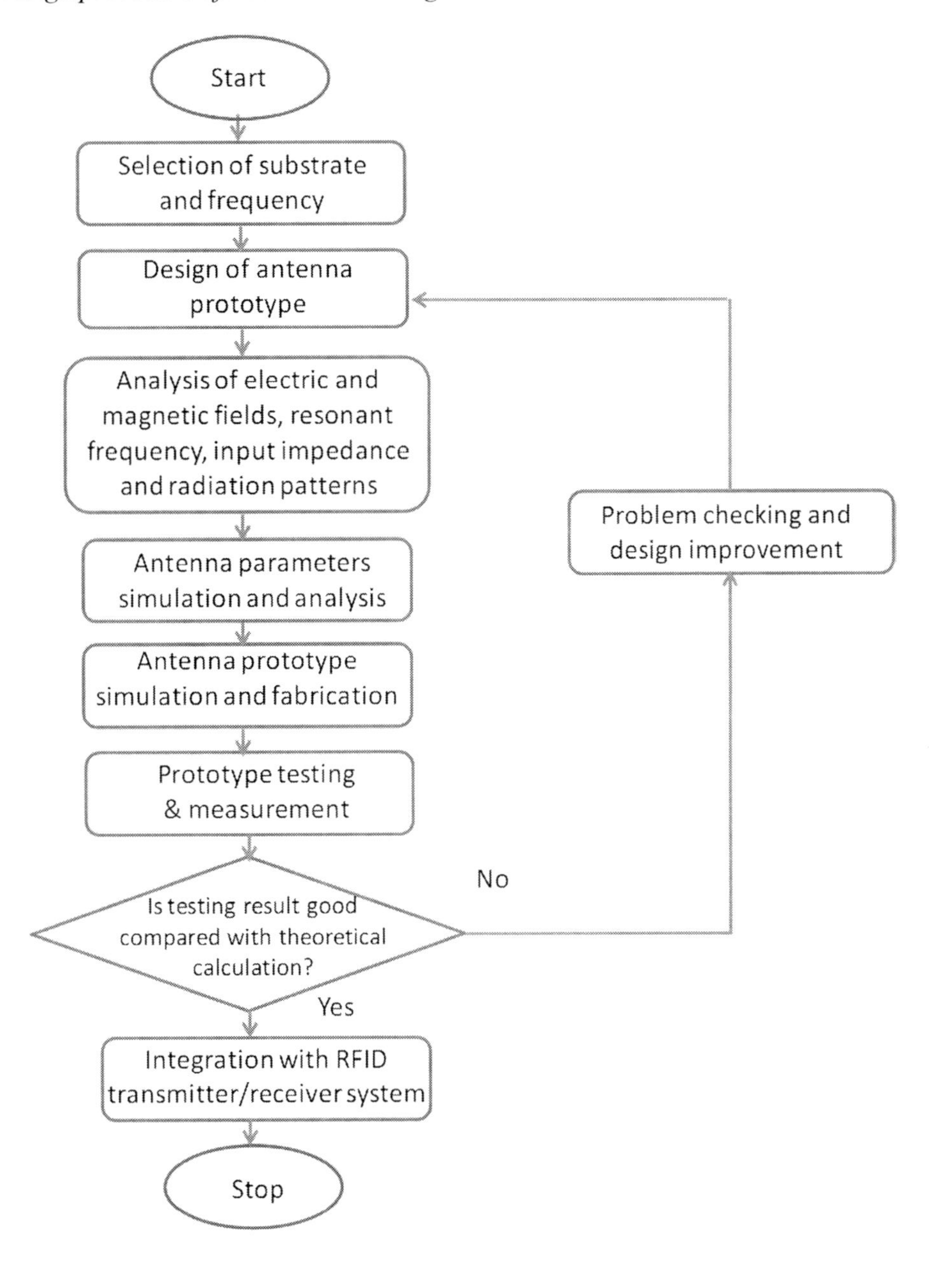

$\hat{\rho}$ is the unit vector. For polarization and reflection-matched antennas, maximum directional radiation and reception can be applied, which means Equation 4 can be simplified to:

$$\frac{P_r}{P_t} = \frac{\lambda^2}{(4\pi R)^2} G_{0t} G_{0r} \tag{5}$$

where G_{0t} and G_{0r} are gains for transmitting and receiving antennas. Due to the fact that two similar microstrip circular patch antennas were used in the testing, the gains for them were the same. So we have:

$$G = G_{0t} = G_{0r} \tag{6}$$

Table 2. Dimensional parameters of 5.8 GHz circular patch antenna tag (mm)

a	b	ρ	h	t
9.232	13.5	3	0.7874	0.0017

Therefore, Equation 5 can be written as:

$$\frac{P_r}{P_t} = \frac{\lambda^2}{(4\pi R)^2} \times (2G) \quad (7)$$

By simplifying Equation 7, the antenna gain can be expressed as:

$$G = \frac{(4\pi R)^2 P_r}{2\lambda^2 P_t} \quad (8)$$

By applying logarithm to both sides of Equation 8, antenna gain can be finally reduced to:

$$G = 0.5(32.45 + 20\log f + 20\log R - P_t + P_r) \quad (9)$$

where f is the in the unit of GHz, R is the distance between the antenna pair in meter and P_t and P_r are in unit of dBm. During the measurement, the twin antennas were placed at the same height with a centre distance of 36 cm.

Figure 9 shows the simulated and measured return loss and antenna gain of the 5.8 GHz circular patch antenna. It can be seen from the graph that the simulated return loss at 5.84 GHz is 20.5 dB by using CST and the measured value is 22.5 dB at 5.77 GHz by using Agilent Performance Network Analyser (PNA). The measured return loss illustrates that the fabricated antenna resonates at 5.79 GHz which shifts 50 MHz lower from the simulated result. The difference between the simulated and measured results is due to the limitation of fabrication and soldering facilities at Monash University (the difference will be eliminated by utilizing a more precise fabrication and soldering machine). The antenna gain in the span from 5.7 GHz to 5.9 GHz gives a strong performance which retains in the range from 7.5 to 9.5 dBi. Figure 10 depicts the simulated and measured radiation pattern in H and E planes respectively. The 3 dB beam-width is about 60° and front to back ratio is 14 dB.

Figure 8. Prototype of microstrip circular patch antenna tag: (a) antenna tag prototypes of 2.4 GHz and 5.8 GHz and (b) dimensions of the proposed prototype

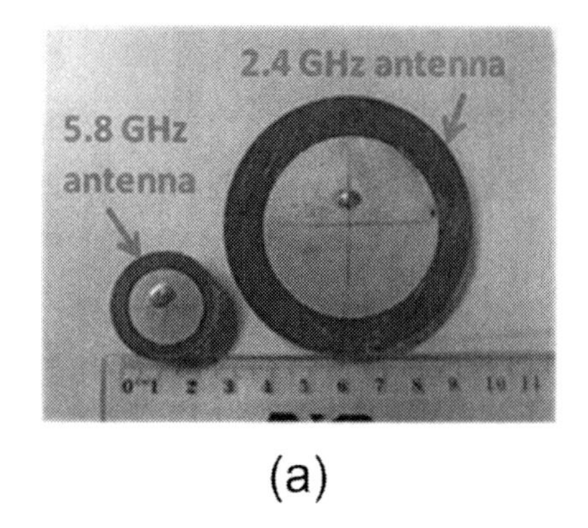

(a)

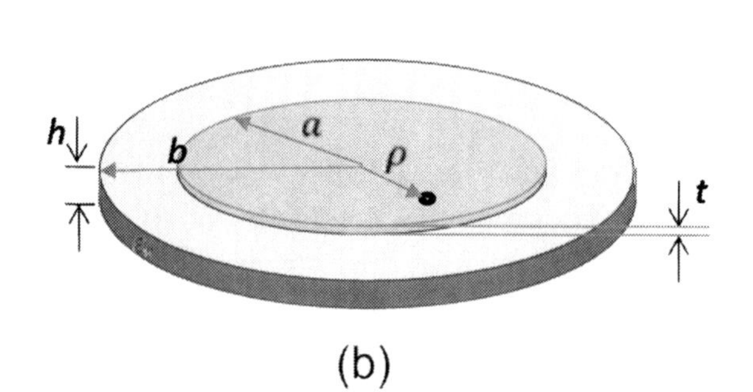

(b)

Figure 9. 5.8 GHz antenna gain and return loss

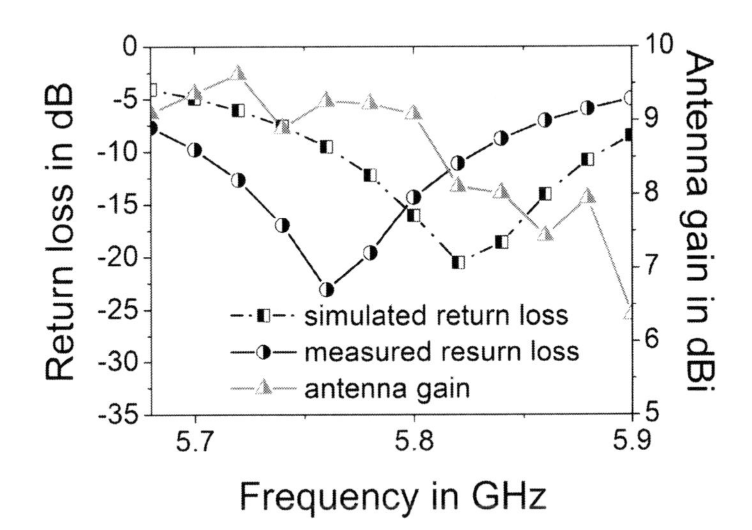

Figure 10. Simulated and measured radiation pattern of the 5.8 GHz antenna tag in (a) H-plane and (b) E-plane

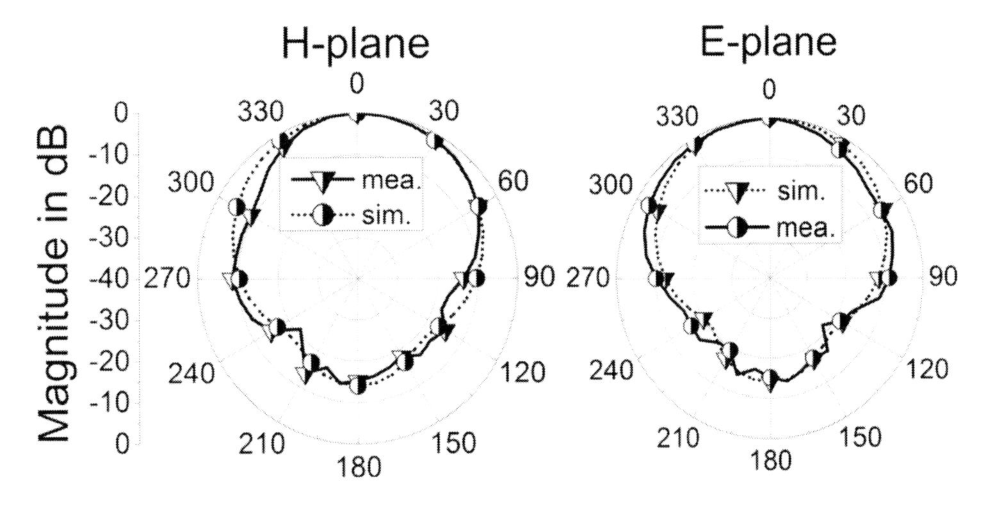

4.2 Propagation Characteristics of 5.8 GHz Radio Wave in WBSN Environment

4.2.1 Introduction

2.4 GHz is one of the most popular radio frequency bands being used in current wireless transmission systems. However, 2.4 GHz channel is over-crowded by technologies such as Bluetooth, Wi-Fi and ZigBee. The 5.8 GHz band is still relatively free for wireless communications compared to 2.4 GHz. By using 5.8 GHz band, the transmission system can properly avoid interference with the wireless devices using Bluetooth, Wi-Fi and ZigBee. Another significant advantage of using 5.8 GHz band in our wireless monitoring of sleep apnea project is that the sizes of antennas will be significantly small. It will augment our research goal of designing compact wireless transducers for various physiological parameters. According our experimental trials, the performance of the 5.8 GHz antenna is even better than the 2.4 GHz one. However, the high frequency signal has short wavelength which can lead to lossy performance and weak propagation compared to the low frequency signal. To prove the feasibility of the 5.8 GHz solution in our project, a series of trials regarding the 5.8 GHz signal propagation are presented in the following sections.

4.2.2 5.8 GHz Signal Propagation Trials

In the real sleep monitoring environment, the patient's body will not block the wireless signal propagation since the receiving antennas will be placed on the ceiling and side wall of the monitoring room to make sure that the transmitting antenna faces the receiving antennas no matter the patient sleeps on his side or back. Some strange sleep positions, such as sleep on stomach or sleep scrunched up, are strictly not allowed during the monitoring period. Moreover, circular patch antenna is utilized as the transmitting antenna with its back sitting on the top layer of the on-body sensor and top plane facing to the ceiling or side wall of the monitoring room. The circular patch antenna has the characteristic of the high front to back ratio which means most of the radiation is in the facing side of the circular patch. Therefore, the impact on signal propagation from skin tissue of human body can be neglected.

Considering the aspect mentioned above, the antenna trials are presented with experimental data and scenario diagrams regarding the antenna performance, antenna return loss, gain and radiation pattern are measured in five different environments (refer to Figure 11): 1) free space; 2) with a cloth covered on the transmitting antenna; 3) with a cardboard and a paper covered on the transmitting antenna; 4) with a cloth, a cardboard and a paper covered on the transmitting antenna; 5) a monitoring objective standing right behind the transmitting antenna with his cloth fully covered on the antenna. Based on the five different testing environments, the measurements of antenna gain and radiation pattern are introduced in the following sections.

4.2.3 Free Space Propagation

Antenna gain and radiation pattern are measured manually in Radio Microwave Laboratory (Monash University). In Figure 12, two identical antennas are fixed at the top of two supporting bars facing each other (rotating start point of 0 degree). The transmitting antenna is connected to the Agilent signal generator E8257D via a coaxial cable. The receiving antenna is connected to the Agilent spectral analyzer E4408B via a

Figure 11. Diagram of the 5.8 GHz signal propagation trials

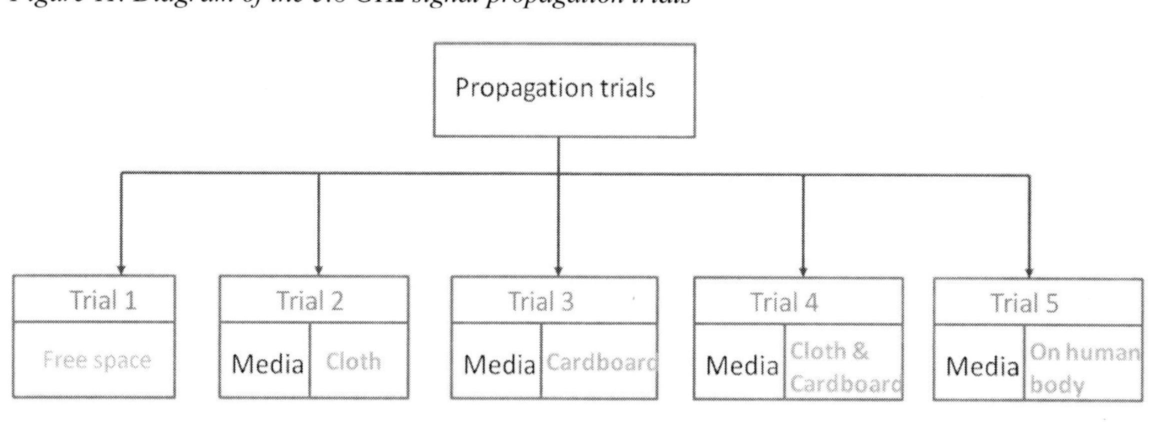

Figure 12. Propagation measurement of 5.8 GHz circular antenna tag in free space

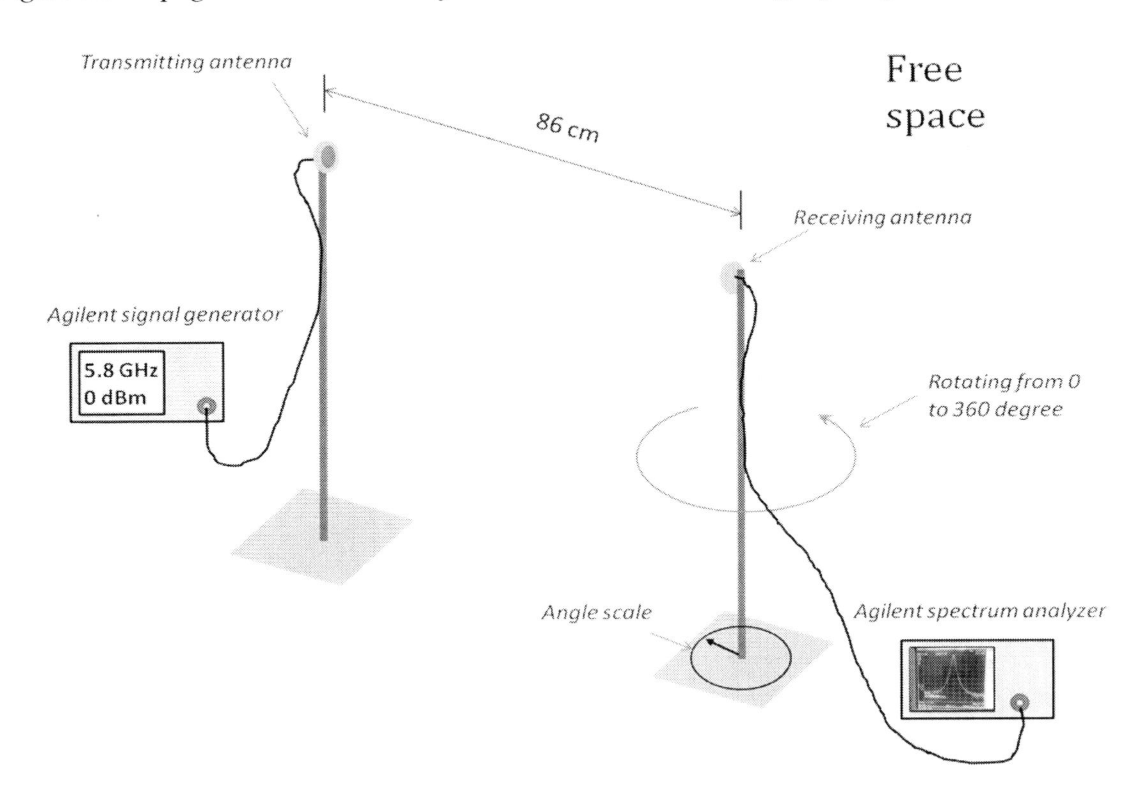

long coaxial cable for easy rotation. The distance between the antenna pair is 86 cm. The angular scale is placed at the base of the supporting bar in receiving side. During the measurement of antenna radiation pattern, the receiving supporting bar with antenna is rotated from 0 to 360 degree with step of 10 degree. The received power is displayed in spectrum analyzer.

During the measurement, the signal generator is sending a 0 dBm signal at the antenna resonant frequency of 5.76 GHz via a coaxial cable to the transmitting antenna facing right towards the receiving antenna. The transmitting antenna is fixed to the supporting bar with the same height as the receiving antenna. In the receiving antenna side, the received 5.76 GHz microwave signal is sent to the signal analyser via another long coaxial cable. By setting the centre frequency of 5.76 GHz with frequency span of 500 MHz, the received the signal is displayed properly in spectrum analyser E4408B. The testing condition and parameter setting are shown in Table 3.

4.2.4 Propagation Test Configuration

The 5.8 GHz circular patch antenna is designed for sleep apnea monitoring. During the monitoring of the patient's bio-parameters, the antenna is usually covered by the patient's pajama and quilt. There are also chances for some obstacles

Table 3. Parameters setting of 5.8 GHz signal propagation

Settings/Conditions	Value
Transmitting power (Tx)	0 dBm
Propagation signal frequency	5.76 GHz
Cable loss in Tx line	1.97 dB
Cable loss in Rx line	3.33 dB
Distance between Tx and Rx antennas	0.86 m

to be occasionally placed between the transmitting and receiving antennas. These obstacles usually are made of nonconductive materials which do not impact the electromagnetic field surrounding the antenna pair. Therefore, high penetrating-capability is required for antenna propagations.

Figure 13 shows the propagation measurements in four different conditions. The setting of the measurements follows the values illustrated in Table 3. Table 4 lists the dielectric properties of skin tissue of human body and the materials used in the trails. The experimental results are presented in next section.

4.2.5 Measurements

Return Loss

As is shown in Figure 14, the five different S_{11} curves are very close to each other. In free space, the 5.8 GHz circular patch antenna has the return loss of 19.6 dB at 5.74 GHz with a bandwidth (S_{11} < 10 dB) of 130 MHz (2.3%). For the case of cloth covered measurement, the resonant frequency is shifted about 300 MHz lower (at 5.71 GHz) with a higher return loss of 23 dB. On the contrary, the measurement with a cardboard covered gives a higher resonant frequency of 5.76 GHz with a re-

Figure 13. Propagation measurements in four different conditions: (a) media of cloth (b) media of cardboard (c) media of both cloth and cardboard (d) on human body (patient)

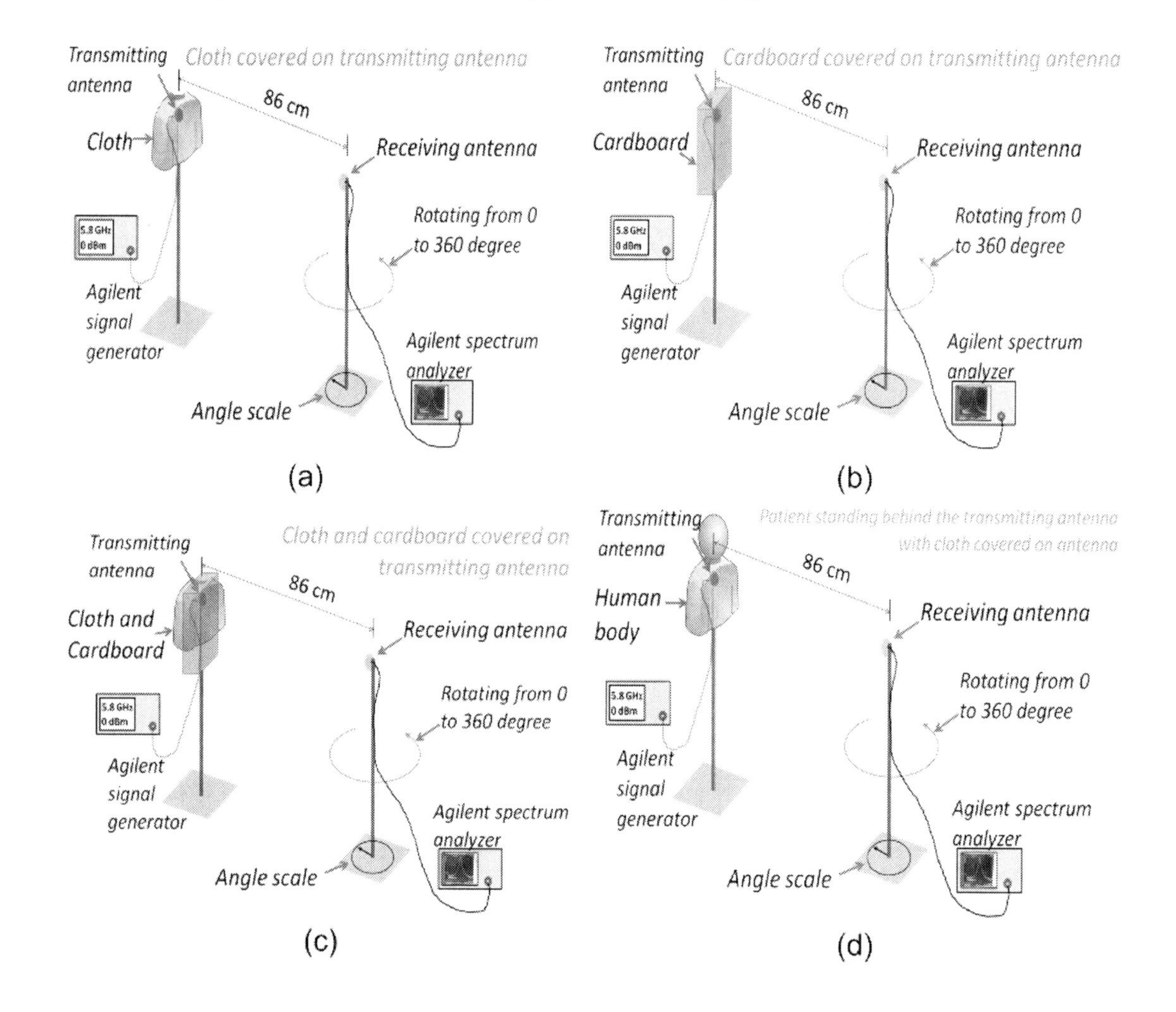

Table 4. Dielectric properties of skin tissue and the materials used in the trails

Dielectric Constant of Skin Tissue	Cardboard		Cloth (100% cotton)	
	Dielectric constant ε_r	Loss tanδ	Dielectric constant ε_r	Loss tanδ
30 – 40	1.66 – 2.97	0.01	4-6	0.057

turn loss of 22 dB. The case of cloth and cardboard covered is similar as the case of measurement with patient except that the latter case shows a wider bandwidth of 130 MHz which is approximately 10 MHz wider than the former one. Both of the two resonate at 5.73 GHz.

According to the experimental results, the 5.8 GHz circular patch antenna is capable of retaining a good performance during the wireless sleep apnea monitoring environment even in case of antenna surface touched to the clothes and cardboard. Compared with the free space environment, the return loss in the other four cases is even higher than in free space. The covered nonconductive materials might make the antenna resonant frequency shifted a few mega-hertz (less than 30 MHz). Due to the wide bandwidth of the antenna (130 MHz), this will not change the antenna performance.

Radiation Pattern

Figure 15 shows the normalized H-plane radiation pattern in terms of five different environments. According to the graph, the 3 dB beam width is about 60 degree with the front to back ratio of 14 dB. The measurement is conducted manually with a sweeping degree of 10 in Radio Microwave Laboratory (Monash University). The extremely high sensitivity in the 300 degree nearby leads to the significant difference during each measurement. The 10 dB difference between the red/brown and others is due to this reason. This problem can be solved by using the Anechoic Chamber which can

Figure 14. Return loss measurements of the 5.8 GHz circular patch antenna under different environments

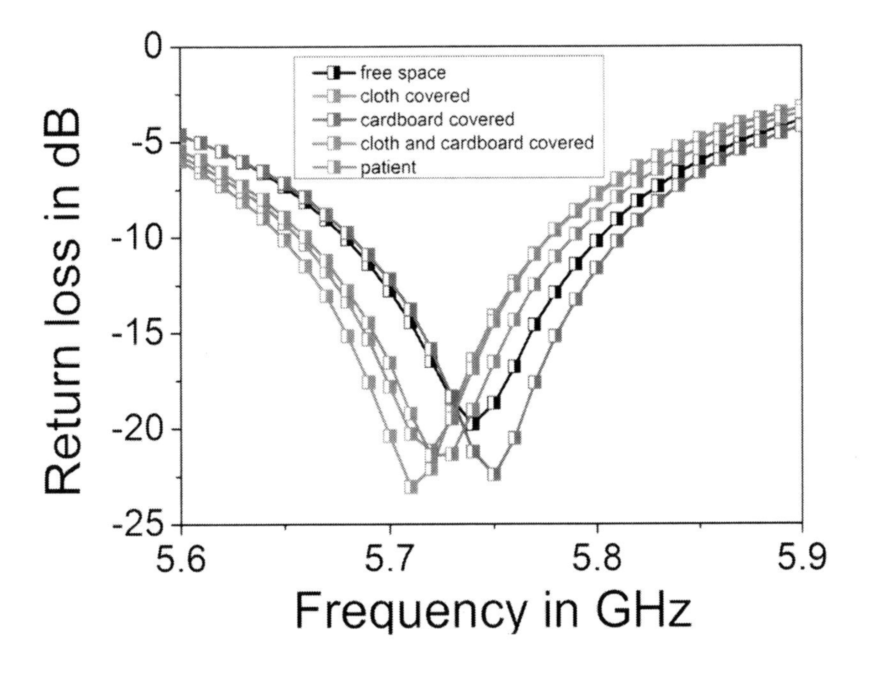

Figure 15. Radiation pattern of the 5.8 GHz circular patch antenna in H-plane

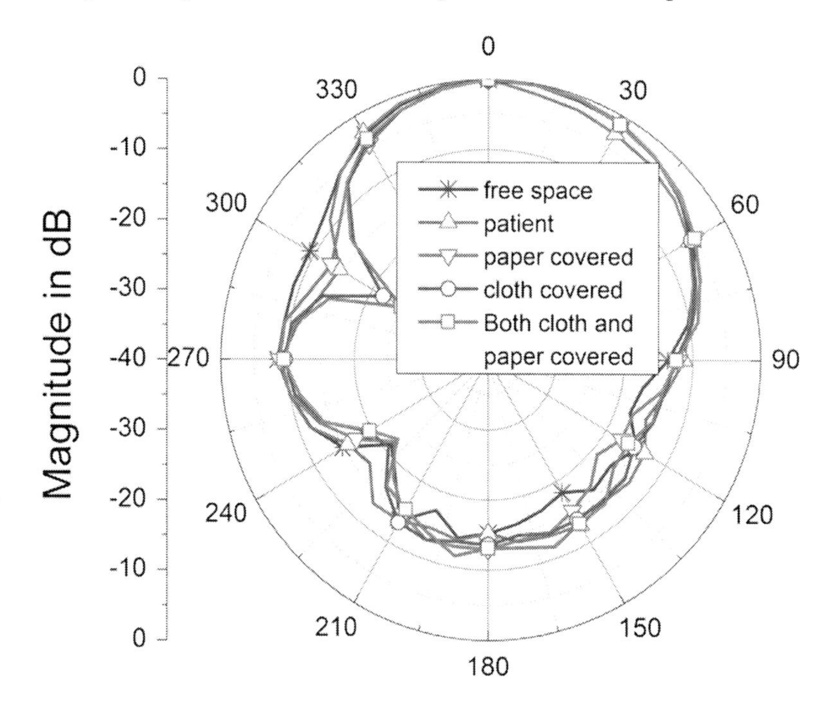

give less environmental noises, accurate control of antenna rotating and precise data reading.

Figure 16 presents the normalized radiation pattern in E-plane. From the graph, the 3 dB beam width is 70 degree and the front to back ratio is 14 dB. Due to the fact that the E-plane is symmetric referring to the antenna feeding point, the E-plane radiation pattern is supposed to be symmetric. In both Figure 15 and 16, the experimental results with four different obstacles applying in the measurement are closely matched to the free space results.

Antenna Gain

The antenna gain is calculated in terms of Equation 9 which is deducted from Friis' transmission (Equation 3). The detailed analysis has been introduced in section 4.1.3. According to Equation 9, the antenna gain in five different environments has been plotted in Figure 17. As is shown in the graph, the five curves representing five different environments are closed matched to each other.

The 5.8 GHz antenna tag yields 9.9 dBi peak gain with a 1 dB gain-drop bandwidth of 2.43% which shows good agreement with the return loss bandwidth (Figure 14). It can be seen from Figure 9 and 17 that the 5.8 GHz circular patch antenna tag has strong penetrating-capability which is feasible to be applied in WBSN applications.

5. MIMO APPLICATION IN WBSN

In WBSN, rician fading channel is considered since there are Line-of-Sight (LOS) as well as Non-Line-of-Sight (NLOS) component. Hence performance of Alamouti Space time block code is investigated and compared with Rayleigh fading channel in this section. Alamouti (Alamouti, 1998) offers Space Time Code, which is a simple method for achieving spatial diversity with two transmit antennas. Suppose there are transmission sequences $x_1, x_2, ..., x_n$. According to Alamouti, x_1 and x_2 should be sent from the first and second

Figure 16. Radiation pattern of the 5.8 GHz circular patch antenna in E-plane

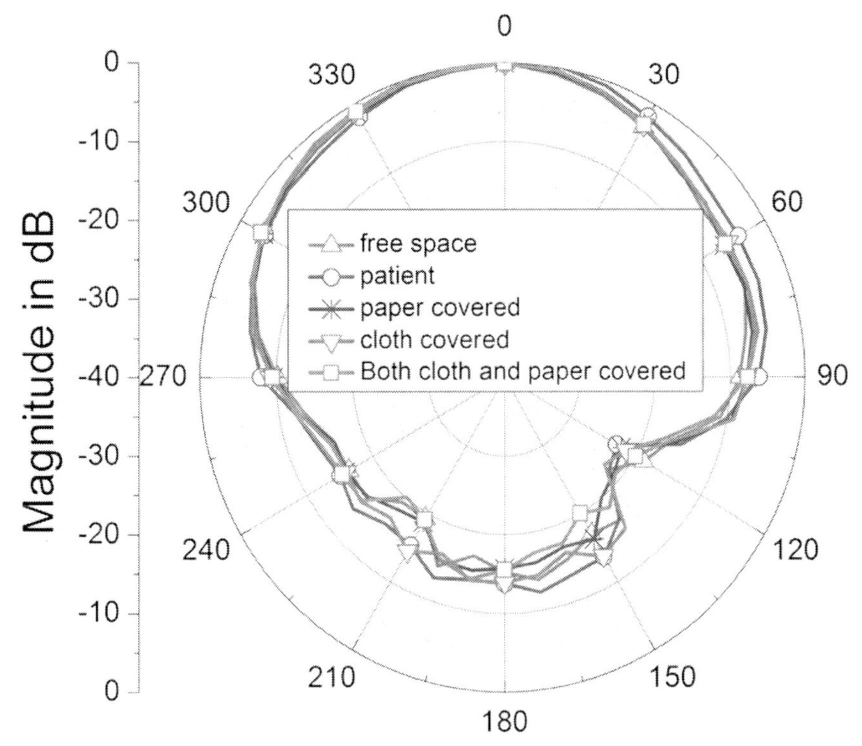

Figure 17. Antenna gain of 5.8 GHz circular patch antenna tag in five different environments

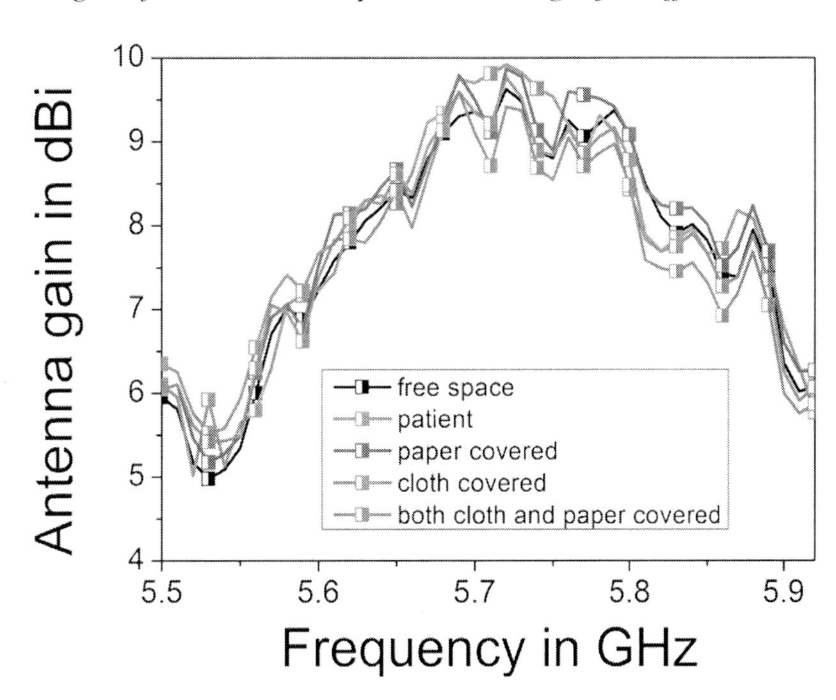

antennas in the first time slot. In the second time slot $-x_2^*$ and x_1^* are sent from the first and second antennas respectively. In the third time slot x_3 and x_4 are sent from the first and second antenna and so on. The general configuration of the 2×N MIMO system for single sensor node is shown in Figure 18.

Table 5 shows the encoding operation of Alamouti scheme. The rows of the table indicate time domain and columns indicate space domain.

The transmit matrix of the scheme is given by Equation 10 (Alamouti, 1998).

$$X = \begin{bmatrix} x_1 & -x_2^* \\ x_2 & x_1^* \end{bmatrix} \quad (10)$$

where x^* represents complex conjugate of x. The two rows and columns of X are orthogonal to each other and the code matrix of Equation 10 is orthogonal (Alamouti, 1998):

$$X.X^H = \begin{bmatrix} |x_1|^2 + |x_2|^2 & 0 \\ 0 & |x_1|^2 + |x_2|^2 \end{bmatrix} = (|x_1|^2 + |x_2|^2)\mathbf{I}_2 \quad (11)$$

where $\mathbf{I}_2$ is a 2x2 identity matrix. This property enables the receiver to detect x_1 and x_2 by a simple linear signal processing operation.

Combining and Maximum Likelihood Decoding is used at the receiver. The Alamouti scheme can be represented in simple form, if the channel fading coefficients are considered constant across two consecutive transmission periods and channels can be perfectly recovered at the receiver (Abdolee, 2008). The received signal for 2×2 antenna configuration in both time intervals can be expressed (Abdolee, 2008) as:

Figure 18. MIMO using Space time coding for WBSN

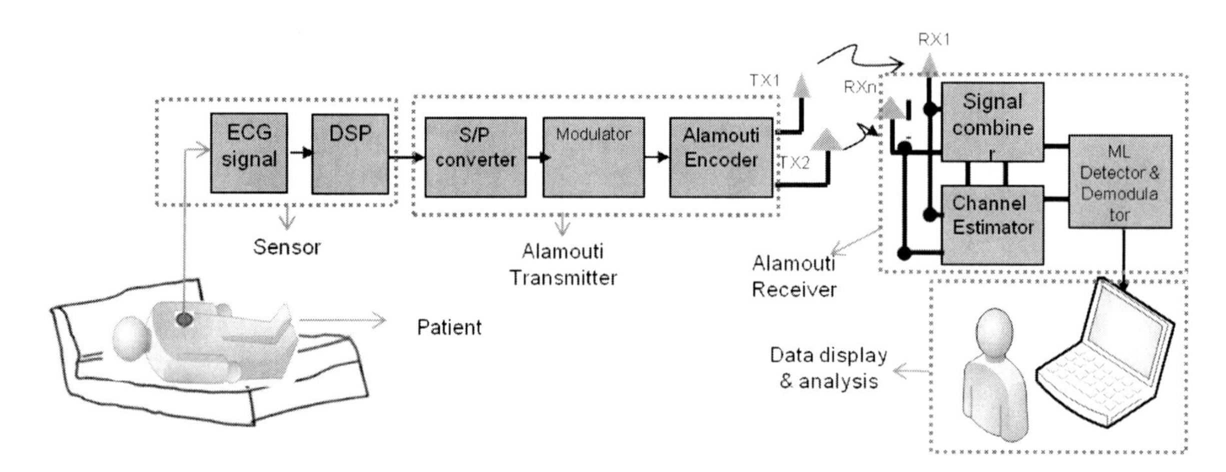

Table 5. Encoding process of Alamouti STBC

	Time Slots							
TX	1st	2nd	3rd	4th	5th	6th	7th	8th
TX-1	x_1	$-x_2^*$	x_3	$-x_4^*$	x_5	$-x_6^*$		
TX-2	x_2	x_1^*	x_4	x_3^*	x_6	x_5^*		

$$\begin{bmatrix} r_1 & r_2 \\ r_3 & r_4 \end{bmatrix} = \begin{bmatrix} h_1 & h_2 \\ h_3 & h_4 \end{bmatrix} \times \begin{bmatrix} x_1 & -x_2^* \\ x_2 & x_1^* \end{bmatrix} + \begin{bmatrix} n_1 & n_2 \\ n_3 & n_4 \end{bmatrix} \tag{12}$$

In Equation 12, r is the received signal, h is the channel matrix, x is the transmitted signal and n is the noise matrix. The output of the combiner can be represented as (Abdolee, 2008):

$$\begin{bmatrix} \tilde{x}_1 \\ \tilde{x}_2 \end{bmatrix} = \begin{bmatrix} h_1^* & h_2 \\ h_2^* & -h_1 \end{bmatrix} \times \begin{bmatrix} r_1 \\ r_2^* \end{bmatrix} \tag{13}$$

These combined signals $\left(\tilde{x}_1, \tilde{x}_2\right)$ of Equation 13 are sent to the maximum likelihood decoder which for each transmitted symbol $x_i, i = 1, 2$ selects an estimated symbol $\hat{x}_i$ from the signal set such that $d^2(\tilde{x}_i, \hat{x}_i)$ is minimum (Abdolee, 2008), where $d^2(\tilde{x}_i, \hat{x}_i)$ is the Euclidian distance between two symbols. Combining and maximum likelihood decoding will be used to detect the signal at the receiver.

In WBSN, the Rician distribution is considered for short term fading since there is Line-of-Sight (LOS) as well as Non-Line-of-Sight (NLOS) component. The MIMO channel model for Rician fading is expressed as (Park, Park, & Lee, 2010),

$$H = \sqrt{\frac{k}{k+1}} H_{LOS} + \sqrt{\frac{1}{k+1}} H_{NLOS} \tag{14}$$

This model is described in terms of a fading parameter '*k*' which is known as the Rician factor. This factor represents the power ratio of the LOS components over other multipath (scattered) components. In Equation 14, H_{LOS} represents the line of sight component of *H*. If there is no correlation at the transmit antenna, H_{NLOS} is i.i.d zero mean circularly symmetric complex Gaussian (ZMCSCG) random variables with unit power and is denoted by H_w. If there is correlated fading channel, the non-line of sight component, H_{NLOS} can be modeled as (Park et al., 2010):

$$H_{NLOS} = H_w R_T^{1/2} \tag{15}$$

where R_T denotes the correlation matrix and is defined as (Loyka, 2001):

$$R_T = \begin{bmatrix} 1 & \rho \\ \rho & 1 \end{bmatrix}, \quad 0 \le \rho \le 1 \tag{16}$$

It is seen from Equation 15 that H_{NLOS} depends on the spatial correlation coefficient (Park et al., 2010). Hence performance of Alamouti STBC is investigated in Rician fading channel.

The BER performance of the Alamouti STBC scheme for both two transmit one receive (Figure 19) and two transmit two receive antenna systems (Figure 20) are compared with the conventional one transmit and one receive antenna scheme. Pillai in (Pillai, 2008) considered flat Rayleigh fading channel. However in WBSN, the Rician distribution is considered for short term fading and this distribution is described in terms of a fading parameter 'k' which is known as the Rician factor. In the simulation, the BPSK modulation scheme has been considered. On the receive antenna, the noise n has the Gaussian probability density function with

$$p(n) = \frac{1}{\sqrt{2\pi\sigma^2}} e^{-\frac{(n-\mu)^2}{2\sigma^2}}$$

with

$$\mu = 0 \text{ and } \sigma^2 = \frac{N_0}{2}$$

Figure 19. Performance comparison of 2×1 Alamouti STBC for Rayleigh and Rician fading for various k-factors

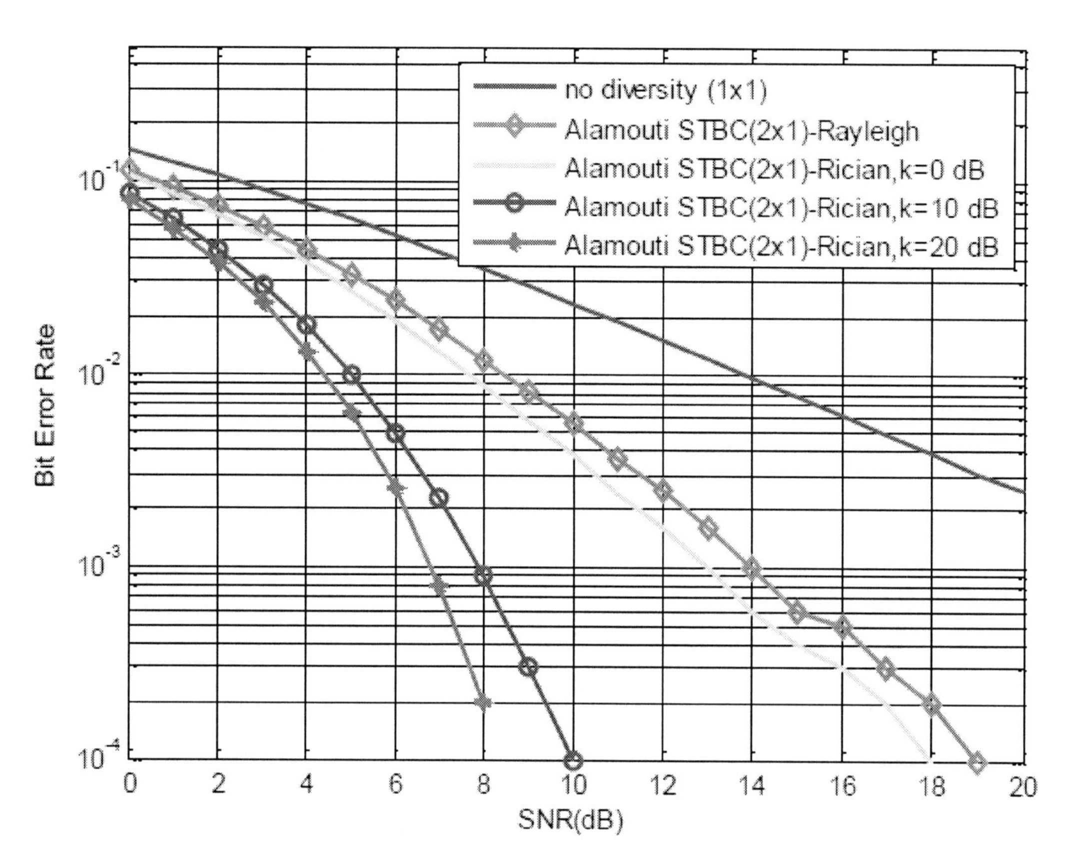

The channel has been assumed to remain constant over two time slots and the channel h_i is known at the receiver.

The Rician distribution is modeled when a line of sight (LOS) between the transmitter and receiver exists in addition to random multiple paths. The simulation is run considering rician fading channel for several values of k-factor. When k-factor is increased, the portion of the LOS signal is increased and therefore probability to encounter a deep fade reduces. It is seen from the simulation result (Figures 19 and 20) that with the increasing value of rician K factor, the LOS path becomes more significant. For example, when K=20 dB, the LOS is the most dominant path and fading is reduced notably. Hence the BER performance improves when the rician K factor increases. Moreover, the BER performance for various rician K factor is observed superior compared to Rayleigh. Also the BER performance is improved with diversity order. MIMO technique can be applied to RFID in which read reliability is increased using space diversity. It will also increase read range and read throughput. The performance of the Alamouti scheme depends greatly on channel characteristics such as Rician k-factor and fading signal correlation. The dependency of Rician k-factor is investigated in this section. Another important factor, correlation coefficient, which influences the performance of Alamouti scheme, is measured in the following section.

Figure 20. Performance comparison of 2×2 Alamouti STBC for Rayleigh and Rician fading for various k-factors

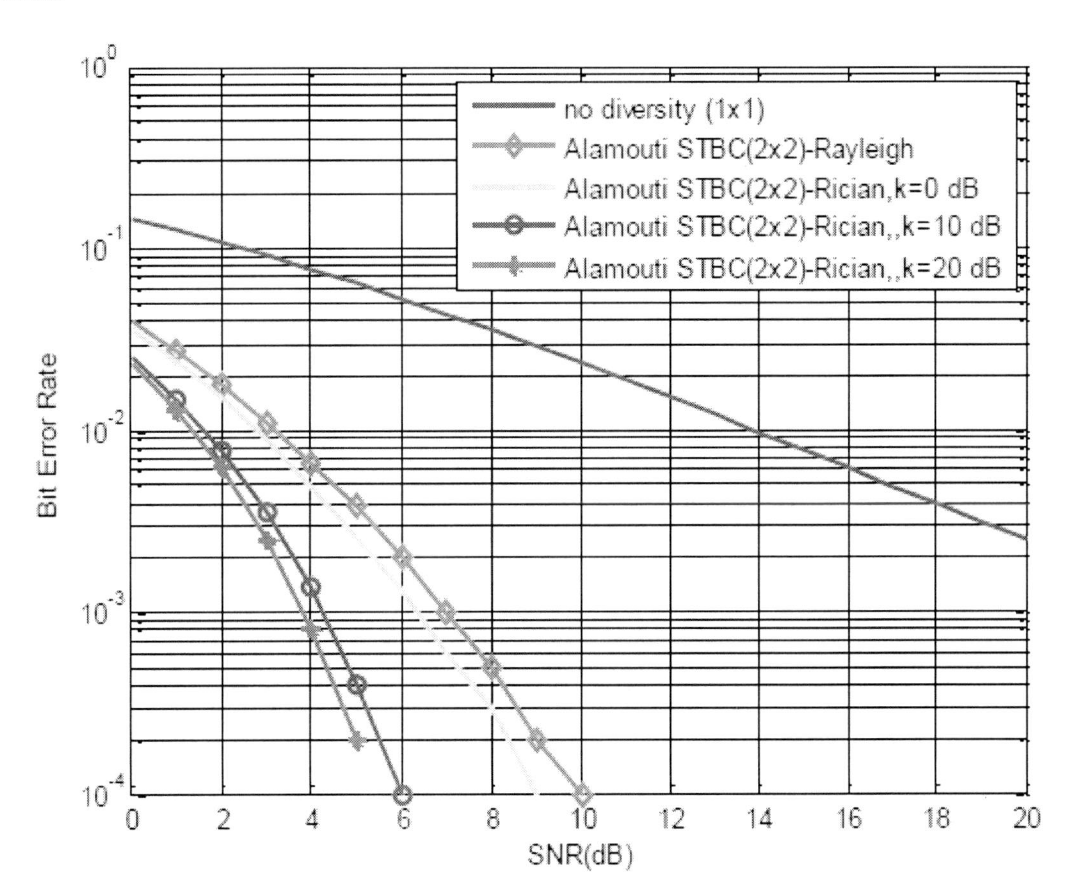

5.1 Correlation Coefficient Measurement for MIMO Application In WBSN

The capacity and reliability of MIMO systems depends on the correlation among the channel transfer functions of different pairs of transmit and receive antennas (Singh & Mohan, 2010). Moreover, the performance of a diversity receiver greatly depends upon the correlation between the received signals at the diversity branches. The essential criterion to quantify the efficiency of MIMO antennas is the correlation between the antennas. A basic requirement to realize the gains from MIMO systems is low correlation (Singh et al., 2010) in the WBAN channel. The diversity gain in Rician fading channel for WBAN is investigated in the previous section. The correlation coefficient is measured using space diversity in this section. Space diversity is achieved by using more than one antenna at the transmitter or receiver side. For measuring the correlation coefficient for WBSN, the following steps are followed.

5.1.1 MIMO Antenna Development

Since the attenuation in the human tissue is very high, propagation around the human body into shadow regions is mostly via creeping waves. Therefore, an antenna which is polarized perpendicularly to the surface of the body, and thus induces strong creeping waves on the body surface, is best suited for on-body communications. The presence of the ground plane behind the antenna is also beneficial as it protects the antenna from the detrimental effects of body proximity. Antennas like monopoles, microstrip patches and planar inverted-F antennas (PIFA) satisfy these conditions and provide higher Propagation Gain (PG) and less fading than other antennas, such as a loop or a dipole (Hall & Hao, 2006). Considering these factors, we shall design MIMO antenna. The general requirements for WBSN antennas are (Hall et al., 2006; Haga, Saito, Takahashi, & Ito, 2009):

- Low mutual coupling between antennas and body for high radiation efficiency and low SAR.
- Polarization normal to the body surface for minimized link loss in on-body communications.
- Small size and low profile.

5.1.2 Patch Antenna Design

The traditional antennas are not suitable for placement on body sensor nodes. For WBSN, the patch antenna is the most suitable device due to its smaller size, conformal configuration, low manufacturing cost, better radiation pattern and higher antenna gain. The operating frequency of the antenna is considered unlicensed ISM 5.8 GHz frequency band for getting due smaller size of the antenna. The single patch 5.8 GHz rectangular antenna has been designed using CST simulation laboratory. Theoretically, the size of the rectangular patch antenna (Figure 21a) is based on the empirical formula below (Balanis, 2005):

$$W = \frac{v_0}{2f_r}\sqrt{\frac{2}{\varepsilon_r + 1}} \quad (17)$$

$$L = \frac{1}{2f_r\sqrt{\varepsilon_{eff}\mu_0\varepsilon_0}} - 2\Delta L \quad (18)$$

where W and L are the width and length of the patch respectively, v_0 is the speed of light in free space, f_r is the resonant frequency of the antenna, ΔL is the incremental length due to fringing field, ε_r is the dielectric constant of the substrate, $\mu_0\left(4\pi \times 10^{-7} H / m\right)$ is the permeability and $\varepsilon_0\left(8.854 \times 10^{-12} F / m\right)$ is the permittiv-

Figure 21. (a) single patch antenna, (b) dual patch antenna

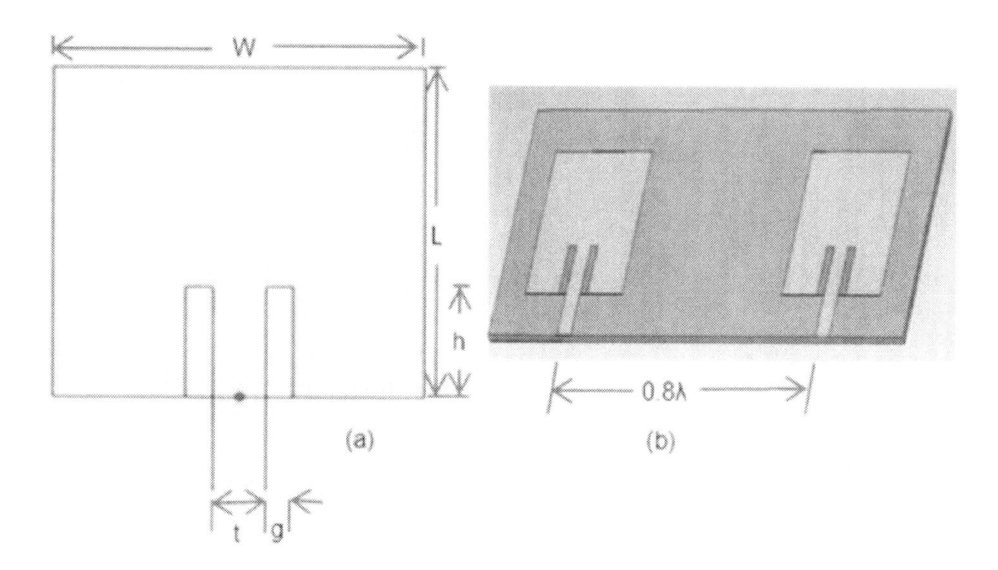

ity of free space. ε_{eff} is the effective dielectric constant of the substrate which accounts for the fringing field effect of the patch antenna.

The rectangular patch antenna is fed by transmission line mode. To adjust the input resistance, the inset feeding method is applied by extending the transmission line into the patch and digging two identical slots along it. The transmission line is set to 50 Ω by adjusting t, the width of the line, to 2.28 mm. To maintain a 50 Ω input resistance for the patch, the width of the slot ***g*** is set to ***t***/2 which is 1.14 mm and the length of the slot is set to ***L***/3, which is 5.28 mm. The material used for antenna is Taconic TXL-0 $\left(\varepsilon_r = 2.45\right)$. The distance between the two patches of the MIMO antenna is $0.8\lambda = 41.36$ mm (Figure 21b), where λ is the wavelength of the carrier signal. This dual patch antenna will be used in each of the proposed wireless on body sensor shown in Figure 4 or Figure 5 to form MIMO in WBSN.

5.1.3 Antenna Radiation Pattern Measurement

The radiation pattern was measured in the anechoic chamber at Monash University which is shown in Figure 22. Based on the graph for a single patch (Figure 22 b), the radiation pattern shows directionality compared with the H plane having a 3 dB beam width of 50 degrees. The H plane has a higher front to back ratio, which is about 11 dB. The two graphs (Figure 22 c and d) below present the radiation pattern of the dual patch antennas (one connected with probe and another to a matched load). Due to the coupling effect between the two patches, the two radiation patterns are slightly shifted. The 3 dB beam width for both of the two is 70 degrees.

5.1.4 Measurement Set-Up and Antennas Placement on Human Body

A number of on-body channels named belt-head, belt-back, belt-wrist, and belt-ankle has been selected for receiving diversity measurement. For each on-body channel, the transmitting antenna is placed at the waist (belt) position on the right side of the body. The receiving antennas are placed on the left side of the head, left side of the chest, left wrist, left ankle, rear side of the body respectively. The belt-back channel represents the Non-Line-of-Sight (NLOS) scenario whereas the other channels represent the partial Line-of-sight (LOS) scenario depending on body movement. Some off-body channels are considered for measurement of diversity. In off-body channels, the transmitting antenna is placed at the chest and the receiving antennas are placed at the base station located at approximately 2-3 m away from the body. Figure 23 represents one of the measurements set-ups for Tx on the chest and Rx at the base station. The received signal envelopes for various WBAN channels are measured using Agilent PNA as shown in Figure 24. The measurement is done in the Antenna and RFID Research laboratory at ECSE department of Monash University.

The received signal envelope for the belt-head on-body channel from the measured result is shown in Figure 25. In this measurement one transmit antenna and two receive antennas are considered. In the receiving side, two adjacent patch antennas as shown in Figure 21b have been used.

5.1.5 Calculation of Correlation Coefficient

In WBSN, propagation is significantly affected by the motion of the body, due to multipath propagation around the body and reflection from the neighboring environment. Fading will occur in WBSN due to various factors such as energy absorption, reflection, diffraction, shadowing by

Figure 22. (a) Photograph of radiation pattern measurement setup at Monash Anechoic chamber, (b) radiation pattern for Single patch antenna, (c) radiation pattern dual patch antenna (port 1) [other antenna is terminated to a matched load], (d) radiation pattern dual patch antenna (port 2) [other antenna is terminated to a matched load]

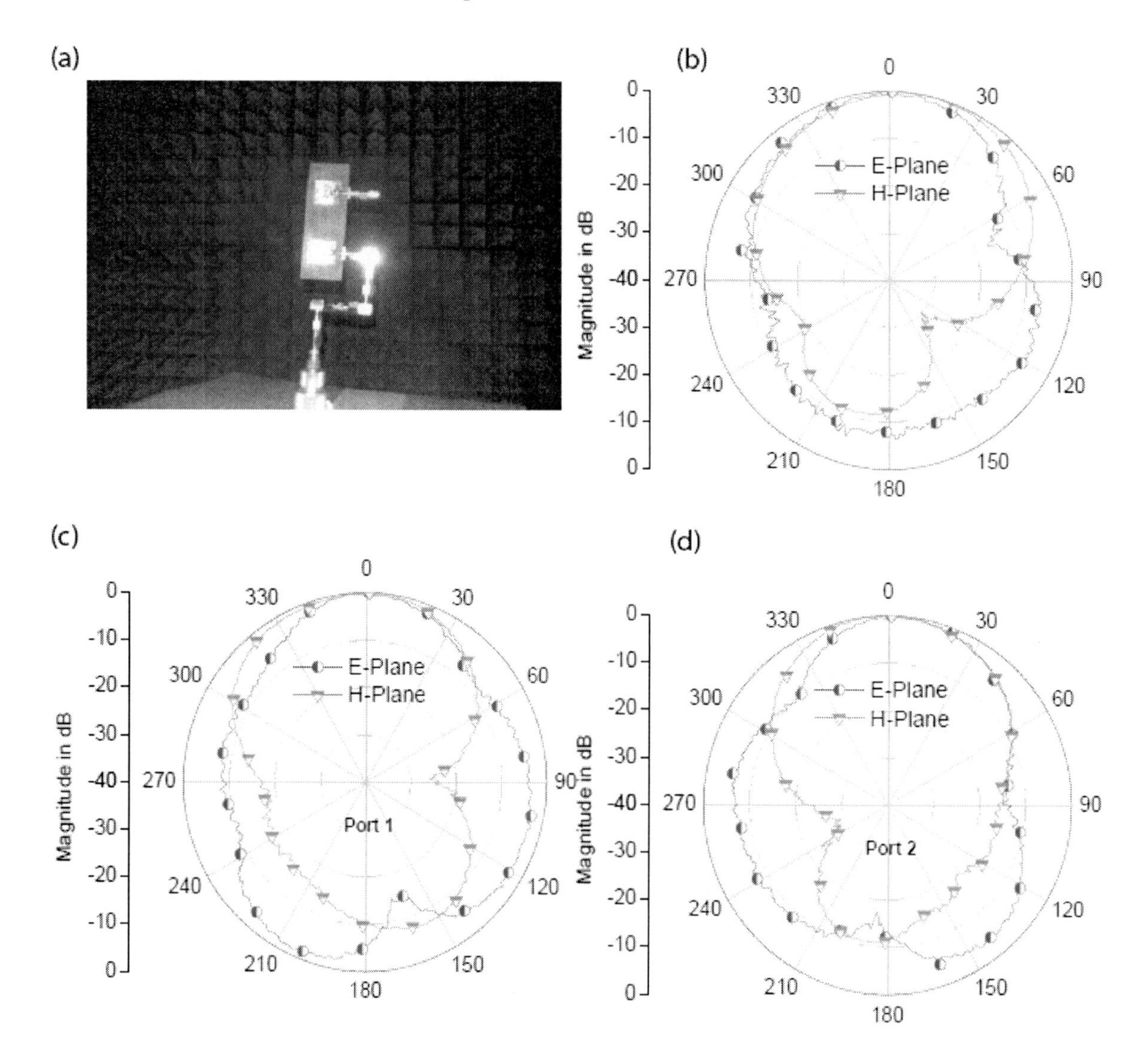

body, body pose (Yazdandoost et al, 2010). To enhance the system performance and overcome fading in WBSN, diversity is a very powerful tool (Khan, 2009). The principle of diversity is the use of more than one independent and thus uncorrelated branch which will each fade independently of each other. Space diversity is achieved by using more than one antenna at the transmitter or receiver side. The correlation comes from three sources, namely the correlated fading channel, correlation among the transmit antennas, and correlation among the receive antennas (Singh et al., 2010). Correlation coefficient was measured in (Votis et al., 2010; Ping, Hynes, van Wonterghem, & Michelson, 2005) using dipole antenna for conventional wireless system whereas patch

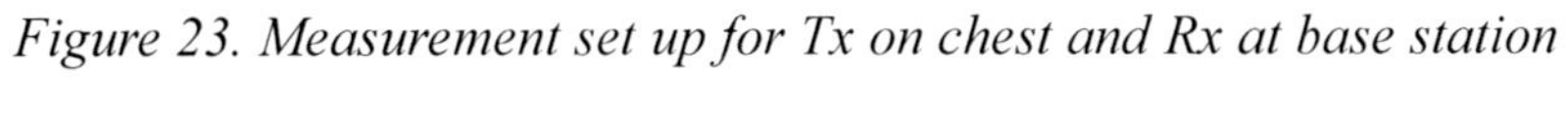

Figure 23. Measurement set up for Tx on chest and Rx at base station

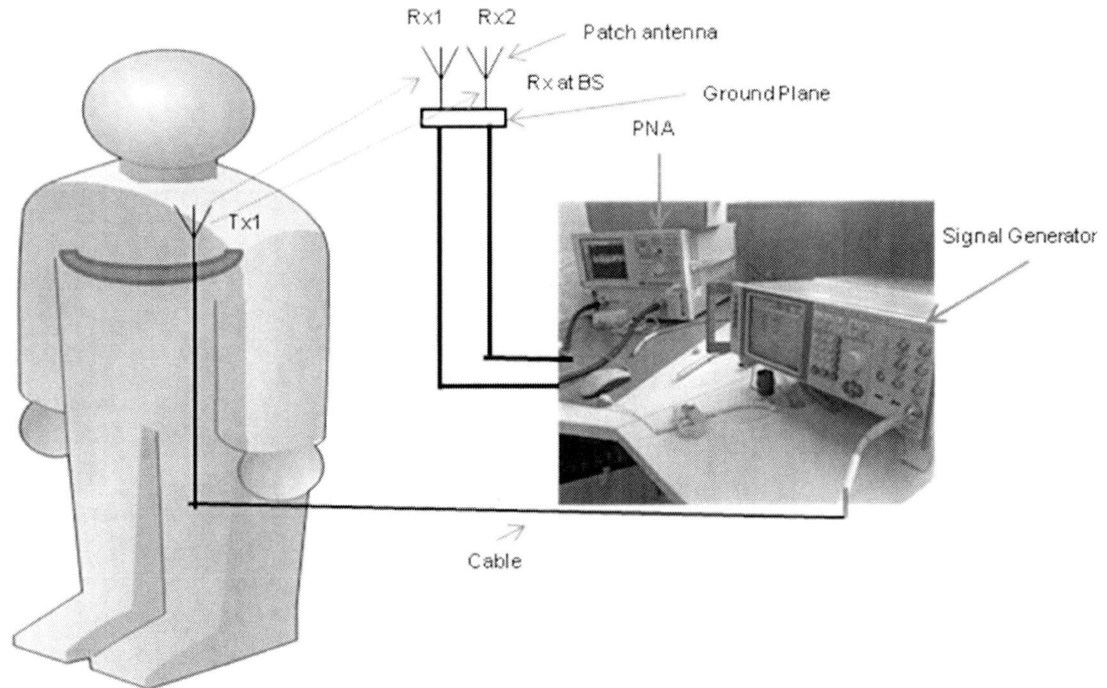

Figure 24. Various on body channels (a) belt-chest channel, (b) belt-head channel

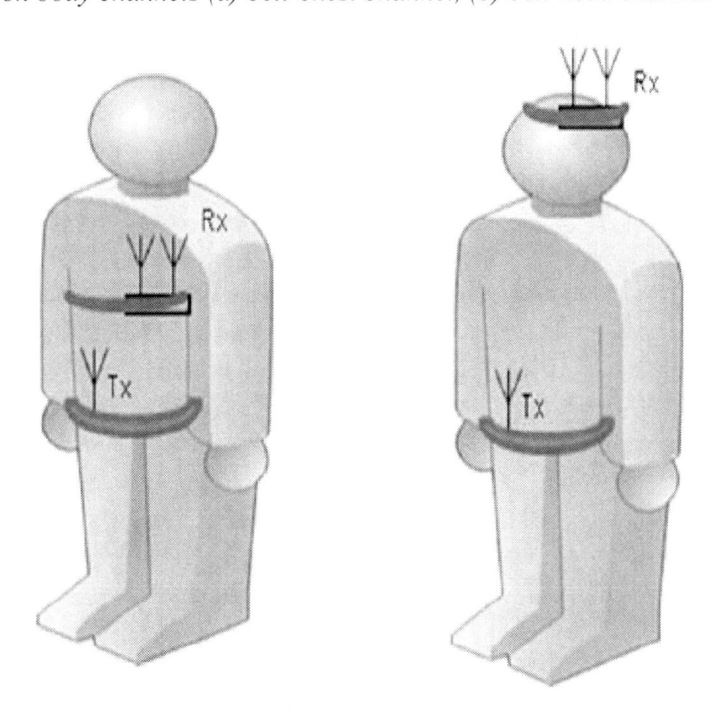

Figure 25. Received signals for belt-head on-body channel for 1 Tx with 0 dBm power at 5.8 GHz carrier signal and 2 Rx antennas

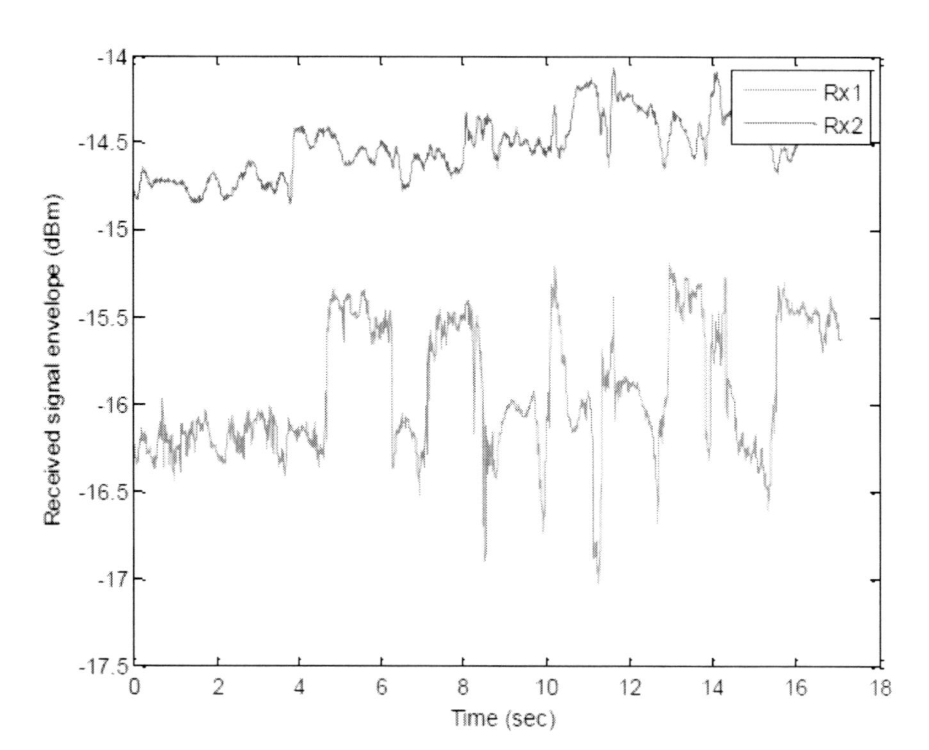

antenna has been used in this paper for WBAN system. The previous work in (Votis et al., 2010) for measuring correlation coefficient has been carried out at the 2.4-GHz ISM band. However we have used 5.8 GHz as the patch antenna is compact at this frequency which is well suited for WBAN. The envelope correlation coefficient (ρ_e) between the two branch signals is calculated using the Equation 19 (Khan, 2009) for various WBSN channels:

$$\rho_e = \frac{\sum_{i=1}^{N}(r_1(i)-\overline{r_1})(r_2(i)-\overline{r_2})}{\sqrt{\sum_{i=1}^{N}(r_1(i)-\overline{r_1})^2}\sqrt{\sum_{i=1}^{N}(r_2(i)-\overline{r_2})^2}} \quad (19)$$

where N is the total samples and $\overline{r_j}$ is the mean value of the fading envelope r_j of the received branch signal. During the received diversity measurement, several body movements, for example, looking right, moving hand etc., were performed in order to create rich scattering environment. The correlation coefficients are shown in Table 6. The correlation coefficient for NLOS dynamic WBAN channel (for example, Belt-Back) is shown lower. Moreover, partial LOS channels (Belt-Head, Belt-Ankle, etc.) show lower correlation due to movement of the body and rich scattering environment. The results conclude that patch antenna is suitable in WBAN channels to get lower correlation coefficient and it can be used to enhance reliability in WBAN. Hence significant diversity gain can be attained using MIMO in WBSN.

Table 6. Correlation coefficients of various WABN channels

WBSN Channel	Correlation Coefficient
Belt Head Looking Right	-0.06
Belt Ankle moving feet while sitting	-0.19
Belt Back sitting and lifting book	0.30
Belt Back moving hand standing	0.03
Belt Back sitting and standing	0.14
Belt Wrist hand moving near Head	-0.22
Belt Wrist hand on chest	0.03

6. CONCLUSION

6.1 Conclusion of the Works

A novel portable wireless monitoring system based on active RFID technology for sleep apnea diagnosis is presented in this chapter. The detailed design specifications of each sub-system are given. The design of on-body sensors is extremely challenging due to the limited power consumption and physical size. In order to minimize the power consumption, the passive channel-selection BPFs are adopted at the RF front-end rather than in the baseband. Moreover, the system is designed to operate in 5.8 GHz ISM band. Therefore, the physical size of passive components can be significantly reduced and the interference with other radios in 2.4 GHz can be effectively eliminated. According to the experimental results presented in previous section, two points can be proved: firstly, the penetrating-capability of the 5.8 GHz circular patch antenna is good enough to let the wireless signal pass through the nonconductive made obstacles; secondly, different nonconductive materials do not change the penetrating-capability of the 5.8 GHz circular patch antenna. Therefore, the strong penetrating performance of 5.8 GHz radio wave proves that the RFID technology using 5.8 GHz ISM band can be applied in WBSN applications. Moreover, the MIMO technique in WBSN for sleep apnea monitoring systems has been introduced in order to overcome the fading effect due to reflection, diffraction, body shadowing, body movement, etc. the performance of MIMO STBC on WBSN channel is investigated for various values of Rician k factor. The performance of STBC depends on Rician k factor which presents the power ratio of the LOS components over NLOS components. Also the performance of MIMO system in WBSN was shown superior compare to SISO. Moreover, the patch antenna suitable for diversity improvement for dynamic WBAN system is designed to measure the correlation coefficient. According to measurement result, the lower correlation coefficient was found for most of the WBAN channels. Hence MIMO can be used in sleep apnea monitoring systems to combat fading effect in dynamic WBAN.

6.2 Overlook of the Future

The future work is to develop the system to carry out the described functions. In spite of the numerous benefits of the presented system, there are many challenges associated with implementation. Practical, compact, and low power on-body sensor design, adaptive system design and cross layer adaptation for RFID need to be considered to maintain quality of service and improve operating lifetime. In sleep apnea monitoring system, several sensors need to be placed on body to transmit physiological signal through WBSN channel. Spatial diversity is achieved by using MIMO systems. However, in case of dynamic WBSN, to achieve diversity gain, multiple antennas cannot be placed for small sensor node, due to the limited physical size. Hence the aim of the future research is to introduce cooperative communication (virtual MIMO) techniques using energy efficient network coding among the sensor nodes mounted on the body for transmitting physiological data to the base station. In cooperative communication, transmitting sensors share their antennas which generate a virtual MIMO scenario to exploit spatial diversity.

Moreover, a key requirement of physical layer network coding (PNC) is synchronization among nodes which has been addressed in other wireless networks such as ad hoc and sensor networks etc. However, unlike them, WBSN sensor nodes pose a number of unique challenges such as very low permitted transmission energy by on-body sensors due to their close proximity to human body, low transmission range compared to traditional sensor network and low processing capabilities at each body sensors. Hence, symbol level time, carrier frequency and carrier phase synchronization issue in PNC for WBSN will be investigated as further research.

ACKNOWLEDGMENT

The project is supported by Australian Research Council Linkage Project Grant LP0776796: Radio Frequency Wireless Monitoring in Sleep Apnoea (Particularly for Paediatric Patients) and Regni Science and Health Pty. Ltd.

REFERENCES

Abd-Elrazzak, M. M., & Al-Nomay, I. S. (2004). A design of a circular microstrip patch antenna for bluetooth and HIPERLAN applications. *The 9th Asia-Pacific Conference on Communications: Vol. 3* (pp. 974-977).

Abdolee, R. (2008). *Performance of MIMO space time coded system and training based channel estimation for MIMO-OFDM system.* Master Thesis, Universiti Teknologi Malaysia.

Alamouti, S. M. (1998). A simple transmitter diversity scheme for wireless communications. *IEEE Journal on Selected Areas in Communications*, *16*, 1451–1458. doi:10.1109/49.730453

Alippi, C., & Vanini, G. (2005). An application-level methodology to guide the design of intelligent-processing, power-aware passive RFIDs. *IEEE International Symposium on Circuits and Systems,* Vol. 6 (pp. 5509-5512).

Baker, J. P., Bones, P. J., & Lim, M. A. (2006). Wireless health monitor. *Electronics New Zealand Conference,* (pp. 7-12).

Balanis, C. A. (2005). *Antenna theory: Analysis design*. John Wiley & Sons, Inc.

Banks, J., Hanny, D., Pachano, M. A., & Thompson, L. G. (2007). *RFID in warehousing and distribution systems*. John Wiley & Sons, Inc.

Bellon, J. S., Cabedo-Fabres, M., Antonino-Daviu, M., Ferrando-Bataller, M., & Penaranda-Foix, F. (2011). Textile MIMO antenna for wireless body area networks. *Proceedings of the 5th European Conference on Antennas and Propagation (EUCAP),* (pp. 428-432).

Brown, L., Grundlehner, B., van de Molengraft, J., Penders, J., & Gyselinckx, B. (2009). *Body area network for monitoring autonomic nervous system responses*. 1st International ICST Workshop on Wireless Pervasive Healthcare.

Cao, H., Leung, V., Chow, C., & Chan, H. (2009). Enabling technologies for wireless body area networks: A survey and outlook. *IEEE Communications Magazine*, *47*, 84–93. doi:10.1109/MCOM.2009.5350373

Cerovic, D. D., Dojcilovic, J. R., Asanovic, K. A., Mihajlidi, T. A., Popovic, D. M., & Spasovic, S. B. (2003). *Investigation of dielectric properties and electric resistance of some textile materials*. The 5th General Conference of the Balkan Physical Union.

Chen, M., Gonzalez, S., Vasilakos, A., Cao, H., & Leung, V. C. M. (2010). Body area networks. *Survey (London, England)*, 1–23.

Fensli, R., Gunnarson, E., & Gundersen, T. (2005). A wearable ECG-recording system for continuous arrhythmia monitoring in a wireless tele-home-care situation. *Proceedings of the 18th IEEE Symposium on Computer-Based Medical Systems,* (pp. 407-412).

Fort, A., Desset, C., De Doncker, P., Wambacq, P., & Biesen, L. V. (2006). An ultra-wideband body area propagation channel model-from statistics to implementation. *IEEE Transactions on Microwave Theory and Techniques, 54,* 1820–1826. doi:10.1109/TMTT.2006.872066

Gabriel, C., Bentall, R. H., & Grant, E. H. (1987). Comparison of the dielectric properties of normal and wounded human skin material. *Bioelectromagnetics, 8,* 23–27. doi:10.1002/bem.2250080104

Garima, B. D., Saxena, V. K., & Saini, J. S. (2009). Broadband circular patch microstrip antenna with diamond shape slot for space communication. *International Conference on Emerging Trends in Electronic and Photonic Devices & Systems,* (pp. 332-335).

Glover, B., & Bhatt, H. (2006). *RFID essentials* (1st ed.). O'Reilly Media, Inc.

Haga, N., Saito, K., Takahashi, M., & Ito, K. (2009). Characteristics of cavity slot antennas for body-area networks. *IEEE Transactions on Antennas and Propagation, 57,* 837–843. doi:10.1109/TAP.2009.2014577

Hall, P. S., & Hao, Y. (2006). *Antennas and propagation for body-centric wireless communications.* Artech House. doi:10.1109/TAP.2009.2018564

Hey-Shipton, G. L., Metthews, P. A., & Mestay, J. (1982). The complex permittivity of human tissue at microwave frequencies. *Physics in Medicine and Biology, 27,* 1067–1071. doi:10.1088/0031-9155/27/8/008

Hu, X., Wang, J., Yu, Q., Liu, W., & Qin, J. (2008). A wireless sensor network based on ZigBee for telemedicine monitoring system. *The 2nd International Conference on Bioinformatics and Biomedical Engineering,* (pp. 1367-1370).

Texas Instruments. (2011). *MSP430FR57x family datasheet.*

Johansson, A., & Karedal, J. Tufvesson, F., & Molisch, A. F. (2005). MIMO channel measurements for personal area networks. *IEEE 61st Vehicular Technology Conference,* Vol. 1 (pp. 171-176).

Jose Tomlal, E., Thomas, P. C., George, K. C., Jayanarayanan, K., & Joseph, K. (2010). Impact, tear, and dielectric properties of cotton/polypropylene commingled composites. *Journal of Reinforced Plastics and Composites, 29,* 1861–1875. doi:10.1177/0731684409338748

Kent, A., & Billiard, M. (2003). *Sleep: physiology, investigations and medicine.* Kluwer Academic / Plenum Publishers.

Khan, I. (2009). *Diversity and MIMO for body-centric wireless communication channels. Unpublished doctoral theses.* UK: School of Electronics, Electrical, & Computer Engineering, University of Birmingham.

Khan, I., & Hall, P. S. (2009). Multiple antenna reception at 5.8 and 10 GHz for body-centric wireless communication channels. *IEEE Transaction on Antennas and Propagation, 57.*

Khan, I., & Hall, P. S. (2010). Experimental evaluation of MIMO capacity and correlation for narrowband body-centric wireless channels. *IEEE Transactions on Antennas and Propagation, 58,* 195–202. doi:10.1109/TAP.2009.2025062

Lam, S. C. K., Wong, K. L., Wong, K. O., Wong, W. & Mow W. H. (2009). A smart phone centric platform for personal health monitoring using wireless wearable biosensors. *The 7th International Conference on Information, Communications and Signal Processing,* (pp. 1-7).

Latre, B., Braem, B., Moerman, I., Blondia, C., & Demeester, P. (2011). A survey on wireless body area networks. *Wireless Networks, 17*, 1–18. doi:10.1007/s11276-010-0252-4

Li, C.-J., Liu, L., Chen, S.-Z., Wu, C. C., Huang, C.-H., & Chen, X.-M. (2004). Mobile healthcare service system using RFID. *IEEE International Conference on Networking, Sensing and Control,* Vol. 2 (pp. 1014-1019).

Lo, B., Thiemjarus, S., King, R., & Yang, G.-Z. (2005). *Body sensor network: A wireless sensor platform for pervasive healthcare monitoring.* Retrieved on September 5, 2011, from http://academic.research.microsoft.com/Publication/11040372/body-sensor-network-a-wireless-sensor-platform-for-pervasive-healthcare-monitoring

Loyka, S. L. (2001). Channel capacity of MIMO architecture using the exponential correlation matrix. *IEEE Communications Letters, 5*, 369–371. doi:10.1109/4234.951380

Marrocco, G. (2007). RFID antennas for the UHF remote monitoring of human subjects. *IEEE Transactions on Antennas and Propagation, 55*, 1862–1870. doi:10.1109/TAP.2007.898626

Neirynck, D., Williams, C., Nix, A., & Beach, M. (2005). *Experimental capacity analysis for virtual array antennas in personal and body area networks.* International Workshop on Wireless Adhoc Networks.

Neirynck, D., Williams, C., Nix, A., & Beach, M. (2006). *Personal area networks with line-of-sight MIMO operation.* IEEE 63rd Vehicular Technology Conference.

Neirynck, D., Williams, C., Nix, A., & Beach, M. (2007). Exploiting MIMO in the personal sphere. *IET Proceedings of Microwaves, Antenna Propagation.*

Occhiuzzi, C., & Marrocco, G. (2010). The RFID technology for neurosciences: feasibility of limbs' monitoring in sleep diseases. *IEEE Transactions on Information Technology in Biomedicine, 14*, 37–43. doi:10.1109/TITB.2009.2028081

Park, J.-H., Seol, J.-A., & Oh, Y.-H. (2005). Design and implementation of an effective mobile healthcare system using mobile and RFID technology. *Proceedings of 7th International Workshop on Enterprise networking and Computing in Healthcare Industry* (pp. 263-266).

Park, Y., Park, S. K., & Lee, H. Y. (2010). Performance of wireless body area network over on-human-body propagation channels. *2010 IEEE Sarnoff Symposium,* (pp. 1-4).

Personal Area Network (PAN). (n.d.). Retreived on May 2, 2011 from http://searchmobilecomputing.techtarget.com/definition/personal-area-network

Philipose, M., Smith, J. R., Jiang, B., Mamishev, A., Roy, S., & Sundara-Rajan, K. (2005). Battery-free wireless identification and sensing. *IEEE Pervasive Computing, 4*, 37–45. doi:10.1109/MPRV.2005.7

Pillai, K. (2008). *Feature of diversity technique.* Retrieved on 15 June, 2010, from http://www.dsplog.com/2008/10/16/alamouti-stbc/

Ping, H., Hynes, C. G., van Wonterghem, J., & Michelson, D. G. (2005). Measurements of correlation coefficients of closely spaced dipoles. *IEEE International Workshop on Antenna Technology: Small Antennas and Novel Metamaterials,* (pp. 474-477).

Rodrigues, J. M., Estevao, M. H., Malaquias, J. L., Santos, P., Gouveia, G., & Simoes, J. B. (2007).Sleep at home – Portable home based system for pediatric sleep apnea diagnosis. *IEEE International Conference on Portable Information Devices* (pp. 1-4).

Sakaguchi, K., Chua, H.-Y.-E., & Araki, K. (2005). MIMO channel capacity in an indoor line-of-sight environment. *IEICE Transactions on Communication, E88-B*.

Sample, A. P., Yeager, D. J., Powledge, P. S., & Smith, J. R. (2007). Design of a passively-powered, programmable sensing platform for UHF RFID systems. *IEEE International Conference on RFID* (pp. 149-156).

Sangwan, R. S., Qiu, R. G., & Jessen, D. (2005). Using RFID tags for tracking patients, charts and medical equipment within an integrated health delivery network. *Proceedings of the IEEE Networking, Sensing and Control* (pp. 1070-1074).

Shi, X., Medard, M., & Lucani, D. E. (2010). *Network coding for energy efficiency in wireless body area networks*. Massachusetts Institute of Technology.

Shnayder, B. C. V., Lorincz, K., Fulford-Jones, T., & Welsh, M. (2005). *Sensor networks for medical care*. Technical Report TR-08-05, Division of Engineering and Applied Sciences, Harvard University.

Singh, C. K., & Mohan, S. (2010). Effect of antennas correlation on the performance of MIMO systems in wireless sensor network. *Wireless Telecommunications Symposium (WTS)*, (pp. 1-5).

Sivakumar, M., & Deavours, D. D. (2006). *A dual-resonant planar microstrip antenna design for UHF RFID using paperboard as a substrate*. Retrieved on 05 September, 2011, from http://www.ittc.ku.edu/~deavours/pubs/rfid06b-submit.pdf

Smith, D., Hanlen, L., Miniutti, D., Zhang, J., Rodda, D., & Gilbert, B. (2008). Statistical characterization of the dynamic narrowband body area channel. *First International Symposium on Applied Sciences on Biomedical and Communication Technologies* (pp. 1-5).

Smith, D., Hanlen, L., Zhang, J., Miniutti, D., Rodda, D., & Gilbert, B. (2009). Characterization of the dynamic narrowband on-body to off-body area channel. *IEEE International Conference on Communications*, (pp. 1-6).

Tafa, Z., & Stojanovic, R. (2006). Bluetooth-based approach to monitoring biomedical signals. *Proceedings of the 5th WSEAS International Conference on Telecommunications and Informatics*, (pp. 415-420).

Takada, J., Aoyagi, T., Takizawa, K., Katayama, N., Kobayashi, T., Yazdandoost, K. Y., et al. (2008). *Static propagation and channel models in body area*. Retrieved September 5, 2011, from http://www.ap.ide.titech.ac.jp/publications/Archive/COST2100_TD%2808%-29639%280810Takada%29.pdf

Theremin, L. (n.d.). *Bio*. Retrieved September 5, 2011, from http://www.absoluteastronomy.com/topics/L%C3%A9on_Theremin

Votis, C., Tatsis, G., & Kostarakis, P. (2010). Envelope correlation parameter measurements in a MIMO antenna array configuration. *International Journal of Communications. Network and System Sciences*, *3*(4), 350–354. doi:10.4236/ijcns.2010.34044

Wang, L., Noel, E., Fong, C., Kamoua, R., & Tang, K. W. (2006). A wireless sensor system for biopotential recording in the treatment of sleep apnea disorder. *Proceedings of the 2006 IEEE International Conference on Networking, Sensing and Control*, (pp. 404-409).

Yang, Y., Karmakar, N. C., & Zhu, X. (2011). A portable wireless monitoring system for sleep apnea diagnosis based on active RFID technology. *Accepted by Asia-Pacific Microwave Conference.*

Yang, Y., Roy, S. M., Karmakar, N. C., & Zhu, X. (2011). A novel high selectivity bandpass filter for wireless monitoring of sleep apnea patients. *Accepted by Asia-Pacific Microwave Conference.*

Yazdandoost, K. Y., & Sayrafian-Pour, K. (2010). *Channel model for body area network*. Retrieved September 5, 2011, from https://mentor.ieee.org/802.15/dcn/08/15-08-0780-12-0006-tg6-channel-model.pdf

Yazicioglu, R. F., Merken, P., Puers, R., & Van Hoof, C. (2007). A 60 μW 60nV/ Hz readout front-end for portable biopotential acquisition systems. *IEEE Journal of Solid-state Circuits*, *42*, 1100–1110. doi:10.1109/JSSC.2007.894804

Zaffaroni, A., de Chazal, P., Heneghan, C., Boyle, P., Mppm, P. R., & McNicholas, W. T. (2009). Sleep minder: An innovative contact-free device for the estimation of the apnea hypopnoea index. *Annual International Conference of the IEEE Engineering in Medicine and Biology Society*, (pp. 7091-9094).

KEY TERMS AND DEFINITIONS

Analog-to-Digital Converter (ADC): It is a device that converts a continuous quantity to a discrete time digital representation.

Antenna Gain: A key performance figure which combines the antenna's directivity and electrical efficiency.

Bandpass Filter: A device that passes frequencies within a certain range and rejects (attenuates) frequencies outside that range.

Correlation Coefficient: A measure of the strength of the linear relationship between two variables that is defined in terms of the (sample) covariance of the variables divided by their (sample) standard deviations.

Diversity: A statistic intended to assess the diversity of any population in which each member belongs to a unique group, type or species.

Electrocardiography (ECG): A transthoracic (across the thorax or chest) interpretation of the electrical activity of the heart over a period of time.

Electroencephalography (EEG): The recording of electrical activity along the scalp. EEG measures voltage fluctuations resulting from ionic current flows within theneurons of the brain.

Electromyography (EMG): A technique for evaluating and recording the electrical activity produced by skeletal muscles.

Electrooculography (EOG): A technique for measuring the resting potential of the retina.

Microcontroller (MCU): It is a small computer on a single integrated circuit containing a processor core, memory, and programmable input/output peripherals.

Multiple Input Multiple Output (MIMO): The use of multiple antennas at both the transmitter and receiver to improve communication performance.

Patch Antenna: It is a type of radio antenna with a low profile, which can be mounted on a flat surface.

Radiation Pattern: The directional (angular) dependence of the strength of the radio waves from the antenna or other source.

Radio Frequency Identification (RFID): A technology that uses radio waves to transfer data from an electronic tag, called RFID tag or label, attached to an object, through a reader for the purpose of identifying and tracking the object.

Radio Telemetry: A radio frequency technology that allows remote measurement and transparent conveyance of remote information.

Return Loss: The loss of signal power resulting from the reflection caused at a discontinuity in a transmission line or optical fiber.

Rician Fading: It is a stochastic model for radio propagation anomaly caused by partial cancellation of a radio signal by itself.

Rician-*k* Factor: The ratio of signal power in dominant component over the (local-mean) scattered power.

Sleep Apnea: A sleep disorder characterized by abnormal pauses in breathing or instances of abnormally low breathing, during sleep.

Space Time Block Code (STBC): A technique used in wireless communications to transmit multiple copies of a data stream across a number of antennas and to exploit the various received versions of the data to improve the reliability of data-transfer.

Transceiver: A device comprising both a transmitter and a receiver which are combined and share common circuitry or a single housing.

Wireless Body Sensor Network (WBSN): It consists of spatially distributed autonomous sensors to monitor physical or environmental conditions.

Wireless Sensor: A sensor communicate with base station or other devices without cables.

ENDNOTES

1. Data acquisition board produced by Crossbow
2. Federal Communications Commission
3. Recording the count of movements per minute

Chapter 13
Chipless RFID Sensor for High Voltage Condition Monitoring

Emran Amin
Monash University, Australia

Nemai C. Karmakar
Monash University, Australia

ABSTRACT

A novel approach for non-invasive radiometric Partial Discharge (PD) detection and localization of faulty power apparatuses in switchyards using Chipless Radio Frequency Identification (RFID) based sensor is presented. The sensor integrates temperature sensing together with PD detection to assist on-line automated condition monitoring of high voltage equipment. The sensor is a multi-resonator based passive circuit with two antennas for reception of PD signal from the source and transmission of the captured PD to the base station. The sensor captures PD signal, processes it with designated spectral signatures as identification data bits, incorporates temperature information, and retransmits the data with PD signals to the base station. Analyzing the PD signal in the base station, both the PD levels and temperature of a particular faulty source can be retrieved. The prototype sensor was designed, fabricated, and tested for performance analysis. Results verify that the sensor is capable of identifying different sources at the events of PD. The proposed low cost passive RFID based PD sensor has a major advantage over existing condition monitoring techniques due to its scalability to large substations for mass deployment.

1. INTRODUCTION

RFID (Radio Frequency Identification) is emerging as a successful alternative to the traditional barcodes due to its embedded security, larger reading distance irrespective of line of sight, greater storage capacity and elevated reliability (Finkenzeller, 2010).The RFID tags are made robust so that they can be used in any harsh environments and temperatures. Each tag has a unique ID which is used to track or localize goods. The tag uses modulated backscattering of RF signals to communicate with the reader.

DOI: 10.4018/978-1-4666-2080-3.ch013

Apart from the tracking an object, RFID tags can monitor the surrounding environmental conditions and act as a sensor. Thus a sensor enabled tag monitors some of the physical parameters (temperature, pressure, humidity etc.) in addition to its identification function (Xia, 2009). This offers an accurate and relatively simple implement solution also. An object providing its own condition and identification simultaneously simplifies the infrastructure and enhances the quality of the information. RFID sensors create a link between the digital networked world and the physical world. Its role is to deliver a unique ID to the networked data related to the product. In addition, RFID acts as the virtual connection between the object and composite decision processes that are based on sensory data inputs from the users. RFID sensor technology has revolutionized the field of healthcare, structural damage detection environmental monitoring, transporting pharmaceuticals, storing goods, promoting security etc. (Want, 2004).

The power industry has introduced RFID for asset management, inventory control and equipment monitoring. In Sen et al (2009), five categories of problems in power system management which can be overcome through RFID and wireless technology is presented. These are 1) Locating asset. 2) Identification and status of asset; 3) Tool tracking; 4) Fleet management; and 5) Access management and infrastructure security. Power facility management using RFID integrated with Wireless Sensor Network (WSN) is proposed in (Kim et al., 2006). The combination of RFID and sensors such as (thermal, wind speed and direction, battery life etc.) provides real time condition monitoring facility and improves transmission reliability. Moreover, active RFID is a key ingredient for the Smart Grid where power distribution system automates generation, delivery and consumption of electrical energy using communication and information technologies. Figure 1 is an illustration of power distribution monitoring system using RFID tag and WSN. This provides real time power system surveillance with minimum cost.

However, chipless RFID is a new dimension in the regime of RFID technology which has not been employed in power management system. Chipless RFID tag circuitry does not have any semiconductor chip or IC in it thus the cost is significantly lower than the chipped tags (S. Preradovic & Karmakar, 2010). It is completely passive having a specific EM structure and operates when the tag is illuminated by a reader antenna. The chipless RFID tag can also incorporate sensing mechanism without utilizing an external sensor within the tag (Bhattacharyya et al., 2010). Chipless RFID sensor can revolutionize the automated condition monitoring of High Voltage (HV) equipment in a power substation by cutting down cost, enhanced dynamic range and longevity. In this chapter a novel application of chipless RFID sensor is proposed for real time condition supervision. Our proposed sensor addresses two aspects of health monitoring for HV equipment which are 1) Partial Discharge (PD) due to insulation deterioration and 2) temperature measurement. We introduce a new era in condition monitoring of electricity substations by integrating these two sensing paradigm on the platform of microstrip based chipless RFID sensor (Figure 2).

In the next section a brief overview and background study of chipless RFID sensor and its applicability to HV condition monitoring is depicted. The section is divided into three major segments. The first two segments illustrate the state- of- art of chipless RFID technology and its integration with temperature sensing. The last segment describes the research undertaken in the area of PD detection and localizing for fault analysis in a switchyard. The section concludes by pointing out substantial pertinence of chipless RFID sensor in the regime of on line switchyard surveillance.

In Section 3, the general theory of our proposed chipless RFID sensor is presented. This section is

Figure 1. A power distribution monitoring system using RFID sensor network

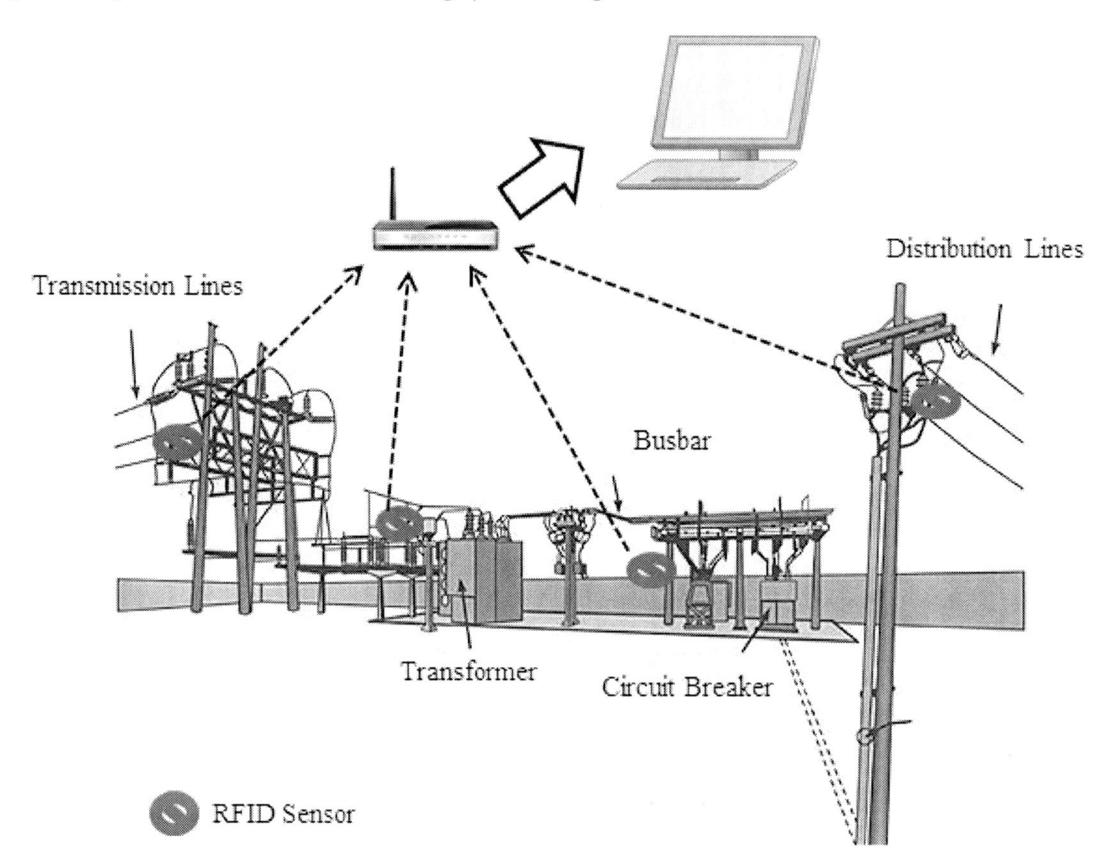

Figure 2. Schemetic block diagram of the functionality and scope of our proposed chipless RFID based PD sensor

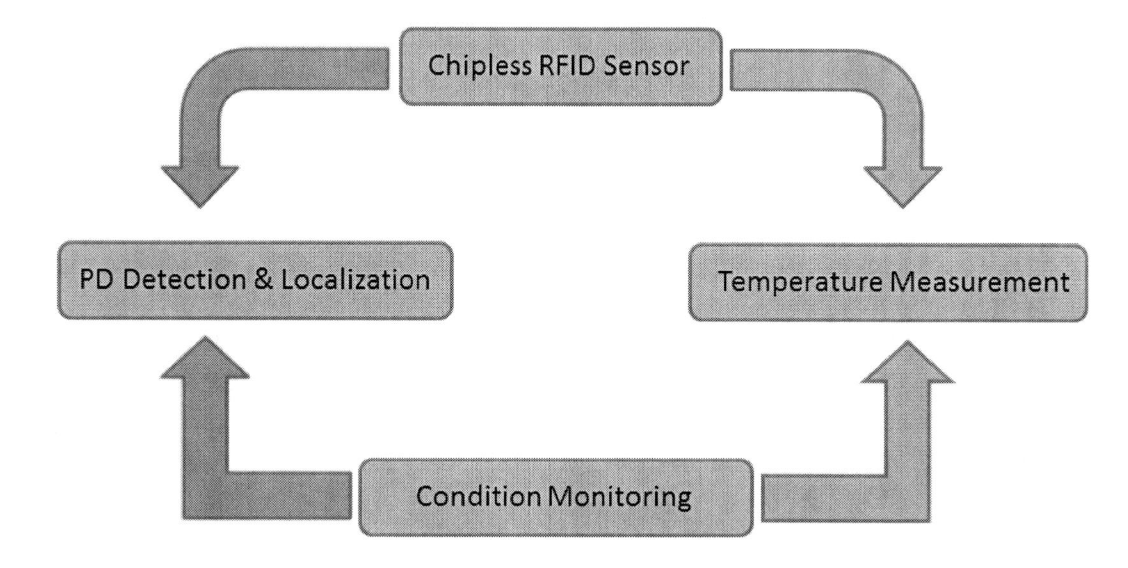

segmented into two parts to demonstrate the system overview followed by the theoretical development and integration of the PD and temperature sensor. Section 4 deals with the guidelines for designing individual components of the sensor prototype. The simulated and experimental results to verify our senor performance is presented in Section 5 which is followed by future research outcomes and challenges to conclude in Section 6.

2. BACKGROUND

2.1 Chipless RFID

RFID is a wireless identification system where the tag is attached to an object to be tracked similar to the existing conventional optical barcodes. The advantage of RFID technique over barcodes are fast and secure reading process which is independent of line of sight, ability to operate in robust or hazardous environment without human intervention thus allowing automated asset management plausible. Moreover, RFID tags can store large amount of data and performs bi-directional communication with the reader (Finkenzeller, 2010).

The cost of a RFID tag is largely on the silicon IC or chip on the tag circuit. In chipless RFID this overhead cost of silicon is excluded from the total budget of the system. The chipless RFID tag is fully printable and passive thus resistant to harsh environment and weather conditions. The potential advantages of this feature allows chipless RFID in applications such as low cost item tagging like banknotes, ID cards, books, consumer goods etc. (Das, 2006; Fletcher, 2002; Plessky & Reindl, 2010). In literature (Preradovic et al., 2009), a spectral signature based multi-resonator chipless RFID tag is presented. The cascaded spiral resonators are coupled to a microstrip line and act as band stop filters modifying the spectrum of the received signal sent by the reader (Figure 3). A continuous wave frequency spectrum from the reader's transmitting antenna illuminates the tag. The receiving antenna of the tag receives the signal and passes through the series of spiral loaded transmission line. Then the signal containing spectral signatures is transmitted back to the reader. Finally the reader decodes data by observing the attenuated frequencies. The number of cascaded spirals corresponds to the number of data bits the tag can store. A time domain based UWB chipless tag is proposed in (Ramos et al., 2011). Here a UWB signal is send through a transmitting antenna which is reflected back from the chipless tag. The receiving antenna captures the backscattered signal and further time domain analysis of the 'antenna mode' can detect the time coding of different tags.

2.2 Chipless RFID Temperature Sensor

The primary challenge of a chipless RFID temperature sensor is to embed the sensing operation within the tags general principle of tracking the object. In active or passive chip based sensors a separate IC or module is dedicated for sensory as well as the tagging purpose. A central control unit or a microcontroller maintains the communication between the sensing unit and the RF link. Conversely, a chipless RFID sensor is designed to integrate the variation in physical parameters in the tags normal operation so that it does not require a separate sensing platform. A number of literatures can be found where the researchers have developed temperature sensor based on the chipless tag technology. A Surface Acoustic Wave (SAW) based RFID temperature sensor is proposed in (Dowling et al., 2009) which uses the Time Domain Reflectrometry (TDR) encoding scheme. It is reported as completely passive still has the drawbacks of not being fully printable. A completely passive LC resonant telemetry scheme, relying on frequency response is applied to develop a temperature sensor in (Wang et al., 2008). Though it is capable of operating in robust environment, it is chipped and therefore is not fully

Figure 3. Basic operation of multiresonator based chipless RFID tag

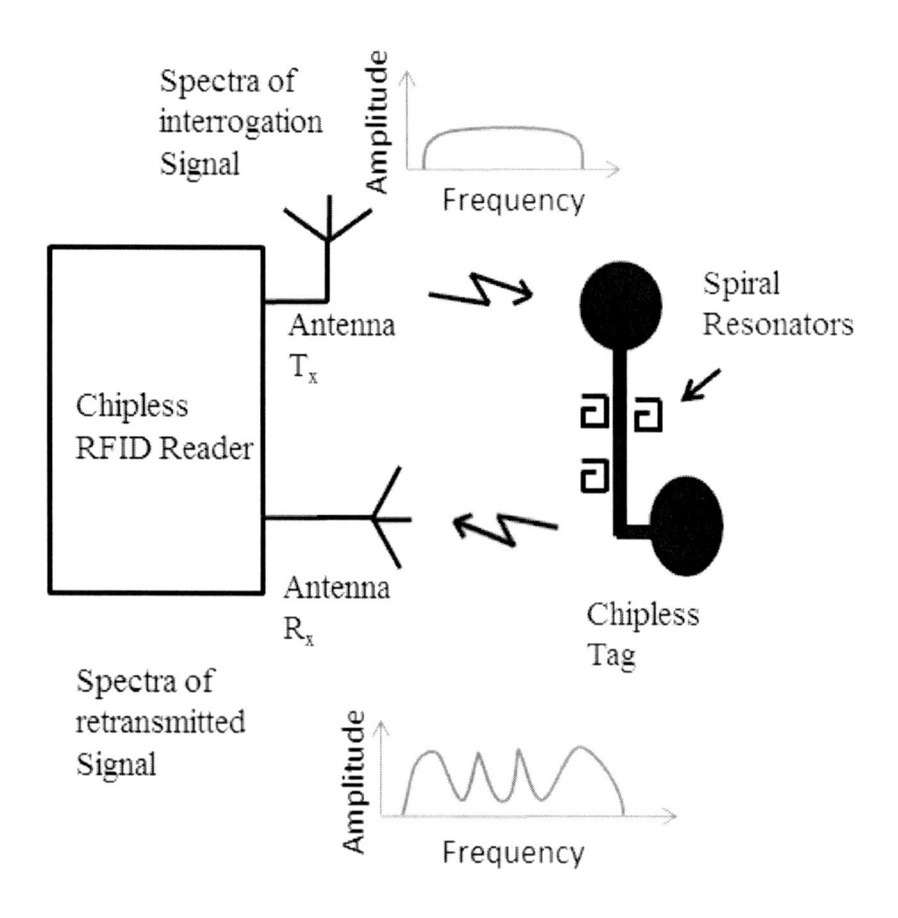

printable. An RFID sensor designed to monitor the change of temperature or other environmental effects using the tags own physical properties is presented in (Bhattacharyya, et al., 2010) which changes mechanical structure of the tag as a temperature threshold is surpassed. In (Shrestha et al., 2009) microstrip transmission line based chipless RFID tag was reported where a single transmission line produces a Pulse Position Modulation (PPM) to denote a definite ID code and the phase change of the reflected pulse is used for detecting sensory information. A spectral signature based multi-resonator chipless RFID tag is modified to incorporate sensing mechanism in (Amin & Karmakar, 2011). Here, the first N-1 spirals of an N bit tag carry the identification data whereas the last spiral contains temperature information. Results verify that the sensing mechanism is independent of identifying the object.

2.3 Power System Condition Monitoring

The condition monitoring of HV equipment in a switchyard or substation is a routine check throughout the whole lifespan of the equipment. The health of insulation in HV equipment's are damaged due to over-voltages, manufacturing defects, or aging. This can eventually result in electrical breakdown causing power failure, equipment casualty, and loss of life and property. Partial Discharge (PD) is the energy dissipation caused when the electric field across a dielectric exceeds a threshold breakdown value (Kreuger, 1989). PD detection is used in power systems to monitor the state of dielectric insulation of HV equipment. As PD is an early warning of insulation deterioration, failure to detect PD efficiently on time, can lead to the catastrophic disruption of the equipment. Reliable online PD detection is of significant interest for power distributors to ensure personnel safety and reduce potential loss of service. Another physical parameter which indicates the health condition of HV apparatus is temperature. In (Chu & Williamson, 1982) thermal imaging system was used to locate faults in a transformer. The temperature rise due to arc discharge was monitored through infra-red camera. Results verify that the sensitivity of fault localization increases as temperature is observed during PD.

2.3.1 Partial Discharge

A partial discharge in an HV transformer (HVT) occurs when the electric field in a localized area changes in such a way that a localized current is created (Boggs, 1990). This localized current manifests itself as an electrical pulse that is measurable at the output of the transformer. PD detection and localization techniques have been developed starting from early 1940s (Austen & Hackett, 1944; Kemp, 1995; Nattrass, 1993). The principle of measuring the level or severity of dielectric degradation is speculating the variation in size and shape of the cavity as well as the change of temperature, pressure, or electrical stress during PD. These variations result in different physically observable signals including electrical (Kreuger, 1989) and acoustic pulses (Lundgaard, 1992), ultraviolet ray, temperature variation (Chu & Williamson, 1982), gas discharge (Duval, 1989), or Radio Frequency (RF) emission (Judd et al., 2005; Qin & Birlasekaran, 2002). According to this, a number of methods are found for detecting PD signals which can be grouped into six categories, based on the PD manifestation that they measure:

- Chemical
- Electrical
- Acoustic
- Optical
- Infra-Red
- Radiometric

2.3.2 Radiometric Detection

In Radiometric detection the RF emission due to PD is captured and analyzed for PD level and location finding. According to electromagnetic theory, The EM wave is produced from the acceleration of free electrons within the voids inside an insulator. These waves last for a very short time ($< 1 \mu s$) (A. J. Reid et al., 2006a). As the duration of consecutive PD pulses are in the range of nanoseconds, the induced RF signals superimpose causing fluctuating amplitudes. Moreover, the wave is distorted due to the reflection and refraction as it propagates through the dielectric (A. J. Reid et al., 2006b). The frequency span of the emitted RF waves can be found between 300 MHz to 3 GHz on UHF band. As mentioned in (A. J. Reid et al., 2011), the electromagnetic emission due to PD is analogous to the acoustic ringing of a bell once struck. This infers that the RF signal emitted from PD source has less attenuation by electrical insulation but contains multiple reflections. The captured PD pulse contains information about the type and severity of insulation damage. In radiometric detection, the frequency spectrum of the PD signal is analyzed to extract information about insulation deterioration, ageing process and other common faults within the equipment. This method makes online PD detection very attractive because it enables real time monitoring of HV systems possible. The major advantage of radiometric detection over the other PD detection techniques is it can measure the severity and source identification in a nonintrusive manner. The equipment under test can be live in the switchyard while detecting PD. Finally, radiometric detection assures safety of personnel while switchyard surveillance.

A major challenge in radiometric PD detection is exact localization of the fault. As mentioned before PD occurrence is very unpredictable and lasts for a short time. Moreover, the environment of a power substation or a switchyard is EM noisy as it has a number of HV equipment emitting RF waves. In most cases the presence of a fault may be detected observing the increased RF level within a switchyard, however, the origin of the RF signal is hard to detect. For analogy, if there is a number of ringing bells in a room, it is difficult to distinguish a specific sound. In practice, researchers have developed a number of locating finding techniques for PD faults which are described in the following section.

2.3.3 Localizing PD Faults

Time Difference of Arrival (TDOA) method is a well-known PD location finding algorithm which has been investigated in (Moore et al., 2005; S. Tenbohlen, 2006; Stewart et al., 2009; Zhiguo et al., 2006). In literature (Moore, et al., 2005), a 4 element wide band antenna array was used to receive PD signal from a distance and TDOA algorithm was developed to localize the faulty source. This method was tested in two situations of on-site testing and a mobile measurement set up. Similar approach was done in (Stewart, et al., 2009) where four smaller sized mobile broadband antennas were used for 3D localization of PD source in a 400KV substation. In (Zhiguo, et al., 2006) a large frequency band (1 GHz-5 GHz) was measured to accurately captured the initial rise time of the PD pulses. An algorithm was proposed to automatically measure the time of arrival which showed reduced location error while experimentation. However, in this localizing algorithm, the antennas have to capture the initial part of wideband radiated PD signal. As it is distorted due to multipath reflection while propagating through the environment, the time between the direct wave and reflected wave is a variable depending on the position and distance of reflecting objects close to the source and antennas (Moore, et al., 2005). Also, the emitted RF signal is distorted due to the EM noisy environment. This largely limits the accuracy of this method. In addition, the fault location precision depends on the sampling rate of the oscilloscope. A high speed oscilloscope generates excessive data which requires additional hardware and computational arrangement to manipulate them. This endues overhead cost and increases system complexity.

In Luo et al (2006), an ultrasonic receiving planar array transducer is developed for detecting PD location in power transformer. The transducer consists of 16 x 16 elements to form digital beam-scanning using planar array theory. It is reported to locate multiple PD within a transformer. A novel VHF-UHF radio interferometer system (VURIS) for PD localization is proposed in (Kawada, 2003). The system captures the emitted EM wave using two identical antennas and calculates the angle of arrival from the phase difference information. This requires fewer antennas for location finding however the feasibility was tested only in laboratory setup. In (Hu et al., 2010; Y. Tian et al., 2008; Ye Tian et al., 2009), FDTD method is used to simulate the propagation of EM waves from PD faults and PD location is measured through the time of flight calculation. FDTD method improves the accuracy of location finding as the optimum 3D positioning of the UHF sensors is analyzed. In (Ishimaru & Kawada, 2010), PD source localization using maximum likelihood estimation is presented. An UHF sensor based PD detection and localization technique is presented in (Reid et al., 2009) where the sensing units are electrically isolated. Each sensing unit is connected to the central monitoring unit or oscilloscope via a cable loop. The cable loop is designed such that the PD signal captured from different units arrives at the oscilloscope at distinct time intervals which identify the faulty unit. The system is envisioned to detect PD in metal enclosed switchgear as the units have to be electrically shielded. In the aforementioned literatures the sensors are not designed for remote localization of PD source.

The above study reveals that radiometric detection has emerged as an inevitable technique for on line non-intrusive PD diagnosis of HV power apparatus. Radiometric detection can automate the condition monitoring of a switchyard as it allows scanning the switchyard using wide band UHF antennas. However, there remain challenges in localizing the faulty source due to the external neighboring interferences, presence of metallic substances, multipath and reflections and above all the PD pulses have very low intensity and short duration. To this extent we propose a chipless RFID based PD sensor for exact localization and condition monitoring of HV apparatus in a distribution switchyard.

2.4 Chipless RFID Sensor for PD Localization and Condition Monitoring

In this chapter a wireless remote PD sensor incorporating frequency modulation based chipless RFID technique is proposed for exact localization of potential PD sources in a switchyard. In addition to localizing PD faults, the sensor can provide the RF level and temperature information of a faulty source to the central base station. The sensor is proposed to have the following features:

- The remote sensor is placed at closed proximity of potential PD sources in a switchyard. It captures PD signal processes it with designated spectral signatures as identification data bits and retransmits the data with PD signals to the base station. Analyzing the PD signal in the base station, both the PD levels and source identifications can be retrieved.
- As discussed before, conventional time domain based localizing techniques encounter multipath and reflection delay as the PD signal travels towards the antenna or UHF sensor. Unlike the conventional RF measurement techniques, our proposed sensor uses frequency modulation based RFID for source identification. This mitigates reflection and multipath effect on source localization in a switchyard.
- Earlier RF detection techniques developed antenna or UHF coupler to capture and analyze PD signal. As PD pulses are impulsive and irregular in nature, the signal processing and denoising requires high speed hardware configuration. This entails a limitation of conventional RF detection in the regime of remote detection. Our PD sensor overcomes this constraint by guiding the PD pulse towards base station with minimal hardware configuration. This ensures low cost remote PD detection conceivable.
- As the sensors are placed close to the potential PD sources, our PD detection and localizing method is more immune to external neighboring interferences. Also, the PD signal transmitted to the base station has a distinct frequency signature which reduces the chance of false detection.
- By integrating microstrip based temperature sensing mechanism, the temperature information of a faulty source is send together with the PD signal. This adds a new dimension in condition monitoring of HV equipment.
- The designed low cost sensor can be deployed in mass scale around a switchyard. It can perform a non-invasive 24/ 7 online condition monitoring in a switchyard.

In the consecutive sections theory and design of proposed PD sensor prototype is depicted. Further experiment was conducted to verify the feasibility of developed sensor.

PD manifests a short transient current which generates EM wave. The current waveform due to PD depends on the type of dielectric fault such as void discharge, surface discharge or discharge relating conducting particles (Niemeyer, 1993). Although the real PD pulse is not measurable, researchers have approximated the current waveform with a Gaussian equation given by(A. J. Reid, et al., 2006b), in Equation 1"

$$I = I_0 e^{-(t/t_0)^2} \quad (1)$$

where

$$t_0 = \frac{T_h}{2\sqrt{\log 2}}$$

Here, I_0 is the peak current and T_h is the half amplitude width of the pulse. Typically T_h has a value in nanoseconds. A list of parametric values of PD current pulse for various dielectric faults is given in (Shibuya et al., 2010). A general waveform of PD current is shown in Figure 4. This is an important parameter for UHF PD detection as the frequency spectrum of the emitted pulse depends on the rise time of PD pulse. The frequency spectrum of a typical PD pulse shown in Figure 5 reveals that it has a relatively flat response over a wide frequency range.

Figure 4. General wave of PD current

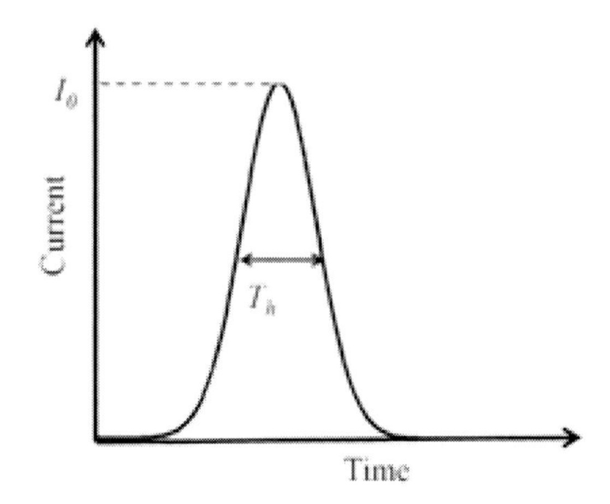

Figure 5. UWB spectrum of PD pulse

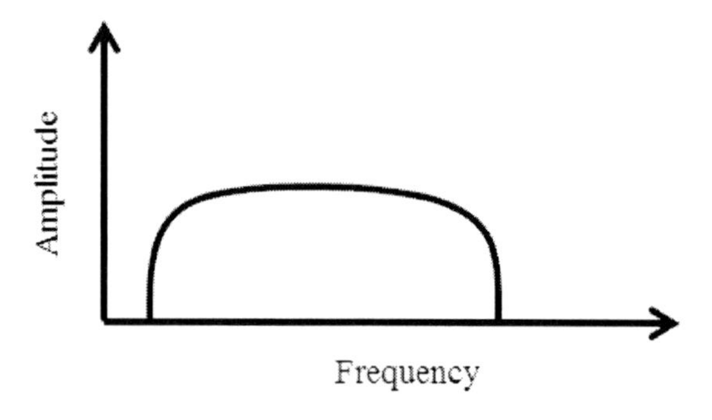

2.5 System Overview

The general working principle of the proposed PD sensor is to capture the UWB RF signal using a wideband receiving antenna R_x and pass it through a number of cascaded microstrip resonators to attenuate distinct frequencies (Preradovic, et al., 2009). This modification embeds a distinct frequency signature within the pulse which is transmitted using another transmitting antenna T_x to the base station for further processing. The block diagram of Figure 6 gives a general overview of the sensor structure. The sensor consists of two identical antennas for capturing and retransmitting the PD signal and a Frequency Modulation Unit (FMU) for PD localization and condition monitoring. The FMU comprises of N passive Stepped Impedance Resonator (SIR) filters to perform frequency attenuation to detect and localize PD and a spiral resonator for monitoring temperature variation. As shown in Figure 6, each SIR filter produces a dip in the magnitude spectrum of the retransmitted signal. This property is used to encode the data into the frequency spectra of received power spectrum. The presence of a magnitude null resembles logic '0' whereas the absence of a magnitude null at a distinct frequency resembles logic '1'. Also the combination of data bits when no filter is present is omitted in our proposed sensor system. Thus, N number of cascaded resonators can produce, $M = 2^N - 1$, different frequency signature to detect

Figure 6. Block diagram of proposed chipless RFID based PD sensor

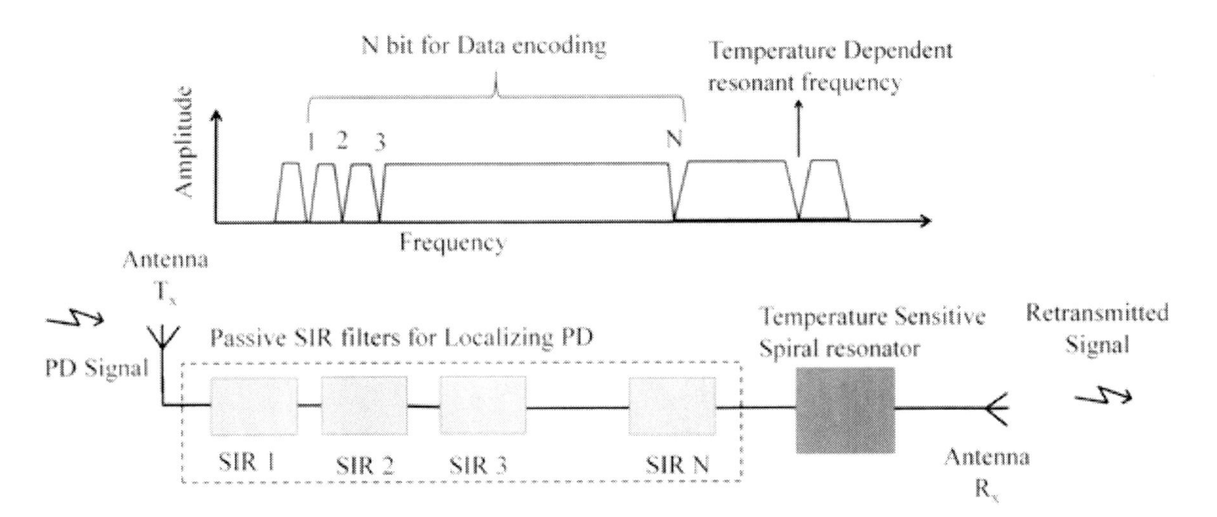

M number of objects. On the other hand, the spiral resonator which operates as an electrically coupled LC filter is modified using a temperature sensitive polyamide to incorporate temperature variation. Thus, the resonant frequency of the spiral resembles surrounding temperature. The basic operation of our proposed PD sensor block can be summarized as follows:

- The sensor is placed closed to a potential PD source. Each individual PD source is identified by the ID or tag of its adjacent sensor.
- The receiving antenna R_x captures the PD signal emitted from the faulty source.
- The captured signal passes through the Frequency Modulation Unit (FMU) which embeds the source ID and temperature information within the signal
- The frequency modulated PD signal is transmitted through the T_x antenna to base station.
- The signal is processed and analyzed in the base station to determine the frequency encoded data. This gives the sensor identification together with RF intensity and temperature.

In the following section a detail example of utilizing remote sensor to detect and localize PD source is illustrated.

The schematic diagram in Figure 7(a) shows three potential PD sources PD_1, PD_2 and PD_3 in a switchyard. These sources are situated apart from one another and also placed at a distance from the base station. The base station has a receiving antenna array and a central control unit to detect and analyze PD signals. There are three RF PD sensors S_1, S_2 and S_3 placed close to PD sources PD_1, PD_2 and PD_3. It is assumed that the receiving antennas R_{x1}, R_{x2} and R_{x3} of the sensors will capture PD signals radiating from the source closest to it. Also due to short duration of PD impulse the probability of multiple PD occurrences at the same instance is extremely small (Moore, et al., 2005)

As depicted in Figure 7(a), at the event of a PD at source PD_1, the signal is captured by receiving antenna R_{x1} and FMU_1 gives a distinct frequency signature after passing through it. This modulated signal is transmitted through T_{x1} antenna to the base station. Likewise, if PD occurs from PD_2 or PD_3, the sensor S_2 or S_3 is activated and analogous retransmitted signal is generated. However, the modulation is dissimilar as FMU_1, FMU_2 and FMU_3 are unique. The FMU is a com-

Figure 7. (a) Proposed PD sensor operating in a switchyard; (b) spectral modification of the retransmitted signals from sensors S_1, S_2 and S_3.

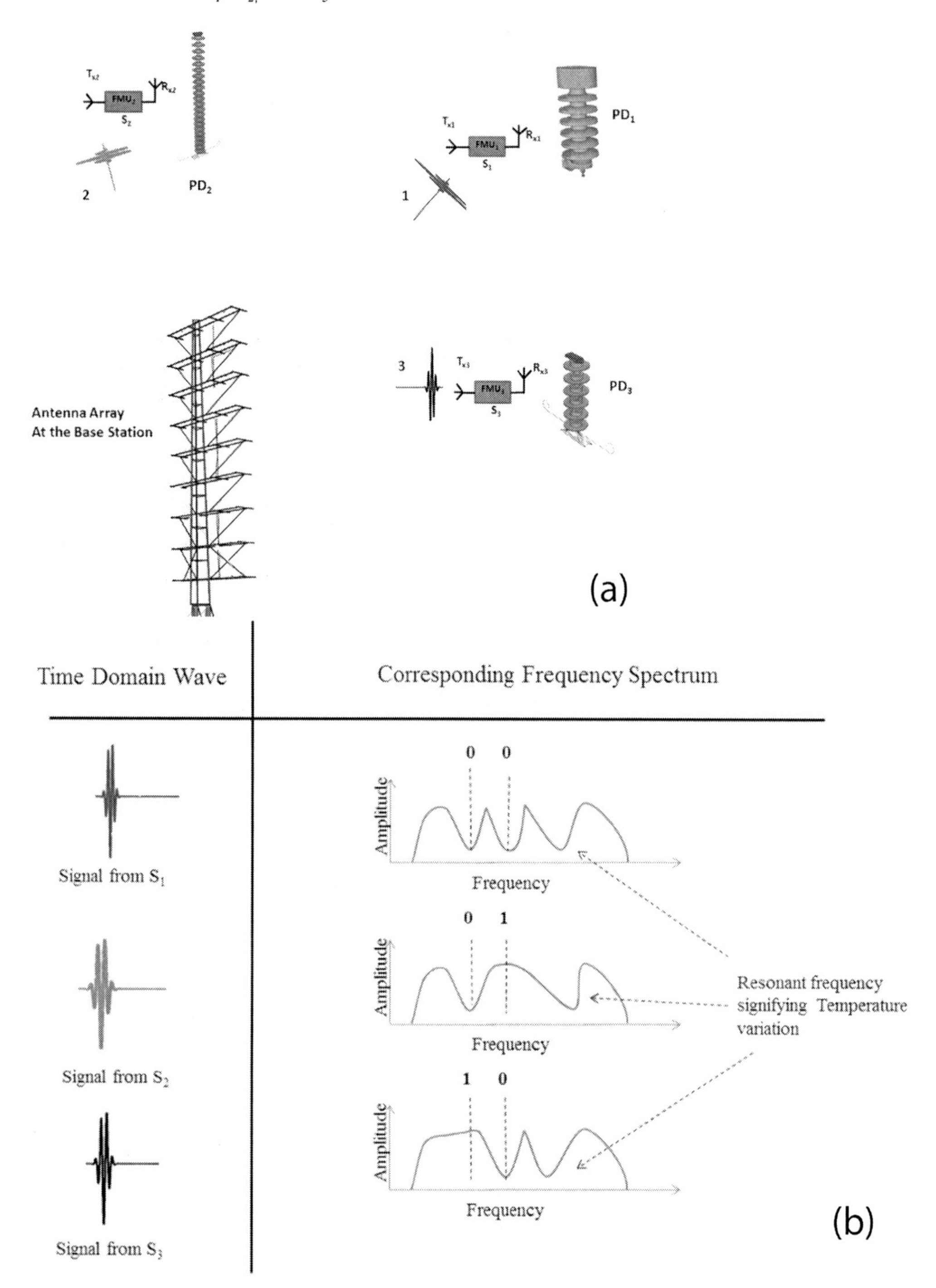

Box 1. Equivalent Impedance

$$Z_s = \frac{j(Z_1Z_2\tan\theta_2 + Z_1Z_3\tan\theta_3 + Z_1^2\tan\theta_1 - \frac{Z_1^2Z_2}{Z_3}\tan\theta_1\tan\theta_2\tan\theta_3)}{(Z_1 - Z_2\tan\theta_1\tan\theta_2 - Z_3\tan\theta_1\tan\theta_3 - \frac{Z_1Z_2}{Z_3}\tan\theta_2\tan\theta_3)} \quad (2)$$

bination of cascaded band stop filters where each filter corresponds to a single data bit. Each FMU has two passive SIR filters for designating identification bits and a spiral resonator to convey temperature information. As a result, there can be three different combination of spectral signature to identify each source.

In Figure 7(b), the spectrum of retransmitted signal from three sensors is shown. If the absence of a filter is denoted by '1' and the presence is denoted by '0' then the combinations of first two bit can represent '00', '01' and '10' bit to tag sources PD_1, PD_2 and PD_3. However, the third bit is present in all the combinations to carry corresponding temperature data.

In the following sections, detailed theory of individual component of the proposed PD sensor and their integration is explained.

2.6 Frequency Modulation Unit

The FMU unit performs a frequency domain modification of the captured PD signal using a number of cascaded passive resonators which are the fundamental component of the PD sensor. The resonators are narrow bandstop filters and operate within the bandwidth of receiving antenna R_x. This band limiting property enables each resonator to function independently while they are cascaded in series. Our proposed sensor combines two types of transmission line based resonators to achieve condition monitoring. Firstly, a tri step SIR filter (Sagawa et al., 1997) is proposed to perform the identification and localization of PD in a switchyard. Secondly, a compact size microstrip spiral resonator is (Young-Taek et al., 2002) modified to incorporate temperature sensing mechanism. The theory behind each of the resonator structures is as follows.

2.6.1 Stepped Impedance Resonator (SIR)

Stepped Impedance Resonator (SIR) is transmission line resonator utilizing quasi TEM modes. The SIR has advantages over the uniform impedance resonators (UIR) in terms of wide degree of freedom in terms of designing, compact size and ease of fabrication etc. (Hong & Lancaster, 2011). The basic structure of a three element half wave SIR is shown in Figure 8(a). This structure is symmetric at the mid center 'O' and comprises of two cascaded quarter wave tri- step SIR as shown in Figure 8(b). The characteristic impedance of the three steps is Z_1, Z_2 and Z_3 having electrical length θ_1, θ_2 and θ_3. Thus, the overall electrical length of the half wave SIR is $2\theta_T = 2(\theta_1+\theta_2+\theta_3)$.

The overall admittance Y looking into the open end in Figure 8(b) is calculated in Box 1 from the equivalent impedance Z_s (Hualiang & Chen, 2005).

Taking $a = Z_1/Z_2$ and $b = Z_3/Z_2$, the resonance condition for the SIR filter is, $Y_s = 0$, thus:

$$b\tan\theta_1\tan\theta_3 + a\tan\theta_1\tan\theta_2 + (\frac{a}{b})\tan\theta_1\tan\theta_3 = 1 \quad (3)$$

From Equation 2 it is found that the resonance condition depends on the three electrical lengths and the impedance ratios *a* and *b*. For, simplicity

Figure 8. (a) Layout of proposed half wave tri-step SIR filter; (b) quarter wave tri- step SIR

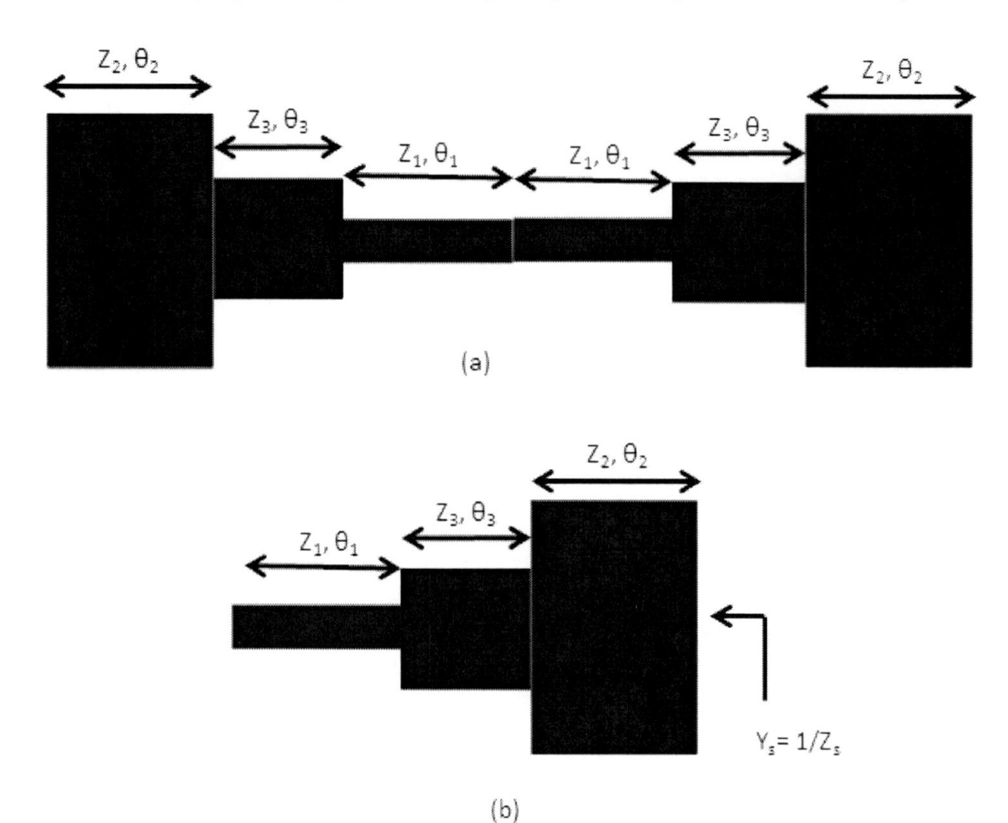

of analysis here we set $\theta_1 = \theta_2$. Thus Z_1 and Z_2 have equal length. Further simplification enables θ_3 to be expressed as (Hualiang & Chen, 2006):

$$\theta_3 = \tan^{-1}\left(\frac{1 - a\tan^2\theta_1}{P\tan\theta_1}\right)$$

where

$$P = (b + \frac{a}{b})$$

So the overall length of quarter wave SIR is, $\theta_T = 2\,\theta_1 + \theta_3$ and is a function of impedance ratios *a* and *b*. It is reported in (Hualiang & Chen, 2006) that the minimum electrical length of a tri step SIR is found when *a* and *b* is within a range of $a > 1 > b$.

2.6.2 Temperature Sensing Spiral Resonator

The microwave compact size spiral resonator is fully printable, planar and exhibits narrow band characteristics. It has great application in wireless and satellite communication services especially in Monolithic Microwave Integrated Circuits (MMIC) as their reduced size and planar 2D configuration. A low cost fully printable tag consisting spiral resonators has been developed in (Preradovic, 2009) for replacing barcodes. To incorporate temperature sensing in multiresonator based chipless tags, a detail analysis of the spectral response for individual spiral resonator is performed. Figure 9(a) shows a single spiral resonator coupled to a microstrip line. The different parameters of the spiral are defined in the Figure 9(a) which affect the filter response.

Figure 9. (a) Layout of a spiral resonator coupled to a strip line; (b) equivalent circuit

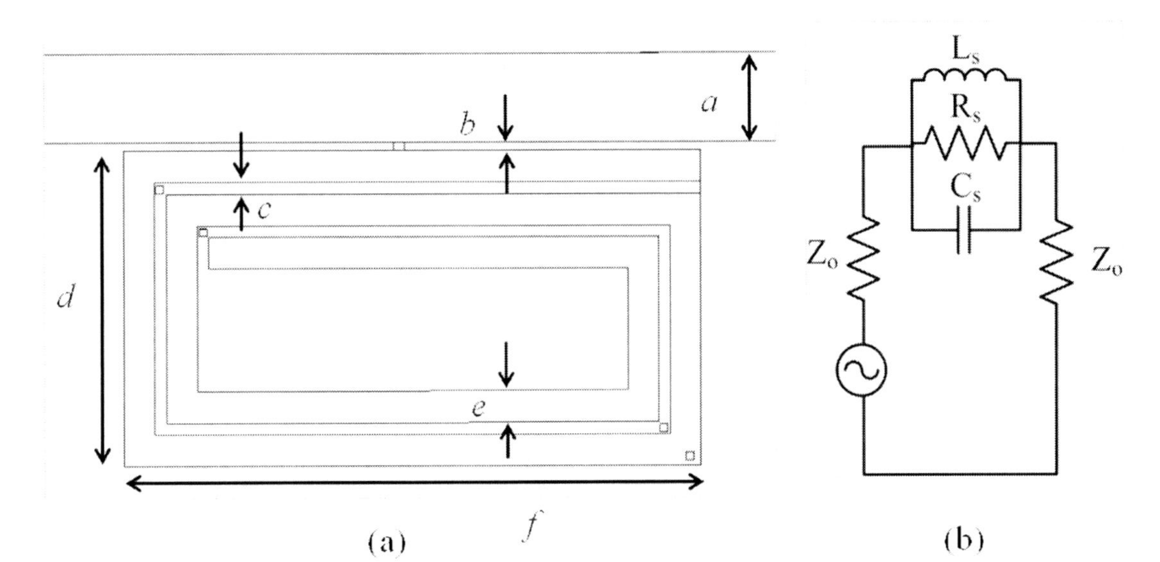

Moreover, a parallel RLC equivalent circuit is presented in Fig. 9(b) (Young-Taek, et al., 2002). In the equivalent circuit, the spiral is modeled by a resistor R_s, an inductor L_s and a capacitor C_s. The value of these lumped elements depends on the distributive parameters described in Fig. 9(a). Thus changing the structural parameters the resonant frequency of a single spiral can be tuned. The resonant frequency of the spiral is given by.

$$f_r = \frac{1}{2\pi\sqrt{L_s C_s}} \tag{4}$$

To incorporate temperature sensing in our proposed sensor, a temperature dependent high K polyamide is used to modify the structure of the spiral. Conventional dielectric substrates do not have temperature dependent dielectric property. As the permittivity of the substrate affects the equivalent capacitance of microstrip resonators, the modified spiral has a temperature dependent capacitance C_T instead of C_s. The resonant frequency of the spiral f_T varies with C_T accordingly thus resembles temperature variation:

$$f_T = \frac{1}{2\pi\sqrt{L_s C_T}} \tag{5}$$

2.6.3 Dipole Antenna

The half wave dipole is the simplest form of dipole antenna having two quarter wavelength wires to form a total length of half the wavelength of operating frequency. Moreover, the dipole antenna has superior transient response than other antennas and its polarization characteristics is analogous to the EM radiation from PD source (Hoshino et al., 2001).

Two identical half wave dipole antennas were used as transmitting and receiving antenna for the PD sensor. A general structure of half wave dipole antenna is shown in Figure 10. One of the wires is connected to ground whereas the other one is powered.

Assuming a dipole antenna of length *l* placed along the z-axis having its center at z=0. The general form of current when a sinusoidal voltage is impressed can be written as (Balanis, 1992):

Figure 10. A half wave dipole antenna

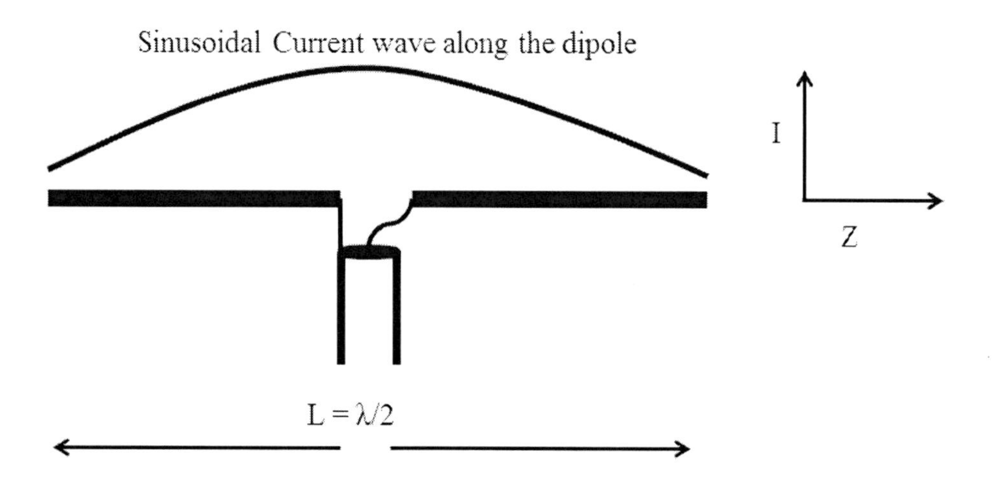

$$I(z) = I_0 \sin[k(\frac{L}{2} \pm z)],$$
$$-L/2 < z < L/2$$

Here, I_0 is the peak of current magnitude. The electric and magnetic fields radiated from the half-wave dipole antenna are given by:

$$E_\theta = \frac{j\eta I_0 e^{-jkr} \cos\left(\frac{\pi \cos\theta}{2}\right)}{2\pi r \sin\theta}$$
$$H_\phi = \frac{E_\theta}{\eta}$$

The half wave dipole antenna has an omni-directional radiation pattern which is symmetric about the z axis. The radiation has a maximum peak at right angles to the dipole axis and reduces to zero at the antenna axis. The impedance of a half wave dipole antenna is Z= 73.2+ j* 42.5 Ohms. By tuning the total wire length slightly less than λ/2, the impedance is calculated as, Z=73.2 ohms. At this impedance the antenna resonates. The gain of a half wave dipole antenna is measured as 2.15 dBi. Inserting a back reflector ground plate the gain can be increased up to 9dBi.

2.7 Integrated PD Sensor for Condition Monitoring

Figure 11 shows overall structure of the proposed RFID PD sensor. The two dipole antennas are kept at 90 degrees apart in spatial domain to maintain isolation between the received PD signal and retransmitted signal. N SIR filters and a modified spiral resonator is cascaded in series to perform frequency modification. As the resonators are band stop filters, each of them operate at a particular band within the overall bandwidth of the antennas. Figure 12 shows an illustration of 2-bit PD sensor with temperature sensing spiral resonator.

A roadmap of chipless RFID based PD sensor is demonstrated in Figure 13. Here, SIR and spiral resonator, two widely used narrowband passive resonator circuits are combined to achieve PD and temperature sensing. PD detection and localization is accomplished by cascading multiple SIR filters. In addition, a modified spiral resonator performs temperature monitoring. Integrating these resonators together with the dipole antennas a novel HV condition monitoring PD sensor is realized. The sensor is envisioned to facilitate online monitoring of switchyard modules through merging PD detection and temperature monitoring.

Figure 11. Schematic diagram of the integrated PD sensor

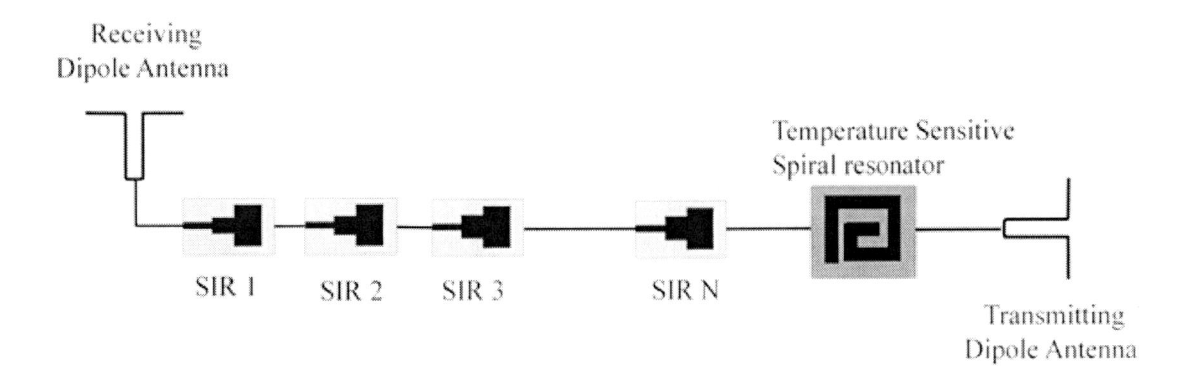

Figure 12. Layout of a 2-bit PD sensor with temperature sensing spiral

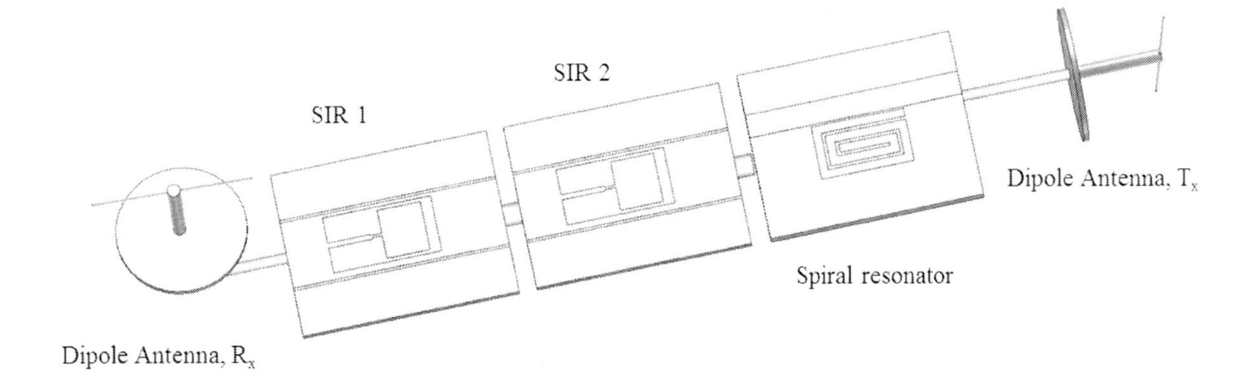

3. DESIGN

In this section, a prototype for the proposed sensor is designed. This section includes detail description of the individual component design procedure and simulation method.

3.1 Frequency Modulation Unit (FMU)

3.1.1 SIR Filter

For the prototype of our PD sensor, a frequency band of 100 MHz with 1GHz center frequency is selected. Two coplanar tri section SIR filters are designed in CST Microwave Studio operating at 965MHz and 1025 MHz. As described before, the condition for minimum electrical length of a tri step SIR is $a > 1 > b$. In this regard, the impedance values of the three sections is taken as $Z_1 = 51.98\Omega$, $Z_2 = 25.68\Omega$ and $Z_3 = 32.1\Omega$. The simulation is performed on Taconic TLX0 substrate having relative permittivity $\varepsilon_r = 2.45$ and $\tan\delta = 0.0019$ and substrate thickness, h= 0.5mm. The CPW line is matched to 50 Ohms. Figure 14 shows the layout of CPW tri step SIR structure designed for 965 MHz. Later, a parametric sweep of the overall length L is performed to attune the structure for 1025 MHz.

Figure 13. Roadmap of chipless RFID based PD sensor development for condition monitoring

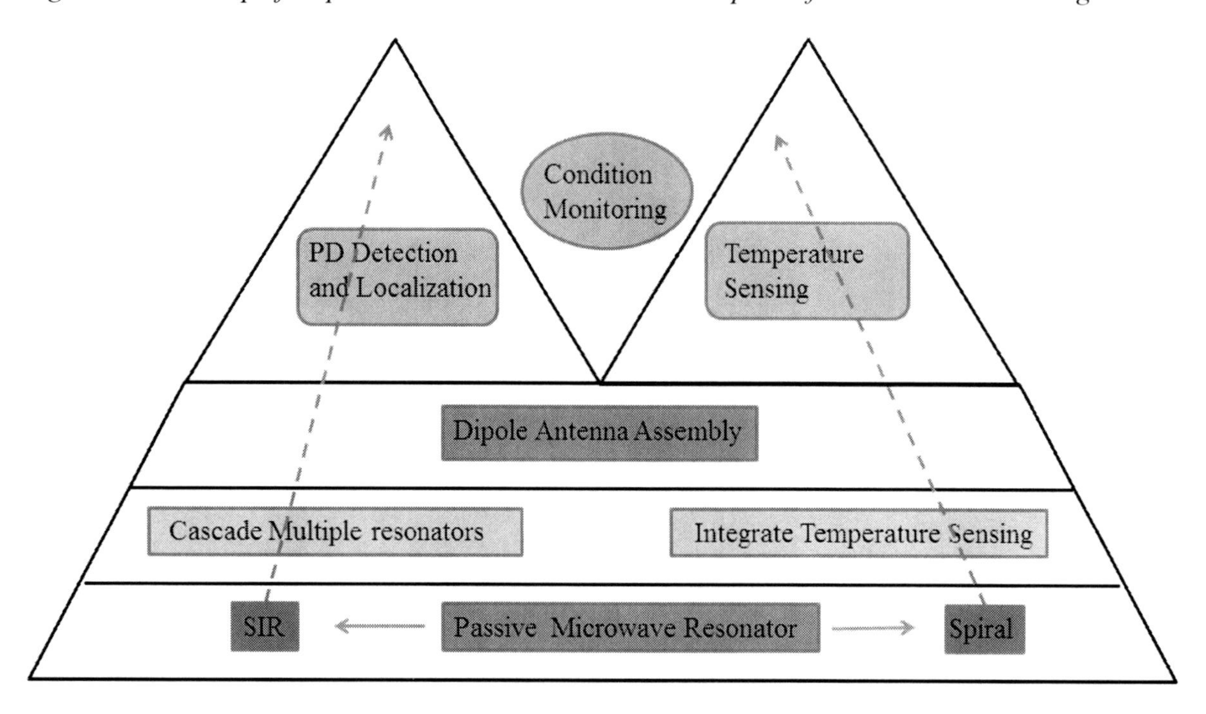

Figure 14. Layout of the tri section SIR filter. The length and width of Z_1, Z_2 and Z_3 sections are 22.4mm and 0.5 mm; 15mm and 4.8 mm; 3.0mm and 0.25mm. Total Length, L = 50mm

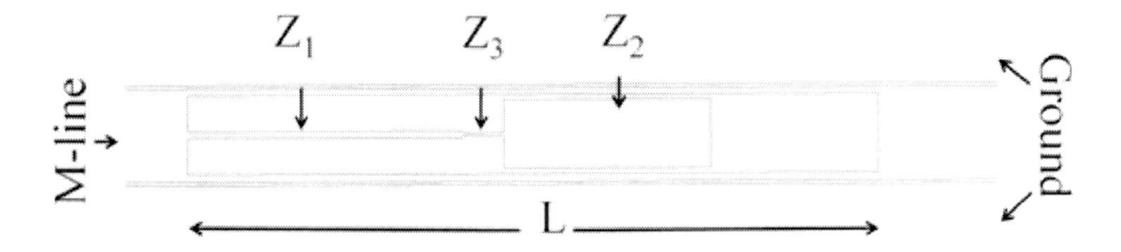

3.1.2 Temperature Sensing Spiral Resonator

A temperature sensing spiral resonator is designed in CST Microwave Studio operating at 2.4 GHz (Figure 15(a)). The spiral is coupled to a 50 ohm microstrip line having Taconic TLX-0 as substrate. The structural parameters of the spiral are given in Table 1. To incorporate temperature sensing, the spiral is modified with Stanyl TE200F6 polyamide (DSM Engineering Plastics) (Pawlikowski, 2009). Stanyl polyamide has a linear variation of dielectric constant with temperature (Figure 16). The top and cross sectional view of the modified spiral is shown in Figure 15. The polyamide is placed between the arms of the spiral. Also, a

Figure 15. Layout of modified spiral resonator for temperature sensing (a) top view; (b) cross sectional view

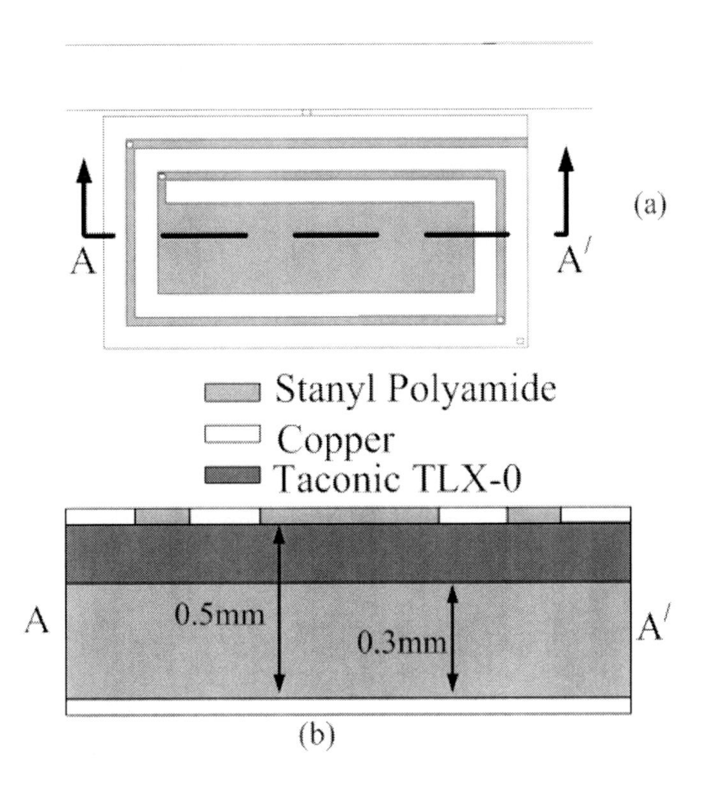

Table 1. Parameter values for temperature sensing spiral

Parameter Description	Parameter Values (mm)
Length of Spiral, *f*	15.3
Width of microstrip line, *a*	1.41
Width of spiral arm, *e*	0.8
Gap between two arms, *c*	0.3
Gap between microstrip and spiral, *b*	0.2
Width of total spiral, *d*	8
Height of substrate, *h*	0.5

part of the Taconic substrate between the ground and top plane was cut away to fill with Stanyl. This structural modification did not affect the adjacent microstrip line thus the resonant condition remained unchanged.

3.2 Dipole Antenna

Two identical half wave dipole antennas are designed having resonant frequency at 1GHz with 100MHz bandwidth. Figure 17 shows the layout of the dipole antenna with back reflector plane. The antenna is simulated in CST Microwave

Figure 16. Relative permittivity vs temperature for Stanyl

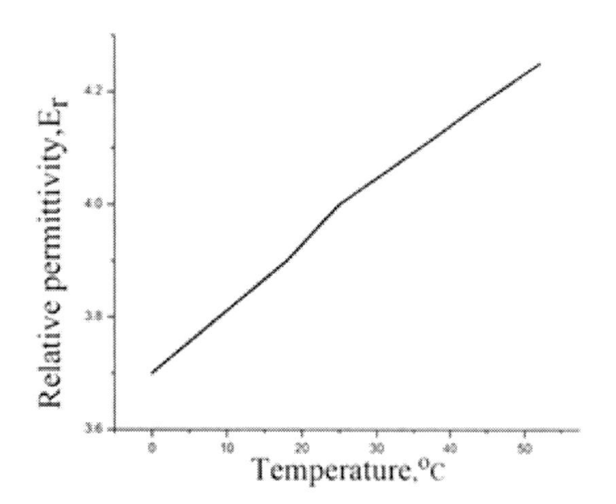

Figure 17. Layout of half wave dipole antenna with back reflector

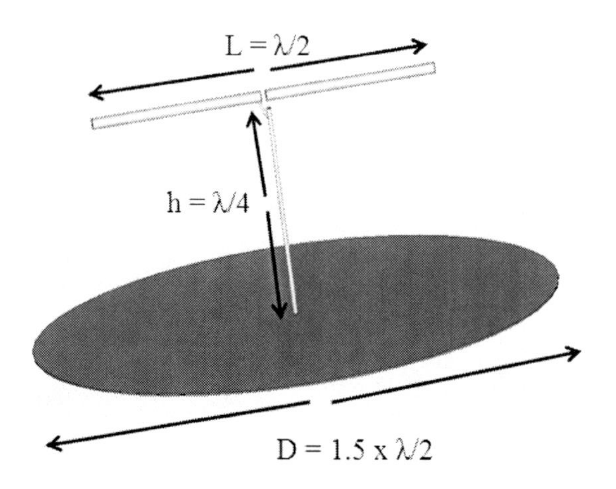

Studio to get the expected frequency response. In simulation two copper wires are energized through a 50Ω coax line using a discrete port. Initially the length of each wire was taken as λ/4, where $\lambda = c/f_c = 300$ mm. Performing a number of parametric sweeps the desired resonant frequency and -10 dB bandwidth was achieved. The length of the coax line is, h= λ/4= 75mm and the diameter of the copper plate is, D= 1.5x λ/2= 225mm. The return loss (S_{11}) and radiation pattern of simulated dipole antenna is shown in the result section.

4. RESULTS

In this section, results obtained from the simulation and experimentation is provided. This section presents the results in two parts. In the first segment, a prototype is developed to test and verify PD detection and localization using cascaded SIR filters. In the second segment, corroborative results for the temperature sensing spiral is depicted.

4.1 PD Detection and Localization

4.1.1 SIR Filter Response

According to the design, SIR filters are fabricated on Taconic TLX-0 substrate. Copper is etched out to create the CPW structure. The photo fabricated SIR filter operating at 965 MHz is shown in Figure 18. The filter response is measured using a Vector Network Analyzer as shown in Figure 19. In Figure 19(a) and Figure 19(b), the insertion loss (S_{21}) vs frequency of the fabricated filters resonating at 965 MHz and 1025 MHz is shown. Both the filters produce approximately 20 dB attenuation at their respective resonant frequencies. Figure 19(c) shows the S_{21} vs frequency of the filters when they are connected in series which provides a different frequency signature than the individual filters.

However, each combination of the filters generates a unique frequency signature that can designate a distinct identification data bit. Here, two SIR filters can generate, $M = 2^2 - 1$ or 3 bit patterns as shown in Table 2. Thus the filters can identify PD signal occurring from three different HV units. According to Table 2, the spectral sig-

Figure 18. Photo of fabricated SIR filter at 965 MHz

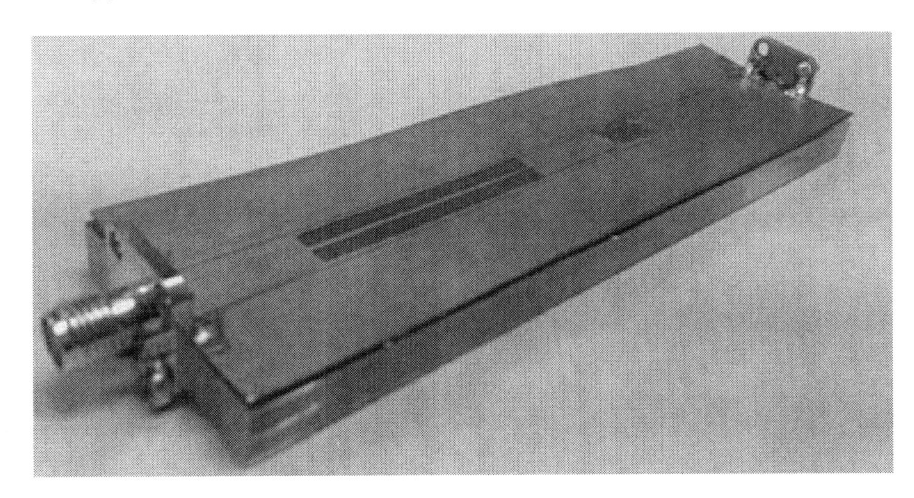

Figure 19. Simulated and measured insertion loss of (a) 965 MHz SIR Filter, (b) 1025 MHz SIR Filter, and (c) cascaded SIR Filters

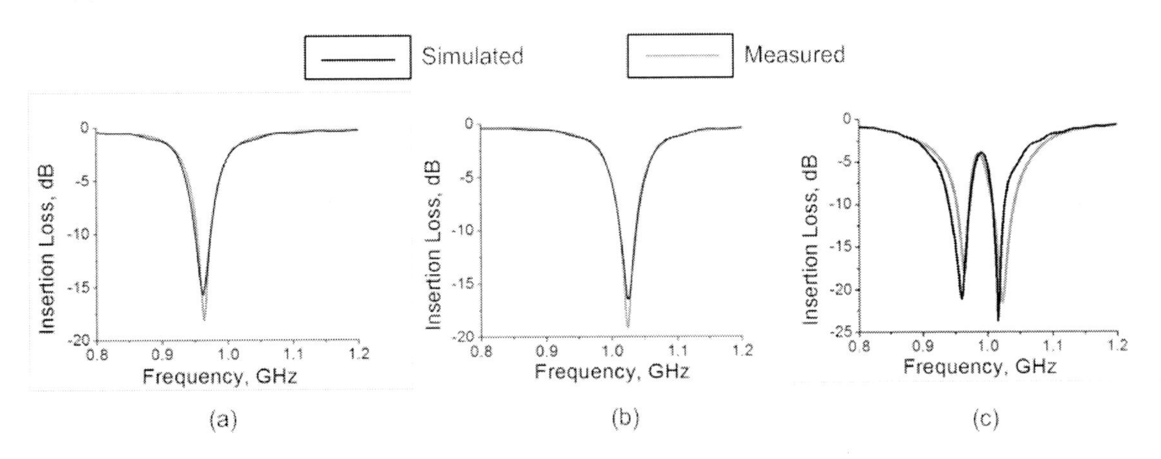

natures of Figure 19(a)-(c) designate corresponding data bits '01', '10' and '00'.

4.1.2 Dipole Antenna Response

A half-wave dipole antenna is constructed using two copper wires of quarter wave length. The wires were connected to a coaxial cable and a circular copper plate was soldered at λ/4 distance from the dipole. The overall antenna is shown in Figure 20. The measured return loss (S_{11}) of the dipole is given in Figure 21. It has a 10dB bandwidth from 950 MHz to 1045 MHz with maximum at-

Table 2. Designated data bits for different combinations of the filters

Resonant Frequency		Data Bit	
965 MHz	1025 MHz		
✓	X	0	1
X	✓	1	0
✓	✓	0	0

Figure 20. Photo of fabricated dipole antenna

Figure 21. Measured return loss (S_{11}) of the dipole antenna

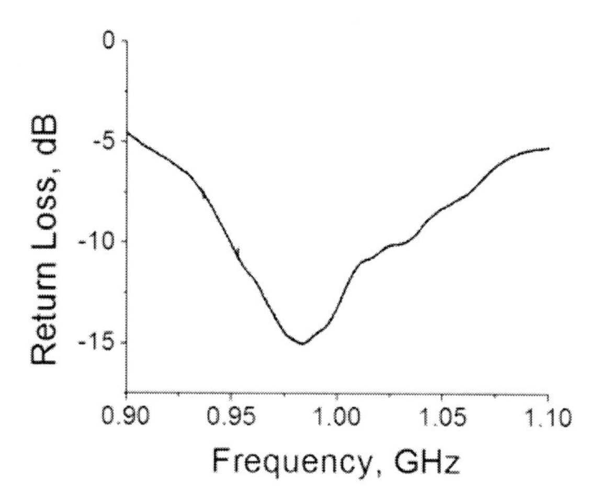

tenuation of 15 dB. The simulated 3D radiation pattern in Figure 22 reveals that maximum gain is vertical to the antenna axis (z- axis). Also, the general torus shape of a half dipole 3D pattern is distorted due to the of back reflector plate.

4.1.3 Experimentation with PD Source

Experimentation is done to validate the response of fabricated PD sensor. The whole experiment was performed in anechoic chamber where the external EM waves do not interfere. For conducting experiment with PD signal, a short duration PD pulse generator is used as a source. This source is a PD calibrator CAL2B from Power Diagnostix (Figure 23) ("Power Diagnostix Systems GmbH,"). The PD source is compliant with IEC 60270 standard and emits PD current pulses with rise time <=200ps and bandwidth >= 1.5 GHz. The voltage is set to 50Volts throughout the whole experiment. Also in our experiment, a high speed oscilloscope displays the captured signal (Figure 24). The oscilloscope used is Tektronix DSA72004 Digital Serial Analyzer. It features high bandwidth up to 20 GHz matched across 4 channels. The sampling frequency was set to 25 GS/sec and the record length was 200ns.

Figure 25(a) shows the time domain PD pulse captured with the antenna without transmitting through a SIR filter. Figure 25(b) shows the power spectrum of the PD pulse over the frequency band of interest. The spectrum is almost constant over the frequency band as expected.

Figure 26 shows the experimental setup conducted in the laboratory to investigate sensor performance. Here the sensor blocks are installed in three EM shielded units. Each unit has a PD calibrator CAL2B which works as a PD source. The cables from each unit is connected to three sampling channels Ch1, Ch2 and Ch3 thus these channels represent the PD signals passing through filters 965 MHz, 1025 MHz and cascaded filters respectively. Figure 27 (a) - (c) shows the time domain signal of the captured PD at Ch1, Ch2 and Ch3. Comparing with the time domain signal presented in Figure 25 (a), these signals spread in time as they are passed through bandstop filters. Also the signal amplitude is less than the direct PD signal.

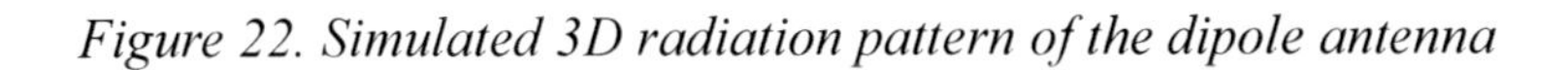

Figure 22. Simulated 3D radiation pattern of the dipole antenna

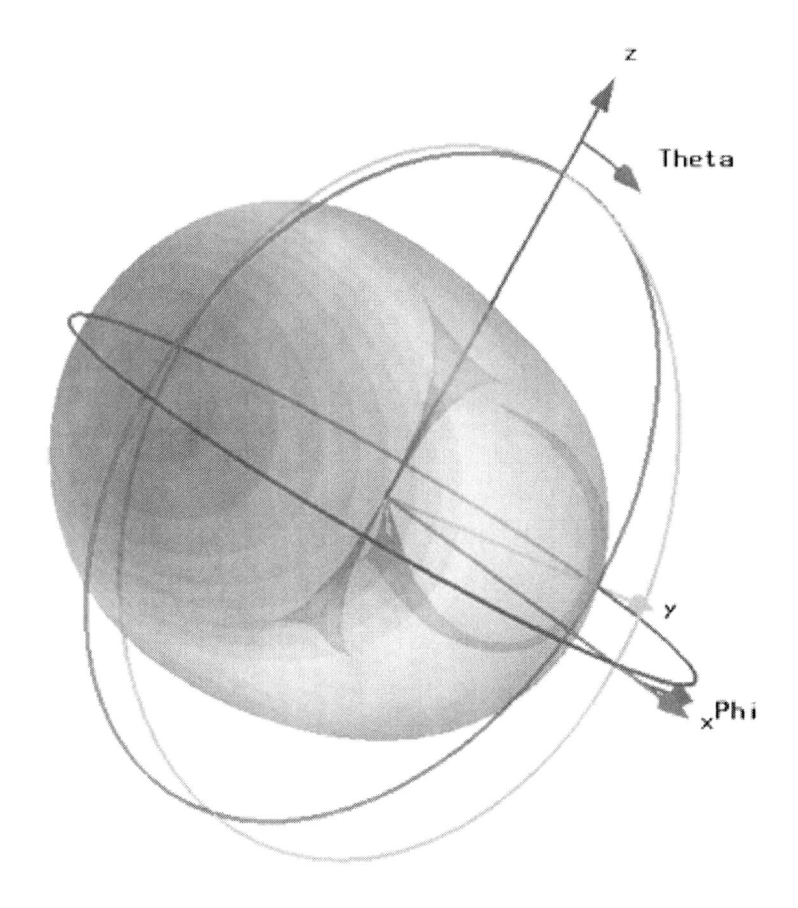

Figure 23. PD calibrator CAL2B

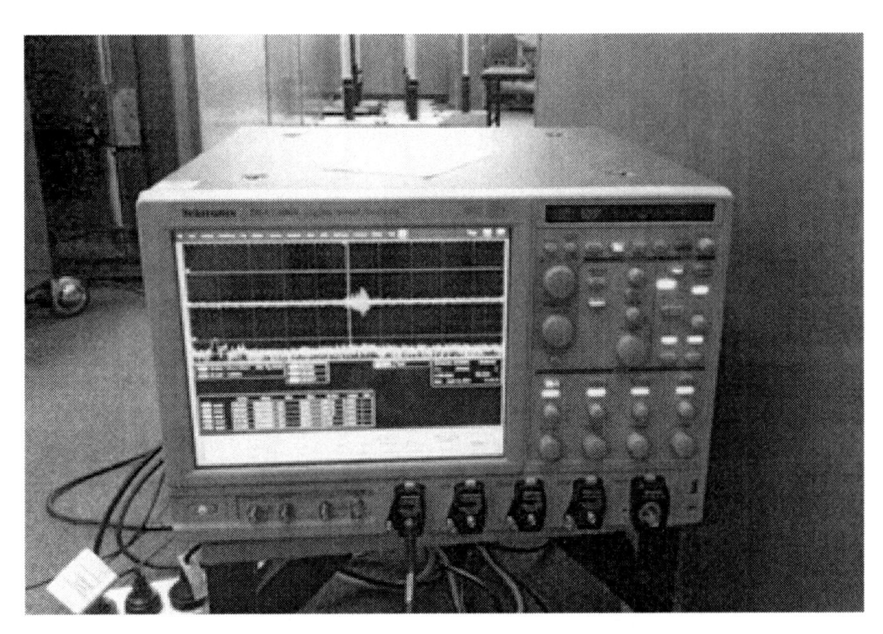

Figure 24. Tektronix DSA72004 digital serial analyzer

In Figure 28 (a) - (c), the power spectrum of captured PD signals transmitting through three sensors is shown. Here, the power received P_{sensor} is normalized by the P_{direct}, where P_{direct} is the signal power received by the antenna without passing through the filter. In Figure 28 (a), the power spectrum has a sharp attenuation at 965 MHz which infers that the signal is originated from source '01' or unit 1. Similarly, the signal in Figure 28 (b) has a dip in power spectrum at 1025 MHz and the signal in Figure 28(c) has two distinct dips at 965 MHz and 1025 MHz. It can be concluded that the PD signals in Figure 28 (b) and Figure 28 (c) are originated from unit 2 and unit 3 respectively. Hence the sensors can identify different PD sources separately.

4.2 Temperature Sensing

The spiral resonator shown in Figure 15 is simulated to verify temperature effect. As shown in Figure 16, the relative permittivity, ε_r of Stanyl varies almost linearly with temperature. A parametric sweep of ε_r is performed from 3.2 to 4.0 and

Figure 25. (a) Time domain response of captured PD signal; (b) power spectrum of the PD signal in (a)

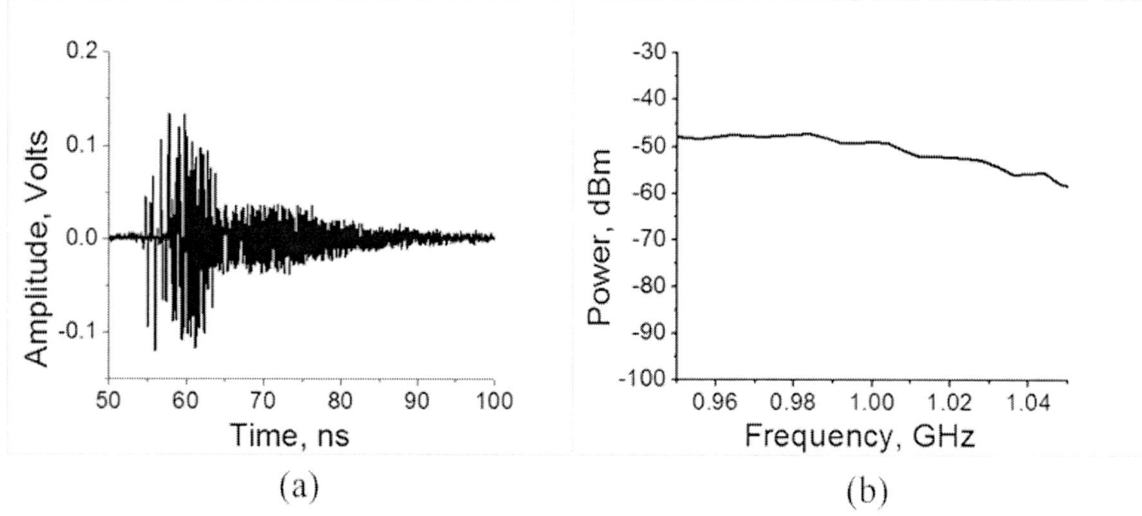

Figure 26. Experimental setup for PD sensor prototype verification

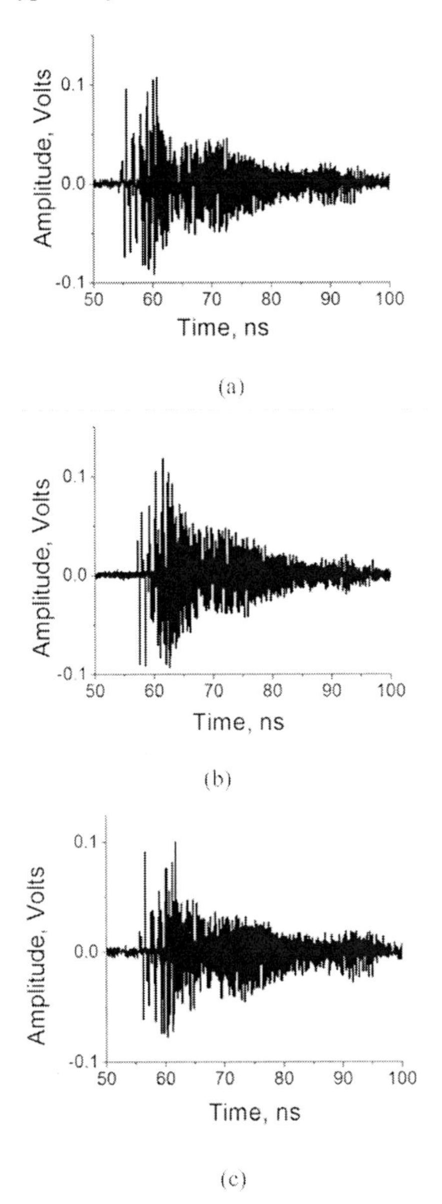

Figure 27. Time domain response of PD signals captured at (a) ch1, (b) ch2, and (c) ch3

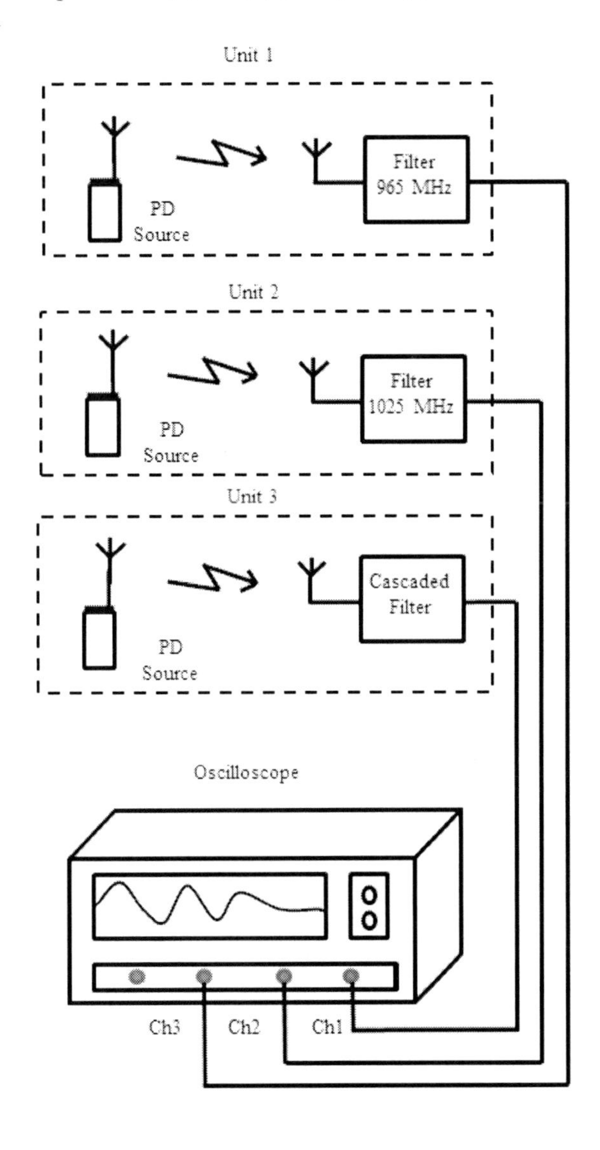

the measured insertion loss (S_{21}) vs frequency is plotted in Figure 29. It is observed that the resonant frequency, f_T shifts with the dielectric property which emulates temperature variation. Figure 30(a) shows the measured resonant frequencies for different ε_r. Correlating this data with Figure 16, resonant frequency, f_T vs temperature is extracted as shown in Figure 30(b). Hence, observing the resonant frequency of the particular spiral resonator, temperature information can be retrieved.

5. CONCLUSION

In recent years chipless RFID has emerged as a low cost reliable solution for item tagging, secured asset management and remote location finding. Especially its resilience to harsh environment has increased the demand for multidimensional

Figure 28. Power spectrum of captured PD signal in (a) unit 1, (b) unit 2, and (c) unit 3

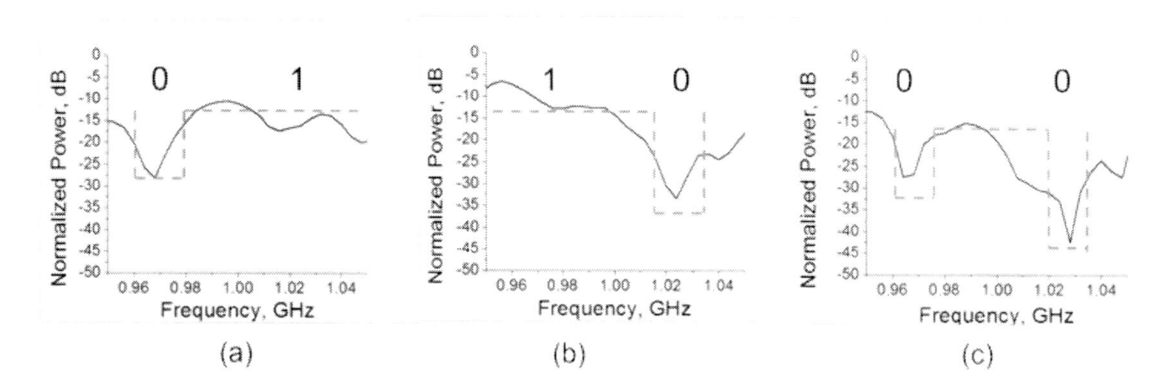

Figure 29. Measured Insertion loss (S_{21}) vs frequency for different dielectric constant of Stanyl

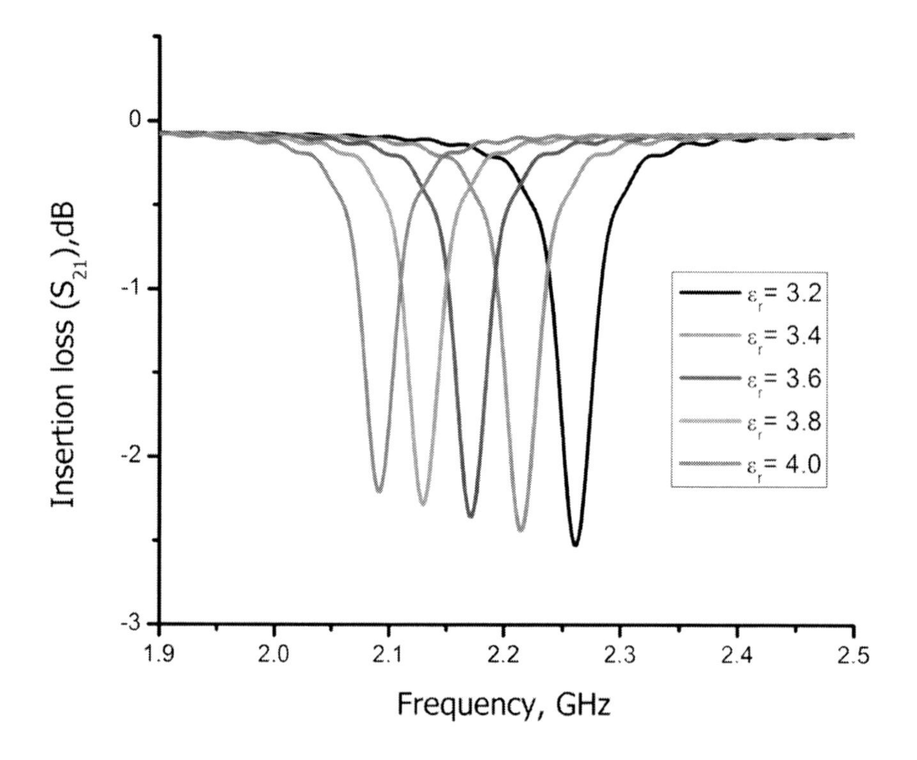

applications. Chipless RFID sensor is such an added feature which has revolutionized the field of automated object identification in conjunction with condition monitoring.

Though RFID technology has been widely explored in power facility management, the intrusion of chipless RFID is still naive in power systems. In this chapter, a novel application of chipless RFID sensor in high voltage condition monitoring is presented. The multiresonator based RFID tag is envisioned to overcome the challenges regarding PD detection and localization in a switchyard. In addition to that, temperature monitoring of individual apparatus is plausible by integrating a unique temperature sensing prototype.

Figure 30. (a) Measured resonant frequency (f_T) for varying relative permittivity (ε_r); (b) Measured resonant frequency (f_T) vs temperature

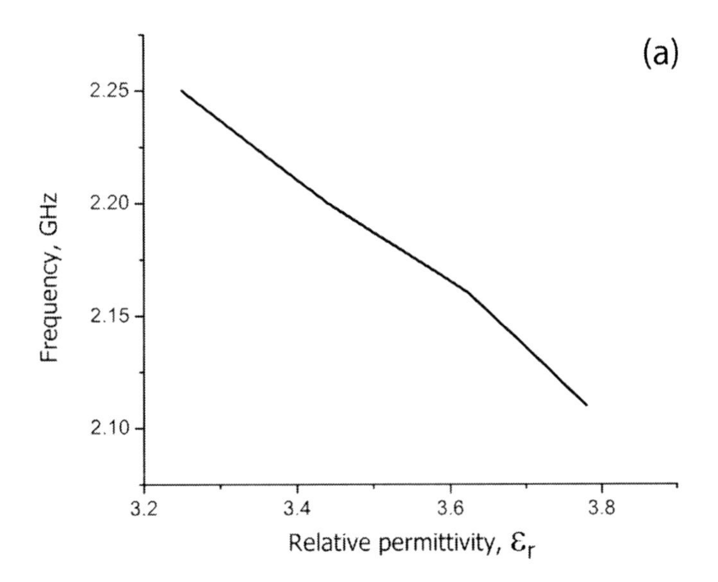

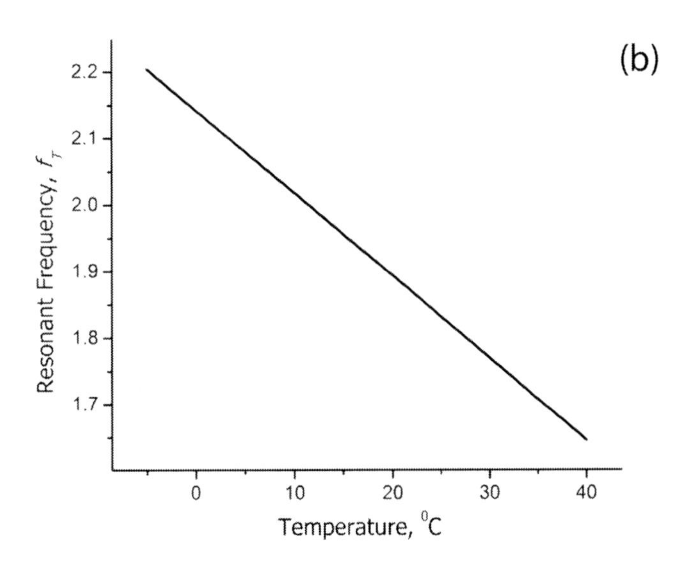

The proposed sensor was developed and tested to verify its applicability in detecting and identifying PD event. Results confirm that the UWB PD pulse is modified to incorporate frequency signature as identification data bits. Also, the temperature sensing spiral denotes noticeable frequency shift with temperature variation. Challenges still remain in integrating the two sensing parameters within the frequency band of operation. Moreover, the retransmission of the modulated PD signal will be investigated in future studies.

Finally, the chipless RFID based PD sensor will provide an inexpensive method for radiometric PD monitoring by utilizing passive sensors. Since a number of HV units within a substation are monitored concurrently, PD severity can be determined by comparing with the threshold level. The proposed system has a major advantage over existing RFI PD measurement techniques due to its scalability to large substations for mass deployment.

ACKNOWLEDGEMENT

The project is supported by Australian Research Council Linkage Project Grant LP0989355: Smart Information Management of Partial Discharge in Switchyards using Smart Antennas and SP-AusNet Pty. Ltd."

REFERENCES

Amin, E. M. A., & Karmakar, N. (2011). *Development of a chipless RFID temperature sensor using cascaded spiral resonators*. Paper presented at the IEEE SENSORs 2011.

Austen, A. E. W., & Hackett, W. (1944). Internal discharges in dielectrics: their observation and analysis. *Journal of the Institute of Electrical Engineers - Part I: General, 91*(44), 298–312.

Balanis, C. A. (1992). Antenna theory: A review. *Proceedings of the IEEE, 80*(1), 7–23. doi:10.1109/5.119564

Bhattacharyya, R., Floerkemeier, C., & Sarma, S. (2010). Low-cost, ubiquitous RFID-tag-antenna-based sensing. *Proceedings of the IEEE, 98*(9), 1593–1600. doi:10.1109/JPROC.2010.2051790

Boggs, S. A. (1990). Partial discharge: Overview and signal generation. *Electrical Insulation Magazine, IEEE, 6*(4), 33–39. doi:10.1109/57.63057

Chu, F. Y., & Williamson, A. (1982). Fault location in SF6 insulated substations using thermal techniques. *IEEE Transactions on Power Apparatus and Systems, PAS-101*(7), 1990–1997. doi:10.1109/TPAS.1982.317446

Das, R. (2006). *Chipless RFID - The end game*. Retrieved from http://www.idtechex.com/research/articles/chipless_rfid_the_end_game_00000435.asp

Dowling, J., Tentzeris, M. M., & Beckett, N. (2009, 24-25 September). *RFID-enabled temperature sensing devices: A major step forward for energy efficiency in home and industrial applications?* IEEE MTT-S International Microwave Workshop on Wireless Sensing, Local Positioning, and RFID, IMWS 2009.

Duval, M. (1989). Dissolved gas analysis: It can save your transformer. *Electrical Insulation Magazine, IEEE, 5*(6), 22–27. doi:10.1109/57.44605

Finkenzeller, K. (2010). *Introduction*. John Wiley & Sons, Ltd.

Fletcher, R. R. (2002). *Low-cost electromagnetic tagging: Design and implementation.*

Hong, J.-S. G., & Lancaster, M. J. (2011). *Microstrip filters for RF/microwave applications* (2nd ed.). Wiley-Interscience. doi:10.1002/9780470937297

Hoshino, T., Kato, K., Hayakawa, N., & Okubo, H. (2001). A novel technique for detecting electromagnetic wave caused by partial discharge in GIS. *IEEE Transactions on Power Delivery, 16*(4), 545–551. doi:10.1109/61.956735

Hu, X., Judd, M. D., & Siew, W. H. (2010, August 31 2010-September 3). *A study of PD location issues in GIS using FDTD simulation.* Paper presented at the 2010 45th International Universities Power Engineering Conference (UPEC).

Hualiang, Z., & Chen, K. J. (2005). A tri-section stepped-impedance resonator for cross-coupled bandpass filters. *Microwave and Wireless Components Letters, IEEE, 15*(6), 401–403. doi:10.1109/LMWC.2005.850475

Hualiang, Z., & Chen, K. J. (2006). Miniaturized coplanar waveguide bandpass filters using multisection stepped-impedance resonators. *IEEE Transactions on Microwave Theory and Techniques*, *54*(3), 1090–1095. doi:10.1109/TMTT.2005.864126

Ishimaru, H., & Kawada, M. (2010). Localization of a partial discharge source using maximum likelihood estimation. *IEEJ Transactions on Electrical and Electronic Engineering*, *5*(5), 516–522. doi:10.1002/tee.20567

Judd, M. D., Li, Y., & Hunter, I. B. B. (2005). Partial discharge monitoring of power transformers using UHF sensors. Part I: Sensors and signal interpretation. *IEEE Electrical Insulation Magazine*, *21*(2), 5–14. doi:10.1109/MEI.2005.1412214

Kawada, M. (2003). Fundamental study on location of a partial discharge source with a VHF-UHF radio interferometer system. *Electrical Engineering in Japan*, *144*(1), 32–41. doi:10.1002/eej.10145

Kemp, I. J. (1995). Partial discharge plant-monitoring technology: Present and future developments. *IEE Proceedings. Science Measurement and Technology*, *142*(1), 4–10. doi:10.1049/ip-smt:19951438

Kim, Y.-I., Bong-Jae, Y., Jae-Ju, S., Jin-Ho, S. A., & Lee, J.-I. (2006). Implementing a prototype system for power facility management using RFID/WSN. *International Journal of Applied Mathematics and Computer Science*, *2*(2).

Kreuger, F. H. (1989). *Partial discharge detection in high-voltage equipment* (1st ed.). Butterworths.

Lundgaard, L. E. (1992). Partial discharge, XIII: Acoustic partial discharge detection-fundamental considerations. *Electrical Insulation Magazine, IEEE*, *8*(4), 25–31. doi:10.1109/57.145095

Luo, Y., Ji, S., & Li, Y. (2006). Phased-ultrasonic receiving-planar array transducer for partial discharge location in transformer. *IEEE Transactions on Ultrasonics, Ferroelectrics, and Frequency Control*, *53*(3), 614–622. doi:10.1109/TUFFC.2006.1610570

Moore, P. J., Portugues, I. E., & Glover, I. A. (2005). Radiometric location of partial discharge sources on energized high-voltage plant. *IEEE Transactions on Power Delivery*, *20*(3), 2264–2272. doi:10.1109/TPWRD.2004.843397

Nattrass, D. A. (1993). Partial discharge, XVII: The early history of partial discharge research. *Electrical Insulation Magazine*, *9*(4), 27–31. doi:10.1109/57.223897

Niemeyer, L. (1993, 28-30 September). *The physics of partial discharges.* Paper presented at the 1993 International Conference on Partial Discharge.

Pawlikowski, G. T. (2009). *Effects of polymer material variations on high frequency dielectric properties*, Vol. 1156. MRS 2009 Spring Meeting.

Plessky, V., & Reindl, L. (2010). Review on SAW RFID tags. *IEEE Transactions on Ultrasonics, Ferroelectrics, and Frequency Control*, *57*(3), 654–668. doi:10.1109/TUFFC.2010.1462

Power Diagnostix Systems GmbH. (n.d.). Retrieved from http://www.pd-systems.com/accessories.html#mail

Preradovic, Balbin, I., Karmakar, N. C., & Swiegers, G. F. (2009). Multiresonator-based chipless RFID system for low-cost item tracking. *IEEE Transactions on Microwave Theory and Techniques*, *57*(5), 1411–1419. doi:10.1109/TMTT.2009.2017323

Preradovic, S. (2009). *Chipless RFID system for barcode replacement.* Monash University.

Preradovic, S., & Karmakar, N. C. (2010). Chipless RFID: Bar code of the future. *Microwave Magazine, IEEE, 11*(7), 87–97. doi:10.1109/MMM.2010.938571

Qin, S., & Birlasekaran, S. (2002, 2002). *The study of propagation characteristics of partial discharge in transformer.* Paper presented at the 2002 Annual Report Conference on Electrical Insulation and Dielectric Phenomena.

Ramos, A. A., Lazaro, D., Girbau, A., & Villarino, R. (2011). Time-domain measurement of time-coded UWB chipless RFID tags. *Progress in Electromagnetics Research, 116*, 313–331.

Reid, A. J., Judd, M. D., Fouracre, R. A., Stewart, B. G., & Hepburn, D. M. (2011). Simultaneous measurement of partial discharges using IEC60270 and radio-frequency techniques. *IEEE Transactions on Dielectrics and Electrical Insulation, 18*(2), 444–455. doi:10.1109/TDEI.2011.5739448

Reid, A. J., Judd, M. D., Stewart, B. G., & Fouracre, R. A. (2006a, 15-18 Oct. 2006). *Frequency distribution of RF energy from PD sources and its application in combined RF and IEC60270 measurements.* Paper presented at the 2006 IEEE Conference on Electrical Insulation and Dielectric Phenomena.

Reid, A. J., Judd, M. D., Stewart, B. G. A., & Fouracre, R. A. (2006b). Partial discharge current pulses in SF 6 and the effect of superposition of their radiometric measurement. *Journal of Physics. D, Applied Physics, 39*(19), 4167. doi:10.1088/0022-3727/39/19/008

Reid, D. I., & Judd, M. (2009, 8-11 June). *UHF monitoring of partial discharge in substation equipment using a novel multi-sensor cable loop.* Paper presented at the 20th International Conference and Exhibition on Electricity Distribution - Part 1, CIRED 2009.

Sagawa, M., Makimoto, M., & Yamashita, S. (1997). Geometrical structures and fundamental characteristics of microwave stepped-impedance resonators. *IEEE Transactions on Microwave Theory and Techniques, 45*(7), 1078–1085. doi:10.1109/22.598444

Sen, D., Sen, P., & Das, A. M. (2009). *RFID for energy and utility industries: PennWell.*

Shibuya, Y., Matsumoto, S., Tanaka, M., Muto, H., & Kaneda, Y. (2010). Electromagnetic waves from partial discharges and their detection using patch antenna. *IEEE Transactions on Dielectrics and Electrical Insulation, 17*(3), 862–871. doi:10.1109/TDEI.2010.5492260

Shrestha, S., Balachandran, M., Agarwal, M., Phoha, V. V., & Varahramyan, K. (2009). A chipless RFID sensor system for cyber centric monitoring applications. *IEEE Transactions on Microwave Theory and Techniques, 57*(5), 1303–1309. doi:10.1109/TMTT.2009.2017298

Stewart, B. G., Nesbitt, A., & Hall, L. (2009, May 31 2009-June 3). *Triangulation and 3D location estimation of RFI and partial discharge sources within a 400kV substation.* Paper presented at the Electrical Insulation Conference, EIC 2009.

Tenbohlen, S. M. H., Denissov, D., Huber, R., Riechert, U., Markalous, S. M., Strehl, T., & Klein, T. (2006). Electromagnetic (UHF) PD diagnosis of GIS, cable accessories and oil-paper insulated power transformers for improved PD detection and localization.

Tian, Y., Kawada, M., & Isaka, K. (2008, 7-11 Sept. 2008). *Visualization of electromagnetic waves emitted from multiple PD sources on distribution line by using FDTD method.* Paper presented at the Electrical Insulating Materials, 2008. (ISEIM 2008). International Symposium on.

Tian, Y., Kawada, M., & Isaka, K. (2009). Locating partial discharge source occurring on distribution line by using FDTD and TDOA methods. *IEEJ Transactions on Fundamentals and Materials*, *129*(2), 89–96. doi:10.1541/ieejfms.129.89

Wang, Y., Jia, Y., Chen, Q., & Wang, Y. (2008). A passive wireless temperature sensor for harsh environment applications. *Sensors (Basel, Switzerland)*, *8*(12), 7982–7995. doi:10.3390/s8127982

Want, R. (2004). Enabling ubiquitous sensing with RFID. *Computer*, *37*(4), 84–86. doi:10.1109/MC.2004.1297315

Xia, F. (2009). Wireless sensor technologies and applications. *Sensors (Basel, Switzerland)*, *9*(11), 8824–8830. doi:10.3390/s91108824

Young-Taek, L., Jong-Sik, L., Chul-Soo, K., Dal, A., & Sangwook, N. (2002). A compact-size microstrip spiral resonator and its application to microwave oscillator. *Microwave and Wireless Components Letters*, *12*(10), 375–377. doi:10.1109/LMWC.2002.804556

Zhiguo, T., Chengrong, L., Xu, C., Wei, W., Jinzhong, L., & Jun, L. (2006). Partial discharge location in power transformers using wideband RF detection. *IEEE Transactions on Dielectrics and Electrical Insulation*, *13*(6), 1193–1199. doi:10.1109/TDEI.2006.258190

KEY TERMS AND DEFINITIONS

Chipless RFID: A passive RFID tag which does not have a silicon IC or chip within its circuit.

Condition Monitoring: To monitor the condition of dielectric degradation of high voltage equipment.

Frequency Modulation Unit: A passive resonator circuit to give spectral signature within PD.

Partial Discharge (PD): A discharge occurring due to high voltage stress across a dielectric.

PD Localization: Locating a faulty source that emits PD.

Radiometric Detection: A PD detection technique where RF emission during PD is monitored.

Stepped Impedance Resonator (SIR): A passive microwave resonator.

Chapter 14
Recent Advancements in Smart Sensors and Sensing Technology

Subhas C. Mukhopadhyay
Massey University, New Zealand

ABSTRACT

The chapter presents the design and development of very low cost planar sensors and sensing systems for measuring fat contents in meat, leather quality assessment, food quality, and biomedical application such as cancer detection, agriculture, and RFID based detection systems. The sensors comprise planar passive microwave integrated circuits in the forms of microstrip meander lines, mesh and inter-digital capacitance. The sensors are excited with voltage controlled oscillators (VCOs) and power supply units. A data acquisition system based on a microcontroller and an op-amp based interfacing circuits complete the sensing system. The results of various characteristics parameters of samples are presented and compared with the results from expensive conventional measurement set up. These low cost sensors bring benefits in the sensing technology with novel and accurate concepts.

INTRODUCTION

The sensors and sensing technology plays a crucial role in our day-to-day life. Sensors and associated measuring instrumentation circuits pose a challenge towards the development of a low-cost intelligent sensing system. Sensors are a significant part of complex and sophisticated systems of modern technology. It will be difficult to achieve the purpose of any modern system or process control without use of any forms of sensors. In recent times, a significant amount of research work is undertaken to develop smart and intelligent sensor system for different novel applications. J. Schmalzel et. al. (Schmalzel, Figueroa, & Morris, 2005) have reported an implementation of a prototype intelligent rocket facility, the results of which established a basis for future advanced development and validation using the rocket test stand facilities at Stennis Space Center. In health

DOI: 10.4018/978-1-4666-2080-3.ch014

management systems, smart sensor components play key roles in providing the distributed intelligence needed to perform diagnosis of overall health of the people. A wireless visual sensor comprising of a low-cost greyscale camera as the sensing hardware and BlueTooth (BT) 100-m slave module as the transmission hardware has been reported in (Ferrigno, Pietrosanto, & Paciello, 2006). The sensor is quite efficient in terms of cost, energy saving and bandwidth for image transfer. Usually the camera based sensor systems are not able to provide internal characteristics of the system though in recent times depth camera has now been reported. A low-cost, high performance displacement sensor has been presented in (Toth & Meijer, 1992). The system has been implemented with simple electrodes, an inexpensive microcontroller and linear capacitance-to-period converter. For the sensing system sometime excitation for the sensors pose design challenges. The excitation part of the sensor as reported in (Toth et al., 1992) is not very clearly explained as this can be challenging in many applications. A low-cost, smart capacitive angular-position sensor with simple, stable and reliable characteristics has been reported in(Li & Meijer, 1995), (Li, Meijer, de Jong, & Spronck, 1996) and (Li, Meijer, & de Jong, 1997). Interfacing the sensors signals to a microprocessor or to a microcontroller pose a challenge. A novel smart interface for voltage-generating sensors has been reported in (Li, Meijer, & Schnitger, 1998). A smart and accurate interface for resistive sensors has been reported in (Li & Meijer, 2001). If the sensors provide DC signals as in for thermocouple types, the problem of interfacing is not so severe. It is important that the signal from the sensors should also have the condition signals, which provide rudimentary information necessary for fault detection and isolation in sensor systems (Amadi-Echendu & Zhu, 1994). Such rudimentary information should be very significant in the development of an intelligent measurement system. An interface circuit for a differential capacitance transducer has been described in Mochizuki, Masuda, & Watanabe (1998) which allows high-accuracy signal processing with standard components. In Patra, Kot, & Panda (2000), a scheme of an intelligent capacitive pressure sensor using an artificial neural network has been proposed. A complete solution to connect an IEEE 802.11-based sensor with a wireless network has been presented in (Ferrari, Flamini, Marioli, & Taroni, 2006). An overview of significant development of methods, structures, manufacturing technologies, and signal processing characterizing today's sensors and sensing systems has been presented in (Kanoun & Trankler, 2004). Most of the common magnetic sensing methods have been described and the underlying principles governing their operation have been highlighted in (Lenz & Edelstein, 2006).

This chapter reports the development work which was carried out at Massey University, New Zealand over the last few years. The idea behind the research works is to design and develop of a low-cost intelligent sensing system for variety of applications. The whole chapter is divided into 4 sections. After the introduction in the first section, the design issues towards a low-cost sensing system will be discussed. The next section discusses the development of different sensing systems for wide variety of applications. The following section deals with the issues of developing intelligent systems. The last section concludes with future possibilities.

DESIGN ISSUES TOWARDS A LOW COST SENSING SYSTEM

A low-cost sensing system is always desirable in many applications. A sensing system is comprised on three main parts: Sensors; an Excitation system; and a Data Acquisition system as shown in Figure 1.

A sensor is a passive device which changes its physical parameters in response to the environment. As for an example, a humidity sensor re-

Figure 1. Block diagram of a wireless sensor

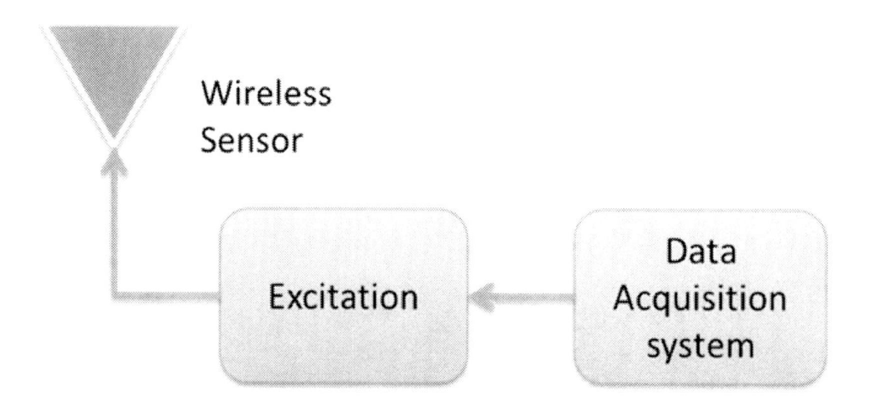

sponds to the change of the humidity in atmosphere with its varying physical parameters. The change is converted to the sensing parameters at convenience. The sensor needs an exciting device to respond to the environmental changes. A data acquisition device or system such as a microcontroller controls the excitation system and well as record the data corresponding to the change of the physical parameters. Following are the detailed description of the three areas at Massey University's Sensor Laboratories

Sensors

The most important elements of any inspection system are the sensors. If the sensors can provide moderate response at relatively low frequency it is always helpful to develop a low-cost sensing system. The developed sensing systems are usually based on planar passive microwave circuits. Three planar type electromagnetic sensors such as (i) meander, (ii) mesh and (iii) interdigital types are shown in Figure 2. The characterization and comparative performance evaluation of the sensors have been performed in the author's laboratories and reported in (Mukhopadhyay, Gooneratne, Sen Gupta, & Yamada, 2005). All of them are suitable for inspection and evaluation of the properties of systems or a structure in a non-destructive and non-invasive ways.

Different sizes of sensors of each meander, mesh and interdigital types have been experimented with using the experimental set-up shown in Figure 3. The sensor has been connected to a high frequency supply while the voltage across and the current through the sensor are recorded. The frequency of excitation has been varied between 1 kHz to 100 MHz. The impedance characteristics of the sensors are shown in Figures 4a, b and c, respectively. It is seen that the transfer impedance for both meander and mesh type increases with the increase in frequency whereas the impedance of interdigital type sensor decreases with frequency. Basically meander and mesh types sensors are inductive type whereas the interdigital type sensor is capacitive type. It has been seen that both meander and mesh type sensors respond well at high frequencies (Mukhopadhyay et al., 2005). The response of the interdigital sensor is very good at low frequencies and doesn't respond well at high frequencies. Thus the operating frequency has to be very carefully selected.

Excitation System

A smart power supply and associated instrumentation to provide the controlled excitation to the exciting coil of the sensor is very important for the intelligent sensing system. A controlled variable

Figure 2. Planar electromagnetic sensors of (a) meander, (b) mesh, and (c) interdigital configuration (Mukhopadhyay et al., 2005)

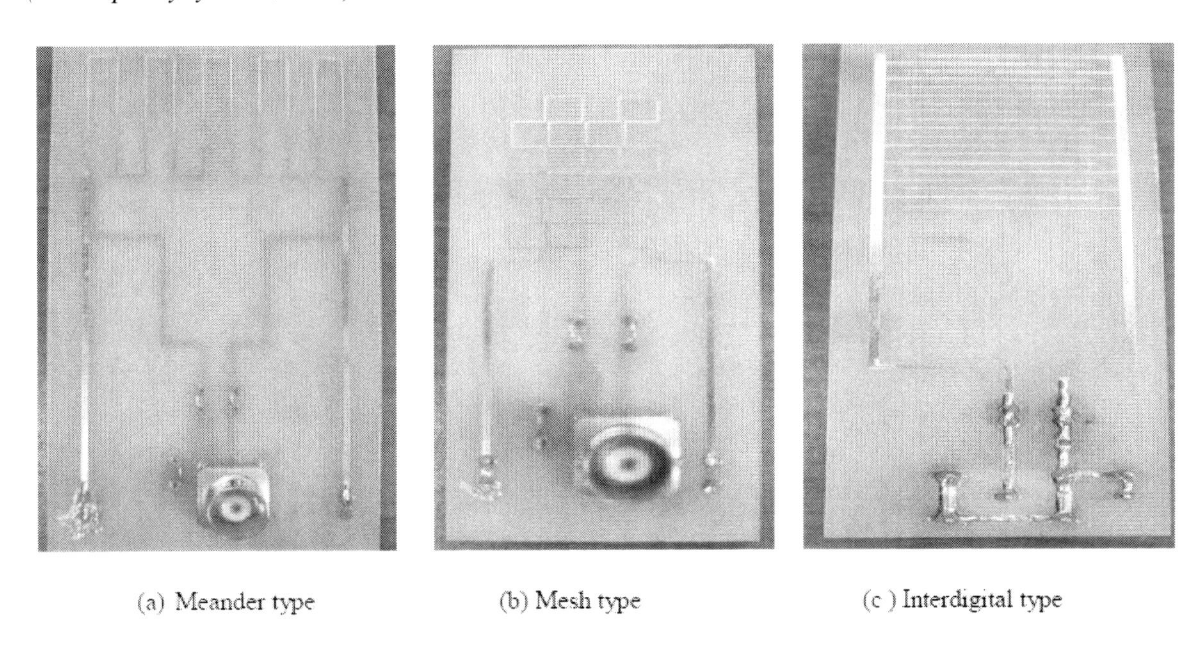

Figure 3. The experimental set-up for sensor characterization

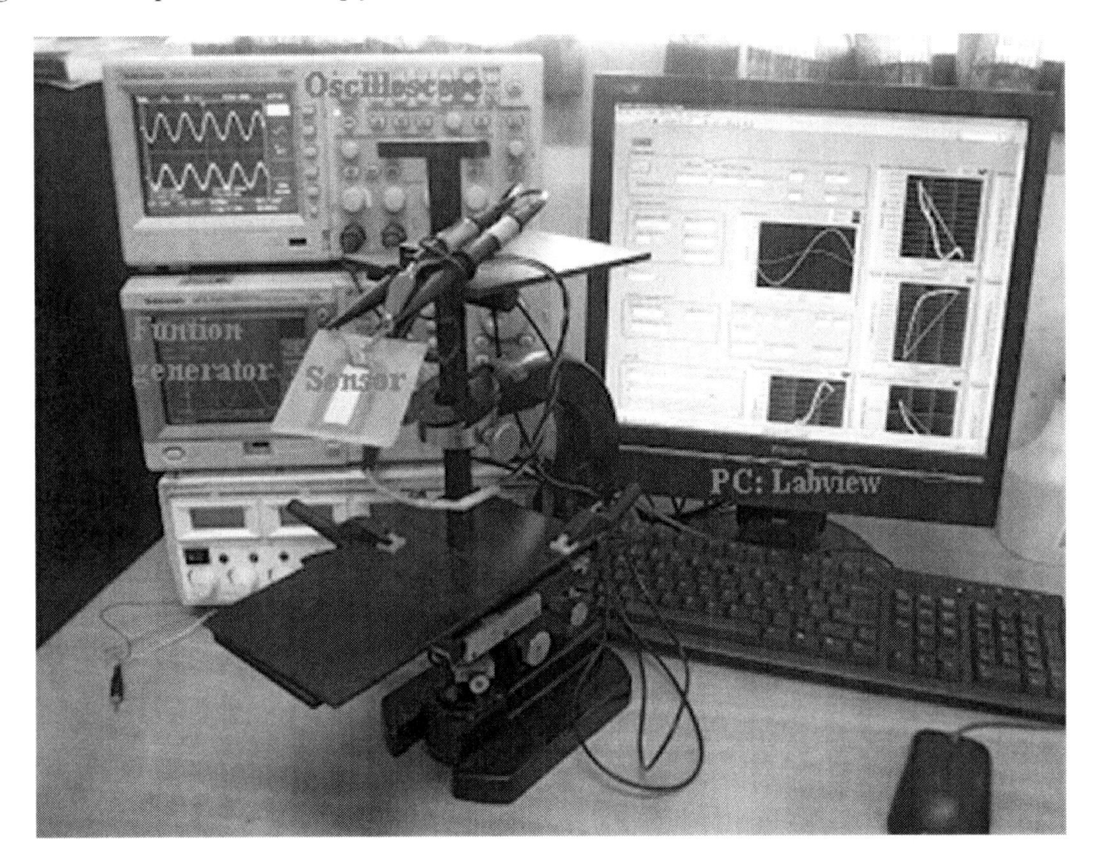

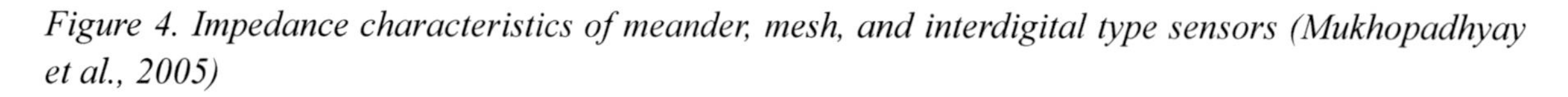

Figure 4. Impedance characteristics of meander, mesh, and interdigital type sensors (Mukhopadhyay et al., 2005)

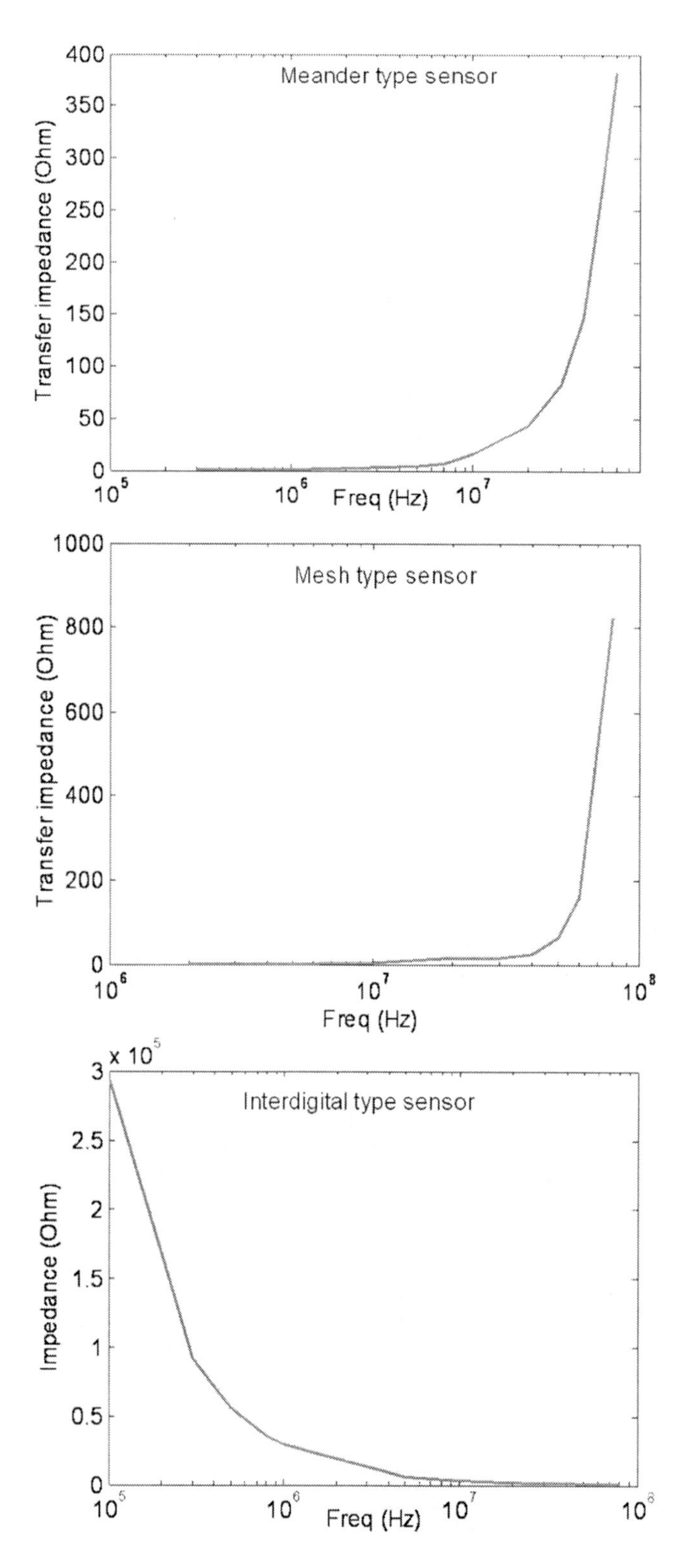

frequency power supply has been designed and developed for the supply of excitation voltage. The frequency can be controlled from tens of kHz to 1 GHz. The important components of the system are briefly discussed here (Mukhopadhyay, 2006).

Figure 4a shows a voltage controlled oscillator (VCO) unit which generates the desirable frequency of operation. Currently there are five VCO units and the output of the VCO is to be manually connected to the power supply unit but later on it can be controlled with the help of the microcontroller used in the system. Figure 5b shows the different component of the power supply unit. The total frequency range has been obtained by switching between different Voltage Controlled Oscillators corresponding to different frequency ranges.

Data Acquisition System

An efficient data acquisition system is very important for the necessary acquisition of voltage and current signals. The analog data is captured with the help an analog-to-digital converter inside the microcontroller. The complexity of acquiring data depend on many factors, the most important one is the data-rate at which the incoming signal is changing. In many transducers the signal is in the form of DC, the problem is not severe at all. The sensors reported here uses alternating current signal. Depending on the application the complete signal may be required to be acquired. Usually the planar meander and mesh types sensors do not respond well at relatively low frequency operation. But the planar interdigital sensor responds very well at low frequency. This chapter has dealt with the development of a low-cost system using planar interdigital sensor. The sensor can be used for the estimation of properties of dielectric materials such as milk, saxophone reeds, meat, leather etc. The electric field lines generated by the sensor penetrate into the system under test and the impedance of the sensor is changed. The sensor behaves as a capacitor in which the capacitive reactance becomes a function of system properties. By measuring the capacitive reactance, the system properties are evaluated. To determine the capacitive impedance both the voltage applied across the sensor and current flowing through the sensor has been acquired by the data acquisition system. A Silicon Lab microcontroller C8051F020 has been considered at the first instant as shown in Figure 6. The SiLab C8051F020 has two ADCs operating at 100 kHz and 500 kHz respectively.

Figure 5. Instrumentation for the excitation systems (a) VCO unit, and (b) power supply unit (Mukhopadhyay, 2006)

(a)

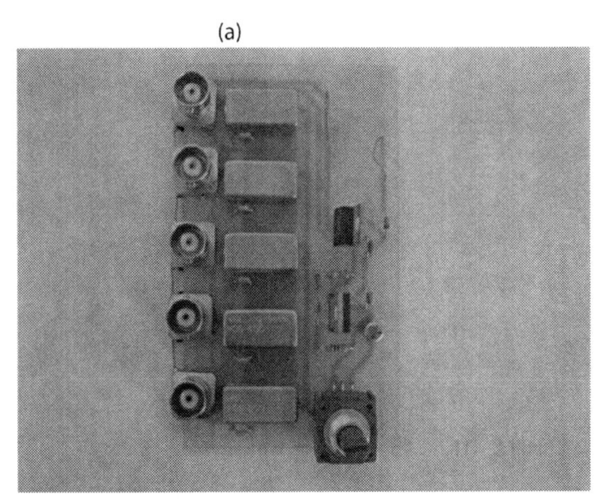

(b)

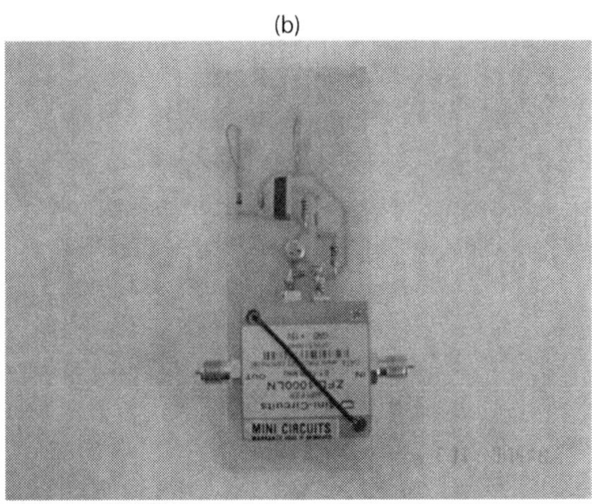

Figure 6. SiLab microcontroller C8051F020 based data acquisition system

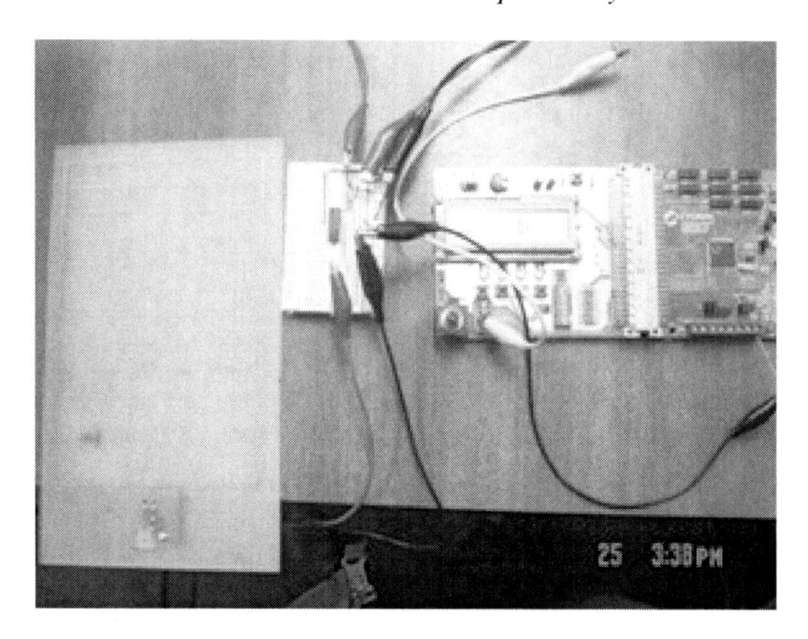

So an operating frequency of up to 50 kHz can be very well utilised using this system. The system has already been used and provides good results for single frequency measurement with an interdigital type sensor.

SMART SENSING SYSTEM FOR DIFFERENT APPLICATIONS

The development of smart sensing systems configured around a low-cost microcontroller. The signal from the sensor is fed to the analog input of the microcontroller. The signal coming from the sensor is alternating in nature. The interfacing circuit used for this setup is shown in Figure 7. The sensor output is fed into the non-inverting input through a capacitor and the voltage divider is used to offset the signal. The sensor input needs to have an offset since the 12-Bit Analog to Digital Converter (ADC) cannot process values less than zero. A LM324 Low Power Quad Operational Amplifier is used in the circuit. The sensor signal is amplified by a gain, the value of which depends on the application. V_{CC} is set to 5V, and the Zener diode makes sure the signal into the ADC input in the microcontroller doesn't exceed 3.3 V as the microcontroller C8051F020 operates at 3.3 V.

The analog signal from the sensor output is fed into the ADC pin in the microcontroller. The program stores the highest and the lowest peak of the signal, hence calculating the peak to peak value. So as the response of the sensor changes with the type of material under test, the output digital value will change accordingly. To measure the phase difference a timer is used to measure the time difference between the zero crossing of the voltage and current signals.

Sensing System for Determination of Fat Content in Meat

The interdigital sensor has been employed to estimate fat contents in pork meat. In the beginning the conventional sensors have been fabricated very similar in size to that of the meat sample as shown in Figure 8. The purpose of doing this is that the sensor can estimate the average quality of the meat by testing once for the whole meat sample. Then the modified designs as shown in

Figure 7. Interfacing circuit used in experimental setup

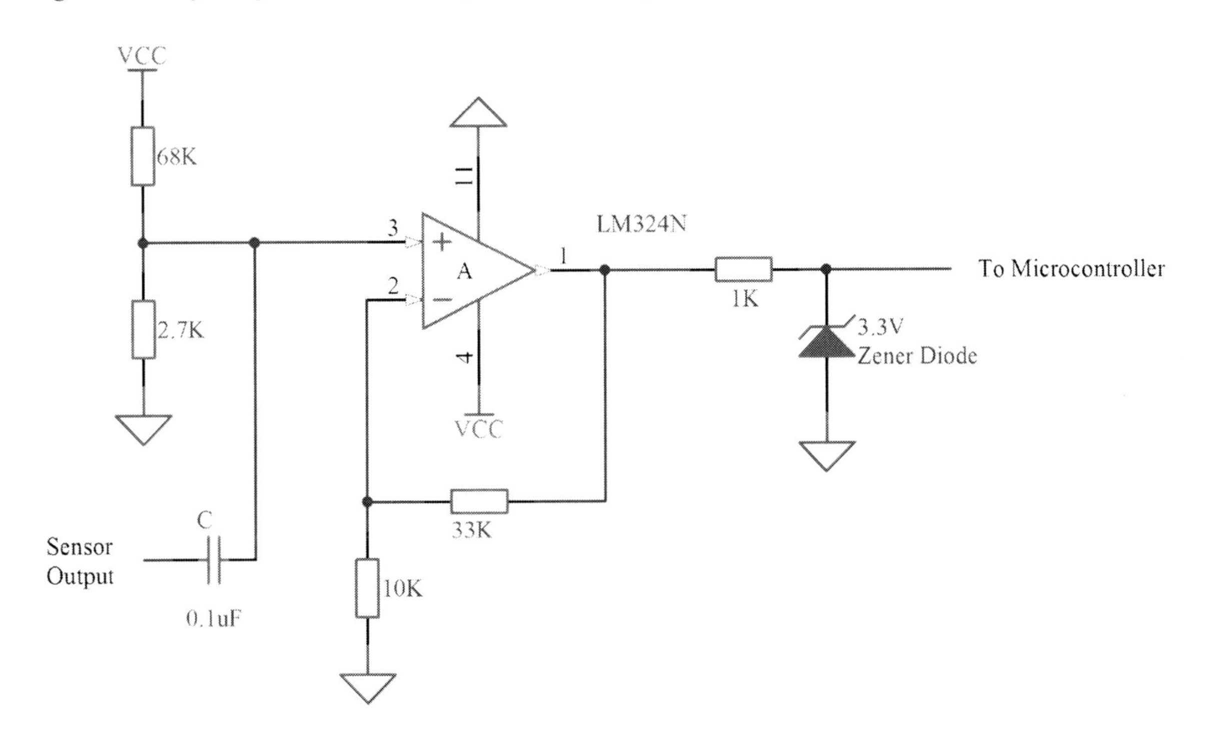

Figure 8. The interdigital sensor and the pork sample

Figure 9 have been designed and fabricated. The labels of Figure 9 are with reference to the Figure 10. Figure 10 shows the experimental comparative results of fat content estimation in meat samples with different types of fabricated sensors (Mukhopadhyay & Gooneratne, 2007).

Leather Quality Assessment Using a Smart Sensing System

Looseness in leather could be determined before and after tanning using planar interdigital sensor sensing system (Mukhopadhyay, Deb Choudhury, Allsop, Kasturi, & Norris, 2008). The looseness in sheep skin is one of the key factors in determining the quality of finished leather. It was found that the looseness could be caused by various factors such as age, bacterial damage, storage or/and processing effects. Also looseness is predominant in some of the breeds. Each of the processing steps in early stages of modern or current tanning procedures can also result in looseness. Therefore, a system that would help us to determine the looseness in sheep skins would be an advantage. So, the skins were brought to the lab for experiments before tanning process then returned to get converted in to finished leather.

Each skin was labelled into five zones as is shown in Figure 11 and sensor voltage was distinctive for each zone of same skin. But as the

Figure 9. The modified design of inter-digital sensors

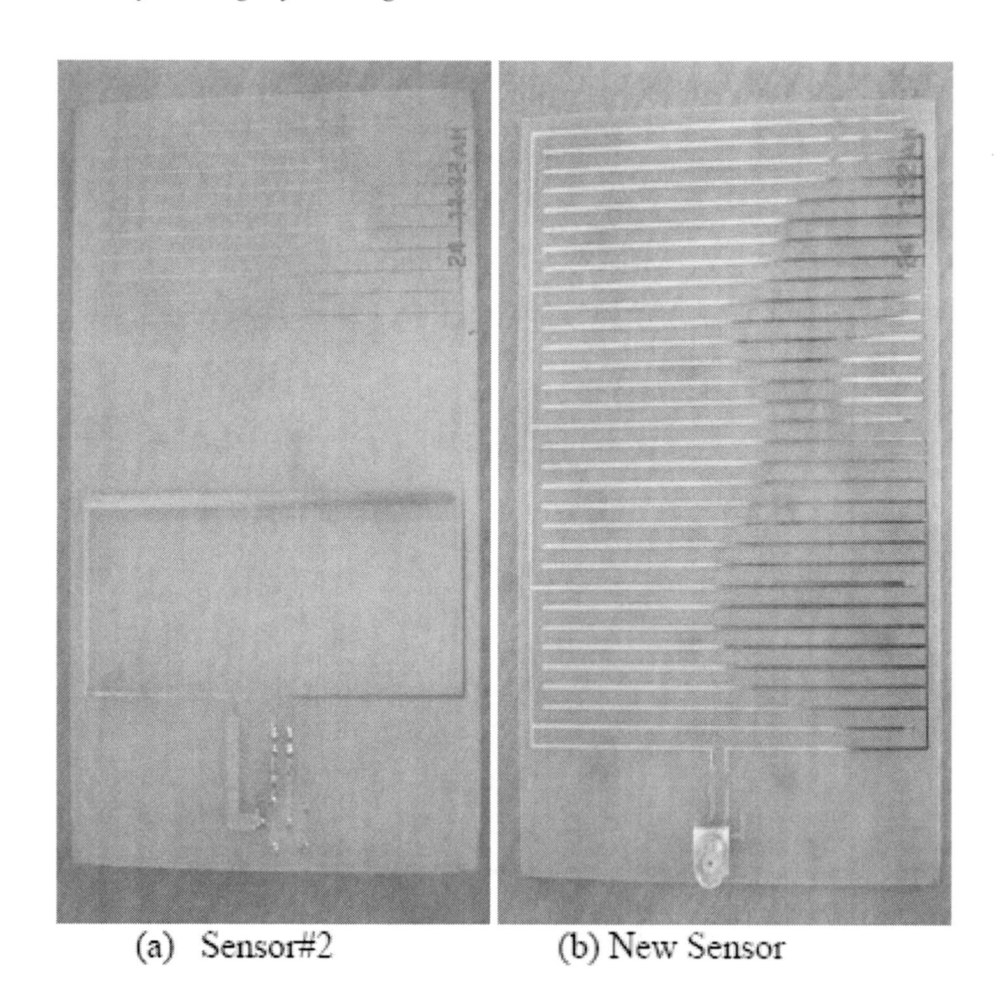

(a) Sensor#2 (b) New Sensor

Figure 10. Estimation of fat content in pork meat (Mukhopadhyay et al., 2007)

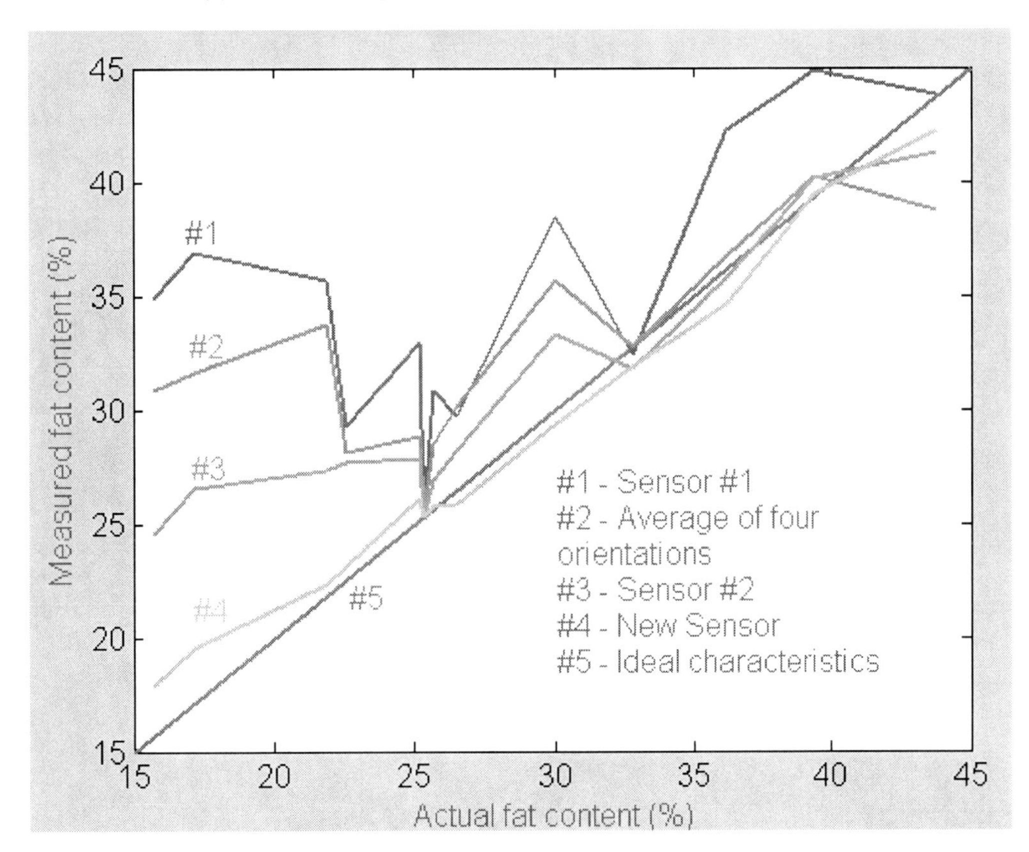

Figure 11. Measurement of looseness of leather at the marked position using the sensing system

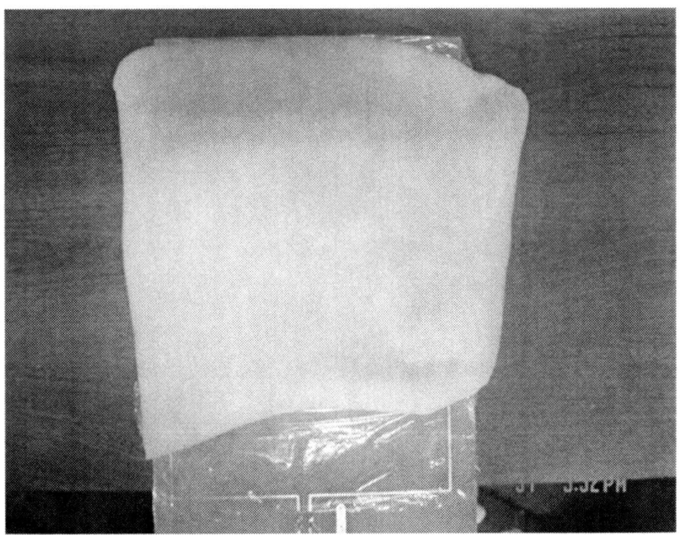

looseness feature is spread all through the skin, an average of voltages of all positions of skin was compared with looseness. Figures 12 and 13 show the comparison of measured values of looseness with the mechanical measurement before and after the tanning process. A good correlation was observed between the looseness values and sensor voltage values. Sensor voltage dropped and increased along with looseness values. It was also observed that some skins had same looseness

Figure 12. Comparison of looseness before tanning process

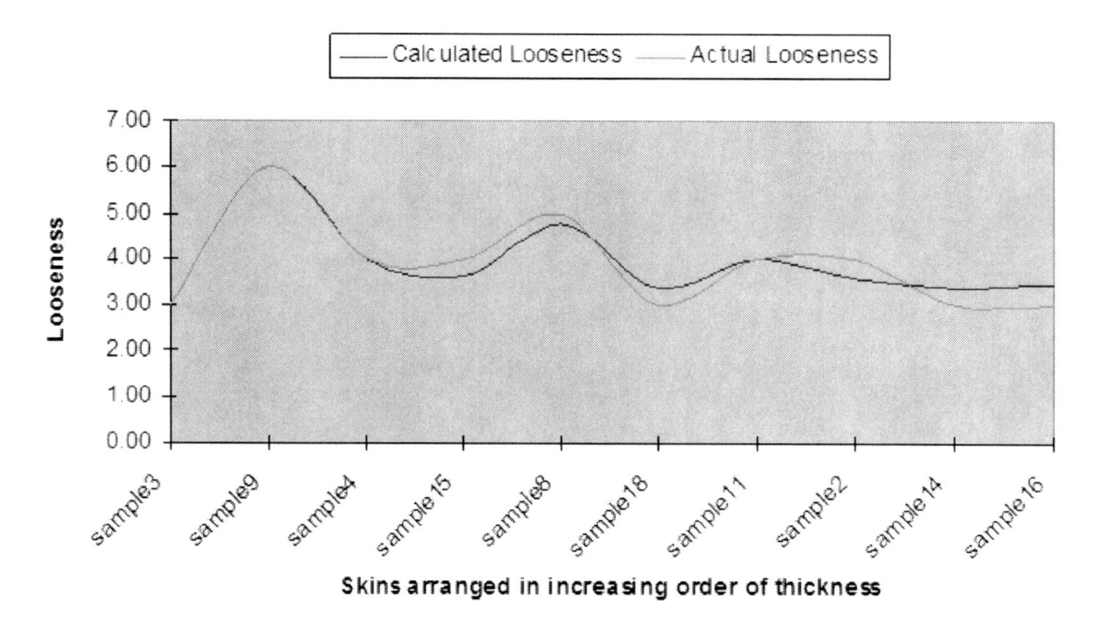

Figure 13. Comparison of looseness after tanning process

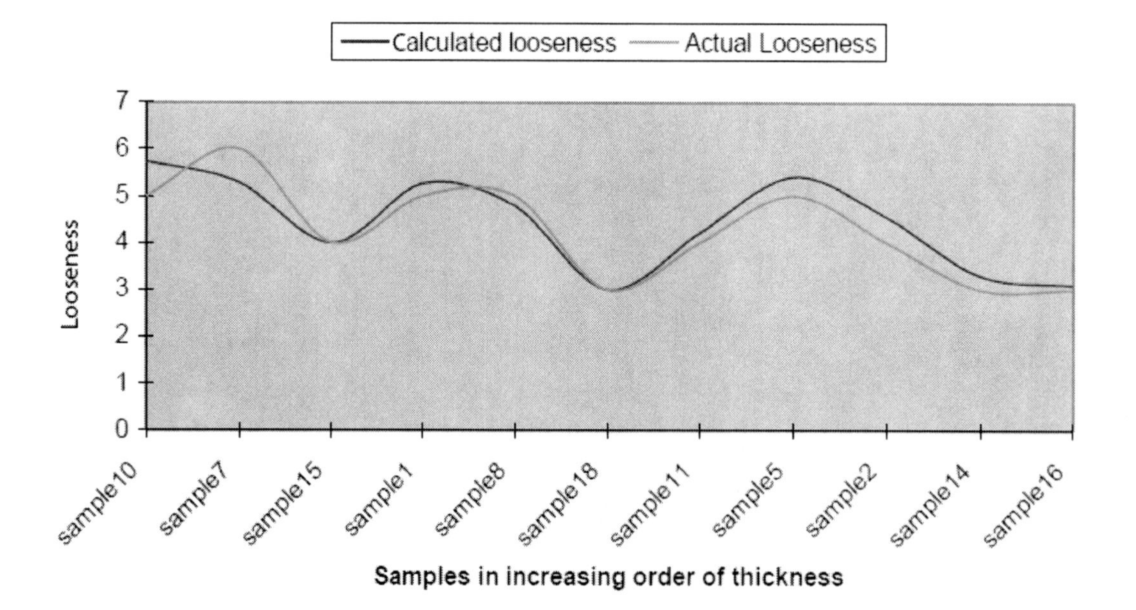

values but different voltage values which was accounted for thickness, presence of fat or also human error. The effect of thickness on sensor voltage was studied by measuring the average thickness of holes made in each of 5 zones of a skin. It was observed that sensor voltage increases along with increase in thickness of skin. Then the sensor voltage was plotted with looseness values for the samples that had same looseness values and a trend of increasing voltage was observed as the samples were also arranged in the increasing order of thickness. After the skins were converted in to finished leather experiments were repeated to verify the repeatability of the sensors. Sensor voltages were measured at the same zones after the skins were tanned and the voltage values followed the same trend as voltage values before tanning process. Hence, the repeatability of sensors was confirmed. A formula for calculation of looseness was developed which was dependent on the scaling factor of skins.

Smart Sensing System for Food Inspection

Food is essential for human life. However, pathogenic bacteria that infect food pose a serious threat to human wellbeing. The public demands a zero tolerance policy with respect to disease causing pathogens in food and is increasingly concerned over food safety. The public health implications of failing to detect pathogens in food can be fatal and the resulting losses due to medical costs and loss productivity can severely affect the world economy. World Health Organization statistics show that 1.8 million people died worldwide from diarrhoeal diseases in 2005 of which 70% were food borne (Buzby & Roberts, 2009). Errors during food production, distribution and preparation allow pathogens to contaminate food. There are many deaths each year due to foodborne illnesses and the medical costs and productivity losses can cost billions of dollars in many countries -(Buzby, Roberts, Jordan Lin, & MacDonald, 1996), (Greig & Ravel, 2009), (Tauxe, 2002), (Humphrey, O'Brien, & Madsen, 2007) and (Nørrung & Buncic, 2008). The most important of the pathogens are zoonotic bacteria, such as Salmonella, Campylobacter, Staphylococcus aureus, Clostridium perfringens, Listeria monocytogenes, Vibrio parahaemolyticus, Escherichia coli and Bacillus cereus(Humphrey & Jørgensen, 2006), (2009) and (McMeekin, 2003). Zoonoses are diseases or infections, which are transmissible from animals to humans (Muhammad-Tahir & Alocilja, 2004). In 2005 zoonotic agents caused illness in over 387,000 people in the European Union (EU) (Damez & Clerjon, 2008). Furthermore, there were many illnesses and deaths due to foodborne diseases in many countries, including New Zealand (Lazcka, Javier Del Campo, & Muñoz, 2007) and (Choi & Chae, 2009). The food poisoning caused by bacteria is a major concern to the food industry around the world (Zhang, Ma, Zhang, Zhang, & Wang, 2009). Hence, it is critical that pathogens are controlled through a complete, continuous food source-consumer system.

Most of the microbiological tests that are being used currently to detect pathogens/bacteria in food are centralized in large stationary laboratories with expensive, complex instrumentation handled by highly qualified personnel. Established methods in pathogen detection include polymerase chain reaction (PCR), culture and colony counting methods, and enzyme-linked immune-sorbent assay (ELISA). Currently ELISA is the most established technique. Even though these microbiological tests are accurate, they are time consuming (up to a week), which is not feasible for both industry and government health agencies that require rapid quality assurance of food materials. Furthermore, the required instruments are expensive, personnel need to be specially trained which is time consuming and costly and the cost of developing the ELISA strip is expensive and the method used to establish the kits is tedious.

A low cost sensing system for inspection of raw seafood was developed, the sensor and the sensing system for meat inspection is shown in Figure 13 (Mohd Syaifudin, Jayasundera, & Mukhopadhyay, 2009) and (Mohd Syaifudin, Mukhopadhyay, & Yu, 2010). The sensing system is based on new design of interdigital sensor.

The first prototype of seafood inspection tool (SIT) was developed to detect the domoic acid (DA) in mussels. SIT consist of ± 9 V power supply, novel planar interdigital sensor, a SiLab C8051F020 microcontroller, a signal processing circuit and an expansion board (for display). A user friendly software was developed to make it ease of use. It can be used by anyone especially by fisherman for pre-screening process at the ranch site. The first prototype of SIT is shown in Figure 14. The response of the sensing system is shown in Figure 15. The sensing system can detect an amount of 12.5 μg/g of domoic acid in meat samples; the regulation from the World Health Organization is 20 μg/g. The developed low cost sensing system can be used for fisherman for pre-screening process. The sensing system analyses the samples from the ranch site and then provide pass or fail analysis. If the results showing certain number of fail analysis (suspicious results), the samples at that ranch site have to be sent to the laboratory for further analysis of contaminated chemicals (DA) in the seafood using expensive techniques. The developed sensing system is easy to be used for the purpose of sampling inspection and can provide fast analysis of DA within shellfish meat for in-situ monitoring. The developed sensing system should be reliable and cost effective (Figure 16).

Smart Sensing System for Biomedical Applications

Cancer is the most deadly disease in the world today. There is a variety of different treatment methods for cancer, including radiotherapy and chemotherapy with anti-cancer drugs that have been in use over a long period of time. Hyperthermia is one of the cancer treatment methods which utilises the property that cancer cells are more sensitive to temperature than normal cells. The control of temperature is an important task in achieving success using this treatment method. The development of a novel needle-type nano-sensor based on the spin-valve giant magneto-resistive (SV-GMR) technique to measure the magnetic flux-density inside the body via pricking the needle has been undertaken (Mukhopadhyay, Chomsuwan, Gooneratne, & Yamada, 2007) and (Goonerate, Mukhopadhyay, & Yamada, 2009). The sensor has been fabricated by TDK (Tokyo Denki Kaisha), Japan. The modelling and experimental results of flux-density measurement have been carried out with an accuracy of around 0.1%. From the information of flux-density the

Figure 14. The sensors and sensing system for inspection of meat quality (Mohd Syaifudin et al., 2009)

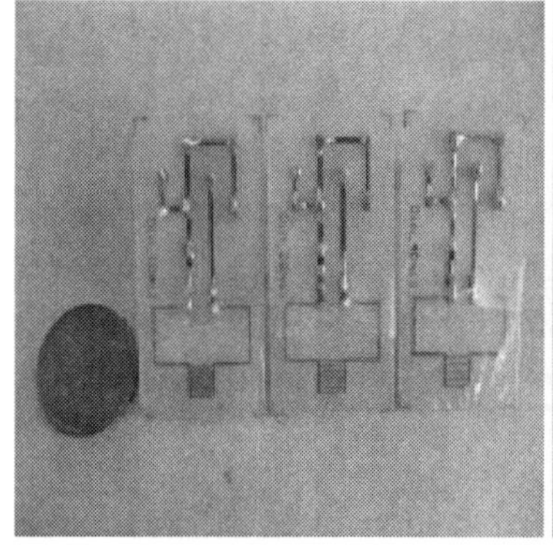

Figure 15. The first prototype of seafood inspection tool (Mohd Syaifudin et al., 2010)

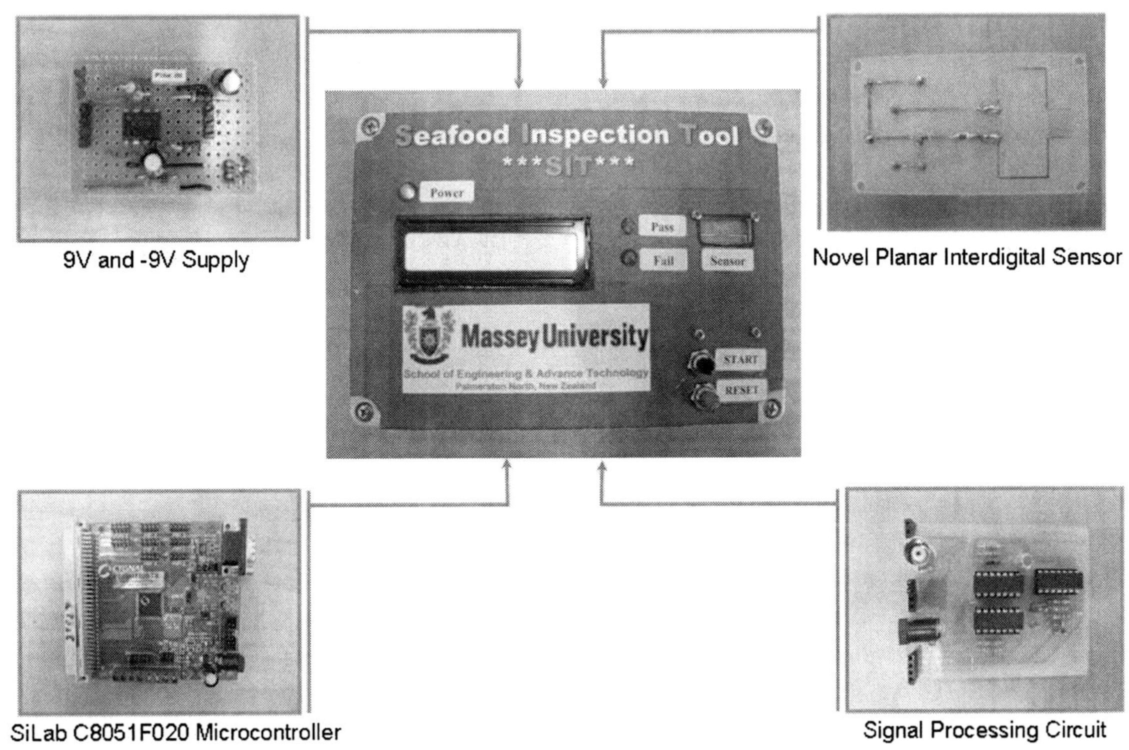

Figure 16. Experimental response of the sensing system as a function of amount of chemicals

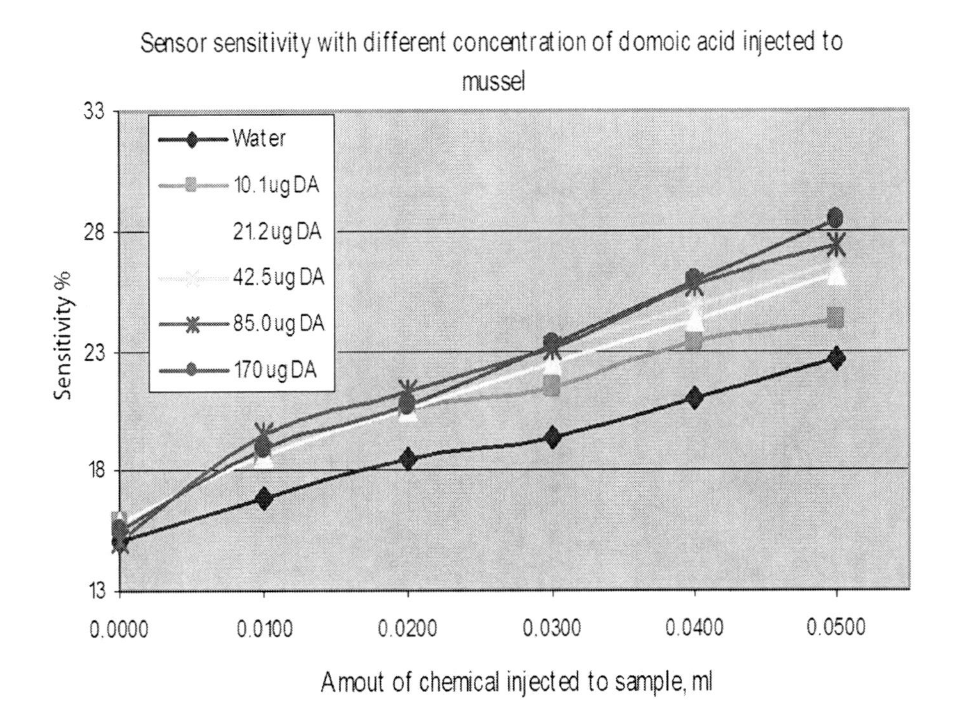

temperature rise can be estimated to permit the delivery of controlled heating to precisely defined locations in controlled hyperthermia cancer treatment. Figure 17 shows the picture of the GMR needle sensor along with the experimental set-up.

Planar Electromagnetic Sensor for Agricultural Applications

A new sensors based on the combination of meander and interdigital sensors for sensing the level of Nitrate in water sources has been designed and fabricated (Yunus, Mendez, & Mukhopadhyay, 2011) and (Mohd. Amri Yunus & Mukhopadhyay, 2011). A series of experiments was conducted involving different type and concentrations of nitrate-based solution. The responses of the sensors were observed and the sensors are viable for Nitrates sensing in water sources applications. The proposed sensors for sensing of Nitrates in water sources were tested with different concentration of Nitrates samples (5 mg, 10 mg, 15 mg and 20 mg) in the form of Sodium Nitrate (NaNO3) and Ammonium Nitrate (NH4NO3) diluted in 1 litre of distilled water. Beforehand, the sensor was sprayed with Wattyl Killrust Incralac to form an acrylic resin-based protective coating. The effect of the samples on the sensor's impedance was examined. The fabricated sensors are shown in Figure 17a and the experimental setup is shown in Figure 17b respectively. The experimental setup (Figure 18) has a frequency waveform generator where standard sinusoidal waveform with 10 Volts peak-to-peak value was set as the input signal for the sensors. An old microscope- was used as a platform for the sample container and the sensors were partially immersed into the water sample. The Agilent 54622D mixed signal oscilloscope was interfaced to a PC where the output signals and the sensor's impedance was recorded and calculated consecutively using LabView software for an infinite period of time (approximately 2000 sec).

Figure 17. Fabricated GMR sensor and associated circuit diagram

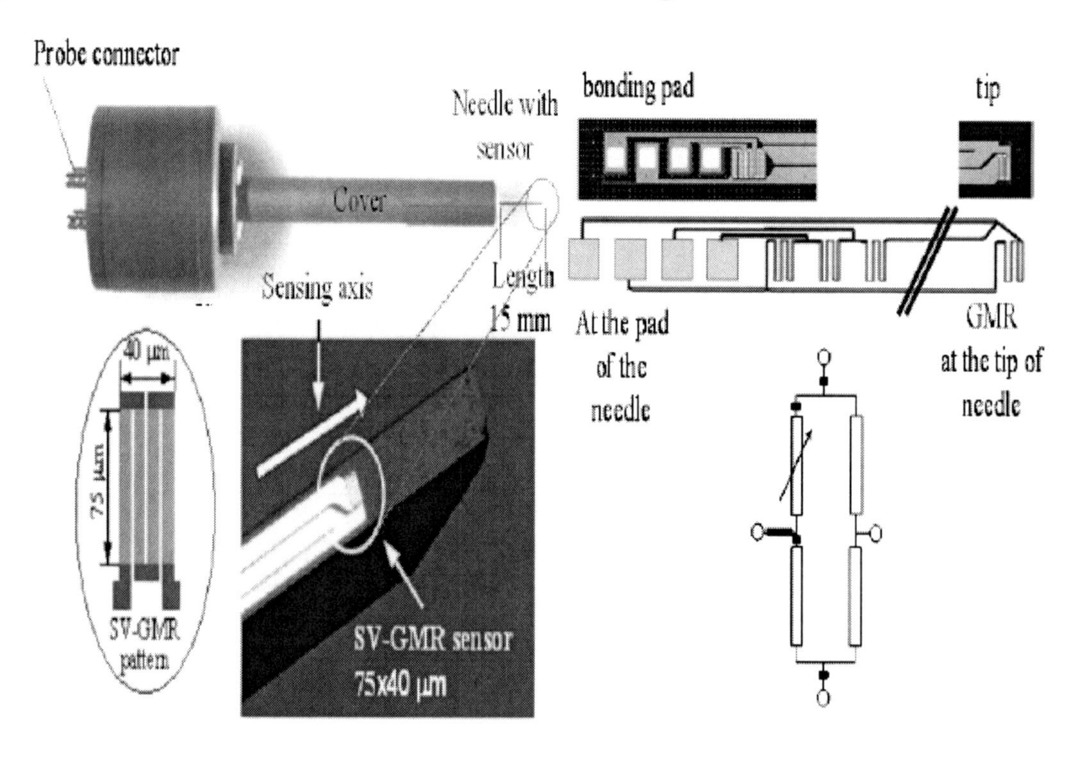

Figure 18. Planar electromagnetic sensor and the experimental set-up for agricultural application

RFID Based Real-Time Tracking System of Hajj Pilgrim in Mecca

Radio-frequency identification (RFID) is a technology that uses communication through the use of radio waves to transfer data between a reader and an electronic tag attached to an object for the purpose of identification and tracking. Radio Frequency Identification (RFID) is a technology that has been used since 1984. In recent years, RFID has drawn more attention and it has been applied and used in many industries. The Department of Defense (DoD) and very well known market stores around the world, including Wal-Mart, Pfizer, Tesco, and Gillette's, have used RFID technology. Furthermore, the European Central Bank is trying to prevent counterfeiting of banknotes by using RFID tags. RFID can also be used in crowded public places where people often lose their family members and children. Therefore, RFID technology will make Hajj season in Mecca much easier for many people. The continuous increase of number of pilgrims made the Hajj more difficult for the pilgrims and the authorities. One of the most difficult tasks during the Hajj season is when someone gets lost in the crowd. Young children and the elderly are the most affected under such situation. Many pilgrims die, get lost and kidnapped every year due the congestion. Therefore, the new technologies have become critical to control the crowds. Recently, a few projects have been conducted in Mecca in order to curb the problem. However, the need for more practical and economical solutions is still required. The chipless RFID as presented in the book in many chapters will alleviate the economy for mass deployment of chipless RFID tags such as human tracking and tracing problems as delineated for pilgrimage.

TOWARDS AN INTELLIGENT SENSING SYSTEM

The future target of this project is to develop a low-cost intelligent sensing system. They are characterized as having significant data processing, storing and analyzing power. These intelligent sensors can be used as autonomous systems or deployed in large numbers to form powerful sensor networks. The sensor networks may depend on multiple embedded processors to simultaneously gather and process information from many sources. They are often flexible, self-organizing and fault-tolerant, thus making them well suited for critical applications.

There has been an increasing interest in the design, development and applications of the intelligent sensors and networks. The sensor described in this paper is based on a Cygnal microcontroller as shown in Figure 6. It has the ability to compensate for systematic errors, drift and random errors generated by system parameters or the characteristics of the sensor. It has the ability to determine the processing parameters automatically. It has the ability to detect corrupted data and self-test its operation. The complete sensor system could determine the type and level of recalibration required by a particular sensor and it has the power of efficient data processing techniques.

Currently the sensor can communicate data with a central computer in the form of wireless communication. It can be asked to carry out necessary operations such as offset measurement and calibration, signal measurement, necessary processing and dispatching of data. Since the sensor system has been developed around a powerful microcontroller so many other features of an intelligent sensing system can be incorporated without much difficulty. The future plan is to integrate the microcontroller and other necessary analog circuit at the back side of the sensor.

CONCLUSION AND FUTURE WORKS

In recent times sensors and sensors network have been widely used in many applications such as monitoring environmental parameters, monitoring and control of industrial situations, structural health monitoring and so on. The advancement of electronics, embedded controller, smart wireless sensors, networking and communication have made the development of low cost and low power smart wireless sensors nodes possible. The sensors data exchange, sharing and interoperability are the major challenges using wireless sensors network for the condition monitoring. It will help to solve majority of the problems faced if a standardized sensors data format and a communication protocol. The electromagnetic sensor based approach proposed in this chapter leads to a low-cost sensing technology and can be an alternative to the existing measurement techniques used in many industries. This sensing technique is non-destructive in nature. It is also safe and cost-effective. In addition, the approach does not involve any fragile sensing elements (such as glass probes) and it is quite stable (requires only infrequent periodic calibration). The design issues towards developing a low-cost and intelligent sensing system have been discussed.

In future, the challenge is to reducing the size of the sensors and sensing system. Moreover, it will be desirable to have the sensors, signal conditioning circuits, embedded controller, along with communication facility supplied with power in the form of photovoltaic cells in one package. Though it looks like a huge challenge in the present scenario, the days are not far ahead to visualise that in practice.

REFERENCES

Amadi-Echendu, J. E., & Zhu, H. (1994). Detecting changes in the condition of process instruments. *IEEE Transactions on Instrumentation and Measurement*, *43*(2), 355–358. doi:10.1109/19.293447

Buzby, J. C., & Roberts, T. (2009). The economics of enteric infections: Human foodborne disease costs. *Gastroenterology*, *136*, 1851–1862. doi:10.1053/j.gastro.2009.01.074

Buzby, J. C., Roberts, T., Jordan Lin, C. T., & MacDonald, J. M. (1996). *Bacterial foodborne disease: Medical costs and productivity losses*. USA: Economic Research Service/USDA Report Number (AER741). Retrieved November 4, 2009 from http://www.ers.usda.gov/Publications/AER741/

Choi, S., & Chae, J. (2009). A microfluidic biosensor based on competitive protein adsorption for thyroglobulin detection. *Biosensors & Bioelectronics*, *25*, 118–123. doi:10.1016/j.bios.2009.06.017

Damez, J. L., & Clerjon, S. (2008). Meat quality assessment using biophysical methods related to meat structure. *Meat Science*, *80*, 132–149. doi:10.1016/j.meatsci.2008.05.039

Ferrari, P., Flamini, A., Marioli, D., & Taroni, A. (2006). IEEE802.11 sensor networking. *IEEE Transactions on Instrumentation and Measurement*, *55*(2), 615–619. doi:10.1109/TIM.2006.870105

Ferrigno, L., Pietrosanto, A., & Paciello, V. (2006). Low-cost visual sensor node for BlueTooth-based measurement networks. *IEEE Transactions on Instrumentation and Measurement*, *55*(2), 521–527. doi:10.1109/TIM.2006.870126

Goonerate, C. P., Mukhopadhyay, S. C., & Yamada, S. (2009, December). Electromagnetic field computation of GMR sensors for hyperthermia based cancer treatment. In *Proceedings of IEEE Applied EletroMagnetic Conference* (AEMC) 2009, 14-16, Kolkata, India. ISBN: 978-1-4244-4819-7

Greig, J. D., & Ravel, A. (2009). Analysis of foodborne outbreak data reported internationally for source attribution. *International Journal of Food Microbiology*, *130*, 77–87. doi:10.1016/j.ijfoodmicro.2008.12.031

Humphrey, T., & Jørgensen, F. (2006). Pathogens on meat and infection in animals – Establishing a relationship using campylobacter and salmonella as examples. *Meat Science*, *74*, 89–97. doi:10.1016/j.meatsci.2006.04.026

Humphrey, T., O'Brien, S., & Madsen, M. (2007). Campylobacters as zoonotic pathogens: A food production perspective. *International Journal of Food Microbiology*, *117*(3), 237–257. doi:10.1016/j.ijfoodmicro.2007.01.006

Kanoun, O., & Trankler, H. R. (2004). Sensor technology advances and future trends. *IEEE Transactions on Instrumentation and Measurement*, *53*(6), 1497–1501. doi:10.1109/TIM.2004.834613

Lazcka, O., Javier Del Campo, F., & Muñoz, F. X. (2007). Pathogen detection: A perspective of traditional methods and biosensors. *Biosensors & Bioelectronics*, *22*, 1205–1217. doi:10.1016/j.bios.2006.06.036

Lenz, J., & Edelstein, A. S. (2006). Magnetic sensors and their applications. *IEEE Sensors Journal*, *6*(3), 631–649. doi:10.1109/JSEN.2006.874493

Li, X., & Meijer, G. C. M. (1995). A novel smart resistive-capacitive position sensor. *IEEE Transactions on Instrumentation and Measurement*, *44*(3), 768–770. doi:10.1109/19.387328

Li, X., & Meijer, G. C. M. (2001). A smart and accurate interface for resistive sensors. *IEEE Transactions on Instrumentation and Measurement*, *50*(6), 1648–1651. doi:10.1109/19.982961

Li, X., Meijer, G. C. M., & de Jong, G. W. (1997). A microcontroller-based self-calibration technique for a smart capacitive angular-position sensor. *IEEE Transactions on Instrumentation and Measurement*, *46*(4), 888–892. doi:10.1109/19.650794

Li, X., Meijer, G. C. M., de Jong, G. W., & Spronck, J. W. (1996). An accurate low-cost capacitive absolute angular-position sensor with a full-circle range. *IEEE Transactions on Instrumentation and Measurement*, *45*(2), 516–520. doi:10.1109/19.492778

Li, X., Meijer, G. C. M., & Schnitger, E. J. (1998). A novel smart interface for voltage-generating sensors. *IEEE Transactions on Instrumentation and Measurement*, *47*(1), 285–288. doi:10.1109/19.728835

McMeekin, T. A. (2003). *Detecting pathogens in food* (1st ed., pp. 241–258). Cambridge, UK: Woodhead.

Mochizuki, K., Masuda, T., & Watanabe, K. (1998). An interface circuit for high-accuracy signal processing of differential-capacitance transducers. *IEEE Transactions on Instrumentation and Measurement*, *47*(4), 823–827. doi:10.1109/19.744628

Mohd Syaifudin, A. R., Jayasundera, K. P., & Mukhopadhyay, S. C. (2009). A low cost novel sensing system for detection of dangerous marine biotoxins in seafood. *Sensors and Actuators. B, Chemical*, *137*, 67–75. doi:10.1016/j.snb.2008.12.053

Mohd Syaifudin, A. R., Mukhopadhyay, S. C., & Yu, P. L. (2010). Novel sensors for food inspection. *Sensors and Transducers Journal*, *108*, 1–44.

Muhammad-Tahir, Z., & Alocilja, E. C. (2004). A disposable biosensor for pathogen detection in fresh produce samples. *Biosystems Engineering*, *88*(2), 145–151. doi:10.1016/j.biosystemseng.2004.03.005

Mukhopadhyay, S. C. (2006, October). Sensing and instrumentation for a low cost intelligent sensing system. *Proceedings of the SICE-ICCAS Joint International Conference 2006*, (pp. 1075-1080). ISBN: 89-950038-5-5 98560

Mukhopadhyay, S. C., Chomsuwan, K., Gooneratne, C., & Yamada, S. (2007). A novel needle-type SV-GMR sensor for biomedical applications. *IEEE Sensors Journal*, *7*(3), 401–408. doi:10.1109/JSEN.2007.891929

Mukhopadhyay, S. C., Deb Choudhury, S., Allsop, T., Kasturi, V., & Norris, G. E. (2008). Assessment of pelt quality in leather making using a novel non-invasive sensing approach. *Journal of Biochemical and Biophysical Methods*, *70*, 809–815. doi:10.1016/j.jbbm.2007.07.003

Mukhopadhyay, S. C., & Gooneratne, C. P. (2007). A novel planar-type biosensor for noninvasive meat inspection. *IEEE Sensors Journal*, *7*(9), 1340–1346. doi:10.1109/JSEN.2007.903335

Mukhopadhyay, S. C., Gooneratne, C. P., Sen Gupta, G., & Yamada, S. (2005). Characterization and comparative evaluation of novel planar electromagnetic sensors. *IEEE Transactions on Magnetics*, *41*(10), 3658–3660. doi:10.1109/TMAG.2005.854792

Nørrung, B., & Buncic, S. (2008). Microbial safety of meat in the European Union. *Meat Science*, *78*, 14–24. doi:10.1016/j.meatsci.2007.07.032

Patra, J. C., Kot, A. C., & Panda, G. (2000). An intelligent pressure sensor using neural networks. *IEEE Transactions on Instrumentation and Measurement*, *49*(4), 829–834. doi:10.1109/19.863933

Public Health Surveillance. (2008). *Annual summary of outbreaks: New Zealand.* Retrieved on October 25, 2009 from www.surv.esr.cri.nz/surveillance/annual_outbreak.php

Schmalzel, J., Figueroa, F., & Morris, J. (2005). An architecture for intelligent systems based on smart sensors. *IEEE Transactions on Instrumentation and Measurement*, *54*(4), 1612–1616. doi:10.1109/TIM.2005.851477

Tauxe, R. V. (2002). Emerging foodborne pathogens. *International Journal of Food Microbiology*, *78*, 31–41. doi:10.1016/S0168-1605(02)00232-5

Toth, F. N., & Meijer, G. C. M. (1992). A low-cost smart capacitive position sensor. *IEEE Transactions on Instrumentation and Measurement*, *41*(6), 1041–1044. doi:10.1109/19.199446

Yunus, M. A. M., Mendez, G. R., & Mukhopadhyay, S. C. (2011, May). Development of a low cost system for nitrate and contamination detections in natural water supply based on a planar electromagnetic sensor. In *Proceedings of IEEE I2MTC 2011* (pp. 1557–1562). Hangzhou, China: Conference. doi:10.1109/IMTC.2011.5944156

Zhang, Y., Ma, H., Zhang, K., Zhang, S., & Wang, J. (2009). An improved DNA biosensor built by layer-by-layer covalent attachment of multi-walled carbon nanotubes and gold nanoparticles. *Electrochimica Acta*, *54*, 2385–2391. doi:10.1016/j.electacta.2008.10.052

KEY TERMS AND DEFINITIONS

Instrumentations: The art and science of measurement and control of systems.

Intelligent Sensors: Sensors incorporated with dedicated signal processing.

Planar Electromagnetic Type Sensors: Sensors that use a propagating electromagnetic wave that has flat planes as surfaces of constant phase.

Sensing: That part of the transducer which reacts directly in response to the input.

Compilation of References

Abd-Elrazzak, M. M., & Al-Nomay, I. S. (2004). A design of a circular microstrip patch antenna for bluetooth and HIPERLAN applications. *The 9th Asia-Pacific Conference on Communications: Vol. 3* (pp. 974-977).

Abdolee, R. (2008). *Performance of MIMO space time coded system and training based channel estimation for MIMO-OFDM system*. Master Thesis, Universiti Teknologi Malaysia.

Aberdeen Group. (2008). *Where RFID meets ROI: Beyond supply chains.*

Aberer, K., Hauswirth, M., & Salehi, A. (2006). A middleware for fast and flexible sensor network deployment. *VLDB, 06*(September), 1215.

Abramson, N. (1970). The ALOHA system - Another alternative for computer communications. *In Proceedings of Fall Joint Computer Conference, AFIPS Conference* (pp. 281–285). Houston, Texas.

Acierno, R., Carata, E., De Pascali, S. A., Fanizzi, F. P., Maffia, M., Mainetti, L., & Patrono, L. (2010a). Potential effects of RFID systems on biotechnology insulin preparation: A study using HPLC and NMR spectroscopy. In (IEEE) *Proceedings of the 5th International Conference on Complex Medical Engineering* (pp. 198-203).

Acierno, R., De Pascali, S. A., Fanizzi, F. P., Maffia, M., Mainetti, L., Patrono, L., & Urso, E. (2010b). Investigating potential effects of RFID systems on the molecular structure of the human insulin. In *Proceedings of the 5th Cairo International Biomedical Engineering Conference* (pp. 192-196).

Acierno, R., De Riccardis, L., Maffia, M., Mainetti, L., Patrono, L., & Urso, E. (2010c). Exposure to electromagnetic fields in UHF band of an insulin preparation: Biological effects. In (IEEE) *Proceedings of Biomedical Circuits and Systems Conference* (pp. 78-81).

Acierno, R., Maffia, M., Mainetti, L., Patrono, L., & Urso, E. (2011). RFID-based tracing systems for drugs: Technological aspects and potential exposure risks. In (IEEE) *Proceedings of Topical Conference on Biomedical Wireless Technologies, Networks, and Sensing Systems* (pp. 87-90).

Aero-plastics, Inc. (n.d.). *Website*. Retrieved from http://www.aero-plastics.tmcsweb.com/

Agilent. (n.d.). *Advanced design system (ADS) simulation environment*. Retrieved from http://www.home.agilent.com/

Alamouti, S. M. (1998). A simple transmitter diversity scheme for wireless communications. *IEEE Journal on Selected Areas in Communications, 16*, 1451–1458. doi:10.1109/49.730453

Alarcon-Aquino, V., Dominguez-Jimenez, M., & Ohms, C. (2008). Design and implimentation of a security layer for RFID. *Journal of Applied Research and Technology, 6*(2), 69–83.

Alien. (2011). *RFID ICs*. Retrieved from http://www.alientechnology.com/tags/rfid_ic.php

Alippi, C., & Vanini, G. (2005). An application-level methodology to guide the design of intelligent-processing, power-aware passive RFIDs. *IEEE International Symposium on Circuits and Systems,* Vol. 6 (pp. 5509-5512).

Almeida, B. E., Oliveira, J. E., Damiani, R., Dalmora, S. L., Bartolini, P., & Ribela, M. T. (2011). A pilot study on potency determination of human follicle-stimulating hormone: A comparison between reversed-phase high-performance liquid chromatography method and the in vivo bioassay. *Journal of Pharmaceutical and Biomedical Analysis*, *54*(4), 681–686. doi:10.1016/j.jpba.2010.10.018

Alomair, B., Clark, A., Cuellart, J., & Poovendran, R. (2010). Scalable RFID systems: A privacy-preserving protocol with constant-time identification. In *the Proceedings of the IEEEI IFIP International Conference on Dependable Systems & Networks* (DSN), (pp. 1-10).

Alotaibi, M., Postula, A., & Portmann, M. (2009). Tag anti-collision algorithms in RFID systems: A new trend. *WSEAS Transactions on Communications*(WTOC). *WSEAS Transactions on Communications*, *8*, 1216–1232.

Altin, N., & Yazgan, E. (2008). A new roof model on randomly placed buildings in mobile communication. *Progress in Electromagnetic Research*, *3*(3), 95–101.

Amadi-Echendu, J. E., & Zhu, H. (1994). Detecting changes in the condition of process instruments. *IEEE Transactions on Instrumentation and Measurement*, *43*(2), 355–358. doi:10.1109/19.293447

Amaral, L. A., Hessel, F. P., Bezerra, E. A., Correa, J. C., Longhi, O. B., & Dias, T. F. O. (2009). An adaptative framework architecture for RFID applications. 33rd Annual IEEE Software Engineering Workshop (SEW), 13-14 October 2009, (pp. 15-24).

Amazon. (2011). *Elastic compute cloud*. Retrieved from http://aws.amazon.com/ec2

Amin, E. M. A., & Karmakar, N. (2011). *Development of a chipless RFID temperature sensor using cascaded spiral resonators*. Paper presented at the IEEE SENSORs 2011.

Anderson, H. A. (1993). A ray tracing propagation model for digital broadcast systems in urban areas. *IEEE Transaction on Broadcasting, 39*(3).

Anderson, A. (1971). *The Raman effect*. M. Dekker.

Antoniou, G., Billington, D., Governatori, G., & Maher, M. (2006). Embedding defeasible logic into logic programming. *Theory and Practice of Logic Programming*, *6*(6), 703–735. doi:10.1017/S1471068406002778

ATA5590 Datasheet. (2005). *1-kbit UHF R/W IDIC with anti-collision function*. ATMEL Corporation.

Austen, A. E. W., & Hackett, W. (1944). Internal discharges in dielectrics: their observation and analysis. *Journal of the Institute of Electrical Engineers - Part I: General*, *91*(44), 298–312.

Avery Dennison. (2011). *About RFID*. Retrieved from http://www.rfid.averydennison.com/us/index.php

Avoine, G. (2004). Privacy issues in RFID banknote protection schemes. In *Proceedings of the 6th International Conference on Smart Card Research and Advanced Applications*, (pp. 33-48).

Avoine, G. (2005). *Radio frequency identification: adversary model and attacks on existing protocols*. Technical Report LASEC-REPORT-2005-001. Lausanne, Switzerland: EPFL. Retrieved from http://www.epcglobalinc.org/standards/tds

Avoine, G., Dysli, E., & Oechslin, P. (2005). Reducing time complexity in RFID systems. *Proceedings of the 12th Annual Workshop Selected Areas in Cryptography* (*SAC*).

Avoine, G., & Oechslin, P. (2005). RFID traceability: A multilayer problem. *Financial Cryptography*, *2005*, 125–140.

Avruch, J. (1998). Insulin signal transduction through protein kinase cascades. *Molecular and Cellular Biochemistry*, *182*(1-2), 31–48. doi:10.1023/A:1006823109415

Bacheldor, B. (2006). *Mitsubishi Electric Asia switches on RFID*. Retrieved March 9, 2011, from http://www.rfidjournal.com/article/view/2644

Bacheldor, B. (2007). *RFID tattoos for livestock*. Retrieved February 6, 2011, from http://www.rfidjournal.com/article/view/3079

Bacheldor, B. (2008a). *Drug pedigree mandate could be expensive*. Retrieved March 12, 2011, from http://www.rfidjournal.com/article/print/4179

Bacheldor, B. (2008b). *Italian alpine resort takes near field communications for ride*. Retrieved March 14, 2011, from http://www.rfidjournal.com/article/view/4463

Baker, J. P., Bones, P. J., & Lim, M. A. (2006). Wireless health monitor. *Electronics New Zealand Conference*, (pp. 7-12).

Balanis, C. A. (2005). *Antenna theory analysis and design*. John Wiley & Son.

Balanis, C. A. (1992). Antenna theory: A review. *Proceedings of the IEEE*, *80*(1), 7–23. doi:10.1109/5.119564

Balanis, C. A. (2005). *Antenna theory - Analysis and design* (3rd ed.). John Wiley & Sons.

Balbin, I., & Karmakar, N. (2009a). Novel chipless RFID tag for conveyor belt tracking using multi-resonant dipole antenna. *European Microwave Conference, 1-3*, 1109-1112.

Balbin, I., & Karmakar, N. (2009b). Phase-encoded chipless RFID transponder for large-scale low-cost applications. *IEEE Microwave and Wireless Components Letters*, *19*(8), 509–511. doi:10.1109/LMWC.2009.2024840

Bancroft, R. (2009). *Microstrip and printed antenna design* (2nd ed.). SciTech Publishing, Inc.

Bang, O., Choi, J. H., Lee, D., & Lee, H. (2009). *Efficient novel anti-collision protocols for passive RFID tags*. Auto-ID Labs White Paper. Retrieved August 11, 2011, from http://www.autoidlabs.org/single- view/dir/article/1/323/page.html

Banks, J., Hanny, D., Pachano, M. A., & Thompson, L. G. (2007). *RFID in warehousing and distribution systems*. John Wiley & Sons, Inc.

Barchetti, U., Bucciero, A., De Blasi, M., Mainetti, L., & Patrono, L. (2009). Implementation and testing of an EPCglobal-aware discovery service for item-level traceability. In (IEEE) *Proceedings of International Conference on Modern Ultra Telecommunications* (pp. 1-8).

Barchetti, U., Bucciero, A., De Blasi, M., Mainetti, L., & Patrono, L. (2010). RFID, EPC and B2B convergence towards an item-level traceability in the pharmaceutical supply chain. In (IEEE) *Proceedings of International Conference on RFID-Technology and Applications* (pp. 194-199).

Bardaki, C., Karagiannaki, A., & Pramatari, K. (2008). A systematic approach for the design of RFID implementations in the supply chain. *Panhellenic Conference on Informatics* (pp. 244-248). IEEE Press.

Baserga, R., Hongo, A., Rubini, M., Prisco, M., & Valentinis, B. (1997). The IGF-I receptor in cell growth, transformation and apoptosis. *Biochimica et Biophysica Acta*, *1332*(3), 105–126.

Bashir, A. K., Chauhdary, S. H., Shah, S. C., & Park, M.-S. (2011). Mobile RFID and its design issues. *IEEE Potential*, *30*(4), 34–38. doi:10.1109/MPOT.2011.940230

Bassen, H. (2005). Liquid pharmaceuticals and 915 MHz radiofrequency identification systems, worst-case heating and induced electric fields. *RFID Journal*. White Paper Library.

Bassen, H., Seidman, S., Rogul, J., Desta, A., & Wolfgang, S. (2007). An exposure system for evaluating possible effects of RFID on various formulations of drug products. In (IEEE) *Proceedings of International Conference on RFID* (pp. 191-198).

Batina, L., Lee, Y. K., Stefaan, S., Singelee, D., & Verbauwhede, I. (2010). Privacy-preserving ECC-based grouping proofs for RFID. In *Proceedings of the 13th International Conference on Information Security* (ISC'2010), (pp. 159-165).

Bauder, D. W. (1983). *An anti-counterfeiting concept for currency systems. Research report*. Albuquerque NM Sandia National Labs.

Bayvel, L. P. (1981). *Electromagnetic scattering and its applications* (Bayvel, L. P., & Jones, A. R., Eds.). London, UK: Applied Science. doi:10.1007/978-94-011-6746-8

Bellon, J. S., Cabedo-Fabres, M., Antonino-Daviu, M., Ferrando-Bataller, M., & Penaranda-Foix, F. (2011). Textile MIMO antenna for wireless body area networks. *Proceedings of the 5th European Conference on Antennas and Propagation (EUCAP)*, (pp. 428-432).

Berthiaume, D. (2004). Standards ease global RFID adoption. *Retail Technology Quarterly*, May 2004.

Bhandari, N., Sahoo, A., & Iyer, S. (2006). Intelligent query tree (IQT) protocol to improve RFID tag read efficiency. In *The 9th International Conference on Information Technology (ICIT'06)* (pp. 46–51).

Bhattacharyya, R., Floerkemeier, C., & Sarma, S. (2010). Low-cost, ubiquitous RFID-tag-antenna-based sensing. *Proceedings of the IEEE*, *98*(9), 1593–1600. doi:10.1109/JPROC.2010.2051790

Bhatt, H., & Glover, B. (2006). *RFID essentials*. Sebastopol, CA: O'Reilly.

Billington, D. (2007). *An introduction to clausal defeasible logic*. David Billington's Home Page. Retrieved August 12, 2011, from http://www.cit.gu.edu.au/~db/research.pdf

Billington, D. (2008). Propositional clausal defeasible logic. In *European Conference on Logics in Artificial Intelligence* (pp. 34-47).

Billington, D., Estivill-Castro, V., Hexel, R., & Rock, A. (2005). Non-monotonic reasoning for localisation in RoboCup. In *Proceedings of the 2005 Australasian Conference on Robotics and Automation.*

Bismuto, E., Mancinelli, F., d'Ambrosio, G., & Massa, R. (2003). Are the conformational dynamics and the ligand binding properties of myoglobin affected by exposure to microwave radiation? *European Biophysics Journal*, *32*(7), 628–634. doi:10.1007/s00249-003-0310-2

Blair, J. (1999). Sine-fitting software for IEEE Standards 1057 and 1241. *Proceedings of the 16th IEEE Instrumentation and Measurement Technology Conference, IMTC/99.*

Blumenstein, M., Liu, X., & Verma, B. (2007). An investigation of the modified direction feature for cursive character recognition. *Pattern Recognition*, *40*(2), 376–388. doi:10.1016/j.patcog.2006.05.017

Blythe, P. (1999). RFID for road tolling, road-use pricing and vehicle access control. *IEEE Colloquium RFID Technology,* London, England, (pp. 8/1-8/16).

Bo, Y., Yehua, H., & Caijiang, Z. (2008). Apparel supply chain management based on RFID. *27th Chinese Control Conference* (pp. 419-423). IEEE Press.

Boggs, S. A. (1990). Partial discharge: Overview and signal generation. *Electrical Insulation Magazine, IEEE*, *6*(4), 33–39. doi:10.1109/57.63057

Bolic, M., Simplot-Ryl, D., & Stojmenovic, I. (Eds.). (2010). *RFID systems: Research trends and challenges*. Wiley Press. doi:10.1002/9780470665251

Braaten, B., Owen, G., Vaselaar, D., Nelson, R., Bauer-Reich, C., & Glower, J. … Reich, M. (2008). A printed rampart line antenna with a dielectric superstrate for UHF RFID applications. *IEEE RFID 2008 Conference Proceedings,* Las Vegas, NV, April 16-17.

Braaten, B., Scheeler, R., Reich, M., Nelson, R., Bauer-Reich, C., Glower, J., & Owen, G. (2010). Compact metamaterial-based UHF RFID antennas: Deformed omega and split-ring resonator structures. *ACES Journal, 25*(6).

Bringer, J., Chabanne, H., & Dottax, E. (2006). HB++: A lightweight authentication protocol secure against some attacks. *Proceedings of IEEE International Conference on Pervasive Service, Workshop Security, Privacy and Trust in Pervasive and Ubiquitous Computing*, 2006.

Brown, L., Grundlehner, B., van de Molengraft, J., Penders, J., & Gyselinckx, B. (2009). *Body area network for monitoring autonomic nervous system responses*. 1st International ICST Workshop on Wireless Pervasive Healthcare.

Bruneo, D., Puliafito, A., Scarpa, M., & Zaia, A. (2005). *Mobile middleware in enterprise systems.*

Burdet, L. A. (2004). *RFID multiple access methods.* Technical Report. Retrieved from http://www.vs.inf.ethz.ch/edu/SS2004 /DS/reports/06 rfid-mac report.pdf

Butner, K. (2010). The smarter supply chain of the future. *Strategy and Leadership*, *38*(1), 22–31. doi:10.1108/10878571011009859

Buzby, J. C., Roberts, T., Jordan Lin, C. T., & MacDonald, J. M. (1996). *Bacterial foodborne disease: Medical costs and productivity losses*. USA: Economic Research Service/USDA Report Number (AER741). Retrieved November 4, 2009 from http://www.ers.usda.gov/Publications/AER741/

Buzby, J. C., & Roberts, T. (2009). The economics of enteric infections: Human foodborne disease costs. *Gastroenterology*, *136*, 1851–1862. doi:10.1053/j.gastro.2009.01.074

Calabrese, C., & Marrocco, G. (2008). Meander-slot antennas for sensor-RFID tags. *IEEE Antennas and Wireless Propagation Letters*, *7*, 5–8. doi:10.1109/LAWP.2007.914123

Cao, H., Leung, V., Chow, C., & Chan, H. (2009). Enabling technologies for wireless body area networks: A survey and outlook. *IEEE Communications Magazine*, *47*, 84–93. doi:10.1109/MCOM.2009.5350373

Cerovic, D. D., Dojcilovic, J. R., Asanovic, K. A., Mihajlidi, T. A., Popovic, D. M., & Spasovic, S. B. (2003). *Investigation of dielectric properties and electric resistance of some textile materials*. The 5th General Conference of the Balkan Physical Union.

Cha, D., Blumenstein, M., Zhang, H., & Jeng, D. (2008). A neural-genetic technique for coastal engineering: Determining wave-induced seabed liquefaction depth. *In Engineering Evolutionary Intelligent Systems,* (pp. 337-351).

Cha, J.-R., & Kim, J.-H. (2005). Novel anti-collision algorithms for fast object identification in RFID system. *The 11th International Conference on Parallel and Distributed Systems*, Korea, (pp. 63–67).

Chalasani, S., & Boppana, B. (2007). Data architectures for RFID transaction. *IEEE Transaction on Industrial Informatics*, *3*(3), 246–257. doi:10.1109/TII.2007.904147

Chawathe, S., Krishnamurthy, V., Ramachandran, S., & Sarma, S. (2004). Managing RFID data. In *Very Large Databases*, (pp. 1189-1195).

Cheng, T., & Jin, L. (2007). Analysis and simulation of RFID anti-collision algorithms. In *The 9th International Conference on Advanced Communication Technology (ICACT), Vol. 1,* (pp. 697–701). Phoenix Park, Korea. IEEE Computer Society.

Chen, M., Gonzalez, S., Vasilakos, A., Cao, H., & Leung, V. C. M. (2010). Body area networks. *Survey (London, England)*, 1–23.

Chen, W. T. (2006). An efficient anti-collision method for tag identification in a RFID system. *IEICE Transactions*, *89-B*(12), 3386–3392. doi:10.1093/ietcom/e89-b.12.3386

Chen, Y., Mihcak, M. K., & Kirovski, D. (2005). Certifying authenticity via fiber-infused paper. *SIGecom Exchange*, *5*(3), 29–37. doi:10.1145/1120680.1120685

Chiang, K. W., Hua, C., & Yum, T. S. (2006). Prefix-randomized query-tree protocol for RFID system. *IEEE International Conference on Communication* (pp. 1653-1657). IEEE Press.

Chien, H.-Y. (2006). Secure access control schemes for RFID systems with anonymity. *Proceedings of the 2006 International Workshop Future Mobile and Ubiquitous Information Technologies* (*FMUIT* '06), 2006.

Chien, H.-Y. (2007). SASI: A new ultralightweight RFID authentication protocol providing strong authentication and strong integrity. *IEEE Transactions on Dependable and Secure Computing*, *4*(4).

Chien, H.-Y., & Chen, C.-H. (2007). Mutual authentication protocol for RFID conforming to EPC class 1 generation 2 standards. *Computer Standards & Interfaces*, *29*(2), 254–259. doi:10.1016/j.csi.2006.04.004

Cho, H., Lee, W., & Baek, Y. (2007). LDFSA: A learning-based dynamic framed slotted ALOHA for collision arbitration in active RFID systems. In *Advances in Grid and Pervasive Computing Second International Conference, Vol. 4459,* (pp. 655–665). Paris, France. Berlin, Germany: Springer.

Cho, J. S., Shin, J. D., & Kim, S. K. (2008). RFID Tag anti-collision protocol: Query tree with reversed IDs. *10th International Conference on* Advanced Communication Technology (pp. 225-230). IEEE Press.

Choi, H., Cha, J. R., & Kim, J.-H. (2004). Fast wireless anti-collision algorithm in ubiquitous ID system. In *Vehicular Technology Conference (VTC2004), IEEE* (pp. 4589–4592).

Choi, J. H., Lee, H. J., Lee, D., Lee, H. S., Youn, Y., & Kim, J. (2008). *Query tree based tag identification method in RFID systems*. Retrieved August 11, 2011, from www.freshpatents.com/Query-tree-based-tag-identification-method-in-rfid-systems-dt20080508ptan20080106383.php

Choi, J., & Lee, W. (2007). Comparative evaluation of probabilistic and deterministic tag anti-collision protocols for RFID networks. In *EUC Workshops '07* (pp. 538–549).

Choi, J. H., Lee, D., & Lee, H. (2007). Query tree-based reservation for efficient RFID tag anti-collision. *IEEE Communications Letters*, *11*(1), 85–87. doi:10.1109/LCOMM.2007.061471

Choi, S., & Chae, J. (2009). A microfluidic biosensor based on competitive protein adsorption for thyroglobulin detection. *Biosensors & Bioelectronics*, *25*, 118–123. doi:10.1016/j.bios.2009.06.017

Cho, J., Shim, Y., Kwon, T., & Choi, Y. (2007). SARIF: A novel framework for integrating wireless sensor and RFID networks. *IEEE Wireless Communications*, *2007*, 50–56. doi:10.1109/MWC.2007.4407227

Chuang, M. L., & Shaw, M. H. (2007). RFID: Integration stages in supply chain management. *IEEE Engineering Management Review*, *35*(2), 80–87. doi:10.1109/EMR.2007.899757

Chu, F. Y., & Williamson, A. (1982). Fault location in SF6 insulated substations using thermal techniques. *IEEE Transactions on Power Apparatus and Systems*, *PAS-101*(7), 1990–1997. doi:10.1109/TPAS.1982.317446

Church, S., & Littman, D. (1991). *Machine reading of visual counterfeit deterrent fea- tures and summary of US research, 1980-90*. Canada: Four Nation Group on Advanced Counterfeit Deterrence.

Clark, R. (2011). *Dataveillance and information privacy*. Retrieved from http://www.rogerclarke.com/DV/

Clauberg, R. (2004). *RFID and sensor networks*. RFID Workshop, University of St. Gallen, Switzerland, Sept. 27, 2004.

Collin, R. E. (1985). *Antennas and radiowave propagation*. New York, NY: McGraw-Hill.

Collins, J. (2004). *RFID fibers for secure applications*. Retrieved February 27, 2011, from http://www.rfidjournal.com/article/view/845

Collins, J. (2005). *New tags use crystal, not silicon*. Retrieved March 1, 2011, from http://www.rfidjournal.com/article/view/1967

Conti, M., Pieto, R. D., Mancini, L. V., & Spognardi, A. (2007a). RIPP-FS: An RFID identification, privacy preserving protocol with forward secrecy. In *the Proceedings of the Fifth IEEE International Conference on Pervasive Computing and Communications Workshops* (PerComW'07), (pp. 229-234).

Conti, M., Pieto, R. D., Mancini, L. V., & Spognardi, A. (2007b). FastRIPP: RFID privacy preserving protocol with forward secrecy and fast resynchronisation. *Proceedings of the 33rd Annual Conference of the IEEE Industrial Electronics Society*, (pp. 52-57).

Cox, F. C., Sharma, V. K., Klibanov, A. M., Wu, B. I., Kong, J. A., & Engels, D. W. (2006). A method to investigate non-thermal effects of radio frequency radiation on pharmaceuticals with relevance to RFID technology. In (IEEE) *Proceedings of International Conference Engineering in Medicine and Biology Society: Vol. 1* (pp. 4340-4343).

Cross, I. D. Inc. (n.d.). *Website*. Retrieved from http://innovya.com/CrossID/

Cui, Y., & Zhao, Y. (2008). Mathematical analysis for binary tree algorithm in RFID. *IEEE Vehicular Technology Conference* (pp. 2725-2729). IEEE Press.

Damez, J. L., & Clerjon, S. (2008). Meat quality assessment using biophysical methods related to meat structure. *Meat Science*, *80*, 132–149. doi:10.1016/j.meatsci.2008.05.039

Daniel Hunt, V., Puglia, A., & Puglia, M. (2007). *RFID - A guide to radio frequency identification*. John Wiley & Sons. doi:10.1002/0470112255

Das, R. (2006). *Chipless RFID - The end game*. Retrieved from http://www.idtechex.com/research/articles/chipless_rfid_the_end_game_00000435.asp

Das, R. (2007). Chipless RFID. *Adhesives & Sealants Industry*, May 2007, 47-48.

De Pomerai, D. I., Smith, B., Dawe, A., North, K., Smith, T., & Archer, D. B. (2003). Microwave radiation can alter protein conformation without bulk heating. *FEBS Letters*, *543*(1-3), 93–97. doi:10.1016/S0014-5793(03)00413-7

Deng, Z., Li, J., & Feng, B. (2008). A taxonomy model of RFID security threats. In *the Proceedings of the IEEE International Conference on Communication Technology*, ICCT 2008, (pp. 765-768).

Derakhshan, R., Orlowska, M. E., & Li, X. (2007). RFID data management: Challenges and opportunities. In *IEEE International Conference on RFID 2007* (pp. 175–182). Texas, USA.

Devadas, S., Suh, E., Paral, S., Sowell, R., Ziola, T., & Khandelwal, V. (2008, 16-17 April). *Design and implementation of PUF-based "unclonable" RFID ICs for anti-counterfeiting and security applications.* Paper presented at the 2008 IEEE International Conference on RFID.

Devarapalli, M. R., Sarangan, V., & Radhakrishnan, S. (2007). AFSA: An efficient framework for fast RFID tag reading in dense environments. In *QSHINE '07: The Fourth International Conference on Heterogeneous Networking for Quality, Reliability, Security and Robustness Workshops,* (pp. 1–7). New York, NY: ACM.

Ding, J., & Liu, F. (2009). Novel tag anti-collision algorithm with adaptive grouping. Wireless Sensor Network (WSN). *Wireless Sensor Network, 1*(5), 475–481. doi:10.4236/wsn.2009.15057

Dolukhanov, M. (1971). *Propagation of radio waves.* Moscow, USSR: MIR Publishers.

Dowling, J., Tentzeris, M. M., & Beckett, N. (2009, 24-25 September). *RFID-enabled temperature sensing devices: A major step forward for energy efficiency in home and industrial applications?* IEEE MTT-S International Microwave Workshop on Wireless Sensing, Local Positioning, and RFID, IMWS 2009.

Driebergen, R., & Baer, G. (2003). Quantification of follicle stimulating hormone (follitropin alfa): Is in vivo bioassay still relevant in the recombinant age? *Current Medical Research and Opinion, 19*(1), 41–46. doi:10.1185/030079902125001344

Duc, D. N., Lee, H., Konidala, D. M., & Kim, K. (2009). Open issues in RFID security. In *the Proceedings of the International Conference on Internet Technology and Secured Transactions*, ICITST 2009, (pp. 1-5).

Duc, D. N., Park, J., Lee, H., & Kim, K. (2006). Enhancing security of EPCglobal Gen-2 RFID tag against traceability and cloning. *Proceedings of the 2006 Symposium on Cryptography and Information Security*, 2006.

Duval, M. (1989). Dissolved gas analysis: It can save your transformer. *Electrical Insulation Magazine, IEEE, 5*(6), 22–27. doi:10.1109/57.44605

Edgeware. (2011). *Syspro ERP software for medical device manufacturers.* Retrieved from http://www.edgeware.net/industries/syspro-erp-software-for-medical-device-manufacturers

Edwards, J. (2011). RFID: The next stage. *RFID Journal*, Nov/Dec 2011, 12-19.

EPCglobal Inc. (2005). *EPC™ radio-frequency identification protocols class-1 generation-2 UHF RFID protocol for communications at 860MHz-960MHz Version 1.0.9.* Specification for RFID air interface. Retrieved July 15, 2010, from http://www.epcglobalinc.org

EPCGlobal. (2008). *EPCGlobal tag data standards version 1.4: Ratified specification.* Retrieved August 11, 2011, from http://www.epcglobalinc.org/standards/tds/

EPCglobal. (2011). *EPCglobal: The EPCglobal network: Overview of design, benefits and security.* Retrieved from http://www.epcglobalinc.org

Europe, M. E. (2001). *CAD benchmark.* Retrieved from http://i.cmpnet.com/edtn/europe/mwee/pdf/CAD.pdf

Fabian, B., Gunther, O., & Spiekermann, S. (2005). *Security analysis of the object name service for RFID.* First International Workshop on Security, Privacy and Trust in Pervasive and Ubiquitous Computing, July 2005.

Fan, X., Song, I., & Chang, K. (2008a). Gen2-based hybrid tag anti-collision Q algorithm using Chebyshev's inequality for passive RFID systems. In *19th International Symposium on Personal, Indoor and Mobile Radio Communications, IEEE* (pp. 1–5). Cannes, France.

Fan, X., Song, I., Chang, K., Shin, D. B., Lee, H. S., Pyo, C. S., & Chae, J. S. (2008b). Gen2-based tag anti-collision algorithms using Chebyshev's inequality and adjustable frame size. *ETRI Journal, 30*(5), 653–662. doi:10.4218/etrij.08.1308.0098

Feng, B., Li, J. T., Guo, J. B., & Ding, Z. H. (2006). ID-binary tree stack anti-collision algorithm for RFID. In *Proceedings of the 11th IEEE Symposium on Computers and Communications, IEEE Computer Society* (pp. 207–212). Washington, DC, USA.

Fensli, R., Gunnarson, E., & Gundersen, T. (2005). A wearable ECG-recording system for continuous arrhythmia monitoring in a wireless tele-home-care situation. *Proceedings of the 18th IEEE Symposium on Computer-Based Medical Systems*, (pp. 407-412).

Ferrari, P., Flamini, A., Marioli, D., & Taroni, A. (2006). IEEE802.11 sensor networking. *IEEE Transactions on Instrumentation and Measurement*, *55*(2), 615–619. doi:10.1109/TIM.2006.870105

Ferrer, G., Dew, N., & Apte, U. (2010). When is RFID right for your service? *International Journal of Production Economics*, *124*(2), 414–425. doi:10.1016/j.ijpe.2009.12.004

Ferrigno, L., Pietrosanto, A., & Paciello, V. (2006). Low-cost visual sensor node for BlueTooth-based measurement networks. *IEEE Transactions on Instrumentation and Measurement*, *55*(2), 521–527. doi:10.1109/TIM.2006.870126

Finkenzeller, K. (2003). *RFID handbook: Fundamentals and applications in contactless smart cards and identification*. John Wiley and Sons Ltd, 2003.

Finkenzeller, K. (2010). *Introduction*. John Wiley & Sons, Ltd.

Finkenzeller, K. (2010). *RFID handbook: Fundamentals and applications in contactless smart cards and identification*. Munich, Germany: John Wiley & Sons, Ltd.

Fletcher, R. R. (2002). *Low-cost electromagnetic tagging: Design and implementation*.

Floerkemeier, C. (2007). Bayesian transmission strategy for framed ALOHA based RFID protocols. In *RFID, 2007 - IEEE International Conference on RFID Gaylord Texan Resort* (pp. 228–235). Grapevine, TX, USA.

Floerkemeier, C., & Lampe, M. (2004). Issues with RFID usage in ubiquitous computing applications. In F. Alois & M. Friedemann (Eds.), *PERVASIVE 2004, Lecture Notes in Computer Science: Vol. 3001*, (pp. 188-193). Berlin, Germany: Springer-Verlag press.

Floerkemeier, C., & Lampe, M. (2004). Issues with RFID usage in ubiquitous computing applications. In *Pervasive Computing, No. 3001* (pp. 188–193). Austria: Springer-Verlag. doi:10.1007/978-3-540-24646-6_13

Foldy, L. L. (1945). The multiple scattering of waves-I: General theory of isotropic scattering by randomly distributed scatterers. *Physical Review*, *67*(3-4), 107. doi:10.1103/PhysRev.67.107

Foreshew, J. (2011). *Chipless tracker to transform the library industry*. Retrieved August 27, 2011, from http://www.theaustralian.com.au/australian-it/chipless-tracker-to-transform-libraries/story-e6frgakx-1226106169778

Fort, A., Desset, C., De Doncker, P., Wambacq, P., & Biesen, L. V. (2006). An ultra-wideband body area propagation channel model-from statistics to implementation. *IEEE Transactions on Microwave Theory and Techniques*, *54*, 1820–1826. doi:10.1109/TMTT.2006.872066

Fortune, J. A., Wu, B. I., & Klibanov, A. M. (2010). Radio frequency radiation causes no non thermal damage in enzymes and living cells. *Biotechnology Progress*, *26*(6), 1772–1776. doi:10.1002/btpr.462

Foster, I. (2002). What is the grid? - A three point checklist. *GRID Today*, *1*(6).

Foster, I. (2005). Service-oriented science. *Science*, *308*(5723), 814–817. doi:10.1126/science.1110411

FP7. (2007). *Resources and services virtualization without barriers*. Retrieved from http://www.reservoir-fp7.eu/

Friedlos, D. (2010). *Stell products maker sees ROI in six months*. Retrieved March 9, 2011, from http://www.rfidjournal.com/article/view/7448/1

Friedlos, D. (2011). *South Korean consortium launches EPC Gen 2 reader for mobile phones*. Retrieved February 5, 2011, from http://www.rfidjournal.com/article/print/8155

GaAs mmIC SP4T Non-reflective positive control switch, DC - 8 GHz. (n.d.). Retrieved from http://www.hittite.com/content/documents/data_sheet/hmc345lp3.pdf

Gabriel, C., Bentall, R. H., & Grant, E. H. (1987). Comparison of the dielectric properties of normal and wounded human skin material. *Bioelectromagnetics*, *8*, 23–27. doi:10.1002/bem.2250080104

García, A., & Chang, Y. (2007). RFID enhanced MAS for warehouse management. *International Journal of Logistics: Research & Applications*, *10*(2), 97–107. doi:10.1080/13675560701427379

Garima, B. D., Saxena, V. K., & Saini, J. S. (2009). Broadband circular patch microstrip antenna with diamond shape slot for space communication. *International Conference on Emerging Trends in Electronic and Photonic Devices & Systems*, (pp. 332-335).

Ghayal, A., Khan, M. Z., & Moona, R. (2008). *SmartRF: A flexible and light-weight RFID middleware*. IEEE International Conference on e-Business Engineering. 978-0-7695-3395-7/08

Glaxo-Smith-Kline. (2009). *Counterfeiting report*. Retrieved from http://www.gsk.com/responsibility/supply-chain/counterfeiting.htm

Global Card Fraud. (2010). *The Nilson report*.

Glover, B., & Bhatt, H. (2006). *RFID essentials* (1st ed.). O'Reilly Media, Inc.

Goldsborough, R. (2010a). *Ancient Fourree counterfeits*. Retrieved from http://rg.ancients.info/fourees/

Goldsborough, R. (2010b). *A case for the world's first coin: The Lydian lion*. Retrieved from http://rg.ancients.info/lion/article.html

Golle, P., Jakobsson, M., Juels, A., & Syverson, P. (2004). Universal re-encryption for mixnets. *The Cryptographers' Track at the RSA Conference-CT-RSA* (pp. 163-178)

Goonerate, C. P., Mukhopadhyay, S. C., & Yamada, S. (2009, December). Electromagnetic field computation of GMR sensors for hyperthermia based cancer treatment. In *Proceedings of IEEE Applied EletroMagnetic Conference* (AEMC) 2009, 14-16, Kolkata, India. ISBN: 978-1-4244-4819-7

Greig, J. D., & Ravel, A. (2009). Analysis of foodborne outbreak data reported internationally for source attribution. *International Journal of Food Microbiology*, *130*, 77–87. doi:10.1016/j.ijfoodmicro.2008.12.031

Haga, N., Saito, K., Takahashi, M., & Ito, K. (2009). Characteristics of cavity slot antennas for body-area networks. *IEEE Transactions on Antennas and Propagation*, *57*, 837–843. doi:10.1109/TAP.2009.2014577

Haller, N. M. (1994). The S/KEY one-time password system. *Proceedings of the Symposium on Network and Distributed System Security*, (pp. 151–157).

Hall, P. S., & Hao, Y. (2006). *Antennas and propagation for body-centric wireless communications*. Artech House. doi:10.1109/TAP.2009.2018564

Harrop, P. (2006). *New advances in RFID help foot traceability*. Retrieved March 7, 2011, from www.idtechex.com/documents/downloadpdf.asp?documentid=1279

Heeseo, C., Taek, L., & Hoh, P. (2006). Situation aware RFID system: Evaluating abnormal behavior detecting approach. *Proceeding of the Fourth Workshop on Software Technologies for Future Embedded and Ubiquitous Systems and Second International Workshop on Collaborative Computing, Integration, and Assurance* (SEUS-WCCIA'06), 2006 IEEE.

Hejn, K., & Pacut, A. (2003, 20-22 May). Sine-wave parameters estimation - The second source of inaccuracy. *Proceedings of the 20th IEEE Instrumentation and Measurement Technology Conference, IMTC '03.*

Henrici, D. (2008). *Security and privacy in large-scale RFID systems - Challenges and solutions*. Msc Thesis, University of Kaiserslautern, Germany.

Henrici, D., & MÄuller, P. (2004). Hash-based en-hancement of location privacy for radio-frequency identification devices using varying identifiers. *IEEE International Workshop on Pervasive Com-puting and Communication Security- PerSec* (pp.149-153)

Hernandez-Castro, J. C., Estevez-Tapiador, J. M., & Ribagorda, A. (2006a). EMAP: An efficient mutual authentication protocol for low-cost RFID tags. *Proceedings OTM Federated Conference and Workshop: IS Workshop.*

Hernandez-Castro, J. C., Estevez-Tapiador, J. M., & Ribagorda, A. (2006b). M2AP: A minimalist mutual-authentication protocol for low-cost RFID tags. *Proceedings of the International Conference on Ubiquitous Intelligence and Computing* (*UIC*'06), (pp. 912-923).

Herzberg, G., & Crawford, B. L. (1946). Infrared and Raman spectra of polyatomic molecules. *Journal of Physical Chemistry*, *50*(3), 288–288. doi:10.1021/j150447a021

Hey-Shipton, G. L., Metthews, P. A., & Mestay, J. (1982). The complex permittivity of human tissue at microwave frequencies. *Physics in Medicine and Biology*, *27*, 1067–1071. doi:10.1088/0031-9155/27/8/008

Holland, J. (1975). *Adaption in natural and artificial systems*. University of Michigan Press.

Holmes, P. (2005). *Will tags get out into the supply chain?* Works Management.

Hong, J.-S. G., & Lancaster, M. J. (2011). *Microstrip filters for RF/microwave applications* (2nd ed.). Wiley-Interscience. doi:10.1002/9780470937297

Hopper, N. J., & Blum, M. (2001). Secure human identification protocols. *Proceedings of the Seventh International Conference on Theory and Application of Cryptology and Information Security*, (pp. 52-66).

Hoshino, T., Kato, K., Hayakawa, N., & Okubo, H. (2001). A novel technique for detecting electromagnetic wave caused by partial discharge in GIS. *IEEE Transactions on Power Delivery*, *16*(4), 545–551. doi:10.1109/61.956735

Hoyer, L., Nolan, P. E. Jr, LeDoux, J. H., & Moore, L. A. (1995). Selective stability-indicating high-performance liquid chromatographic assay for recombinant human regular insulin. *Journal of Chromatography. A*, *699*(1-2), 383–388. doi:10.1016/0021-9673(95)00086-3

Hsu, C. H., Yu, C. H., Huang, Y. P., & Ha, K. J. (2008). An enhanced query tree (EQT) protocol for memoryless tag anti-collision in RFID systems. *2008 Second International Conference on Future Generation Communication and Networking* (pp. 427 - 432). IEEE Press.

Hu, X., Judd, M. D., & Siew, W. H. (2010, August 31 2010-September 3). *A study of PD location issues in GIS using FDTD simulation.* Paper presented at the 2010 45th International Universities Power Engineering Conference (UPEC).

Hu, X., Wang, J., Yu, Q., Liu, W., & Qin, J. (2008). A wireless sensor network based on ZigBee for telemedicine monitoring system. *The 2nd International Conference on Bioinformatics and Biomedical Engineering,* (pp. 1367-1370).

Hualiang, Z., & Chen, K. J. (2005). A tri-section stepped-impedance resonator for cross-coupled bandpass filters. *Microwave and Wireless Components Letters, IEEE*, *15*(6), 401–403. doi:10.1109/LMWC.2005.850475

Hualiang, Z., & Chen, K. J. (2006). Miniaturized coplanar waveguide bandpass filters using multisection stepped-impedance resonators. *IEEE Transactions on Microwave Theory and Techniques*, *54*(3), 1090–1095. doi:10.1109/TMTT.2005.864126

Huang, Y., Hua, K., Zhou, X., Jin, H., Chen, X., & Lu, X. (2008). Activation of the PI3K/AKT pathway mediates FSH-stimulated VEGF expression in ovarian serous cystadenocarcinoma. *Cell Research*, *18*(7), 780–791. doi:10.1038/cr.2008.70

Huang, Y., Zhao, Y. Q., Su, M., Gao, S. J., Jin, H. Y., & Feng, Y. J. (2007). Effects of follicle stimulating hormone on proliferation, apoptosis, migration and invasion of ovarian carcinoma cells: An in vitro experiment. *Zhonghua Yi Xue Za Zhi*, *87*(35), 2512–2514.

Humphrey, T., & Jørgensen, F. (2006). Pathogens on meat and infection in animals – Establishing a relationship using campylobacter and salmonella as examples. *Meat Science*, *74*, 89–97. doi:10.1016/j.meatsci.2006.04.026

Humphrey, T., O'Brien, S., & Madsen, M. (2007). Campylobacters as zoonotic pathogens: A food production perspective. *International Journal of Food Microbiology*, *117*(3), 237–257. doi:10.1016/j.ijfoodmicro.2007.01.006

Hu, S. M., Zhou, Y., Law, C. L., & Dou, W. B. (2010). Study of a uniplanar monopole antenna for passive chipless UWB-RFID locatization system. *IEEE Transactions on Antennas and Propagation*, *58*(2), 271–278. doi:10.1109/TAP.2009.2037760

Hush, D. R., & Wood, C. (1998). Analysis of tree algorithms for RFID arbitration. *The IEEE International Symposium on Information Theory*, (pp. 107–107).

Igure, V. M., & Williams, R. D. (2008). Taxonomies of attacks and vulnerabilities in computer systems. *IEEE Communications Surveys & Tutorials*, *10*(1), 6–19. doi:10.1109/COMST.2008.4483667

Iizuka, H., & Hall, P. (2007). Left-handed dipole antennas and their implementations. *IEEE Transactions on Antennas and Propagation*, *55*(5), 1246–1253. doi:10.1109/TAP.2007.895568

InSync. (2011). *Auto-ID development platform, 2011*. Retrieved from http://www.insyncinfo.com/INSYNCsensoredgew.asp

Ishimaru, H., & Kawada, M. (2010). Localization of a partial discharge source using maximum likelihood estimation. *IEEJ Transactions on Electrical and Electronic Engineering*, *5*(5), 516–522. doi:10.1002/tee.20567

ISO. (2005). *ISO/IEC 17799: Information technology – Security techniques – Code of practice for information security management*. International Organization for Standardization.

ISO/IEC 18000-6. (2010). *Radio frequency identification for item management -- Part 6: Parameters for air interface communications at 860 MHz to 960 MHz*.

Jacomet, M., Ehrsam, A., & Gehrig, U. (1999). *Contactless identification device with anti-collision algorithm*. In IEEE Computer Society, Conference on Circuits, Systems, Computers and Communications. Athens, Greece.

Jain, S., & Das, S. R. (2006). Collision avoidance in a dense RFID network. *In WiNTECH'06: Proceedings of the 1st International Workshop on Wireless Network Testbeds, Experimental Evaluation & Characterization*, (pp. 49–56). New York, NY: ACM.

Jakobsson, M. (2002). Fractal hash sequence representation and traversal. In *Proceedings of the 2002 IEEE International Symposium on Information Theory* (ISIT '02), (pp. 437-444).

Jari, P., Pascal, C., Declereq, M., Dehollain, C., & Joehi, N. (2007). *Design and optimization of passive UHF RFID system*. Springer.

Jea, K. F., & Wang, J. Y. (2008). An RFID encoding method for supply chain management. *Asia-Pacific Services Computing Conference* (pp. 601-606). IEEE Press.

Jeffery, S., Garofalakis, M., & Franklin, M. (2006). Adaptive cleaning for RFID data streams. In *Very Large Databases*, (pp. 163-174).

Jia, X. L., Feng, Q. Y., & Ma, C. Z. (2010). An efficient anti-collision protocol for RFID tag identification. *IEEE Communications Letters*, *14*(11), 1014–1016. doi:10.1109/LCOMM.2010.091710.100793

Johansson, A., & Karedal, J. Tufvesson, F., & Molisch, A. F. (2005). MIMO channel measurements for personal area networks. *IEEE 61st Vehicular Technology Conference*, Vol. 1 (pp. 171-176).

Johnson, D., Menezes, A., & Vanstone, S. (2001). The elliptic curve digital signature algorithm (ECDSA). *International Journal of Information Security*, *1*(1), 36–63. doi:doi:10.1007/s102070100002

Jong, R. D., Helmond, D. J. V., & Koot, M. (2005). *An exploration of RFID technology*. Technical report, 2005.

Jordan, E. C., & Balmain, K. G. (1968). *Electromagnetic waves and radiating systems*. Englewood Cliffs, NJ: Prentice Hall.

Jose Tomlal, E., Thomas, P. C., George, K. C., Jayanarayanan, K., & Joseph, K. (2010). Impact, tear, and dielectric properties of cotton/polypropylene commingled composites. *Journal of Reinforced Plastics and Composites*, *29*, 1861–1875. doi:10.1177/0731684409338748

Journal, R. F. I. D. (2003). *1-cent RFID tags for supermarkets*. Retrieved February 27, 2011, from http://www.rfidjournal.com/article/view/363

Journal, R. F. I. D. (2004). *Firewall protection for paper documents*. Retrieved February 27, 2011, from http://www.rfidjournal.com/article/view/790

Journal, R. F. I. D. (2006). *PolyIC announces printable RFID prototypes*. Retrieved February 27, 2011, from http://www.rfidjournal.com/article/print/6589

Judd, M. D., Li, Y., & Hunter, I. B. B. (2005). Partial discharge monitoring of power transformers using UHF sensors. Part I: Sensors and signal interpretation. *IEEE Electrical Insulation Magazine*, *21*(2), 5–14. doi:10.1109/MEI.2005.1412214

Judson, R., & Porter, R. (2010). *Estimating the volume of counterfeit U.S. currency in circulation worldwide: Data and extrapolation* (F. R. B. o. Chicago & F. M. Group, Trans.). Policy Discussion Paper Series.

Juels, A. (2004). "Yoking-Proofs" for RFID Tags. In *Proceedings of the Second IEEE Annual Conference on Pervasive Computing and Communications Workshops* (Per-comw '04), (pp. 138–143).

Juels, A. (2005). Strengthening EPC tag against cloning. *Proceedings of the ACM Workshop Wireless Security* (*WiSe* '05), (pp. 67-76).

Juels, A., Molner, D., & Wagner, D. (2005). Security and privacy issues in e-passports. *Proceedings of the First International Conference Security and Privacy for Emerging Areas in Communication Networks (SecureComm '05)*, 2005.

Juels, A. (2006). RFID security and privacy: A research survey. *IEEE Journal on Selected Areas in Communications*, *24*(2), 381–394. doi:10.1109/JSAC.2005.861395

Juels, A., & Pappu, R. (2003). Squealing Euros: Privacy protection in RFID–enabled banknotes. *Financial Cryptography 2003*. *LNCS*, *2742*, 103–121.

Kanoc, T. (1999). *Mobile middleware: The next frontier in enterprise application integration*. White paper Nettech Systems Inc.

Kanoun, O., & Trankler, H. R. (2004). Sensor technology advances and future trends. *IEEE Transactions on Instrumentation and Measurement*, *53*(6), 1497–1501. doi:10.1109/TIM.2004.834613

Kantor, A. (2003, December 19). Tiny transmitters give retailers, privacy advocates goose bumps. *USAToday.com*.

Kapoor, G., Wei, Z., & Piramuthu, S. (2008). RFID and information security in supply chains. *4th International Conference on Mobile Ad-hoc and Sensor Networks* (pp. 59-62). IEEE Press.

Karthikeyan, S., & Nesterenko, M. (2005). RFID security without extensive cryptography. *Proceedings of the Third ACM Workshop Security of Ad Hoc and Sensor Networks*, (pp. 63-67).

Kawada, M. (2003). Fundamental study on location of a partial discharge source with a VHF-UHF radio interferometer system. *Electrical Engineering in Japan*, *144*(1), 32–41. doi:10.1002/eej.10145

Kemp, I. J. (1995). Partial discharge plant-monitoring technology: Present and future developments. *IEE Proceedings. Science Measurement and Technology*, *142*(1), 4–10. doi:10.1049/ip-smt:19951438

Kent, A., & Billiard, M. (2003). *Sleep: physiology, investigations and medicine*. Kluwer Academic / Plenum Publishers.

Kephart, J. O., & Chess, D. M. (2003). The vision of autonomic computing. *Computer*, *36*(1), 41–50. doi:10.1109/MC.2003.1160055

Khan, I., & Hall, P. S. (2009). Multiple antenna reception at 5.8 and 10 GHz for body-centric wireless communication channels. *IEEE Transaction on Antennas and Propagation, 57*.

Khan, I. (2009). *Diversity and MIMO for body-centric wireless communication channels. Unpublished doctoral theses*. UK: School of Electronics, Electrical, & Computer Engineering, University of Birmingham.

Khan, I., & Hall, P. S. (2010). Experimental evaluation of MIMO capacity and correlation for narrowband body-centric wireless channels. *IEEE Transactions on Antennas and Propagation*, *58*, 195–202. doi:10.1109/TAP.2009.2025062

Khoussainova, N., Balazinska, M., & Suciu, D. (2008). Probabilistic event extraction from RFID data. In *International Conference on Data Engineering*, (pp. 1480-1482).

Kim, M., Lee, J. W., Lee, Y. J., & Ryou, J. (2008). COSMOS: A middleware for integrated data processing over heterogeneous sensor networks. *ETRI Journal*, *30*(5).

Kim, J. G. (2008). A divide-and-conquer technique for throughput enhancement of RFID anti-collision protocol. *IEEE Communications Letters*, *12*(6), 474–476. doi:10.1109/LCOMM.2008.080277

Kim, Y.-I., Bong-Jae, Y., Jae-Ju, S., Jin-Ho, S. A., & Lee, J.-I. (2006). Implementing a prototype system for power facility management using RFID/WSN. *International Journal of Applied Mathematics and Computer Science*, *2*(2).

Klair, D. K., Chin, K. W., & Raad, R. (2007). On the suitability of framed slotted Aloha based RFID anti-collision protocols for use in RFID-enhanced WSNs. In *Proceedings of 16th International Conference on Computer Communications and Networks 2007 (ICCCN 2007)* (pp. 583–590).

Klair, D. K., Chin, K.-W., & Raad, R. (2007). *An investigation into the energy efficiency of pure and slotted Aloha based RFID anti-collision protocols*. IEEE International Symposium on a World of Wireless, Mobile and Multimedia Networks (IEEE WoWMoM'07), 2007.

Klair, D. K., Chin, K. W., & Raad, R. (2010). A survey and tutorial of RFID anti- collision protocols. *IEEE Communications Surveys Tutorials*, *12*(3), 400–421. doi:10.1109/SURV.2010.031810.00037

Konidala, M., & Divyan, K. W.-S. (2006). *Security assessment of EPCglobal architecture framework* (pp. 13–16). AUTO-ID Labs.

Kreuger, F. H. (1989). *Partial discharge detection in high-voltage equipment* (1st ed.). Butterworths.

Kumar, S. S., & Paar, C. (2006). Are standards compliant elliptic curve cryptosystems feasible on RFID? *Proceedings Workshop RFID Security*, July 2006.

Kumar, A., & Parkash, D. (2010). Planar antennas for passive UHF RFID tag. *Progress in Electromagnetics Research*, *19*, 305–327. doi:10.2528/PIERB09121609

Kurokkawa, K. (1965). Power waves and scattering matrix. *IEEE Transactions on Microwave Theory and Techniques*, *13*(2), 194–202. doi:10.1109/TMTT.1965.1125964

Lai, Y. C., & Lin, C. C. (2009). Two blocking algorithms on adaptive binary splitting: Single and pair resolutions for RFID tag identification. *IEEE/ACM Transactions on Networking*, *17*(3), 962–975. doi:10.1109/TNET.2008.2002558

Lakafosis, V., Traille, A., Lee, H., Orecchini, G., Gebara, E., Tentzeris, M. M., et al. (2010, 23-28 May 2010). *An RFID system with enhanced hardware-enabled authentication and anti-counterfeiting capabilities.* 2010 IEEE MTT-S International Microwave Symposium Digest (MTT).

Lakafosis, V., Traille, A., Hoseon, L., Gebara, E., Tentzeris, M. M., DeJean, G. R., & Kirovski, D. (2011). RF fingerprinting physical objects for anticounterfeiting applications. *IEEE Transactions on Microwave Theory and Techniques*, *59*(2), 504–514. doi:10.1109/TMTT.2010.2095030

Lam, S. C. K., Wong, K. L., Wong, K. O., Wong, W. & Mow W. H. (2009). A smart phone centric platform for personal health monitoring using wireless wearable biosensors. *The 7th International Conference on Information, Communications and Signal Processing,* (pp. 1-7).

Landt, J. (2005). The history of RFID. *IEEE Potentials*, *24*(4), 8–11. doi:10.1109/MP.2005.1549751

Langheinrich, M. (2009). A survey of RFID privacy approaches. *Personal and Ubiquitous Computing*, *13*(6), 413–421. doi:10.1007/s00779-008-0213-4

Latre, B., Braem, B., Moerman, I., Blondia, C., & Demeester, P. (2011). A survey on wireless body area networks. *Wireless Networks*, *17*, 1–18. doi:10.1007/s11276-010-0252-4

Law, C., Lee, K., & Siu, K. Y. (2000). Efficient memoryless protocol for tag identification. In *Proceedings of the 4th International Workshop on Discrete Algorithms and Methods for Mobile Computing and Communications, DIALM '00, ACM* (pp. 75–84). New York, NY, USA.

Lax, M. (1951). Multiple scattering of waves. *Reviews of Modern Physics*, *23*(4), 287. doi:10.1103/RevModPhys.23.287

Lazcka, O., Javier Del Campo, F., & Muñoz, F. X. (2007). Pathogen detection: A perspective of traditional methods and biosensors. *Biosensors & Bioelectronics*, *22*, 1205–1217. doi:10.1016/j.bios.2006.06.036

Le, T. V., Burnmester, M., & Medeiros, B. (2007). Universally composable and forward secure RFID authentication and authenticated key exchange. In *the Proceedings of the 2nd ACM Symposium on Information, Computer and Communications Security*, (pp. 242-252).

Lee, J. G., Hwang, S. J., & Kim, S. W. (2008b). Performance study of *anti-collision* algorithms for EPC-C1 Gen2 RFID protocol. In *Information Networking. Towards Ubiquitous Networking and Services, Vol. 5200,* (pp. 523–532). Estoril, Portugal. Berlin, Germany: Springer.

Lee, S. (2005). *Mutual authentication of RFID system using synchronized secret information.* MSc Thesis, School of Engineering, Information and Communications University, 2005.

Lee, S. R., & Lee, C. W. (2006). An enhanced dynamic framed slotted ALOHA anti- collision algorithm. In *Emerging Directions in Embedded and Ubiquitous Computing, Vol. 4097,* (pp. 403–412). Seoul, Korea. Berlin, Germany: Springer.

Lee, S. R., Joo, S. D., & Lee, C. W. (2005). An enhanced dynamic framed slotted ALOHA algorithm for RFID tag identification. *2nd Annual International Conference on Mobile and Ubiquitous Systems: Networking and Services* (pp. 166-172). IEEE Press.

Lee, S.-M., Hwang, Y. J., Lee, D. H., & Lim, J. I. (2005). Efficient authentication for low-cost RFID systems. *International Conference on Computational Science and its Applications* (pp. 619-627)

Lee, Y. K., Batina, L., Stefaan, S., Singelee, D., & Verbauwhede, I. (2010). Low-cost untraceable authentication protocols for RFID (extended version). In S. Wetzel, C. N. Ro-taru, & F. Stajano (Eds.), *Proceedings of the 3rd ACM Conference on Wireless Network Security* (WiSec '10), (pp. 55–64).

Lee, C. W., Cho, H., & Kim, S. W. (2008a). An adaptive RFID anti-collision algorithm based on dynamic framed ALOHA. *IEICE Transactions*, *91-B*(2), 641–645. doi:10.1093/ietcom/e91-b.2.641

Lee, D. (2008). RFID–based traceability in the supply chain. *Industrial Management & Data Systems*, *108*(6), 713–725. doi:10.1108/02635570810883978

Lee, Y. K. (2008). Elliptic-curve-based security processor for RFID. *IEEE Transactions on Computers*, *57*(11), 1514–1527. doi:10.1109/TC.2008.148

Lenz, J., & Edelstein, A. S. (2006). Magnetic sensors and their applications. *IEEE Sensors Journal*, *6*(3), 631–649. doi:10.1109/JSEN.2006.874493

Leong, K. S., & Ng, M. L. (2004). *A simple EPC enterprise model*. Auto-ID Labs Workshop, Zurich 2004. Retrieved from http://www.m-lab.ch

Li, B., Yang, Y., & Wang, J. (2009). *Anti-collision issue analysis in Gen2 protocol - Anti-collision issue analysis considering capture effect*. Auto-ID Labs White Paper. Retrieved August 11, 2011, from http://www.autoidlabs.org/single-view/dir/article/6/320/page.html

Li, C.-J., Liu, L., Chen, S.-Z., Wu, C. C., Huang, C.-H., & Chen, X.-M. (2004). Mobile healthcare service system using RFID. *IEEE International Conference on Networking, Sensing and Control,* Vol. 2 (pp. 1014-1019).

Li, H., Lin, J., & Wu, H. (2008). *Effect of antenna mutual coupling on the UHF passive RFID tag detection*. IEEE Antennas and Propagation Society International Symposium, San Diego, CA.

Li, T., & Deng, R. H. (2007). Vulnerability analysis of EMAP-An efficient RFID mutual authentication protocol. *Proceedings of the Second International Conference on Availability, Reliability, and Security* (*AReS* '07), 2007.

Li, T., & Wang, G. (2007). Security analysis of two ultra-lightweight RFID authentication protocols. *Proceedings of the 22nd IFIP TC-11 International Information Security Conference*, May 2007.

Li, Y., & Ding, X. (2007). *Protecting RFID communications in supply chains*. Presented at the 2nd ACM Symposium on Information, Computer & Communication Security, Singapore, 2007.

Liang, Y., & Li, L. (2007). Integration of intelligent supply chain management (SCM) system. *International Conference on Service Systems and Service Management* (pp. 1-4). IEEE Press.

Linear. (n.d.). *6GHz RMS power detector*. Retrieved from http://cds.linear.com/docs/Datasheet/5581fa.pdf

Lin, X., Lu, R., Kwan, D., & Shen, X. (2010). REACT: An RFID-based privacy-preserving children tracking scheme for large amusement parks. *Computer Networks*, *54*, 2744–2755. doi:10.1016/j.comnet.2010.05.005

Liu, F., & Hu, L. (2008). ROAD: An RFID offline authentication, privacy preserving protocol with DoS resilience. In *Proceedings of the 2008 IFIP International Conference on Network and Parallel Computing*, (pp. 139-146).

Liu, S., Wang, F., & Liu, P. (2007). *A temporal RFID data model for querying physical objects.* Technical Report TR-88, TimeCenter.

Liu, D. S., & Zou, X. C. (2006). Embeded EEPROM memory achieving lower power - New design of EEPROM memory for RFID tag IC. *IEEE Circuits and Devices Magazine*, *22*(6), 53–59. doi:10.1109/MCD.2006.307277

Li, X., & Meijer, G. C. M. (1995). A novel smart resistive-capacitive position sensor. *IEEE Transactions on Instrumentation and Measurement*, *44*(3), 768–770. doi:10.1109/19.387328

Li, X., & Meijer, G. C. M. (2001). A smart and accurate interface for resistive sensors. *IEEE Transactions on Instrumentation and Measurement*, *50*(6), 1648–1651. doi:10.1109/19.982961

Li, X., Meijer, G. C. M., & de Jong, G. W. (1997). A microcontroller-based self-calibration technique for a smart capacitive angular-position sensor. *IEEE Transactions on Instrumentation and Measurement*, *46*(4), 888–892. doi:10.1109/19.650794

Li, X., Meijer, G. C. M., de Jong, G. W., & Spronck, J. W. (1996). An accurate low-cost capacitive absolute angular-position sensor with a full-circle range. *IEEE Transactions on Instrumentation and Measurement*, *45*(2), 516–520. doi:10.1109/19.492778

Li, X., Meijer, G. C. M., & Schnitger, E. J. (1998). A novel smart interface for voltage-generating sensors. *IEEE Transactions on Instrumentation and Measurement*, *47*(1), 285–288. doi:10.1109/19.728835

Li, Y., & Ding, X. (2007). *Protecting RFID communications in supply chains. 2nd ACM syMposium on Information, Computer and Communications Security* (pp. 234–241). ACM Press.

Li, Y., Rida, A., Vyas, R., & Tentzeris, M. M. (2007). RFID tag and RF structures on a paper substrate using inkjet-printing technology. *IEEE Transactions on Microwave Theory and Techniques*, *55*(12), 2894–2901. doi:10.1109/TMTT.2007.909886

Lo, B., Thiemjarus, S., King, R., & Yang, G.-Z. (2005). *Body sensor network: A wireless sensor platform for pervasive healthcare monitoring*. Retrieved on September 5, 2011, from http://academic.research.microsoft.com/Publication/11040372/body-sensor-network-a-wireless-sensor-platform-for-pervasive-healthcare-monitoring

Loo, H., Elmahgoub, K., Elsherbeni, A., & Kajfez, D. (2008). Chip impedance matching for UHF RFID tag antenna design. *Progress in Electromagnetics Research, PIER*, *81*, 359–370. doi:10.2528/PIER08011804

Lòpez, T. S., & Kim, D. (2007). A context middleware based on sensor and RFID information. *Proceedings of the Fifth Annual IEEE International Conference on Pervasive Computing and Communications Workshops* (PerComW'07) IEEE.

Loureiro, R. F., de Oliveira, J. E., Torjesen, P. A., Bartolini, P., & Ribela, M. T. (2006). Analysis of intact human follicle-stimulating hormone preparations by reversed-phase high-performance liquid chromatography. *Journal of Chromatography. A*, *1136*(1), 10–18. doi:10.1016/j.chroma.2006.09.037

Loyka, S. L. (2001). Channel capacity of MIMO architecture using the exponential correlation matrix. *IEEE Communications Letters*, *5*, 369–371. doi:10.1109/4234.951380

Luckett, D. (2004). The supply chain. *BT Technology Journal*, *22*(3), 50–55. doi:10.1023/B:BTTJ.0000047119.22852.38

Lundgaard, L. E. (1992). Partial discharge, XIII: Acoustic partial discharge detection-fundamental considerations. *Electrical Insulation Magazine, IEEE*, *8*(4), 25–31. doi:10.1109/57.145095

Luo, Q., & Fei, Y. (2011). Algorithmic collision analysis for evaluating cryptographic systems and side-channel attacks. In *the Proceedings of the IEEE International Symposium on Hardware-Oriented Security and Trust* (HOST), (pp. 75-80).

Luo, Y., Ji, S., & Li, Y. (2006). Phased-ultrasonic receiving-planar array transducer for partial discharge location in transformer. *IEEE Transactions on Ultrasonics, Ferroelectrics, and Frequency Control*, *53*(3), 614–622. doi:10.1109/TUFFC.2006.1610570

MacKenzie, C. M., Laskey, K., McCabe, F., Brown, P. F., Metz, R., & Hamilton, B. A. (2006). *Reference model for service oriented architecture 1.0*. OASIS SOA Reference Model Technical Committee, 2006.

Marcuse, D. (1969). Mode conversion caused by surface imperfections of a dielectric slab waveguide. *The Bell System Technical Journal*, 3187–3215.

Marrocco, G. (2007). Gain-optimized self-resonant meander line antennas for RFID applications. *IEEE Antennas and Wireless Propagation Letters*, *2*, 302–305. doi:10.1109/LAWP.2003.822198

Marrocco, G. (2007). RFID antennas for the UHF remote monitoring of human subjects. *IEEE Transactions on Antennas and Propagation*, *55*, 1862–1870. doi:10.1109/TAP.2007.898626

Marrocco, G. (2008). The art of UHF RFID antenna design: Impedance matching and size reduction techniques. *IEEE Antennas and Propagation Magazine*, *50*(1), 66–79. doi:10.1109/MAP.2008.4494504

Mathys, P., & Flajolet, P. (1985). Q-ary collision resolution algorithms in random-access systems with free or blocked channel access. *IEEE Transactions on Information Theory*, *31*(2), 217–243. doi:10.1109/TIT.1985.1057013

McCulloch, W., & Pitts, W. (1943). A logical calculus of the ideas immanent in nervous activity. *The Bulletin of Mathematical Biophysics*, *5*, 115–133. doi:10.1007/BF02478259

McFarlane, D., & Sheffi, Y. (2003). The impact of automatic identification on supply chain operations. *International Journal of Logistics Management*, *14*, 1–17. doi:10.1108/09574090310806503

McMeekin, T. A. (2003). *Detecting pathogens in food* (1st ed., pp. 241–258). Cambridge, UK: Woodhead.

Mead, P. S., Slutsker, L., Dietz, V., et al. (2000). *Food-related illness and death in the United States*. Retrieved March 9, 2011, from http://www.cdc.gov/ncidod/eid/vol5no5/mead.htm

Menezes, A. J., Oorschot, P. C. v., & Vanstone, S. A. (1996). *Handbook of applied cryptography*. CRC Press. doi:10.1201/9781439821916

Milligan, T. A. (2005). *Modern antenna design*. John Wiley-IEEE Press. doi:10.1002/0471720615

Mochizuki, K., Masuda, T., & Watanabe, K. (1998). An interface circuit for high-accuracy signal processing of differential-capacitance transducers. *IEEE Transactions on Instrumentation and Measurement*, *47*(4), 823–827. doi:10.1109/19.744628

Mohd Syaifudin, A. R., Jayasundera, K. P., & Mukhopadhyay, S. C. (2009). A low cost novel sensing system for detection of dangerous marine biotoxins in seafood. *Sensors and Actuators. B, Chemical*, *137*, 67–75. doi:10.1016/j.snb.2008.12.053

Mohd Syaifudin, A. R., Mukhopadhyay, S. C., & Yu, P. L. (2010). Novel sensors for food inspection. *Sensors and Transducers Journal*, *108*, 1–44.

Molnar, D., & Wagner, D. (2004). Privacy and security in library RFID: Issues, practices, and architectures. *ACM Conference on Computer and Communications Security - ACM CCS* (pp. 210-219)

Monti, G., Catarinucci, L., & Tarricone, L. (2010). Broad band dipole for RFID applications. *Progress in Electromagnetic Research C*, *12*, 163–172. doi:10.2528/PIERC10012606

Moore, B. (2002). RFID: The chips are down. *Material Handling Management*, *57*(4), 17.

Moore, P. J., Portugues, I. E., & Glover, I. A. (2005). Radiometric location of partial discharge sources on energized high-voltage plant. *IEEE Transactions on Power Delivery*, *20*(3), 2264–2272. doi:10.1109/TPWRD.2004.843397

MoreRFID. (2011). *FASTTag™- RFID edgeware, middleware, tagging and tracking*. Retrieved from http://www.morerfid.com/details.php?subdetail=Product&action=-details&product_id=84&display=RFID

Morris, T. S., Singer, R., Ambrose, M. R., & Hauck, W. W. (2009). Biological potency assays are key to assessing product consistency. *BioPharm International*, *22*(6).

Mosmann, T. (1983). Rapid colorimetric assay for cellular growth and survival: Application to proliferation and cytotoxicity assays. *Journal of Immunological Methods*, *65*(1-2), 55–63. doi:10.1016/0022-1759(83)90303-4

Muhammad-Tahir, Z., & Alocilja, E. C. (2004). A disposable biosensor for pathogen detection in fresh produce samples. *Biosystems Engineering*, *88*(2), 145–151. doi:10.1016/j.biosystemseng.2004.03.005

Mukherjee, S. (2008). Antennas for chipless tags based on remote measurement of complex impedance. *European Microwave Conference, 1-3,* 1728-1731.

Mukhopadhyay, S. C. (2006, October). Sensing and instrumentation for a low cost intelligent sensing system. *Proceedings of the SICE-ICCAS Joint International Conference 2006,* (pp. 1075-1080). ISBN: 89-950038-5-5 98560

Mukhopadhyay, S. C., Chomsuwan, K., Gooneratne, C., & Yamada, S. (2007). A novel needle-type SV-GMR sensor for biomedical applications. *IEEE Sensors Journal*, *7*(3), 401–408. doi:10.1109/JSEN.2007.891929

Mukhopadhyay, S. C., Deb Choudhury, S., Allsop, T., Kasturi, V., & Norris, G. E. (2008). Assessment of pelt quality in leather making using a novel non-invasive sensing approach. *Journal of Biochemical and Biophysical Methods*, *70*, 809–815. doi:10.1016/j.jbbm.2007.07.003

Mukhopadhyay, S. C., & Gooneratne, C. P. (2007). A novel planar-type biosensor for noninvasive meat inspection. *IEEE Sensors Journal*, *7*(9), 1340–1346. doi:10.1109/JSEN.2007.903335

Mukhopadhyay, S. C., Gooneratne, C. P., Sen Gupta, G., & Yamada, S. (2005). Characterization and comparative evaluation of novel planar electromagnetic sensors. *IEEE Transactions on Magnetics*, *41*(10), 3658–3660. doi:10.1109/TMAG.2005.854792

Murugesan, S. (2008). Harnessing green it: Principles and practices. *IT Professional*, *10*(1), 24–33. doi:10.1109/MITP.2008.10

Myung, J., & Lee, W. (2005). Adaptive binary splitting: A RFID tag collision arbitration protocol for tag identification. *IEEE BROADNETs*, (pp. 347–355).

Myung, J., & Lee, W. (2006b). Adaptive splitting protocols for RFID tag collision arbitration. In *MobiHoc '06: Proceedings of the 7th ACM International Symposium on Mobile Ad Hoc Networking and Computing,* (pp. 202–213). New York, NY: ACM.

Myung, J., & Lee, W. (2006a). Adaptive binary splitting: A RFID tag collision arbitration protocol for tag identification. *Mobile Networks and Applications*, *11*(5), 711–722. doi:10.1007/s11036-006-7797-6

Myung, J., Lee, W., & Shih, T. (2006). An adaptive memoryless protocol for RFID tag collision arbitration. *IEEE Transactions on Multimedia*, *8*(5), 1096–1101. doi:10.1109/TMM.2006.879817

Myung, J., Lee, W., Srivastava, J., & Shih, T. K. (2007). Tag-splitting: adaptive collision arbitration protocols for RFID tag identification.. *IEEE Transactions on Parallel and Distributed Systems*, *18*(6), 763–775. doi:10.1109/TPDS.2007.1098

National Institute of Standards and Technology (U.S.). (2008). *Secure hash standard* (SHS). Federal information processing standards publication FIPS PUB 180-3. Retrieved from http://purl.access.gpo.gov/GPO/LPS121031

Nattrass, D. A. (1993). Partial discharge, XVII: The early history of partial discharge research. *Electrical Insulation Magazine*, *9*(4), 27–31. doi:10.1109/57.223897

Neirynck, D., Williams, C., Nix, A., & Beach, M. (2005). *Experimental capacity analysis for virtual array antennas in personal and body area networks*. International Workshop on Wireless Adhoc Networks.

Neirynck, D., Williams, C., Nix, A., & Beach, M. (2006). *Personal area networks with line-of-sight MIMO operation*. IEEE 63rd Vehicular Technology Conference.

Neirynck, D., Williams, C., Nix, A., & Beach, M. (2007). Exploiting MIMO in the personal sphere. *IET Proceedings of Microwaves, Antenna Propagation*.

Ngai, E. (2008). RFID: Technology, applications, and impact on business operations. *International Journal of Production Economics*, *112*(2), 507–509. doi:10.1016/j.ijpe.2007.05.003

Niederman, F. (2007). Examining RFID applications in supply chain. *Communications of the ACM*, *50*(7), 93–101. doi:10.1145/1272516.1272520

Niemeyer, L. (1993, 28-30 September). *The physics of partial discharges.* Paper presented at the 1993 International Conference on Partial Discharge.

Nikitin, P., & Rao, K. V. S. (2008). Antennas and propagation in UHF RFID systems. *IEEE RFID Conference Proceedings,* Las Vegas, NV, April 16-17.

Nikitin, P. (2005). Power reflection coefficient analysis for complex impedances in RFID tag design. *IEEE Transactions on Microwave Theory and Techniques*, *53*(9), 2721–2725. doi:10.1109/TMTT.2005.854191

NIST Special Publication. (2005). *PIV middleware and PIV card application conformance test guidelines,* (pp. 800-85). Retrieved from http://csrc.nist.gov/piv-project

Nørrung, B., & Buncic, S. (2008). Microbial safety of meat in the European Union. *Meat Science*, *78*, 14–24. doi:10.1016/j.meatsci.2007.07.032

O'Connor, M. C. (2006). *GlaxoSmithKline tests RFID on HIV drugs*. Retrieved March 12, 2011, from http://www.rfidjournal.com/article/view/2219

O'Connor, M. C. (2008a). *Wal-Mart, DOD point to sustained progress*. Retrieved February 5, 2011, from http://www.rfidjournal.com/article/articleview/4407/

O'Connor, M. C. (2008b). *American Apparel expands RFID to additional stores*. Retrieved March 8, 2011, from http://www.rfidjournal.com/article/view/4510/1

O'Connor, M. C. (2009a). *RFID trims costs for retailer of Lacoste, CK, Burberry*. Retrieved March 7, 2011, from http://www.rfidjournal.com/article/view/4626

O'Connor, M. C. (2009b). *Marigold industrial gets a better grip on glove production, inventory*. Retrieved March 9, 2011, from http://www.rfidjournal.com/article/print/5004

O'Connor, M. C. (2010). *RFID finds flavor at Izzy's ice cream shop*. Retrieved February 5, 2011, from http://www.rfidjournal.com/article/view/7651

Occhiuzzi, C., & Marrocco, G. (2010). The RFID technology for neurosciences: feasibility of limbs' monitoring in sleep diseases. *IEEE Transactions on Information Technology in Biomedicine*, *14*, 37–43. doi:10.1109/TITB.2009.2028081

OCED. (2011). *The OECD principles*. Retrieved from http://www.anu.edu.au/people/Roger.Clarke/DV/OECDPs.html

OpenNebula. (2002). *The open source toolkit for cloud computing*. Retrieved from http://www.opennebula.org/

Oren, Y., & Shamir, A. (2007). Remote password extraction from RFID tags. *IEEE Transactions on Computers*, *56*(9), 1292–1296. doi:10.1109/TC.2007.1050

Oswald, D., Kasper, T., & Paar, C. (2011). *Side-channel analysis of cryptographic RFIDs with analog demodulation*. RFIDSec'11, June 2011.

Pais, S., & Symonds, J. (2011). Data storage on a RFID tag for a distributed system. *International Journal of UbiComp (IJU)*, *2*(2), 26–39. doi:10.5121/iju.2011.2203

Pappu, R., Recht, B., Taylor, J., & Gershenfeld, N. (2002). Physical one-way functions. *Science*, *297*(5589), 2026–2030. doi:10.1126/science.1074376

Park, J.-H., Seol, J.-A., & Oh, Y.-H. (2005). Design and implementation of an effective mobile healthcare system using mobile and RFID technology. *Proceedings of 7th International Workshop on Enterprise networking and Computing in Healthcare Industry* (pp. 263-266).

Park, Y., Heo, P., & Rim, M. (2008). Measurement of a customer satisfaction index for improvement of mobile RFID services in Korea. *ETRI Journal, 30*(5).

Park, Y., Park, S. K., & Lee, H. Y. (2010). Performance of wireless body area network over on-human-body propagation channels. *2010 IEEE Sarnoff Symposium*, (pp. 1-4).

Patra, J. C., Kot, A. C., & Panda, G. (2000). An intelligent pressure sensor using neural networks. *IEEE Transactions on Instrumentation and Measurement*, *49*(4), 829–834. doi:10.1109/19.863933

Pautasso, C., Zimmermann, O., & Leymann, F. (2008). RESTful web services vs. big web services: Making the right architectural decision. *Proceedings of the 17th International World Wide Web Conference* (WWW2008) Beijing, China.

Pawlikowski, G. T. (2009). *Effects of polymer material variations on high frequency dielectric properties*, Vol. 1156. MRS 2009 Spring Meeting.

Peris-Lopez, P., Hernández-Castro, J. C., Estévez-Tapiador, J. M., & Ribagorda, A. (2006). RFID systems: A survey on security threats and proposed solutions. *PWC*, *2006*, 159–170.

Peris-Lopez, P., Hernandez-Castro, J. C., Estevez-Tapiador, J. M., & Ribagorda, A. (2009). LAMED – A PRNG for EPC class-1 generation-2 RFID specification. *Computer Standards & Interfaces*, *31*(1), 88–97. doi:10.1016/j.csi.2007.11.013

Personal Area Network (PAN). (n.d.). Retreived on May 2, 2011 from http://searchmobilecomputing.techtarget.com/definition/personal-area-network

Philipose, M., Smith, J. R., Jiang, B., Mamishev, A., Roy, S., & Sundara-Rajan, K. (2005). Battery-free wireless identification and sensing. *IEEE Pervasive Computing*, *4*, 37–45. doi:10.1109/MPRV.2005.7

Phillips Semiconductor. (2006). *Application note*

Pillai, K. (2008). *Feature of diversity technique*. Retrieved on 15 June, 2010, from http://www.dsplog.com/2008/10/16/alamouti-stbc/

Ping, H., Hynes, C. G., van Wonterghem, J., & Michelson, D. G. (2005). Measurements of correlation coefficients of closely spaced dipoles. *IEEE International Workshop on Antenna Technology: Small Antennas and Novel Metamaterials,* (pp. 474-477).

Piramuthu, S. (2006). HB and related lightweight authentication protocols for secure RFID tag/reader authentication. *Proceedings of CollECTeR Europe Conference*, June 2006.

Piramuthu, S. (2007). Protocols for RFID tag/reader authentication. *Decision Support Systems*, *43*(3), 897–914. doi:10.1016/j.dss.2007.01.003

Plessky, V., & Reindl, L. (2010). Review on SAW RFID tags. *IEEE Transactions on Ultrasonics, Ferroelectrics, and Frequency Control*, *57*(3), 654–668. doi:10.1109/TUFFC.2010.1462

Plos, T. (2008). Susceptibility of UHF RFID tags to electromagnetic analysis. *CT-RSA 2008. LNCS*, *4964*, 288–300.

Pope, A. (1998). *The Corba reference guide: Understanding the common object request broker architecture.* Addison-Wesley, 1998.

Popovic, N. (2011). UHF RFID antenna: Printed dipole antenna with a CPS matching circuit and inductively coupled feed. *International Journal of Radio Frequency Identification and Wireless Sensor Networks*, *1*(1), 28–33.

Power Diagnostix Systems GmbH. (n.d.). Retrieved from http://www.pd-systems.com/accessories.html#mail

Preradovic, S., & Karmakar, N. (2009). Design of fully printable planar chipless RFID transponder with 35-bit data capacity. *European Microwave Conference, 1-3*, 13-16.

Preradovic, S., & Karmakar, N. C. (2009, 20-22 Aug. 2009). *Design of fully printable chipless RFID tag on flexible substrate for secure banknote applications.* Paper presented at the 3rd International Conference on Anti-counterfeiting, Security, and Identification in Communication, ASID 2009.

Preradovic, S., Balbin, I., Karmakar, N. C., & Swiegers, G. F. (2009). Multiresonator-based chipless RFID system for low-cost item tracking. *IEEE Transactions on Microwave Theory and Techniques, 57*(5:2), 1411–1419.

Preradovic, Balbin, I., Karmakar, N. C., & Swiegers, G. F. (2009). Multiresonator-based chipless RFID system for low-cost item tracking. *IEEE Transactions on Microwave Theory and Techniques*, *57*(5), 1411–1419. doi:10.1109/TMTT.2009.2017323

Preradovic, S. (2009). *Chipless RFID system for barcode replacement*. Monash University.

Preradovic, S., Balbin, I., Karmakar, N. C., & Swiegers, G. F. (2009). Multiresonator-based chipless RFID system for low cost item tracking. *IEEE Transactions on Microwave Theory and Techniques*, *57*(5), 1411–1419. doi:10.1109/TMTT.2009.2017323

Preradovic, S., & Karmakar, N. C. (2010). Chipless RFID: Bar code of the future. *IEEE Microwave Magazine*, (December): 87–97. doi:10.1109/MMM.2010.938571

Public Health Surveillance. (2008). *Annual summary of outbreaks: New Zealand.* Retrieved on October 25, 2009 from www.surv.esr.cri.nz/surveillance/annual_outbreak.php

Pupunwiwat, P., & Stantic, B. (2009). Unified q-ary tree for RFID tag anti-collision resolution. In B. Athman & L. Xuemin (Eds.), *20th Australasian Database Conference: Vol. 92, Conferences in Research Research and Practice in Information Technology* (pp. 49-58). Australian Computer Society Press.

Pupunwiwat, P., & Stantic, B. (2009a). Performance analysis of enhanced Q-ary tree anti-collision protocols. In *The First Malaysian Joint Conference on Artificial Intelligence (MJCAI), Vol. 1* (pp. 229–238). Kuala Lumpur, Malaysia.

Pupunwiwat, P., & Stantic, B. (2009b). Unified Q-ary tree for RFID tag anti-collision resolution. In A. Bouguettaya & X. Lin (Eds.), *The Twentieth Australasian Database Conference (ADC), Vol. 92 of CRPIT, ACS* (pp. 47–56). Wellington, New Zealand.

Pupunwiwat, P., & Stantic, B. (2010b). Dynamic framed-slot ALOHA anti-collision using precise tag estimation scheme. In H. T. Shen & A. Bouguettaya (Eds.), *The Twenty- First Australasian Database Conference (ADC), Vol. 104 of CRPIT, ACS* (pp. 19–28). Brisbane, Australia.

Pupunwiwat, P., & Stantic, B. (2010c). Joined Q-ary Tree anti-collision for massive tag movement distribution. In B. Mans & M. Reynolds (Eds.), *The Thirty-Third Australasian Computer Science Conference (ACSC), Vol. 102 of CRPIT, ACS* (pp. 99–108). Brisbane, Australia.

Pupunwiwat, P., & Stantic, B. (2010d). Resolving RFID data stream collisions using set- based approach. In *The Sixth International Conference on Intelligent Sensors, Sensor Networks and Information Processing (ISSNIP), IEEE* (pp. 61–66). Brisbane, Australia.

Pupunwiwat, P., & Stantic, B. (2010a). A RFID explicit tag estimation scheme for dynamic framed-slot ALOHA anti-collision. In *The Sixth Wireless Communications, Networking and Mobile Computing (WiCOM)* (pp. 1–4). Chengdu, China: IEEE. doi:10.1109/WICOM.2010.5601080

Qin, S., & Birlasekaran, S. (2002, 2002). *The study of propagation characteristics of partial discharge in transformer.* Paper presented at the 2002 Annual Report Conference on Electrical Insulation and Dielectric Phenomena.

Quan, C. H., Hong, W. K., & Kim, H. C. (2006). Performance analysis of tag anti-collision algorithms for RFID systems. In *Emerging Directions in Embedded and Ubiquitous Computing, Vol. 4097*, (pp. 382–391). Seoul, Korea. Berlin, Germany: Springer.

Rach, N. M. (2008). RFID applications spread in upstream operations. *Oil & Gas Journal*, *106*(27), 37–44.

Rahmat-Samii, Y., Williams, L. I., & Yaccarino, R. G. (1995). The UCLA bi-polar planar-near-field antenna-measurement and diagnostics range. *Antennas and Propagation Magazine, IEEE*, *37*(6), 16–35. doi:10.1109/74.482029

Ramos, A. A., Lazaro, D., Girbau, A., & Villarino, R. (2011). Time-domain measurement of time-coded UWB chipless RFID tags. *Progress in Electromagnetics Research*, *116*, 313–331.

Rao, J., Doraiswamy, S., Thakkar, H., & Colby, L. (2006). A deferred cleansing method for RFID data analytics. In *Very Large Databases*, (pp. 175-186).

Ray, B., Chowdhury, M. U., & Pham, T. (2010). Mutual authentication with malware protection for RFID system. *Annual International Conference on Information Technology Security* (pp. I-24 - I-29)

Reid, A. J., Judd, M. D., Stewart, B. G., & Fouracre, R. A. (2006a, 15-18 Oct. 2006). *Frequency distribution of RF energy from PD sources and its application in combined RF and IEC60270 measurements.* Paper presented at the 2006 IEEE Conference on Electrical Insulation and Dielectric Phenomena.

Reid, D. I., & Judd, M. (2009, 8-11 June). *UHF monitoring of partial discharge in substation equipment using a novel multi-sensor cable loop.* Paper presented at the 20th International Conference and Exhibition on Electricity Distribution - Part 1, CIRED 2009.

Reid, A. J., Judd, M. D., Fouracre, R. A., Stewart, B. G., & Hepburn, D. M. (2011). Simultaneous measurement of partial discharges using IEC60270 and radio-frequency techniques. *IEEE Transactions on Dielectrics and Electrical Insulation*, *18*(2), 444–455. doi:10.1109/TDEI.2011.5739448

Reid, A. J., Judd, M. D., Stewart, B. G. A., & Fouracre, R. A. (2006b). Partial discharge current pulses in SF 6 and the effect of superposition of their radiometric measurement. *Journal of Physics. D, Applied Physics*, *39*(19), 4167. doi:10.1088/0022-3727/39/19/008

Research, A. B. I. (2009). *Market for item-level RFID in fashion apparel and footwear will nearly triple by 2014.* Retrieved March 7, 2011, from http://www.abiresearch.com/press/1489-Market+for+Item-Level+RFID+in+Fashion+Apparel+and+Footwear+Will+Nearly+Triple+by+2014

RFMD. (n.d.). *General purpose amplifier*. Retrieved from http://www.rfmd.com/CS/Documents/3378DS.pdf

Rhino. (2011). *JavaScript for Java*. Retrieved from http://www.mozilla.org/rhino/

Rieback, M. R., Crispo, B., & Tanenbaum, A. S. (2006). Is your cat infected with a computer virus? In *Proceedings of PerCom* (pp.169-179)

Rieback, M. R., Crispo, B., & Tanenbaum, A. S. (2006). *The evolution of RFID security* (pp. 62–69). Pervasive Computing.

Rieback, M. R., Simpson, P. N. D., Crispo, B., & Tanenbaum, A. S. (2006). RFID malware: Design principles and examples. *Pervasive and Mobile Computing*, 405–426. doi:10.1016/j.pmcj.2006.07.008

Rivest, R. L., Shamir, A., & Adleman, L. (1978). A method for obtaining digital signatures and public-key cryptosystems. *Communications of the ACM*, *21*(2), 120–126. doi:10.1145/359340.359342

Roberti, M. (2003). Big brother's enemy. *RFID Journal*, July 2003.

Roberti, M. (2004a). *Wal-Mart begins RFID rollout*. Retrieved February 6, 2011, from http://www.rfidjournal.com/article/view/926

Roberti, M. (2004b). *DOD releases final RFID policy*. Retrieved February 6, 2011, from http://www.rfidjournal.com/article/view/1080

Roberti, M. (2006). *RFID is fit to track clothes*. Retrieved March 7, 2011, from http://www.rfidjournal.com/article/view/2195

Roberti, M. (2009). *Restoring confidence in the food chain*. Retrieved March 8, 2011, from http://www.rfidjournal.com/article/view/4621/1

Roberti, M. (2010). *Wal-Mart relaunches EPC RFID effort, starting with men's jeans and basics*. Retrieved February 6, 2011, from http://www.rfidjournal.com/article/view/7753

Robyn, M. (2008). *Market-driven fraud: The impact and consequences of counterfeit products and intellectual property violations.* ASC Annual Meeting, St. Louis, Missouri

Rodrigues, J. M., Estevao, M. H., Malaquias, J. L., Santos, P., Gouveia, G., & Simoes, J. B. (2007).Sleep at home – Portable home based system for pediatric sleep apnea diagnosis. *IEEE International Conference on Portable Information Devices* (pp. 1-4).

Romanov, V. G., & Kabanikhin, S. I. (1994). *Inverse problems for Maxwell's equations*. Utrecht, The Netherlands: VSP.

Romero, H. P., Remley, K. A., Williams, D. F., & Chih-Ming, W. (2009). Electromagnetic measurements for counterfeit detection of radio frequency identification cards. *IEEE Transactions on Microwave Theory and Techniques*, *57*(5), 1383–1387. doi:10.1109/TMTT.2009.2017318

RongLin, L., DeJean, G., Tentzeris, M. M., & Laskar, J. (2004). Development and analysis of a folded shorted-patch antenna with reduced size. *IEEE Transactions on Antennas and Propagation*, *52*(2), 555–562. doi:10.1109/TAP.2004.823884

Rooij, A., Johnson, R., & Jain, L. (1996). *Neural network training using genetic algorithms*. River Edge, NJ: World Scientific Publishing Company Incorporated.

Ross, J. W., & Westerman, G. (2004). Preparing for utility computing: The role of it architecture and relationship management. *IBM Systems Journal*, *43*(1), 5–19. doi:10.1147/sj.431.0005

Rumelhart, D., Hinton, G., & Williams, R. (1986). Learning representations by back-propagating errors. *Nature*, *323*, 533–536. doi:10.1038/323533a0

Russell, W. M. S., & Burch, R. L. (1959). *The principles of humane experimental technique*. London, UK: Methuen.

Ryu, J., Lee, H., Seok, Y., Kwon, T., & Chio, Y. (2007). A hybrid query tree protocol for tag collision arbitration in RFID system. *IEEE International Conference on Communications* (pp. 5981-5986). IEEE Press.

Sagawa, M., Makimoto, M., & Yamashita, S. (1997). Geometrical structures and fundamental characteristics of microwave stepped-impedance resonators. *IEEE Transactions on Microwave Theory and Techniques*, *45*(7), 1078–1085. doi:10.1109/22.598444

Saito, J., Ryou, J.-C., & Sakurai, K. (2004). *Enhancing privacy of universal re-encryption scheme for RFID tags* (pp. 879–890). Embedded and Ubiquitous Computing. doi:10.1007/978-3-540-30121-9_84

Sakaguchi, K., Chua, H.-Y.-E., & Araki, K. (2005). MIMO channel capacity in an indoor line-of-sight environment. *IEICE Transactions on Communication, E88-B*.

Sample, A. P., Yeager, D. J., Powledge, P. S., & Smith, J. R. (2007). Design of a passively-powered, programmable sensing platform for UHF RFID systems. *IEEE International Conference on RFID* (pp. 149-156).

Sangwan, R. S., Qiu, R. G., & Jessen, D. (2005). Using RFID tags for tracking patients, charts and medical equipment within an integrated health delivery network. *Proceedings of the IEEE Networking, Sensing and Control* (pp. 1070-1074).

Sarma, S., Weis, S., & Engels, D. (2003). RFID systems and security and privacy implications. *Workshop on Cryptographic Hardware and Embedded Systems: Vol. 2523, Lecture Notes in Computer Science* (pp. 454-470). Springer-Verlag Press.

Sasaoka, T., Ishiki, M., Sawa, T., Ishihara, H., Takata, Y., & Imamura, T. (1996). Comparison of the insulin and insulin-like growth factor 1 mitogenic intracellular signaling pathways. *Endocrinology*, *137*(10), 4427–4434. doi:10.1210/en.137.10.4427

Schmalzel, J., Figueroa, F., & Morris, J. (2005). An architecture for intelligent systems based on smart sensors. *IEEE Transactions on Instrumentation and Measurement*, *54*(4), 1612–1616. doi:10.1109/TIM.2005.851477

Schoenberger, C. R. (2002). The internet of things. *Forbes Magazine, 6*.

Schoute, F. C. (1983). Dynamic frame length ALOHA. *IEEE Transactions on Communications*, *31*(4), 565–568. doi:10.1109/TCOM.1983.1095854

Schussler, M., Mandel, C., Maasch, M., Giere, A., & Jalcoby, R. (2009). Phase modulation scheme for chipless RFID and wireless sensor tags. *Asia Pacific Microwave Conference,* Vol. 1, (pp. 5229-5232).

Schwartz, M. (1988). *Telecommunication networks protocols, modeling and analysis*. USA: Addison-Wesley.

Seevers, R. H. (2005). *Report to FDA of PQRI RFID working group. Office of regulatory affairs, guidance for industry, prescription drug marketing act* (pp. 2-4/14).

Sen, D., Sen, P., & Das, A. M. (2009). *RFID for energy and utility industries: PennWell.*

Shang, J. Q., Umana, J. A., Bartlett, F. M., & Rossiter, J. R. (1999). Measurement of complex permittivity of asphalt pavement materials. *Journal of Transportation Engineering*, *125*(4), 347–356. doi:10.1061/(ASCE)0733-947X(1999)125:4(347)

Shelby, R. A., Smith, D. R., & Schultz, S. (2001). Experimental verification of a negative index of refraction. *Science*, *292*(5514), 77–79. doi:10.1126/science.1058847

Shibuya, Y., Matsumoto, S., Tanaka, M., Muto, H., & Kaneda, Y. (2010). Electromagnetic waves from partial discharges and their detection using patch antenna. *IEEE Transactions on Dielectrics and Electrical Insulation*, *17*(3), 862–871. doi:10.1109/TDEI.2010.5492260

Shih, D. H., Sun, P. L., Yen, D. C., & Huang, S. M. (2006). Taxonomy and survey of RFID anti-collision protocols. *Computer Communications*, *29*(11), 2150–2166. doi:10.1016/j.comcom.2005.12.011

Shin, W. J., & Kim, J. G. (2007). Partitioning of tags for near-optimum RFID anti-collision performance. In *IEEE Wireless Communications and Networking Conference, WCNC 2007,* (pp. 1673–1678).

Shi, X., Medard, M., & Lucani, D. E. (2010). *Network coding for energy efficiency in wireless body area networks*. Massachusetts Institute of Technology.

Shnayder, B. C. V., Lorincz, K., Fulford-Jones, T., & Welsh, M. (2005). *Sensor networks for medical care.* Technical Report TR-08-05, Division of Engineering and Applied Sciences, Harvard University.

Shrestha, S., Balachandran, M., Agarwal, M., Phoha, V. V., & Varahramyan, K. (2009). A chipless RFID sensor system for cyber centric monitoring applications. *IEEE Transactions on Microwave Theory and Techniques*, *57*(5), 1303–1309. doi:10.1109/TMTT.2009.2017298

Shukla, A., Enzmann, H., & Mayer, D. (2009). Proliferative effect of Apidra (insulin glulisine), a rapid-acting insulin analogue on mammary epithelial cells. *Archives of Physiology and Biochemistry*, *115*(3), 119–126. doi:10.1080/13813450903008628

Sigala, M. (2007). RFID applications for integrating and informationalizing the supply chain of foodservice operators: Perspectives from Greek operators. *Journal of Foodservice Business Research*, *10*(1), 7–29. doi:10.1300/J369v10n01_02

Simmons, G. J. (1991, 1-3 Oct 1991). *Identification of data, devices, documents and individuals.* Paper presented at the 25th Annual 1991 IEEE International Carnahan Conference on Security Technology.

Singh, C. K., & Mohan, S. (2010). Effect of antennas correlation on the performance of MIMO systems in wireless sensor network. *Wireless Telecommunications Symposium (WTS)*, (pp. 1-5).

Sivakumar, M., & Deavours, D. D. (2006). *A dual-resonant planar microstrip antenna design for UHF RFID using paperboard as a substrate*. Retrieved on 05 September, 2011, from http://www.ittc.ku.edu/~deavours/pubs/rfid06b-submit.pdf

Skoric, B. (2007). *The entropy of keys derived from laser speckle*. ArXiv e-prints.

Smith, D., Hanlen, L., Miniutti, D., Zhang, J., Rodda, D., & Gilbert, B. (2008). Statistical characterization of the dynamic narrowband body area channel. *First International Symposium on Applied Sciences on Biomedical and Communication Technologies* (pp. 1-5).

Smith, D., Hanlen, L., Zhang, J., Miniutti, D., Rodda, D., & Gilbert, B. (2009). Characterization of the dynamic narrowband on-body to off-body area channel. *IEEE International Conference on Communications,* (pp. 1-6).

Smith, D. R., Pendry, J. B., & Wiltshire, M. C. K. (2004). Metamaterials and negative refractive index. *Science, 305*(5685), 788–792. doi:10.1126/science.1096796

Son, H. W., & Pyo, C. (2005). Design of RFID tag antenna using an inductively coupled feed. *Electronics Letters, 41*(18), 994–996. doi:10.1049/el:20051536

Steelman, S. L., & Pohley, F. M. (1953). Assay of the follicle stimulating hormone based on the augmentation with human chorionic gonadotropin. *Endocrinology, 53*(6), 604–616. doi:10.1210/endo-53-6-604

Stern, B. (1996). Warning! Bogus parts have turned up in commercial jets. Where's the FAA? *Business Week.* Retrieved from http://www.businessweek.com/1996/24/b34791.htm

Stewart, B. G., Nesbitt, A., & Hall, L. (2009, May 31 2009-June 3). *Triangulation and 3D location estimation of RFI and partial discharge sources within a 400kV substation.* Paper presented at the Electrical Insulation Conference, EIC 2009.

Suhl, H. (1956). The nonlinear behavior of ferrites at high microwave signal levels. *Proceedings of the IRE, 44*(10), 1270–1284. doi:10.1109/JRPROC.1956.274950

Swedberg, C. (2005). Hospital uses RFID for surgical patients. *RFID Journal*. Retrieved August 12, 2011, from http://www.rfidjournal.com/article/articleview/1714/1/1/

Swedberg, C. (2006). *Marks & Spencer to tag items at 120 stores*. Retrieved February 5, 2011, from ttp://www.rfidjournal.com/article/articleview/2829/1/1/

Swedberg, C. (2007a). *Growers and grocers get into plastic pallet pool*. Retrieved March 9, 2011, from http://www.rfidjournal.com/article/view/3821

Swedberg, C. (2007b). *RFID sweetens imperial's shipping process*. Retrieved March 9, 2011, from http://www.rfidjournal.com/article/view/3720/1/1

Swedberg, C. (2008a). *Norwegian food group Nortura to track meat*. Retrieved March 8, 2011, from http://www.rfidjournal.com/article/view/4208/1.

Swedberg, C. (2008b). *Cosmetics and liquor companies assess Toppan Printing's holographic RFID labels*. Retrieved March 7, 2011, from http://www.rfidjournal.com/article/view/4356

Swedberg, C. (2009a). *New Belgium brewing rolls out RFID to track kegs*. Retrieved March 7, 2011, from http://www.rfidjournal.com/article/view/4925

Swedberg, C. (2009b). *PLS uses RFID to track pallets, containers.* Retrieved March 9, 2011, from http://www.rfidjournal.com/article/view/5043

Swedberg, C. (2009c). *BMW finds the right tool*. Retrieved March 12, 2011, from http://www.rfidjournal.com/article/view/5104/1

Swedberg, C. (2010a). *American Apparel adds RFID to two more stores, switches RFID software*. Retrieved March 7, 2011, from http://www.rfidjournal.com/article/view/7313

Swedberg, C. (2010b). *Korean clothing company adds RFID to its supply chain.* Retrieved March 7, 2011, from http://www.rfidjournal.com/article/view/7360

Swedberg, C. (2010c). *Axios MA launches tagged pallets and real-time tracking solution.* Retrieved March 9, 2011 from http://www.rfidjournal.com/article/view/8014/1

Swedberg, C. (2010d). *RFID illuminates Lithuanian lamp manufacturer*. Retrieved March 9, 2011, from http://www.rfidjournal.com/article/view/7419/1

Swedberg, C. (2010e). *Lithuanian manufacturer tracks IKEA-bound furniture*. Retrieved March 9, 2011, from http://www.rfidjournal.com/article/view/7978/1

Swedberg, C. (2010f). *Nigerian drug agency opts for RFID anticounterfeiting technology*. Retrieved March 12, 2011, from http://www.rfidjournal.com/article/view/7856

Swedberg, C. (2010g). *RFID boosts profit margin, safety for Axxa Parma*. Retrieved March 12, 2011, from http://www.rfidjournal.com/article/view/7823/1

Swedberg, C. (2010h). *Parexel tests system to track temperature of test drugs*. Retrieved March 12, 2011, from http://www.rfidjournal.com/article/view/7848/1

Swedberg, C. (2010i). *RFID to take the chill out offrozen plasma tracking*. Retrieved March 13, 2011, from http://www.rfidjournal.com/article/view/7632/1

Swedberg, C. (2010j). *Hospital robot tracks controlled substances, high-value meds*. Retrieved March 13, 2011, from http://www.rfidjournal.com/article/view/7825

Swedberg, C. (2010k). *Cityzi seeks to spur adoption of NFC RFID technology*. Retrieved March 14, 2011, from http://www.rfidjournal.com/article/view/7650

Swedberg, C. (2011a). *Beverage metrics serves up drink-management solution*. Retrieved March 7, 2011, from http://www.rfidjournal.com/article/view/8237/1

Swedberg, C. (2011b). *Norsk Lastbaerer Pool inserts RIFD into Norwegian food chain*. Retrieved March 9, 2011, from http://www.rfidjournal.com/article/view/8137/1

Swedberg, C. (2011c). *Payback could be a lollapalooza for concert promoters*. Retrieved March 9, 2011, from http://www.rfidjournal.com/article/view/8199/1

Tafa, Z., & Stojanovic, R. (2006). Bluetooth-based approach to monitoring biomedical signals. *Proceedings of the 5th WSEAS International Conference on Telecommunications and Informatics*, (pp. 415-420).

Takada, J., Aoyagi, T., Takizawa, K., Katayama, N., Kobayashi, T., Yazdandoost, K. Y., et al. (2008). *Static propagation and channel models in body area*. Retrieved September 5, 2011, from http://www.ap.ide.titech.ac.jp/publications/Archive/COST2100_TD%2808%-29639%280810Takada%29.pdf

Tang, Z., & He, Y. (2007). Research of multi-access and anti-collision protocols in RFID systems. *IEEE International Workshop on Anticounterfeiting, Security, Identification*, China, (pp. 377–380).

Tauxe, R. V. (2002). Emerging foodborne pathogens. *International Journal of Food Microbiology*, *78*, 31–41. doi:10.1016/S0168-1605(02)00232-5

Tenbohlen, S. M. H., Denissov, D., Huber, R., Riechert, U., Markalous, S. M., Strehl, T., & Klein, T. (2006). Electromagnetic (UHF) PD diagnosis of GIS, cable accessories and oil-paper insulated power transformers for improved PD detection and localization.

Texas Instruments. (2011). *MSP430FR57x family datasheet*.

Texas Instruments. (n.d.). *MSP-EXP430F5438 experimenter board user's guide (Rev. E)*.

Theremin, L. (n.d.). *Bio*. Retrieved September 5, 2011, from http://www.absoluteastronomy.com/topics/L%C3%A9on_Theremin

Thiesse, F., Floerkemeier, C., Harrison, M., Michahelles, F., & Roduner, C. (2009). Technology, standards, and real-world deployments of the EPC network. *IEEE Internet Computing Magazine*, *13*(2), 36–43. doi:10.1109/MIC.2009.46

Thuemmier, C., Buchanan, W., Fekri, A., & Lawson, A. (2009). Radio frequency identification (RFID) in pervasive healthcare. *International Journal of Healthcare Technology and Management*, *10*(1/2), 119. doi:10.1504/IJHTM.2009.023731

Tian, Y., Kawada, M., & Isaka, K. (2008, 7-11 Sept. 2008). *Visualization of electromagnetic waves emitted from multiple PD sources on distribution line by using FDTD method.* Paper presented at the Electrical Insulating Materials, 2008. (ISEIM 2008). International Symposium on.

Tian, Y., Kawada, M., & Isaka, K. (2009). Locating partial discharge source occurring on distribution line by using FDTD and TDOA methods. *IEEJ Transactions on Fundamentals and Materials*, *129*(2), 89–96. doi:10.1541/ieejfms.129.89

Tingting Mao, T., Williams, J., & Sanchez, A. (2008). Interoperable internet scale security framework for RFID networks. In *the Proceedings of the IEEE 24th International Conference on Data Engineering Workshop*, ICDEW'08.

Toth, F. N., & Meijer, G. C. M. (1992). A low-cost smart capacitive position sensor. *IEEE Transactions on Instrumentation and Measurement*, *41*(6), 1041–1044. doi:10.1109/19.199446

Tsang, L., Kong, J. A., & Ding, K.-H. (2002). *Scattering of electromagnetic waves: Theories and applications*. John Wiley & Sons, Inc.doi:10.1002/0471224286

Tsang, L., Kong, J. A., & Shin, R. T. (1985). *Theory of microwave remote sensing*. New York, NY: Wiley.

Tuyls, P., & Batina, L. (2006). RFID-tags for anti-counterfeiting. In Pointcheval, D. (Ed.), *Topics in Cryptology – CT-RSA 2006* (*Vol. 3860*, pp. 115–131). Berlin, Germany: Springer. doi:10.1007/11605805_8

U.S. FDA. CPG. Sec. 400.210. (2004). *Radiofrequency identification feasibility studies and pilot programs for drugs*. November 2004.

Uysal, I., De Hay, P. W., Altunbas, E., Emond, J. P., Rasmussen, R. S., & Ulrich, D. (2010). Non-thermal effects of radio frequency exposure on biologic pharmaceuticals for RFID applications. In *Proceedings of IEEE International Conference on RFID* (pp. 266-273).

Vaudenay, S. (2007). On privacy models for RFID. In *Advances in Cryptology* (ASI-ACRYPT'07), *Lecture Notes in Computer Science, LNCS 4833*, (pp. 68–87). Springer-Verlag.

Vogt, H. (2002). Efficient object identification with passive RFID tags. *In Pervasive '02: Proceedings of the First International Conference on Pervasive Computing,* (pp. 98–113). London, UK: Springer.

Vogt, H. (2002). Multiple object identification with passive RFID tags. *The IEEE International Conference on Man and Cybernetics*, (pp. 6–13).

Votis, C., Tatsis, G., & Kostarakis, P. (2010). Envelope correlation parameter measurements in a MIMO antenna array configuration. *International Journal of Communications. Network and System Sciences*, *3*(4), 350–354. doi:10.4236/ijcns.2010.34044

Vyas, R., Lakafosis, V., Rida, A., Chaisilwattana, N., Travis, S., Pan, J., & Tentzeris, M. M. (2009). Paper-based RFID-enabled wireless platforms for sensing applications. *IEEE Transactions on Microwave Theory and Techniques*, *57*(5), 1370–1382. doi:10.1109/TMTT.2009.2017317

Waggoner, M. (2008). Application of RFID technology in the manufacturing process. *Plant Engineering*, *62*(4), 45–47.

Wang, L., Noel, E., Fong, C., Kamoua, R., & Tang, K. W. (2006). A wireless sensor system for biopotential recording in the treatment of sleep apnea disorder. *Proceedings of the 2006 IEEE International Conference on Networking, Sensing and Control,* (pp. 404-409).

Wang, Z., Liu, D., Zhou, X., Tan, X., Wang, J., & Min, H. (2007). *Anti-collision scheme analysis of RFID system*. Auto-ID Labs White Paper. Retrieved August 11, 2011, from http://www.autoidlabs.org/single-view/dir/article/6/281/page.html

Wang, Y., Jia, Y., Chen, Q., & Wang, Y. (2008). A passive wireless temperature sensor for harsh environment applications. *Sensors (Basel, Switzerland)*, *8*(12), 7982–7995. doi:10.3390/s8127982

Want, R. (2004). Enabling ubiquitous sensing with RFID. *Computer*, *37*(4), 84–86. doi:10.1109/MC.2004.1297315

Wasserman, E. (2006). *Keeping fresh foods fresh*. Retrieved March 8, 2011, from http://www.rfidjournal.com/article/print/2137

Wasserman, E. (2011). RFID serves up benefits for guests and hosts. *RFID Journal*, Nov/Dec 2011, 20-23.

Weijie, C., & Weiping, L. (2008). *Study of integrating RFID middleware with enterprise applications based on SOA*. IEEE.

Weinstein, R. (2005). RFID: A technical overview and its application to the enterprise. *IT Prof Magazine,* May/June 2005.

Weis, S. A., Sarma, S. E., Rivest, R. L., & Engels, D. W. (2004). *Security and privacy aspects of low-cost radio frequency identification systems. Security in Pervasive Computing* (pp. 201–212). Springer.

Wei, Z., Kapoor, G., & Piramuthu, S. (2009). RFID-enabled item-level product information revelation. *European Journal of Information Systems*, *18*(6), 570–577. doi:10.1057/ejis.2009.45

Wessel, R. (2007). *Frankfurt widens its NFC-enabled transit network*. Retrieved March 14, 2011, from http://www.rfidjournal.com/article/view/3755

Wessel, R. (2008a). *Dutch Forensic Institute uses RFID to control crime evidence*. Retrieved March 13, 2011, from http://www.rfidjournal.com/article/view/4410

Wessel, R. (2010a). *RFID helps Medlog monitor pharmaceutical cold chain*. Retrieved February 5, 2011, from http://www.rfidjournal.com/article/view/7494

Wessel, R. (2010b). *German clothing company s.Oliver puts RFID to the test*. Retrieved March 7, 2011, from http://www.rfidjournal.com/article/view/8013/1

Wessel, R. (2010c). *Mars, Rewe, Deustsche Post and Luthansa cargo work on SmaRTI*. Retrieved March 9, 2011 from http://www.rfidjournal.com/article/view/8095/1

Wikipedia. (n.d.). *Enterprise software*. Retrieved September 12, 2011, from http://en.wikipedia.org/wiki/Enterprise_software

Williams, R., & Herrup, K. (1988). The control of neuron number. *Annual Review of Neuroscience*, *11*(1), 423–453. doi:10.1146/annurev.ne.11.030188.002231

Windows Azure. (2011). *Website*. Retrieved from http://www.microsoft.com/azure/default.mspx

Wu, B., Liu, Z., George, R., & Shujaee, K. A. (2005). eWellness: Building a smart hospital by leveraging RFID networks. *IEEE Proceedings of the 27th Annual Conference on Engineering in Medicine and Biology*, Shanghai, China, September 1-4, 2005.

Wu, J., Wang, D., & Sheng, H. (2005). *Design an OSGi extension service for mobile RFID applications*. IEEE International Conference on e-Business Engineering 2007.

Wu, H., & Zeng, Y. (2010). Bayesian tag estimate and optimal frame length for anti- collision Aloha RFID system. *IEEE Transactions on Automation Science and Engineering*, 7(4), 963–969. doi:10.1109/TASE.2010.2042957

Wu, N. C., Nystrom, M. A., & Lin, T. R. (2005). Challenges to global RFID adoption. *Technovation*, *26*, 1317–1323. doi:10.1016/j.technovation.2005.08.012

Xia, F. (2009). Wireless sensor technologies and applications. *Sensors (Basel, Switzerland)*, *9*(11), 8824–8830. doi:10.3390/s91108824

Xia, H. H., Bertoni, H. L., Maciel, L. R., Lindsay-Stewart, A., & Rowe, R. (1993). Radio propagation characteristics for line-of-sight microcellular and personal communications. *IEEE Transactions on Antennas and Propagation*, *41*, 1439–1446. doi:10.1109/8.247785

Xiao, Q., Boulet, C., & Gibbons, T. (2007). RFID security issues in military supply chains. *ARES*, *2007*, 599–605.

XRA00-SBN18I datasheet. (2005). *ST microelectronics*.

Yamaguchi, Y., Abe, T., & Sekiguchi, T. (1989). Radio wave propagation loss in the VHF to microwave region due to vehicle in tunnels. *IEEE Transactions on Electromagnetic Compatibility*, *31*(1). doi:10.1109/15.19912

Yan, B., Chen, Y., & Meng, X. (2008). RFID technology applied in warehouse management system. *ISECS International Colloquium on Computing, Communication, Control, and Management* (pp. 362-367). IEEE Press.

Yang, C. N., Kun, Y. C., Chiu, C. Y., & Chu, Y. Y. (2010). A new adaptive query tree on resolving RFID tag collision. *IEEE International Conference on RFID-Technology and Applications* (pp. 153-158). IEEE Press.

Yang, C. N., Kun, Y. C., He, J. Y., & Wu, C. C. (2010). A practical implementation of ternary query tree for RFID tag anti-collision. *IEEE International Conference on Information Theory and Information Security* (pp. 283-286). IEEE Press.

Yang, C.-N., Chen, J.-R., Chiu, C.-Y., Wu, G.-C., & Wu, C.-C. (2009). Enhancing privacy and security in RFID-enabled banknotes. In *the Proceedings of the IEEE International Symposium on Parallel and Distributed Processing with Applications*, (pp. 439-444).

Yang, J., Park, J., Lee, H., Ren, K., & Kim, K. (2005). Mutual authentication protocol for low-cost RFID. *Proceedings of the Ecrypt Workshop RFID and Lightweight Crypto*, 2005.

Yang, L., Staiculescu, D., Zhang, R., Wong, C. P., & Tenzeris, M. M. (2009). A novel "green" fully-integrated ultrasensitive RFID-enabled gas sensor utilizing inkjet-printed antennas and carbon nanotubes. *IEEE Antennas and Propagation Society: International Symposium and USNC/URSI National Radio Science Meeting,* Vol. 1, (pp. 6804-6807).

Yang, Y., Karmakar, N. C., & Zhu, X. (2011). A portable wireless monitoring system for sleep apnea diagnosis based on active RFID technology. *Accepted by Asia-Pacific Microwave Conference.*

Yang, Y., Roy, S. M., Karmakar, N. C., & Zhu, X. (2011). A novel high selectivity bandpass filter for wireless monitoring of sleep apnea patients. *Accepted by Asia-Pacific Microwave Conference.*

Yang, C. N., & He, J. Y. (2011). An effective 16-bit random number aided query tree algorithm for RFID tag anti-collision. *IEEE Communications Letters, 15*(5). doi:10.1109/LCOMM.2011.031411.110213

Yang, L., Zhang, R. W., Staivulescu, D., Wong, C. P., & Tentzeris, M. M. (2009). A novel conformal RFID-enabled module utilizing inkjet-printed antennas and carbon nanotubes for gas-detection applications. *IEEE Antennas and Wireless Propagation Letters*, 8653–8656.

Yan, L., & Rong, C. (2008). Strengthen RFID tags security using new data structure. *International Journal of Control and Automation, 1*, 51–58.

Yazdandoost, K. Y., & Sayrafian-Pour, K. (2010). *Channel model for body area network.* Retrieved September 5, 2011, from https://mentor.ieee.org/802.15/dcn/08/15-08-0780-12-0006-tg6-channel-model.pdf

Yazicioglu, R. F., Merken, P., Puers, R., & Van Hoof, C. (2007). A 60 µW 60nV/ Hz readout front-end for portable biopotential acquisition systems. *IEEE Journal of Solid-state Circuits, 42*, 1100–1110. doi:10.1109/JSSC.2007.894804

Young-Taek, L., Jong-Sik, L., Chul-Soo, K., Dal, A., & Sangwook, N. (2002). A compact-size microstrip spiral resonator and its application to microwave oscillator. *Microwave and Wireless Components Letters, 12*(10), 375–377. doi:10.1109/LMWC.2002.804556

Yunus, M. A. M., Mendez, G. R., & Mukhopadhyay, S. C. (2011, May). Development of a low cost system for nitrate and contamination detections in natural water supply based on a planar electromagnetic sensor. In *Proceedings of IEEE I2MTC 2011* (pp. 1557–1562). Hangzhou, China: Conference. doi:10.1109/IMTC.2011.5944156

Zaffaroni, A., de Chazal, P., Heneghan, C., Boyle, P., Mppm, P. R., & McNicholas, W. T. (2009). Sleep minder: An innovative contact-free device for the estimation of the apnea hypopnoea index. *Annual International Conference of the IEEE Engineering in Medicine and Biology Society*, (pp. 7091-9094).

Zeidler, M. (2003, January 8). RFID: Der Schnuffelchip im Joghurtbecher. *Monitor-Magazine.*

Zernike, F., & Midwinter, J. E. (1973). *Applied nonlinear optics.* Wiley.

Zhang, C., & Jacobsen, H. A. (2003). Refactoring middleware with aspects. *IEEE Transactions on Parallel and Distributed Systems, 14*(11). doi:10.1109/TPDS.2003.1247668

Zhang, Y., Ma, H., Zhang, K., Zhang, S., & Wang, J. (2009). An improved DNA biosensor built by layer-by-layer covalent attachment of multi-walled carbon nanotubes and gold nanoparticles. *Electrochimica Acta, 54*, 2385–2391. doi:10.1016/j.electacta.2008.10.052

Zhen, B., Kobayashi, M., & Shimizu, M. (2005). Framed Aloha for multiple RFID objects identification. *IEICE-Transactions on Communications. E (Norwalk, Conn.), 88-B*, 991–999.

Zheng, L. L., Rodriguez, S., Zhang, L., Shao, B. T., & Zhang, L. R. (2008). Design and implementation of a fully reconfigurable chipless RFID tag using inkjet printing technology. *IEEE International Symposium on Circuits and Systems,* Vol. 1-10, (pp. 1524-1527).

Zhiguo, T., Chengrong, L., Xu, C., Wei, W., Jinzhong, L., & Jun, L. (2006). Partial discharge location in power transformers using wideband RF detection. *IEEE Transactions on Dielectrics and Electrical Insulation, 13*(6), 1193–1199. doi:10.1109/TDEI.2006.258190

Zhou, F., Chen, C., Jin, D., Huang, C., & Min, H. (2004). Evaluating and optimizing power consumption of anti-collision protocols for applications in RFID systems. In *Proceedings of the 2004 International Symposium on Low Power Electronics and Design (ISLPED04),* (pp. 357–362). New York, NY: ACM.

Zhu, L., & Yum, P. T. (2009). The optimization of framed Aloha based RFID algorithms. In *MSWiM '09: Proceedings of the 12th ACM International Conference on Modeling, Analysis and Simulation of Wireless and Mobile Systems,* (pp. 221–228). New York, NY: ACM.

Zhu, L., & Yum, P. T. (2011). A critical survey and analysis of RFID anti-collision mechanisms. *IEEE Communications Magazine*, *49*, 214–221. doi:10.1109/MCOM.2011.5762820

Ziolkowski, R. W., & Lin, C. (2008). Metamaterial-inspired magnetic based UHF and VHF antennas. *Proceedings of IEEE Antennas and Propagation Society International Symposium Digest,* San Diego, CA.

Zuo, Y. (2010). Survivable RFID systems: Issues, challenges, and techniques. *IEEE Transactions on Systems, Man and Cybernetics. Part C, Applications and Reviews*, *40*(4), 406–418. doi:10.1109/TSMCC.2010.2043949

About the Contributors

Nemai Chandra Karmakar obtained his PhD in Information Technology and Electrical Engineering from the University of Queensland, St. Lucia, Australia, in 1999. He has about twenty years of teaching, design, and research experience in smart antennas, microwave active and passive circuits, and chipless RFIDs in both industry and academia in Australia, Canada, Singapore, and Bangladesh. He has published more than 180 refereed journal and conference papers and many book chapters. He holds two patents in the field. Currently, he is an Associate Professor in the Department of Electrical and Computer Systems Engineering at Monash University.

***

Emran Amin received the B.Eng. degree in Electrical and Electronics Engineering Department from Bangladesh University of Engineering and Technology (BUET) in 2009. Currently he is pursuing his PhD at Electrical and Computer Systems Engineering Department of Monash University. His research area is Chipless RFID Sensors and Radiometric Partial Discharge detection of High Voltage Equipments. He is working on a project titled, "Smart Information Management of Partial Discharge in Switchyards using Smart Antennas," funded by ARC and SP-AusNet.

A.K.M. Azad completed his PhD from the Faculty of Information Technology, Monash University, Australia, on May, 2010, and Master's and Bachelor's of Science degrees in Computer Science and Engineering from the Dept. of Computer Science and Engineering, Bangladesh University of Engineering and Technology (BUET), Dhaka, Bangladesh, on 1998 and 2005, respectively. Dr. Azad is currently working as a Research Engineer at the Antenna and RFID research lab of the department of Electrical and Computer Systems Engineering, Monash University, Australia. His research interest includes RFID and Sensors, ad hoc and sensor networks, signal processing, control system design, graph theory, et cetera.

Morshed Uddin Chowdhury is an academic staff member in School of Information Technology, Deakin University, Melbourne, Australia since July 1999. Prior to joining Deakin University, he was an academic staff in Gippsland School of Computing and Information Technology, Monash University, Australia. He has published and reviewed a number of papers in the area of RFID security, sensor network security, suppression of video coding distortion, multimedia synchronisation, multimedia documents retrieval, data mining, computer networks, e-commerce and online distance learning, et cetera. He has more than 12 years of industry experience in Bangladesh and Australia.

Angelo Cucinotta received the degree in Computer Science from the University of Messina, Italy. As a PhD student, he has been engaged in research on RFID systems and sensor network. His main research interests include distributed systems programming and wireless systems.

Peter Darcy received both his Bachelor's of Information Technology (BIT) (2006) and BIT (honours, first class) (2007) at Griffith University, Australia. Mr. Darcy is currently in the final stages of his PhD in Radio Frequency Identification (RFID) focusing on Correcting Ambiguous Anomalies in Captured Data. He has had numerous conference, journal and book chapter publications since the commencing of his PhD in 2008. His main research interests are in radio frequency identification, Bayesian networks, neural networks, and non-monotonic reasoning.

Gerald R. DeJean is a researcher at Microsoft Research in the field of RF and antenna design. He conducts research in antenna design, RF/microwave design and characterization, and 3D system-on-package (SOP) integration of embedded functions that focuses largely on modern commercial RF systems and devices within Microsoft Corporation. He is also an Adjunct Assistant Professor at the Georgia Institute of Technology. Gerald received the B.S. degree in Electrical and Computer Engineering from Michigan State University in 2001 and the M.S. and Ph.D. degrees in Electrical and Computer engineering from the Georgia Institute of Technology in 2005 and 2007, respectively. He has authored and co-authored over 60 papers in refereed journals and conference proceedings.

Edward Gebara received his B.S. (with highest honors), M.S., and Ph.D. degrees in Electrical and Computer Engineering from the Georgia Institute of Technology, Atlanta, in 1996, 1999, and 2003, respectively. From 1999 to 2000 he was an invited scientist at Chalmers University, Sweden. In 2001, Dr. Gebara applied the results of his research to define the core technology for Quellan, Inc as its initial employee, which is now part of Intersil (NASDQ: ISIL). These technologies served as the basis for signal integrity solutions developed for the enterprise, video, storage and wireless markets. At Georgia Tech Dr. Gebara has supported the mixed- signal team research efforts since 2003 on a part-time basis before he became a full time researcher in spring of 2008. Since October 1, 2010, Dr. Gebara has been appointed as Adjunct Assistant Professor. The team's research interest is to develop the foundation of alternate modulation schemes, equalization techniques, and crosstalk cancellation techniques on pure CMOS applied to next generation optic, wired and wireless communication systems. Dr. Gebara has authored or co-authored over 70 papers and has 5 patents issued. He is a reviewer for the IEEE International Symposium and Circuits and Systems (ISCAS) and for the IEEE MTT-S International Microwave Symposium. Additionally, Dr. Gebara served as Workshops and Tutorials Chair of the Technical Program for the IEEE MTT-S International Microwave Symposium 2008.

Salvador Ricardo Meneses González is a Communications and Electronic Engineer from ESIME IPN. He earned his M. S. E. degree in Electrical Engineering from Centro de Investigación y de Estudio-sAvanzadosdel IPN (CINVESTAV IPN). He is a Professor and the departmental head of Communications and Electronic Engineering Department of ESIME Zacatenco, IPN. He is currently a PhD student in the Electromagnetic Compatibility Laboratory.

Jyun-Yan He was a graduate student in the Department of Computer Science and Information Engineering at National Dong Hwa University, Taiwan. His research includes wireless network and cryptography.

Joarder Kamruzzaman received a B.Sc. and M.Sc. in Electrical Engineering from Bangladesh University of Engineering & Technology, Dhaka, Bangladesh in 1986 and 1989, respectively, and a PhD in Information System Engineering from Muroran Institute of Technology, Japan, in 1993. Currently, he is a faculty member in the Faculty of Information Technology, Monash University, Australia. His research interest includes computer networks, computational intelligence, and bioinformatics. He has published over 150 peer-reviewed publications which include 40 journal papers and 6 book chapters, and edited two reference books on computational intelligence theory and applications. He is the recipient of best paper award in two IEEE sponsored international conferences. He is currently serving as a program committee member of a number of international conferences and an editor of international journal.

Gour Chandra Karmakar received the B.Sc. Eng. degree in Computer Science and Engineering from Bangladesh University of Engineering and Technology in 1993 and Master's and Ph.D. degrees in Information Technology from the Faculty of Information Technology, Monash University, in 1999 and 2003, respectively. He is currently a Senior Lecturer at the Gippsland School of Information Technology, Monash University. He has published over 90 peer-reviewed research publications including 13 international peer reviewed reputed journal papers and was awarded 3 best papers in reputed international conferences. His research interest includes image and video processing, mobile ad hoc and wireless sensor networks.

Darko Kirovski, since joining Microsoft Research in 2000, has split time between the Crypto and Machine Learning groups there, flirting with the freedom that the lab offers to its researchers to work on a wide variety of systems projects that span authentication, anti-counterfeiting, mobile payments, and Kinect games, among others. He has published over 100 research articles and co-invented over 100 filed patents; all that thanks to a 2001 PhD in CS from UCLA. His hobby is statistical arbitrage.

Yu-Ching Kun was a graduate student in the Department of Computer Science and Information Engineering at National Dong Hwa University, Taiwan. Her research includes network security and RFID EPCglobal network.

Vasileios Lakafosis received the Diploma degree in Electrical and Computer Engineering from the National Technical University, Athens, Greece, in 2006 and the M.Sc. degree in Electrical and Computer Engineering from the Georgia Institute of Technology, Atlanta, in 2009. He is currently pursuing his PhD degree at Georgia Institute of Technology. In the past, Vasileios has worked for Microsoft Research, the Cisco Research Center and the University of Tokyo. His cross-disciplinary research interests range from novel networking protocols in the areas of wireless mobile ad hoc, mesh, and sensor networks to perpetual, ambient RF energy harvesting communication systems, security authentication, wireless localization techniques, distributed computing in delay tolerant and opportunistic networks, and ubiquitous computing applications. He has authored and coauthored two book chapters and over 15 papers in peer-reviewed journals and conference proceedings. Vasileios is a member of IEEE, ACM, HKN and the Technical Chamber of Greece. He serves as reviewer for a number of IEEE journal transactions and conference proceedings.

Ming K. Lim obtained his PhD degree from the University of Exeter, U.K. He is currently Head of RFID Advanced Research and Lecturer in Logistics Management at Aston University, UK. His research expertise is in the areas of supply chain management, logistics and manufacturing systems and management. Ming has been providing consultancy to companies from the manufacturing and logistics sectors in the implementation of RFID technology, as well as in enhancing production and supply chain competitiveness. Aston RFID Advanced Research is aimed at exploring new RFID functions to enhance the competitiveness in the manufacturing and supply chain to achieve rapid responsiveness, cost optimisation and system efficiency under dynamic changing environment. There are 8 researchers currently working under this research group. Ming's collaborative partners include DHL, UK health sector NHS, Caterpillars, Norbert Dentressangle, and a range of SMEs from different industry sectors.

Michele Maffia is an Associate Professor of Physiology in the Department of Biological and Environmental Sciences and Technologies at the University of Salento, Lecce, Italy. He obtained a PhD in Physiology from the University of Naples in 1989 and since 2008 he is Director of the Laboratory of Clinical Proteomics at "Vito Fazzi" Hospital, Lecce. He is author of more than 70 articles on international scientific journals. His main research activities concern proteomic applications in oncological, cardiovascular and neurodegenerative diseases; the structural and functional analysis of transport proteins in cell membranes and epithelia; dismetabolisms of micro-nutrients and essential metals and their relationship with neurodegenerative and oncological diseases; the functional immobilization of proteins and enzymes for nano-biosensor applications.

Luca Mainetti is an Associate Professor of Software Engineering and Computer Graphics in the Department of Innovation Engineering at the University of Salento (Italy). His research interests include web design methodologies, notations and tools, web and services oriented architectures and applications, collaborative computer graphics. He is scientific coordinator of the GSA Lab – Graphics and Software Architectures Lab (www.gsalab.unisalento.it), and IDA Lab – Identification Atomation Lab (www.idalab.unisalento.it). He is Rector's delegate to ICT. Mainetti received a PhD in computer science from Politecnico di Milano (Italy), where he has been supply teacher of Hypermedia Applications and kaddman Computer Interaction, and he contributed to create the HOC (Hypermedia Open Center) laboratory. He is a member of the IEEE and ACM. He is author of more than 90 international scientific papers.

Antonino Longo Minnolo received the degree in Electronic Engineering from the University of Messina, Italy. As a PhD student, he has been engaged in research on RFID systems and sensor network. His main research interests include distributed systems programming and wireless systems.

Roberto Linares y Miranda received his Ph. D. degree from Centro de Investigación y de Estudios Avanzadosdel IPN (CINVESTAV- IPN). He is a Professor in the Post-graduate Section of the ESIME Zacatenco and the head of the Electromagnetic Compatibility Laboratory.

Luigi Patrono received the Laurea degree (*cum laude*) in Computer Engineering from the University of Lecce, Italy, in 1999. He received Ph.D. in Innovative Materials and Technologies (area Satellite Networks) at the ISUFI of Lecce, Italy, in 2003. He is currently an Assistant Professor at the Faculty of Engineering of the University of Salento (Italy) of "Network Design." His main research interests include design and performance evaluation of wireless communication protocols, RFID technologies, EPCGlobal architecture, and Wireless Sensor Networks. He contributed to create the IDAutomation Laboratory (IDA Lab) of the Department of Innovation Engineering at the University of Salento (since 2008). He is author of about 50 scientific papers published on international journals and conferences and four chapters of books with international diffusion. He is chair of the international symposium on "RFID Technologies & Internet of Things" organized within the IEEE conference on Software in Telecommunications and Computer Networks.

Antonio Puliafito is a full Professor of Computer Engineering at the University of Messina, Italy. His interests include parallel and distributed systems, networking, wireless and GRID computing. He is the coordinator of the Ph.D. course in Advanced Technologies for Information Engineering currently available at the University of Messina and the responsible for the course of study in computers engineering. He was a referee for the European community for the projects of the fourth, fifth and sixth Framework Program and he is currently acting as a referee also in the seventh FP. He is currently the director of the RFIDLab, a joint research lab with Oracle and Intel on RFID and wireless. He is currently a member of the general assembly and of the technical committee of the Reservoir project, an IP project funded from the European Commission.

Prapassara Pupunwiwat received her Bachelor of Information Technology (BIT) in 2006 and a BIT with First Class Honours in 2007 from Griffith University, Australia. She also received a University medal for outstanding academic excellence. Miss Pupunwiwat is currently in final stages of her Ph.D. at Griffith University. Her research interests include various aspects of Radio-Frequency Identification (RFID) technology, such as Anti-Collision Algorithms, Data Streams Errors Management, and RFID Business Model Strategies.

Geoffrey Ramadan from Monash University (Clayton Campus) in 1981 with a Bachelor of Engineering (Electrical) degree, and specialised in microprocessor systems and electronics. Geoffrey undertook his first commercial venture during his last year of university in the design and development of a microprocessor based production monitoring system for a major carpet weaving company. In 1983, he founded Unique Micro Design (UMD) with partners Harry Ramadan and Alan Walker, and has become its lead solution architect and Managing Director, and is supported by a highly skilled team of engineering and software professionals. Initially focused on the design and manufacturing of microprocessor-based devices and interfaces, the company has now evolved into an engineering ICT solutions company in data capture and supply chain applications including RFID.

Abdur Rahim received B.Sc. Eng. (Electrical & Electronic) from Chittagong University of Engineering & Technology, Bangladesh in 1999 and Master of Engineering (Telecommunication) from Asian Institute of Technology (AIT), Bangkok, Thailand in 2005. He worked as a faculty member at International Islamic University Chittagong, Bangladesh from 2000 to 2003 and 2005 to 2009. Currently he is working toward the Ph.D. degree in Electrical and Computer Systems Engineering at Monash University, Australia. His research area is MIMO-based wireless RF monitoring of sleep apnea patient.

Biplob Ranjan Ray is working as an academic staff in school of information technology, Melbourne Institute of Technology, Australia. He is also working as an academic staff in school of School of Science Information Technology and Engineering, University of Ballarat, Australia since March 2009. He has more than 5 year's industry experience in Philippines and Australia. He has published many academic papers. He also worked as a committee members and reviewed many conferences papers. He is also member of Australian computer society and Golden Key International Honour Society. His key research areas are network security, RFID security, and network performance optimization.

Abdul Sattar is the founding Director of the Institute for Integrated and Intelligent Systems (IIIS) and a Professor of Computer Science and Artificial Intelligence at Griffith University. He is also a Research Leader at National ICT Australia (NICTA) Queensland Research Lab. He has been an academic staff member at Griffith University since 1992. His research interests include knowledge representation and reasoning, constraint satisfaction, rational agents, temporal reasoning, and bioinformatics.

Bala Srinivasan is a Professor of Information Technology and Head of Clayton School of Information Technology in the Faculty of Information Technology, Monash University, Australia. He was formerly an academic staff member of the Department of Computer Science and Information Systems at the National University of Singapore, Singapore and the Indian Institute of Technology, Kanpur, India. He has more than 30 years of experience in academia, industries and research organizations. He has authored and jointly edited 7 technical books and more than 300 refereed publications in international journals and conferences in the areas of multimedia databases, data communications, data mining and distributed systems, and has attracted a number of research grants. His substantial contribution towards research and training has been recognized by Monash University by awarding him the Vice-Chancellor's Medal for excellence in supervision. He is a founding chairman of the Australasian Database Conference.

Bela Stantic is a Senior Lecturer at the School of Information and Communication Technology, Griffith University, and member of the Institute for Integrated and Intelligent Systems (IIIS). He is also a Senior Researcher at National ICT Australia (NICTA) Queensland Research Lab (QR). He has been an academic staff member at Griffith University since 2001. His research interests include RFID systems, temporal and spatio-temporal databases, efficient management of complex data structures, bioinformatics, and database systems.

Manos M. Tentzeris received the Diploma Degree in Electrical and Computer Engineering from the National Technical University of Athens (*magna cum laude*) in Greece and the M.S. and Ph.D. degrees in Electrical Engineering and Computer Science from the University of Michigan, Ann Arbor, MI and he is currently a Professor with the School of ECE, Georgia Tech, Atlanta, and GA. He has published more than 420 papers in refereed journals and conference proceedings, 5 books, and 19 book chapters. Dr. Tentzeris has helped develop academic programs in Highly Integrated/Multilayer Packaging for RF and Wireless Applications using ceramic and organic flexible materials, paper-based RFIDs and sensors, biosensors, wearable electronics, inkjet-printed electronics, "Green" electronics and power scavenging, nanotechnology applications in RF, Microwave MEMs, SOP-integrated (UWB, multiband, mmW, conformal) antennas, and Adaptive Numerical Electromagnetics (FDTD, MultiResolution Algorithms) and heads the ATHENA research group (20 researchers). He is currently the Head of the GT-ECE Electromagnetics Technical Interest Group and he has served as the Georgia Electronic Design Center Associate Director for RFID/Sensors research from 2006-2010 and as the Georgia Tech NSF-Packaging Research Center Associate Director for RF Research and the RF Alliance Leader from 2003-2006. He was the recipient/co-recipient of the 2010 IEEE Antennas and Propagation Society Piergiorgio L. E. Uslenghi Letters Prize Paper Award, the 2010 Georgia Tech Senior Faculty Outstanding Undergraduate Research Mentor Award, the 2009 IEEE Transactions on Components and Packaging Technologies Best Paper Award, the 2009 E.T.S.Walton Award from the Irish Science Foundation, the 2007 IEEE APS Symposium Best Student Paper Award, the 2007 IEEE IMS Third Best Student Paper Award, the 2007 ISAP 2007 Poster Presentation Award, the 2006 IEEE MTT Outstanding Young Engineer Award, the 2006 Asian-Pacific Microwave Conference Award, the 2004 IEEE Transactions on Advanced Packaging Commendable Paper Award, the 2003 NASA Godfrey "Art" Anzic Collaborative Distinguished Publication Award, the 2003 IBC International Educator of the Year Award, the 2003 IEEE CPMT Outstanding Young Engineer Award, the 2002 international Conference on Microwave and Millimeter-Wave Technology Best Paper Award (Beijing, CHINA), the 2002 Georgia Tech-ECE Outstanding Junior Faculty Award, the 2001 ACES Conference Best Paper Award, the 2000 NSF CAREER Award, and the 1997 Best Paper Award of the International Hybrid Microelectronics and Packaging Society.

Emanuela Urso got a Laurea Degree (*cum laude*) in Biological Sciences from the University of Salento, Lecce, Italy, in 2003. In 2004 she joined the Department of Biological and Environmental Sciences and Technologies, Laboratory of General Physiology, at the same university, where she has been working in the basic research in the field of metal dismetabolisms. Since February 2011 she is a PhD student of the "Biology and Biotechnologies" course in the same department. Her research interests include the functional characterization of copper ion transporters in cultured cells and tissues, the role of copper in angiogenesis and cancerogenesis, the relationship between metal ion dismetabolisms and neurodegenerative disorders.

Ching-Nung Yang was born on May 9, 1961 in Kaohsiung, Taiwan. He received the B.S. degree in 1983 and the M.S. degree in 1985, both from the Department of Telecommunication Engineering at National Chiao Tung University. He received the Ph.D. degree in Electrical Engineering from National Cheng Kung University in 1997. During 1987-1989 and 1990-1999, he worked at the Telecommunication Lab. and Training Institute Kaohsiung Center, Chunghwa Telecom Co., Ltd., respectively. He is presently a Professor in the Department of Computer Science and Information Engineering at National Dong Hwa University. He is also an IEEE Senior Member. His research interests include coding theory, information security, and cryptography.

Yang Yang was born in Inner Mongolia, China, in 1982. He received B.E. degree from Dalian Nationalities University, Dalian, China, in 2005, M.E. degree in Telecommunication Eng. from Monash University in 2007, M.Sc. Degree in Digital Communications and Protocols from Monash University in 2008 and is currently working toward the Ph.D. degree in Electrical and Computer Systems Engineering at Monash University, Melbourne, Victoria, Australia. His research interests include microstrip passive filter and antenna designs for portable wireless communications system, transceivers designs, MMIC designs and bio-sensors development. His publications mainly focused on microstrip filter designs and portable wireless monitoring system designs.

Index

R

S

T

U

V

W

X

CPSIA information can be obtained at www.ICGtesting.com
Printed in the USA
BVOW051711281112

306516BV00007B/124/P

9 781466 620803